CAMBRIDGE TRACTS IN MATHEMATICS
General Editors
B. BOLLOBAS, P. SARNAK, C.T.C. WALL

113 Spectral Decomposition and Eisenstein Series

C. Moeglin, J.-L. Waldspurger
Université de Paris VII

Spectral Decomposition and Eisenstein Series

Une Paraphrase de l'Écriture

CAMBRIDGE
UNIVERSITY PRESS

CAMBRIDGE UNIVERSITY PRESS
Cambridge, New York, Melbourne, Madrid, Cape Town, Singapore, São Paulo

Cambridge University Press
The Edinburgh Building, Cambridge CB2 8RU, UK

Published in the United States of America by Cambridge University Press, New York

www.cambridge.org
Information on this title: www.cambridge.org/9780521418935

First published 1995
This digitally printed version 2008

A catalogue record for this publication is available from the British Library

Library of Congress Cataloguing in Publication data

Moeglin, Colette, 1953–
Spectral decomposition and Eisenstein series / C. Moeglin,
J.-L. Waldspurger.
p. cm. – (Cambridge tracts in mathematics : 113)
Includes bibliographical references.
ISBN 0 521 41893 3
1. Eisenstein series. 2. Automorphic forms. 3. Spectral theory
(Mathematics) 4. Decomposition (Mathematics) I. Waldspurger,
Jean-Loup, 1953– II. Title. III. Series.
QA295.M62 1995
515′.243 – dc20 94-11429 CIP

ISBN 978-0-521-41893-5 hardback
ISBN 978-0-521-07035-5 paperback

Contents

Preamble

Over the last few years, Arthur has formulated some extremely precise conjectures generalising Jacquet's conjectures for $GL(n)$, on the description of square integrable automorphic forms (see [J], [A2], [A3], [A4]). The origin of these conjectures can be found in his work on the trace formula and their point of departure goes back to Langlands' work (see Arthur's Corvallis talk [A1]). It is probable that the full force of the constructions of these automorphic forms as residues of Eisenstein series has not been exploited. It is thus important to fully understand the basic book [L]; this was our main motivation. Since Langlands wrote the material of [L] (around 1967), several authors have already given personal presentations (Godement [G], Harish-Chandra [HC], Osborne-Warner [OW]). Morris extended Langlands' results to the function field case ([M1] and [M2]).

The following notes are a reworking of the book [L] and an ameliorated (and unified) version of talks which we gave in the 'automorphic' seminar (Paris 7/ENS). A seminar cannot exist without critical auditors: we wish to thank P. Barrat, J.-P. Labesse, P. Perrin (who also gave some talks on this subject), A.-M. Aubert, C. Blondel, L. Clozel, G. Henniart, G. Laumon (for whom the goal of the seminar was to render entirely obscure what was already not particularly clear), M.-F. Vignéras and D. Wigner. We also thank Y. Colin de Verdières who gave a talk on his proof of the meromorphic continuation of Eisenstein series [C], K. Lai for his remarks on a first version of the manuscript, P. Deligne who communicated to us the proof of Appendix I and finally, most particularly, H. Jacquet who explained to us the proof of the meromorphic continuation of Eisenstein series given in Chapter IV.

Here is a more precise description of the different chapters.

The first chapter is a general discussion of automorphic forms and their constant terms. Our goal is not to give a complete description of their properties, but only to formulate these properties in the framework which we need, with particular emphasis on those properties resulting from the principle according to which an automorphic form is determined by its constant terms. In particular, we use without proof some results of reduction theory and of the theory of cuspidal automorphic forms. We are mainly inspired by Harish-Chandra [HC], Godement [G], Borel and Jacquet [BJ] and of course Langlands [L], Chapters 1, 2, 3 and 5.

Let k be a global field, G a connected reductive group defined over k, $G(\mathbb{A})$ the group of its adelic points. We take a group $\mathbf{G}$ which is a finite central covering of $G(\mathbb{A})$ in which the group $G(k)$ of rational points lifts to a group also denoted by $G(k)$. This framework allows us to include the metaplectic groups. After a long paragraph devoted to the properties of these objects, we introduce the notion of constant term of a function on $G(k)\backslash\mathbf{G}$: let $P = MU$ be a standard parabolic subgroup of G and let ϕ be a function, let us say continuous, on $G(k)\backslash\mathbf{G}$; its constant term along P is the function ϕ_P on $U(\mathbb{A})M(k)\backslash\mathbf{G}$ defined by

$$\phi_P(g) = \int_{U(k)\backslash U(\mathbb{A})} \phi(ug)du,$$

see I.2.6. We explain how a function with moderate growth can be approximated by a linear combination of its constant terms along the different standard parabolic subgroups of G (see I.2.7 to I.2.16; the paragraphs I.2.13 to I.2.16 come from [A2]). We then introduce the notions of automorphic form and cuspidal automorphic form on $G(k)\backslash\mathbf{G}$, or more generally on $U(\mathbb{A})M(k)\backslash\mathbf{G}$, $P = MU$ being a standard parabolic subgroup, see I.2.17, I.2.18.

Suppose for a moment that the center $Z_{\mathbf{G}}$ of $\mathbf{G}$ is trivial, and let ϕ be an automorphic form on $G(k)\backslash\mathbf{G}$. We easily define the 'cuspidal component' of ϕ: it is the unique cuspidal automorphic form ϕ^{cusp} on $G(k)\backslash\mathbf{G}$ such that

$$\langle\phi_0, \phi^{\text{cusp}}\rangle = \langle\phi_0, \phi\rangle$$

for every cuspidal automorphic form ϕ_0, where we have set

$$\langle\phi_0, \phi\rangle = \int_{G(k)\backslash\mathbf{G}} \overline{\phi}_0(g)\phi(g)dg.$$

This definition extends to the case of an arbitrary group $\mathbf{G}$ and an automorphic form on $U(\mathbb{A})M(k)\backslash\mathbf{G}$, see I.2.18 and [L], p. 55. In particular, if ϕ is an automorphic form on $G(k)\backslash\mathbf{G}$, we define the cuspidal component ϕ_P^{cusp} of the constant term of ϕ along P. When P runs through the standard parabolic subgroups, these functions determine ϕ (Proposition I.3.4, [L] Lemma 3.7).

Let $P = MU$ be a standard parabolic subgroup and π an irreducible cuspidal automorphic representation of $\mathbf{M}$ (:= the inverse image of $M(\mathbb{A}) \in \mathbf{G}$). We define the space $A_0\big(U(\mathbb{A})M(k)\backslash\mathbf{G}\big)_\pi$ of cuspidal forms of type π on $U(\mathbb{A})M(k)\backslash\mathbf{G}$, see I.3.3.

Denote by $\mathrm{Rat}(M)$ the group of rational characters of M,

$$M(\mathbb{A})^1 := \bigcap_{\chi\in\mathrm{Rat}(M)} \mathrm{Ker}\,|\chi|$$

and by $\mathbf{M}^1$ its inverse image in $\mathbf{G}$. The abelian group $\mathbf{M}/\mathbf{M}^1$ is a real finite-dimensional vector space if k is a number field, a finitely-generated free $\mathbb{Z}$-module if k is a function field. This allows us to define the space of polynomials on $\mathbf{M}/\mathbf{M}^1$, which we have reasons to denote by $\mathbb{C}[\mathrm{Re}\,\mathfrak{a}_M]$. Such a polynomial defines a function on $\mathbf{G}$, left invariant under $\mathbf{M}^1U(\mathbb{A})$.

This being so, for π as above,

(1) $$\mathbb{C}[\mathrm{Re}\,\mathfrak{a}_M] \otimes_{\mathbb{C}} A_0\big(U(\mathbb{A})M(k)\backslash\mathbf{G}\big)_\pi$$

can be identified with a subspace of the space of cuspidal automorphic forms on $U(\mathbb{A})M(k)\backslash\mathbf{G}$. This last is in fact the sum of these subspaces, when π runs through all the irreducible cuspidal representations of $\mathbf{M}$, see I.3.3.

This allows us to define the support of a cuspidal form ϕ as the smallest set Π of irreducible cuspidal representations of $\mathbf{M}$ such that ϕ belongs to the sum of the spaces (1) for $\pi \in \Pi$. In particular, if ϕ is automorphic on $G(k)\backslash\mathbf{G}$, ϕ_P^{cusp} has a support which we denote by $\Pi_0(\mathbf{M}, \phi)$.

Knowledge of this support, now as P runs through the standard parabolic subgroups, determines upper bounds of the function ϕ (see Lemma I.4.1). Using arguments of [L], Lemmas 5.1 and 5.2, and an argument of Waldschmidt (see Lemma I.4.2), we obtain relatively precise criteria for the convergence of a sequence of automorphic forms (Lemma I.4.4) or for a function $z \mapsto \phi_z$ with values in the space of automorphic forms (Lemma I.4.10) to be holomorphic. These results are somewhat stronger than those in Chapter 5 of [L]. Finally, if we suppose that ϕ has a unitary central character, the fact that ϕ is or is not square integrable (modulo $Z_{\mathbf{G}}$) can be read off the central characters of the elements of its supports $\Pi_0(\mathbf{M}, \phi)$ (see I.4.11, [L] pp. 104 and 186).

In Chapter II, we introduce the basic objects, Eisenstein series and what we call pseudo-Eisenstein seris (Godement calls them theta series and Langlands does not really call them by any precise name); they are in fact integrals of Eisenstein series: their properties which we use are proved in Chapter 4 of [L]. These objects have become classical and we were largely inspired by Godement's Bourbaki talk presenting Langlands' work. They depend first and foremost on the given standard Levi subgroup M of G whose inverse image in $\mathbf{G}$ we denote by $\mathbf{M}$. We denote by $X_M^{\mathbf{G}}$ the set of (continuous) characters of $\mathbf{M}$ trivial on $\mathbf{M}^1$ (see above) and on the centre of $\mathbf{G}$ (a precise description of this group is given in I.3; if k is a number field, $X_M^{\mathbf{G}}$ is naturally a vector subspace of $\mathrm{Rat}\, M \otimes_{\mathbb{Z}} \mathbb{C}$). They also depend on an orbit of $X_M^{\mathbf{G}}$ in the set of (not indispensably) irreducible automorphic subrepresentations of $\mathbf{M}$, denoted by $\mathfrak{P}$, i.e. $\mathfrak{P}$ is of the form:

$$\mathfrak{P} := \{\lambda \otimes \pi_0,\ \lambda \in X_M^{\mathbf{G}}\},$$

where π_0 is a representation of $\mathbf{M}$ into the space of automorphic forms on $\mathbf{M}$. Finally, they depend on sections of the fibre bundle over $\mathfrak{P}$ whose fibre over a point π of $\mathfrak{P}$ is the space of automorphic forms on $M(k)U(\mathbb{A})\backslash\mathbf{G}$ of type π (see II.1.1) (where U is the unipotent radical of the standard parabolic of Levi M).

Note that if k is a number field, it is possible to fix π_0 canonically (requiring it to have a central character which is trivial on a subtorus (in the sense of Lie groups) of $\mathbf{M}$ which is a supplementary of $Z_{\mathbf{G}}\mathbf{M}^1$). Thus, in this case, we obtain a canonical identification of $X_M^{\mathbf{G}}$ with $\mathfrak{P}$, and the fibre bundle which interests us is canonically isomorphic to the trivial fibre bundle on $X_M^{\mathbf{G}}$ whose fibre is the space of automorphic forms on $M(k)U(\mathbb{A})\backslash\mathbf{G}$ of type π_0. On the other hand, if k is a function field, there is in general no canonical choice of π_0 and the fibre bundle which interests us is isomorphic to a principal fibre bundle over $X_M^{\mathbf{G}}/\mathrm{Fix}_{X_M^{\mathbf{G}}}\mathfrak{P}$ of fibre the space of automorphic forms on $M(k)U(\mathbb{A})\backslash\mathbf{G}$ of type π_0.

It is not difficult to equip $X_M^{\mathbf{G}}$ and $\mathfrak{P}$ with structures of complex analytic manifolds such that for any choice of $\pi_0 \in \mathfrak{P}$ the map:

$$\lambda \in X_M^{\mathbf{G}} \mapsto \lambda \otimes \pi_0 \in \mathfrak{P}$$

is holomorphic. We then easily define the notion of holomorphic, meromorphic, Paley–Wiener, etc. sections of the above fibre bundle. Let ϕ be a section: we classically define an Eisenstein series depending on $\mathfrak{P}$ by setting, when the series converges absolutely,

$$E(\phi, \pi)(g) = \sum_{\gamma' in P(k)\backslash G(k)} \phi(\pi)(\gamma g);$$

closely following Godement, we give the classical sufficient conditions which ensure the absolute and uniform convergence on all compact subsets of **G**, conditions on the absolute value of the central character of π, denoted by $\mathrm{Re}\,\pi$.

In what follows, we fix a unitary character of the centre of **G**, denoted by ξ, and we suppose that $\mathfrak{P}$ consists of cuspidal representations of **M**, the restriction of whose central character to the centre of **G** is ξ, and we consider only the sections ϕ as above with values in the space of cuspidal automorphic forms on $M(k)U(\mathbb{A})\backslash\mathbf{G}$. We then say that $(M, \mathfrak{P})$ is a cuspidal datum relative to ξ. For $\mathfrak{P}$ satisfying these hypotheses and for $\pi \in \mathfrak{P}$, the character $\mathrm{Re}\,\pi$ mentioned above is in fact an element of $X_M^{\mathbf{G}}$. We can define the pseudo-Eisenstein series for $(M, \mathfrak{P})$ fixed as above and for ϕ a Paley-Wiener section, by setting:

$$\theta_\phi := \int_{\substack{\pi\in\mathfrak{P}\\ \mathrm{Re}\ \pi=\lambda_0}} E(\phi,\pi)\mathrm{d}\pi,$$

where λ_0 is a very positive real element of $X_M^{\mathbf{G}}$. We check that θ_ϕ is well-defined (i.e. that convergence of the integral does not depend on the choice of λ_0 and that the integral is a rapidly decreasing function on $G(k)\backslash\mathbf{G}$ (with compact support if k is a function field). One of the properties of pseudo-Eisenstein series is the following 'density' theorem (see II.1.12):

Let ψ be a function on $G(k)\backslash\mathbf{G}$ on which the center of **G** *acts via ξ, and which is either slowly increasing or L^2 modulo the centre, such that:*

$$\int_{Z_{\mathbf{G}}G(k)\backslash\mathbf{G}} \overline{\psi}(g)\theta_\phi(g)\mathrm{d}g = 0,$$

for every pseudo-Eisenstein series (we obviously allow any cuspidal datum relative to ξ). Then $\psi = 0$.

This result is in fact a more or less immediate consequence of Chapter I. The second part of Chapter II is devoted to the calculation of the scalar product of two pseudo-Eisenstein series. This calculation is standard: it is done by calculating the constant terms of either Eisenstein series ([L], 4.6(ii)) or of pseudo-Eisenstein series. It is the second point of view which we adopt (II.2.1). The calculation of the constant terms of Eisenstein series is done in II.1.7 and that of the constant terms of pseudo-Eisenstein series in II.2.2; these calculations are similar and are done in a relatively general framework, already used by Arthur. The calculation of the scalar product gives in particular the first orthogonal decomposition of $L^2(G(k)\backslash\mathbf{G})_\xi$ (see [L], 4.6(i)):

Let $(M, \mathfrak{P})$ and $(M', \mathfrak{P}')$ be cuspidal data relative to ξ; we say that $(M, \mathfrak{P})$ is equivalent to $(M', \mathfrak{P}')$ if there exists $\gamma \in G(k)$ such that $\gamma M \gamma^{-1} = M'$ and $\gamma\mathfrak{P} = \mathfrak{P}'$. Let $\mathfrak{X}$ be an equivalence class of cuspidal data: we denote by

$L^2(G(k)\backslash\mathbf{G})_{\mathfrak{X}}$ the closure of the subspace of $L^2(G(k)\backslash\mathbf{G})_\xi$ generated by the pseudo-Eisenstein series θ_ϕ corresponding to all the elements of $\mathfrak{X}$. Then if $\mathfrak{X}$, $\mathfrak{X}'$ are two distinct equivalence classes, $L^2(G(k)\backslash\mathfrak{G})_{\mathfrak{X}}$ is orthogonal to $L^2(G(k)\backslash\mathfrak{G})_{\mathfrak{X}'}$ and $L^2(G(k)\backslash\mathbf{G})_\xi$ is the completion of the orthogonal sum $\oplus_{\mathfrak{X}} L^2(G(k)\backslash\mathfrak{G})_{\mathfrak{X}}$ where $\mathfrak{X}$ runs through the set of equivalence classes of cuspidal data.

In fact, for what follows and essentially to solve convergence problems in the case where k is a number field, Langlands remarked that it is too restrictive to work only with pseudo-Eisenstein series coming from Paley-Wiener sections ϕ. Following Langlands, we introduce the notions of R-Paley-Wiener (see II.1.4), and for R large enough, we define θ_ϕ as above. We then show that θ_ϕ is still square integrable modulo the centre (II.1.10): there are no major obstacles to calculating the constant terms of θ_ϕ under this hypothesis or to extending the scalar product formula to this situation. Thanks to the density theorem, we immediately see that $\theta_\phi \in L^2(G(k)\backslash\mathbf{G})_{\mathfrak{X}}$ if ϕ is R-Paley-Wiener and is defined from a cuspidal datum belonging to $\mathfrak{X}$.

Chapter III introduces an algebra of operators. Let $\mathfrak{X}$ be an equivalence class of cuspidal data. For $(M, \mathfrak{P}) \in \mathfrak{X}$, let $f_{M,\mathfrak{P}}$ be a function on $\mathfrak{P}$, say with values in $\mathbb{C}$, to simplify. If certain properties of growth, of regularity and of invariance are satisfied, the family $f = (f_{M,\mathfrak{P}})_{(M,\mathfrak{P})\in\mathfrak{X}}$ determines an operator $\Delta(f)$ on $L^2(G(k)\backslash\mathbf{G})_{\mathfrak{X}}$. In particular, if $(M, \mathfrak{P}) \in \mathfrak{X}$ and ϕ is a Paley-Wiener function on $\mathfrak{P}$, we have $\Delta(f)\theta_\phi = \theta_{\phi'}$, where $\phi' = f_{M,\mathfrak{P}}\phi$. The space $H^R_{\mathfrak{X}}$ of these operators appears as a global analogue of the centre of the enveloping algebra of $\mathfrak{G}_\infty$. In particular, it contains an auto-adjoint operator which plays an essential role in Chapter IV to eliminate questions about convergence at infinity and whose introduction by Langlands is one of his subtlest contributions.

These operators allow us to decompose the space of automorphic forms in the following way. Let h be a conjugacy class of pairs (M, π) where M is a standard Levi and π an irreducible cuspidal representation of $\mathbf{M}$. Let us denote by $A(G(k)\backslash\mathbf{G})_{\mathfrak{y}}$ the space of automorphic forms ϕ on $G(k)\backslash\mathbf{G}$ such that the set consisting of pairs (M, π), where M is a standard Levi and $\pi \in |Pi_0(\mathbf{M}, \phi)$ is contained in $\mathfrak{y}$. Then when we let $\mathfrak{y}$ vary, the spaces $A(G(k)\backslash\mathbf{G})_{\mathfrak{y}}$ generate the space of automorphic forms on $G(k)\backslash\mathbf{G}$ (see III.2.6).

The chapter ends with a statement reformulating Lemma 7.3 of [L], according to which if ϕ is an automorphic form on $G(k)\backslash\mathbf{G}$, with unitary central character and square integrable modulo $Z_{\mathbf{G}}$, the central characters χ_π of the elements π of the cuspidal supports $\Pi_0(\mathbf{M}, \phi)$ are not arbitrary. To state the result simply, suppose that $\mathbf{G} = G(\mathbb{A})$ and $Z_{\mathbf{G}} = \{1\}$. Denote by T_M the largest split torus contained in the centre of M. We show that there exists an integer $N(G)$, depending only on G as indicated by the notation, such that for π as above, the restriction to $T_M(\mathbb{A})$ of $(\chi_\pi)^{N(G)}$ has positive real values.

In Chapter IV we prove that the Eisenstein series coming from a pair $(M, \mathfrak{P})$ consisting of a standard Levi and an orbit of irreducible cuspidal automorphic representations of **M** can be meromorphically continued to all of $\mathfrak{P}$. This is also true for the intertwining operators. The proof was communicated to us by Jacquet. He actually attributes it to Selberg. The idea of this proof can also be found in [E].

We start with the case where M is a proper maximal Levi of G. The principle of the proof can then be applied, via induction, to all of M. Suppose thus that M is proper and maximal, and suppose to simplify things that k is a number field and $\mathrm{Stab}(M, \mathfrak{P})$ consists of two elements 1 and w. We first show that the truncated Eisenstein series $\wedge^T E(\phi, \pi)$, defined in a positive half-plane, are solutions of functional equations in which only compact operators occur (Lemma IV.3.4). The usual theory of resolvents of compact operators then allows us to construct a function $\tilde{E}$ which is meromorphic on (nearly) all of $\mathfrak{P}$, such that in a positive half-plane, we have the equality:

$$(**) \qquad E(\phi, \pi) = \tilde{E}(\phi, \pi) + \tilde{E}(M(w, \pi)\phi, w\pi),$$

where $M(w, \pi)$ is the intertwining operator. But E itself satisfies a functional equation which $\tilde{E}$ does not satisfy. Applying this equation, we deduce from (**) the equality

$$0 = R(\phi, \pi) + R(M(w, \pi)\phi, w\pi),$$

see IV.3.9. From this, we deduce by inversion the meromorphic continuation of $M(w, \pi)$. We then continue $E(\phi, \pi)$ by the equality (**).

Using the lemmas of I.4 and the operators of Chapter III, we prove diverse properties of the singularities of Eisenstein series and intertwining operators. We also obtain the functional equations which in the above situation can be written

$$E(\phi, \pi) = E(M(w, \pi), w\pi)$$

$$M(w, w\pi)M(w, \pi) = 1.$$

Finally, we reproduce an argument of Harder ([H] I.6.6) which shows that, if k is a function field, the operators $M(w, \pi)$ and the values $E(\phi, \pi)(g)$ (for fixed g and suitable ϕ) are rational functions of π.

The objective of Chapters V and VI is to give the 'spectral' decomposition of $L^2(G(k)\backslash \mathbf{G})_{\mathfrak{X}}$ where $\mathfrak{X}$ is an equivalence class of cuspidal data relative to the character ξ (see above). For these chapters, we follow practically word for word Chapter VII and Appendix II of Langlands. Thus we fix $\mathfrak{X}$ and for $(M, \mathfrak{P}) \in \mathfrak{X}$, we define $S_{(M,\mathfrak{P})}$ (see V.1.1) to be a set of 'affine subspaces' of $\mathfrak{P}$ which contains all the singular hyperplanes of the intertwining operators:

$$\pi \mapsto M(w^{-1}, -w\overline{\pi})\phi(-w\overline{\pi}),$$

where w is an element of the Weyl group of G of minimal length in its right coset modulo the Weyl group of M and such that wMw^{-1} is still a standard Levi of G (we denote this set by $W(M)$; it is evidently not a group), $-\overline{\pi}$ is the Hermitian contragredient of π and ϕ is a holomorphic section of the fibre bundle over $-w\overline{\mathfrak{P}}$ which is the obvious analogue of the one described above.

We suppose moreover that $S_{(M,\mathfrak{P})}$ is stable under intersection and under conjugation in the following sense:

$$\forall w \in W(M),\ wS_{(M,\mathfrak{P})} = S_{(wMw^{-1}, w\mathfrak{P})}.$$

We set $S_{\mathfrak{X}} = \cup_{(M,\mathfrak{P})\in\mathfrak{X}} S_{(M,\mathfrak{P})}$. Let $\mathfrak{S}', \mathfrak{S}'' \in S_{\mathfrak{X}}$; we say that $\mathfrak{S}'$ and $\mathfrak{S}''$ are equivalent if $\mathfrak{S}' \in S_{(M',\mathfrak{P}')}$, $\mathfrak{S}'' \in S_{(M'',\mathfrak{P}'')}$ with $(M', \mathfrak{P}'), (M'', \mathfrak{P}'') \in \mathfrak{X}$ and if there exists $w \in W(M')$ such that $w\mathfrak{S}' = \mathfrak{S}''$ (which also implies that $wM'w^{-1} = M''$ and $w\mathfrak{P}' = \mathfrak{P}''$). We denote by $[S_{\mathfrak{X}}]$ the set of equivalence classes. We note that $S_{\mathfrak{X}}$ is a locally countable set and that it becomes locally finite if we fix a finite number of $\mathbf{K}$-types $\mathfrak{F}$, and we require that the action on ϕ above be via one of these $\mathbf{K}$-types. This is the point of view of Langlands, which we have also adopted, following him, but we do not develop it in this introduction in order to keep the notation reasonable. The goal is to associate with every element $\mathfrak{C}$ of $[S_{\mathfrak{X}}]$ a closed subspace of $L^2(G(k)\backslash\mathbf{G})_{\mathfrak{X}}$, denoted by $L^2(G(k)\backslash\mathbf{G})_{\mathfrak{C}}$ and possibly zero, such that

(1) $L^2(G(k)\backslash\mathbf{G})_{\mathfrak{C}}$ is orthogonal to $L^2(G(k)\backslash\mathbf{G})_{\mathfrak{C}'}$ if $\mathfrak{C}$ is different from $\mathfrak{C}'$,

(2) $L^2(G(k)\backslash\mathbf{G})_{\mathfrak{X}}$ is the completion of the orthogonal sum of the

$$L^2(G(k)\backslash\mathbf{G})_{\mathfrak{C}},$$

the discrete spectrum of $L^2(G(k)\backslash\mathbf{G})_{\mathfrak{X}}$ is the completion of the orthogonal sum of $L^2(G(k)\backslash\mathbf{G})_{\mathfrak{C}}$ when $\mathfrak{C}$ consits of elements of dimension 0, i.e. points.

These results from [L] do not have the abstract form given here, but [L] gives at the same time a very explicit method for constructing the orthogonal projection of $L^2(G(k)\backslash\mathbf{G})_{\mathfrak{X}}$ onto $L^2(G(k)\backslash\mathbf{G})_{\mathfrak{C}}$. We denote this projection by $\mathrm{proj}_{\mathfrak{C}}$. For this, we describe $\mathrm{proj}_{\mathfrak{C}}\,\theta_\phi$ when ϕ is R-Paley-Wiener. This projection is an integral of residues of Eisenstein series. Thus there are several problems to solve:

(a) These residues must be defined: we do this with, as our only tool, the classical residue theorem generalised to several variables. For us the definitions and the residue theorem form the object of V.1.5 and the description of the residues is in V.2.2. These residues are denoted by $\mathrm{Res}^G_{\mathfrak{G}'} E(\phi, \pi)$ for $\mathfrak{G}' \in \mathfrak{C}$ and $\pi \in \mathfrak{G}'$. (These references correspond to [L], 7.1, which is more precise than V.1.5 as we explain there, and to [L], proof of 7.7).

(b) The set over which we integrate must be defined: for this we fix $\mathfrak{G} \in \mathfrak{C}$ (everything that follows is independent of this choice) and we integrate over the imaginary axis of $\mathfrak{G}$ defined as follows: we recall that $\mathrm{Re}\,\mathfrak{G} := \{\mathrm{Re}\,\pi, \pi \in \mathfrak{G}\}$ is an affine subspace (in the usual sense) of $\mathrm{Rat}\,M \otimes_{\mathbf{Z}} \mathbb{R}$; we write $(\mathrm{Re}\,\mathfrak{G})^0$ for its vector part and we set:

$$o(\mathfrak{G}) = (\mathrm{Re}\,\mathfrak{G})^{0\perp} \cap \mathrm{Re}\,\mathfrak{G},$$

where the orthogonal is taken for an invariant scalar product with which we equip $\mathrm{Rat}\,M \otimes_{\mathbf{Z}} \mathbb{R}$. The imaginary axis is then:

$$\{\pi \in \mathfrak{G} \mid \mathrm{Re}\,\pi = o(\mathfrak{G})\}.$$

(c) The function to be integrated must be defined: we must show that it is holomorphic at every point of the set of integration and that the integral converges in an appropriate sense. For the function, let $\mathfrak{G}$ be as above and $(M, \mathfrak{P}) \in \mathfrak{C}$ such that $\mathfrak{G} \in S_{(M,\mathfrak{P})}$: we take

$$\{\pi \in \mathfrak{G} \mid \mathrm{Re}\,\pi = o(\mathfrak{G})\} \mapsto \sum_{w \in W(M)} \left(\mathrm{Res}^G_{w\mathfrak{G}} E(\phi, \pi')\right)_{\pi' = w\pi} =: {*}e_{\mathfrak{C}}(\phi, \pi),$$

where $*$ is a suitable scalar.

The study of this integral is delicate in the number field case because it does not seem to converge absolutely. Suppose thus that k is a number field. We consider the operator $\Delta(f_0)$ (see Chapter III) associated with the function f_0, element of $H^R_{\mathfrak{X}}$, defined by $f_0(\pi) = (\lambda_\pi, \lambda_\pi)$ where λ_π is the unique element of $\mathrm{Rat}\,M \otimes_{\mathbf{Z}} \mathbb{C}$ such that $\lambda_\pi \otimes \pi_0 \simeq \pi$ (π_0 is the canonical element of $\mathfrak{P}$) and where (,) is the $\mathbb{C}$-linear continuation of an invariant scalar product on $\mathrm{Rat}\,M \otimes_{\mathbf{Z}} \mathbb{R}$. The spectrum of this operator is real, and for a point of continuity T of this spectrum we write q_T for the projection onto the spectral part greater than or equal to $-T$. In fact we begin by describing:

$$L^2(G(k)\backslash\mathbf{G})_{\mathfrak{C}} \cap q_T L^2(G(k)\backslash\mathbf{G})_{\xi} =: L^2(G(k)\backslash\mathbf{G})_{\mathfrak{C},T}$$

which is neither more nor less than $q_T L^2(G(k)\backslash\mathbf{G})_{\mathfrak{C}}$. For this, we give an *ad hoc* definition ([L] first notation in 7.6 and V.2.3) of $L^2(G(k)\backslash\mathbf{G})_{\mathfrak{C},T}$:

we write $\mathrm{proj}_{\mathfrak{C},T}$ for the orthogonal projection onto this space. We then easily check that removing a subset of measure zero from

$$(**) \qquad \{\pi \in \mathfrak{G} \mid \mathrm{Re}\,\pi = o(\mathfrak{G}) \text{ and } \|\,\mathrm{Im}\,\lambda_\pi\|^2 < T + \|o(\mathfrak{G})\|^2\},$$

the map $\pi \mapsto e_{\mathfrak{C}}(\phi, \pi)$, for a fixed, R-Paley-Wiener ϕ, is holomorphic: we write $A_{\mathfrak{C},\pi}$ for the vector space generated as ϕ varies. Still removing a set of measure zero, we show that $A_{\mathfrak{C},\pi}$ is equipped with a Hilbertian scalar product characterised by the formula:

$$\langle e_{\mathfrak{C}}(\phi, \pi), e_{\mathfrak{C}}(\phi', \pi)\rangle = \langle \theta_\phi, e_{\mathfrak{C}}(\phi', \pi)\rangle,$$

for every Paley-Wiener ϕ and R-Paley-Wiener ϕ' (see V.3.9 which is 'greatly inspired' by [L], 7.5).

We thus show that $\pi \mapsto A_{\mathfrak{C},\pi}$ is a preHilbertian field which we easily transform, by completion, into a Hilbertian field. We then have a 'spectral' description of $L^2(G(k)\backslash\mathbf{G})_{\mathfrak{C},T}$ (see V.3.11), and it is with the help of this description and simple arguments from linear algebra that we show that $\pi \mapsto e_{\mathfrak{C}}(\phi, \pi)$ is holomorphic at every point of the set (**) above (see [L], proof of 7.6 and V.3.11). It is then no longer very difficult (see V.3.11) to prove the formula:

$$\mathrm{proj}_{\mathfrak{C},T}\,\theta_\phi = \int_{\substack{\pi\in\mathfrak{G},\,\mathrm{Re}\,\pi=o(\mathfrak{G})\\ \|\,\mathrm{Im}\,\lambda_\pi\|^2<T+\|o(\mathfrak{G})\|^2}} e_{\mathfrak{C}}(\phi, \pi)\,\mathrm{d}\pi.$$

We then prove that:

$$q_T\theta_\phi = \sum_{\mathfrak{C}\in[S_{\mathfrak{X}}]} \mathrm{proj}_{\mathfrak{C},T}\,\theta_\phi,$$

almost all the terms which occur are zero since we are supposing that ϕ is $\mathbf{K}$-finite in the definition of R-Paley-Wiener. For this, it is necessary to compare, and thus in particular to compute, the scalar product of $q_T\theta_\phi$ and $\mathrm{proj}_{\mathfrak{C},T}\,\theta_\phi$ against all the pseudo-Eisenstein series (see V.2.9 and V.3.4): this uses Corollary 1 of [L], 7.4, which becomes Lemmas V.2.10 and V.3.5 below. We thus decompose not $L^2(G(k)\backslash\mathbf{G})_{\mathfrak{X}}$ but $q_TL^2(G(k)\backslash\mathbf{G})_{\mathfrak{X}}$ (see [L] 7.6 and V.3.12). Finally, letting T tend to infinity, we check that the slowly decreasing functions $\mathrm{proj}_{\mathfrak{C},T}\,\theta_\phi$ (for ϕ fixed and R-Paley-Wiener) on $G(k)\backslash\mathbf{G}$, possess a limit in $L^2(G(k)\backslash\mathbf{G})_{\mathfrak{X}}$ ([L] 7.6 and 7.7 and V.3.14); the closed space generated by these limits is $L^2(G(k)\backslash\mathbf{G})_{\mathfrak{C}}$ and we obtain the desired decomposition. It is more or less immediate that $L^2(G(k)\backslash\mathbf{G})_{\mathfrak{C}}$ meets the discrete spectrum if and only if the elements of $\mathfrak{C}$ are reduced to a point (i.e. $\mathrm{proj}_{\mathfrak{C},T}$ is defined without integration). This space is then entirely contained in the discrete spectrum and we say that $\mathfrak{C}$ is the singular class attached to this space. In general, as a representation of $\mathbf{G}$, this space is not irreducible, it is even of infinite length (see the example of G_2).

The goal of Chapter VI is to describe $L^2(G(k)\backslash\mathbf{G})_{\mathfrak{C}}$ when this space does not belong to the discrete spectrum, with the help of a part of the discrete spectrum of a standard Levi of G. To avoid non-conceptual difficulties, we continue to suppose here that k is a number field. The result is then as follows.

We show that there exists a pair (L, δ), unique up to association, where $\mathbf{L}$ is a standard Levi of G and δ is a subspace of the spectrum of $\mathbf{L}$ of the form $L^2(L(k)\backslash\mathbf{L})_{\mathfrak{C}_L}$ (where $\mathfrak{C}_L$ is a singular class, as above) characterised by:

> $\forall \mathfrak{G}_L \in \mathfrak{C}_L$, $\mathfrak{G}_L$ is reduced to a point (see above), denoted by π_L, and $\pi_L \otimes X_L^G$ is an element of $\mathfrak{C}$ (where X_L^G is the analogue of X_M^G defined at the beginning of the introduction) (see ([L], 7.2) and VI.1.9).

Then we show that $L^2(G(k)\backslash\mathbf{G})_{\mathfrak{C}}$ is generated by the integrals over the imaginary axis of X_L^G of Eisenstein series of $\mathbf{L}$ to $\mathbf{G}$ associated with automorphic forms for $\mathbf{L}$ contained in δ. The integral is just a limit in the L^2 sense explained above (see VI.2.2 and the formulation given by Arthur at Corvallis ([A1]) of Langlands' results). In fact, this description of $L^2(G(k)\backslash\mathbf{G})_{\mathfrak{C}}$ is, as Langlands explains in Appendix II of his description, a consequence for all general π of the spaces $A_{\mathfrak{C},\pi}$ introduced above (see [L], 7.4 and VI.2.4) and of the functional equation for the residues of Eisenstein series (see VI.1.5). We show in particular that $A_{\mathfrak{C},\pi}$ (for general π) can be identified with the space of Eisenstein series from $\mathbf{L}$ to $\mathbf{G}$ constructed from automorphic forms contained in δ. For this, we use relatively precise information about the locations of the poles of the residues of Eisenstein series, $\mathrm{Res}_{\mathfrak{G}}^{G} E(\phi, \pi)$ defined above, which is [L], corollary to 7.6 and VI.1.2. Langlands uses this information in the proof of 7.7; we use it in the 'reduction' to the Levi (see VI.1.4) which is implicit in ([L], 7.4 and after).

In Appendix I, we show that the covering $\mathbf{G} \to G(A)$ is split over every subgroup $U(\mathbb{A})$, where U is the unipotent radical of a parabolic subgroup of G. This is easy to prove if the characteristic of k does not divide the number of sheets of the covering. It is much less easy in the general case. The proof was communicated to us by P. Deligne.

In Appendix II we show that if k is a function field, every automorphic form on $G(k)\backslash\mathbf{G}$ is a linear combination of derivatives of Eisenstein series (see 1.4). The reasoning is as follows. Let ϕ be an automorphic form on $G(k)\backslash\mathbf{G}$. For every pseudo-Eisenstein series θ_Φ, the integral

$$I(\phi, \Phi) = \int_{G(k)\backslash\mathbf{G}} \phi(g)\theta_\Phi(g)\,dg$$

is convergent. It is easy enough to express this by means of cuspidal components of constant terms of ϕ (see 2.2). Then the implication

$$\theta_\Phi = 0 \Rightarrow I(\phi, \Phi) = 0$$

imposes on these components the condition that they satisfy a system of linear equations. In particular, the cuspidal components of derivatives of Eisenstein series are solutions of this system. We show that every solution of this system comes from such a function. The argument is algebraic: it is here that we use the hypothesis that k is a function field. The final result is an easy consequence of this.

Appendix III studies the part of the discrete spectrum of a group of type G_2 coming from a minimal Levi and its trivial representation. Langlands showed that the subspace of vectors of this spectrum invariant under the maximal compact subgroup is of dimension 2. We remove this hypothesis of invariance (we do keep it for the archimedean places; we suppose k is a number field). The result is bizarre...

In Appendix IV, we study the modifications necessary to the theory when G is no longer supposed (algebraically) connected. We are thinking, for example, of the case of the orthogonal group. Essentially, we can say that the results are exactly the same as in the connected case, as long as the split ranks of the center of G and of its connected component of the identity G^0 are equal. In order to perform the inductions, we are led to define the notion of Levi subgroup in such a way that these subgroups satisfy these conditions. For this, if M^0 is a Levi subgroup of G^0, we associate with it a Levi subgroup M of G which is by definition the commutator in G of the largest split central torus in M^0. With these definitions, the results of Chapters 5 and 6 can be extended to the non-connected case. There are however some problems which must be resolved. For example, over a local field, there does not always exist a compact maximal subgroup intersecting every connected component. There is no Iwasawa decomposition in this case. But we show that this does not seriously perturb the proofs. If G does not satisfy the condition mentioned above, we must introduce the group G', the commutator in G of the largest split central torus in G^0. This group satisfies the desired condition and the spectral decomposition for the group G can be deduced by induction from the one for G'. Note, for example, that for such a group, the 'discrete spectrum' is reduced to $\{0\}$!

Notation

k, $\mathbb{A}$, $\mathbb{A}_\infty$, $\mathbb{A}_f$, q, G, K, $\mathbf{G}$, i_G I.1.1
P_0, M_0, Z_G, $\mathbf{K}$, $\mathbf{M}$, $Z_{\mathbf{G}}$, $\mathrm{Rat}(M)$, M^1, $\mathbf{M}^1$, $|\chi|$, $\mathfrak{a}_M^*$, $\mathrm{Re}\,\mathfrak{a}_M^*$, $\mathfrak{a}_M$, $\mathrm{Re}\,\mathfrak{a}_M$, X_M, $X_M^{\mathbf{G}}$, $\mathrm{Re}\,X_M$, $\mathrm{Im}\,X_M$, $\kappa : \mathfrak{a}_M^* \to X_M$, $\log_M$,
$m_P : \mathbf{G} \to \mathbf{M}^1\backslash\mathbf{M}$ I.1.4
$Z_{\mathbf{G}}^1$ I.1.5
T_0, $R(T_0, G)$, $R^+(T_0, G)$, Δ_0, $R(T_0, M)$, Δ_0^M, T_M, $R(T_M, G)$, Δ_M, $(\mathfrak{a}_M^{M'})^*$, I.1.6
W, W_M, $W(M)$ I.1.7
$\check{\alpha}$, α^* I.1.11
ρ_0, ρ_P I.1.13
$A_{\mathbf{M}}$, $\mathbf{M}^c$, S, S^P I.2.1
$\|g\|$ I.2.2
δ, $\mathcal{U}$ I.2.3
$L^2(G(k)\backslash\mathbf{G})_\xi$ I.2.5
ϕ_P I.2.6
$\mathbf{M}_0(P, t)$ I.2.7
$s\phi$ I.2.9
$\hat{\tau}_P$, $\Lambda^T\phi$ I.2.13
$\mathfrak{z}$, $\mathfrak{z}^{\mathbf{M}}$, ϕ_k, $A(U(\mathbb{A})M(k)\backslash\mathbf{G})$, $A(U(\mathbb{A})M(k)\backslash\mathbf{G})_\xi$ I.2.17
ϕ^{cusp}, $A_0(U(\mathbb{A})M(k)\backslash\mathbf{G})_\xi$ I.2.18
$X_M(A_{\mathbf{M}})$, $\mathfrak{z}(A_{\mathbf{M}}; \Lambda, N)$, $\mathbb{C}[\mathrm{Re}\,\mathfrak{a}_M]$, $\mathfrak{q}_M$ I.3.1
$A(U(\mathbb{A})M(k)\backslash\mathbf{G})_Z$ I.3.2
$\Pi_0(\mathbf{M})_\xi$, $\Pi_0(\mathbf{M})$, $A_0(U(\mathbb{A})M(k)\backslash\mathbf{G})_\pi$, χ_π, $\mathrm{Re}\,\pi$, $\mathrm{Im}\,\pi$, $-\pi$, $-\overline{\pi}$ I.3.3
$D(M, \phi)$, $\Pi_0(\mathbf{M}, \phi)$ I.3.3 and I.3.5

$C_0(U(\mathbb{A})M(k)\backslash\mathbf{G})$ I.3.4
ϕ_P^{cusp} I.3.5
$A(d,R,Y)$ I.4.2
$s(\varphi,r)$ I.4.3
$A((V_P,\Gamma_P,N_P)_{P_0\subset P\subset G})$ I.4.4
$\mathfrak{w}_{\check{\alpha}}$ I.4.11
$A(U(\mathbb{A})M(k)\backslash\mathbf{G})_\pi$ II.1.1
$\mathrm{Fix}_{X_M^{\mathbf{G}}}\mathfrak{P}$ II.1.1
$P_{(M,\mathfrak{P})}$ II.1.2
$\epsilon F(\phi)$ II.1.3
$P_{(M,\mathfrak{P})}^R$ II.1.4
$E(\phi,\pi)$ II.1.5
$M(w,\pi)$ II.1.6
$W(M,M')$ II.1.7
θ_ϕ II.1.10
$(M,\mathfrak{P})\sim(M',\mathfrak{P}')$ II.2.1
$\mathfrak{P}_{M'}$, $P_{(M,\mathfrak{P}_{M'})}^R$ II.2.2
$\Theta_{\mathfrak{X}}^R$, $L_{\mathfrak{X}}^2$, $H_{\mathfrak{X}}^R$, $\tilde{\mathfrak{X}}^R$, $H_{\mathfrak{X},\mathbb{C}}^R$, $f_{M,\pi}$ III.1.1
$\Delta_\theta(f)$ III.1.2
$H_{\mathfrak{X},b}^R$ III.1.4
$\Delta(f)$ III.1.4 and III.2.1
λ_π, Δ III.1.5
p_T, q_T III.1.6 and V.2.8
Θ_ξ^R, H_ξ^R, Θ_ξ, $H_{\xi,b}^R$, $H_{\xi,\mathbb{C}}^R$ III.1.7
$\Theta_{\eta^\bullet}^{M,G}$ III.2.3
$H_\xi^{\exp}$ III.2.4
$A^2(G(k)\backslash\mathbf{G})_\xi$ III.3.1
$N(G)$ III.3.2
$\mathfrak{F}$, A_ξ, L_ξ^2, $L_{\xi,\mathrm{loc}}^2$, $\mathscr{H}^{\mathfrak{F}}$ IV.1.1
$\mathscr{C}_{\mathfrak{P}}$ IV.1.7
$\mathscr{C}$ IV.2.1
k_h, $\Lambda_1^T k_h$ IV.2.5
E^T IV.3.2
F^T IV.3.3
V_h, e^T, $V_{w,h}$ IV.3.5
$\tilde{E}$ IV.3.6
$\mathfrak{X}$, ξ V.1.1
$\mathrm{Stab}(M,\mathfrak{P})$ V.1.1
$\mathfrak{G}^0$, $\tilde{\mathfrak{G}}^0$ V.1.1
$H_{\alpha^\bullet}$ V.1.1
$S_{(M,\mathfrak{P})}$, $S_{\mathfrak{X}}$ V.1.1

$S_{\mathfrak{x}}^{\mathfrak{F}}$ V.1.1

$\mathfrak{G}_{\le T} = T$ V.1.5(a)

λ_π V.1.5(a)

$P_{\mathfrak{x}}^{\mathfrak{F}}$ V.2.1

$R,\; P_{(M,\mathfrak{P})}^{R,\mathfrak{F}}$ V.2.1

$o(\mathfrak{G})$ V.2.1

$T - \mathfrak{F}$-general point V.2.1

$A(\phi',\phi)$ V.2.1 and V.3.1

$\mathrm{Norm}\,\mathfrak{G},\; \mathrm{Res}_{\mathfrak{G}}^{G}$ V.2.2

q_T V.2.8

H^R V.2.9

$\mathfrak{G}_{\mathfrak{C}},\; z(\mathfrak{G})$ V.3.1

$\mathfrak{G} \sim \mathfrak{G}'$ V.3.1

$[S_{\mathfrak{x}}],\; [S_{\mathfrak{x}}^{\mathfrak{F}}]$ V.3.1

$r_{\mathfrak{C}}(\phi',\phi),\; m_{\mathfrak{C},T}(\phi',\phi)$ V.3.1

$\mathrm{proj}_{\mathfrak{C},T}^{\mathfrak{F}}$ V.3.2 and V.3.3

$e_{\mathfrak{C}}(\phi,\pi),\; A_{\mathfrak{C},\pi}^{\mathfrak{F}}$ V.3.2

$L_{\mathfrak{C},T}^{2,\mathfrak{F}}$ V.3.2

$P_{\mathfrak{C},T'}$ V.3.3

$P_{\mathfrak{C},T'}^{\mathfrak{F}},\; P_{\mathfrak{C},T'}^{R,\mathfrak{F}}$ V.3.3

$\mathfrak{C} > \mathfrak{C}'$ V.3.3

$[S_{\mathfrak{x}}']$ V.3.3

$\mathrm{Hilb}_{\mathfrak{C},T}^{\mathfrak{F}},\; F_\phi$ V.3.10 or 11

$\mathrm{Sing}_T^{G,\mathfrak{F}},\; \mathrm{Sing}^{G,\mathfrak{F}}$ V.3.12

$L^2(G(k)\backslash \mathbf{G})_{\mathfrak{x},\mathfrak{F}}$ V.3.12 corollary

Sing^G V.3.14

$\mathrm{Hilb}_{\mathfrak{C}}$ V.3.14 corollary

$\xi_L,\; \mathfrak{P}_L$, VI.1 introduction

$M_{\mathfrak{G}}$ VI.1.1

$\mathfrak{X}_L,\; [S_{\mathfrak{x}_L}]$ VI.1.4(a)

$\mathrm{Res}_{\mathfrak{G}_L}^{L}$ VI.1.4(b)

W^L VI.2.1(iv)

$\mathrm{Der}\,X_M$ Appendix II, 1.3

$Y,\; P^{\mathfrak{F}}$ Appendix II, 2.1

$N(w,\pi)$ Appendix II, 2.3

$S_M,\; \hat{S}_M,\; \hat{S}_M^{\ge N},\; S_{M,\pi},\; \hat{S}_{M,\pi},\; \hat{S}_{M,\pi}^{\ge N},\; s_{M,\pi}$ Appendix II, 2.4

$\hat{I}(.,.)$ Appendix II, 2.5

I

Hypotheses, Automorphic Forms, Constant Terms

I.1. Hypotheses and general notation

I.1.1. Definitions

Let k be a global field and $\mathbb{A}$ be the ring of adeles of k. For a finite place v of k, we write ϑ_v for the ring of integers. Let $\mathbb{A}_f$ be the ring of finite adeles of k and $\mathbb{A}_\infty = \prod k_v$, the product being over the archimedean places. If k is a function field, let q be the number of elements of its field of constants.

Let G be a connected reductive algebraic group defined over k. Fix an embedding into a linear group as follows. First choose an embedding $i'_G : G \hookrightarrow GL_n$, defined over k, with closed image. Then $i_G : G \hookrightarrow GL_{2n}$ is defined by

$$i_G(g) = \begin{pmatrix} i'_G(g) & 0 \\ 0 & {}^t i'_G(g)^{-1} \end{pmatrix}.$$

There exists a finite set S of places of k, containing the archimedean places, such that the image of i_G is defined and smooth over $\vartheta^S := \prod_{v \notin S} \vartheta_v$ (see [Sp] §4.9). For $v \notin S$, this allows us to define the group $G(\vartheta_v)$ of points with values in ϑ_v. For almost all $v \notin S$, this is a maximal compact subgroup of $G(k_v)$ (see [Sp] p.18, 1.3 and what follows). We fix a compact maximal subgroup K of $G(\mathbb{A})$ such that $K = \prod_v K_v$, product over all places of k, where K_v is a maximal compact subgroup of $G(k_v)$. We suppose, as we may, that $K_v = G(\vartheta_v)$ for almost all finite places. We will impose further properties on K in I.1.4.

Let $\mathbf{G}$ be a topological group which is a finite central covering of $G(\mathbb{A})$ (see [M3]), i.e. there exists a surjective morphism pr of topological groups, whose kernel $\mathbf{N}$ is a finite subgroup of the centre of $\mathbf{G}$, pr being a topological covering. This last condition can be interpreted as follows: there exists an open neighborhood U of the unit element of $G(\mathbb{A})$ and an isomorphism j of $\mathrm{pr}^{-1}(U)$ onto $U \times \mathbf{N}$ which makes the following diagram commute:

$$\begin{array}{ccc} \mathrm{pr}^{-1}(U) & \xrightarrow{j} & U \times \mathbf{N} \\ & \mathrm{pr} \searrow \quad \swarrow \mathrm{pr}_1 & \\ & U & \end{array}$$

where pr_1 denotes projection onto the first factor.

I.1.2. Description of G

We begin with a remark.

Remark *There exists a finite set S of places of k such that for all $v \notin S$, K_v lifts into* $\mathbf{G}$.

Proof Fix U as above. We require that j satisfy the condition $j(1_{\mathbf{G}}) = 1_{\mathbf{N}} \times 1_{G(\mathbb{A})}$, where 1 with an index indicates the unit element of the group in the index. Set $\tilde{U} = j^{-1}(U \times 1_{\mathbf{N}})$. As j is a homeomorphism, $\tilde{U}$ is an open set of $\mathrm{pr}^{-1}(U)$ containing $1_{\mathbf{G}}$; thus it is a neighborhood of $1_{\mathbf{G}}$ on which the restriction of pr coincides with j and induces an isomorphism onto U. Fix an open neighborhood $\tilde{U}'$ of $1_{\mathbf{G}}$ in $\mathbf{G}$, contained in $\tilde{U}$, such that $\tilde{U}'\tilde{U}' \subset \tilde{U}$. Set $U' = \mathrm{pr}(\tilde{U}')$. Let S be a finite set of places of k and U'_S a neighborhood of the unit element of $\prod_{v \in S} G(k_v)$ such that

$$U' \supset U'_S \times \prod_{v \notin S} K_v;$$

the existence of S and U'_S results from the description of the adelic topology of $G(\mathbb{A})$. For all $v \notin S$, K_v can be identified with a subgroup of U' and we set $\tilde{K}_v = \mathrm{pr}^{-1}(K_v) \cap \tilde{U}$; we note that $\tilde{K}_v$ is contained in $\tilde{U}'$. By the preceding discussion, pr induces an isomorphism of $\tilde{K}_v$ onto K_v. In particular, if v is finite, $\tilde{K}_v$ is open and compact. The property of $\tilde{U}'$ then implies that $\tilde{K}_v$ is a subgroup of $\mathbf{G}$, which gives the desired result. □

The lifting defined above is clearly not canonical and has no reason to be unique. It does however satisfy the following property:

(1) for every open set $\mathbf{U}$ of $\mathbf{G}$ containing $1_{\mathbf{G}}$, there exists a finite set of places S such that $\mathbf{U}$ contains $\prod_{v \notin S} \tilde{K}_v$.

Two systems of liftings (i.e. liftings of K_v for almost all v) satisfying this hypothesis coincide almost everywhere.

For a place v of k, we denote by $\mathbf{G}_v$ the inverse image of $G(k_v)$ under pr. It is clear that $\mathbf{G}_v$ equipped with the topology induced by that of $\mathbf{G}$ is a topological group. There is an exact sequence realising $\mathbf{G}_v$ as a covering of $G(k_v)$:

$$1 \longrightarrow \mathbf{N} \longrightarrow \mathbf{G}_v \xrightarrow{\text{pr}} G(k_v) \longrightarrow 1$$

For almost all v, fix a lifting $\tilde{K}_v$ of K_v such that we obtain a system of liftings satisfying (1) above, and set

$$\prod_v{}' \mathbf{G}_v = \{(g_v)_{v \text{ a place of } k}; g_v \in \mathbf{G}_v \text{ and for almost all } v, g_v \in \tilde{K}_v\}.$$

This is independent of the choice of liftings. Put the adelic topology on $\prod_v' \mathbf{G}_v$. This group contains in its centre the group

$$\bigoplus_v \mathbf{N} = \{(n_v)_{v \text{ a place of } k}; n_v \in \mathbf{N} \text{ and } n_v = 1 \text{ for almost all } v\}.$$

Define

$$(\bigoplus_v \mathbf{N})^1 = \{(n_v) \in \bigoplus_v \mathbf{N};\ \prod n_v = 1\}.$$

We then have the isomorphism

$$\prod_v{}' \mathbf{G}_v/(\bigoplus_v \mathbf{N})^1 \xrightarrow{\sim} \mathbf{G}. \tag{2}$$

Conversely, let $\mathbf{N}$ be a finite abelian group and for every place v of k, fix a topological group $\mathbf{G}_v$ which is a central extension of $G(k_v)$ by $\mathbf{N}$, i.e. such that

$$1 \longrightarrow \mathbf{N} \longrightarrow \mathbf{G}_v \xrightarrow{\text{pr}_v} G(k_v) \longrightarrow 1.$$

We suppose that for almost all finite places of k, there exists a compact open subgroup $\tilde{K}_v$ of $\mathbf{G}_v$ such that the restriction of pr_v to $\tilde{K}_v$ is an isomorphism from $\tilde{K}_v$ onto K_v. We define the topological group $\prod_v{}' \mathbf{G}_v$ as above, as well as its subgroup $(\bigoplus_v \mathbf{N})^1$. Then

$$\mathbf{G} := (\prod_v{}' \mathbf{G}_v)/(\bigoplus_v \mathbf{N})^1$$

is a natural extension of $G(\mathbb{A})$ by $\mathbf{N}$.

I.1.3. The covering $\mathbf{G}_v \xrightarrow{\text{pr}} G(\mathbb{A})$: hypotheses and properties

Suppose that

(1) $G(k)$ lifts to a subgroup of $\mathbf{G}$.

Fix once and for all a lifting of it which we also denote by $G(k)$. Note that:

(2) if C is a commutative subgroup of $G(\mathbb{A})$ and $\mathbf{C}$ is its inverse image, then the centre of $\mathbf{C}$ contains the inverse image of

$$C^{|\mathbf{N}|}(:= \{c^{|\mathbf{N}|}; c \in C\}),$$

where $|\mathbf{N}|$ is the number of elements of $\mathbf{N}$.

The proof of this is elementary.

We need to lift the unipotent subgroups of $G(\mathbb{A})$. More precisely, let P be a parabolic subgroup of G (defined over k) and U its unipotent radical. In the appendix we prove that

(3) $U(\mathbb{A})$ lifts canonically into $\mathbf{G}$.

We still use $U(\mathbb{A})$ to denote the image of this lifting. Then $U(\mathbb{A})$ is the adelic product of subgroups defined for every place v of k and isomorphic to $U(k_v)$. Moreover (3) implies that

(4) The inverse image of the normaliser of $U(\mathbb{A})$ in $G(\mathbb{A})$ is the normaliser of $U(\mathbb{A})$ in $\mathbf{G}$.

We will return to the properties of the lifting $\mathrm{pr} : \mathbf{G} \to G(\mathbb{A})$ later on (see I.1.5).

I.1.4. Levi subgroups and characters

Let $k, A, G, \mathbf{G}, \mathrm{pr}, K$ be as previously introduced (see I.1.1). We denote by $\mathbf{K}$ the inverse image of K in $\mathbf{G}$. Let P be a parabolic subgroup of G (i.e. G/P is a complete algebraic variety), defined over k, of unipotent radical U, and let M be a Levi subgroup of P. Let $\mathbf{M}$ denote the inverse image of $M(\mathbb{A})$ in $\mathbf{G}$. We still use $U(\mathbb{A})$ to denote the canonical lifting of $U(\mathbb{A})$ into $\mathbf{G}$ (see I.1.3 (3)). Fix a parabolic subgroup P_0 of G, defined over k and minimal, and a Levi subgroup M_0 of P_0, defined over k. We use the phrase 'standard parabolic subgroup of G' to denote any parabolic subgroup of G defined over k and containing P_0. By a 'standard Levi subgroup of G' we mean any Levi subgroup, containing M_0, of a standard parabolic subgroup of G. Every standard parabolic subgroup possesses a unique standard Levi subgroup.

It is possible to require that the group K have the following properties:

(i) $G(\mathbb{A}) = P_0(\mathbb{A})K$;

(ii) for every standard parabolic subgroup $P = MU$ of G, the equality $P(\mathbb{A}) \cap K = (M(\mathbb{A}) \cap K)(U(\mathbb{A}) \cap K)$ is satisfied, and $M(\mathbb{A}) \cap K$ is still a compact maximal subgroup of $M(\mathbb{A})$.

Thanks to (ii), the choice of K also fixes a choice of compact maximal subgroup of $M(\mathbb{A})$ for every standard Levi M.

Remark Hypothesis (ii) is useful but not indispensable. If it was not satisfied, we would have to choose a compact maximal subgroup of $M(\mathbb{A})$ containing the image in $M(\mathbb{A}) \simeq P(\mathbb{A})/U(\mathbb{A})$ of $P(\mathbb{A}) \cap K$.

Denote by Z_G the centre of G, by $Z_{\mathbf{G}}$ that of $\mathbf{G}$ and for M a Levi subgroup of G, denote by Z_M the centre of M and by $Z_{\mathbf{M}}$ that of $\mathbf{M}$.

Remark In general, $Z_{\mathbf{M}}$ is not equal to the inverse image of $Z_M(\mathbb{A})$ under pr. However, we will show as in I.1.3 (2), that $Z_{\mathbf{M}}$ contains $\mathrm{pr}^{-1}(Z_M(\mathbb{A})^{|\mathbf{N}|})$.

Fix a standard Levi M of G; let $\mathrm{Rat}(M)$ denote the group of rational characters of M (i.e. the homomorphisms as algebraic groups of M into the multiplicative group $\mathbb{G}_m$). Set

$$\mathfrak{a}_M^* = \mathrm{Rat}(M) \bigotimes_{\mathbb{Z}} \mathbb{C}, \quad \mathfrak{a}_M = \mathrm{Hom}_{\mathbb{Z}}(\mathrm{Rat}(M), \mathbb{C}).$$

The complex vector spaces $\mathfrak{a}_M$ and $\mathfrak{a}_M^*$ are duals of each other. Note that they are naturally equipped with $\mathbb{Q}$-structures: if we set

(1) $$\mathfrak{a}_{M,\mathbb{Q}}^* = \mathrm{Rat}(M) \bigotimes_{\mathbb{Z}} \mathbb{Q}, \quad \mathfrak{a}_{M,\mathbb{Q}} = \mathrm{Hom}_{\mathbb{Z}}(\mathrm{Rat}(M), \mathbb{Q}),$$

then obviously

$$\mathfrak{a}_M^* = \mathfrak{a}_{M,\mathbb{Q}}^* \bigotimes_{\mathbb{Q}} \mathbb{C}, \quad \mathfrak{a}_M = \mathfrak{a}_{M,\mathbb{Q}} \bigotimes_{\mathbb{Q}} \mathbb{C}.$$

We will also use the real spaces

$$\mathrm{Re}\,\mathfrak{a}_M^* = \mathrm{Rat}(M) \bigotimes_{\mathbb{Z}} \mathbb{R}, \quad \mathrm{Re}\,\mathfrak{a}_M = \mathrm{Hom}_{\mathbb{Z}}(\mathrm{Rat}(M), \mathbb{R}).$$

Let $\chi \in \mathrm{Rat}(M)$: for every place v of k, χ defines an algebraic character denoted by χ_v of $M(k_v)$ into k_v^*. We define $|\chi|$, a continuous homomorphism of $M(\mathbb{A})$ into $\mathbb{C}^*$, by

$$\forall m = (m_v) \in M(\mathbb{A}), \; m^{|\chi|} = \prod_v |m_v^{\chi_v}|_v,$$

where $|\;|_v$ is the absolute value of k_v (in general, if H is a group, h an element of H and χ a character of H, we write h^χ for the value of χ at the point h, which induces us to write the group law on the set of characters of H additively). Let

$$M^1 = \bigcap_{\chi \in \mathrm{Rat}(M)} \mathrm{Ker}|\chi|;$$

it is a normal subgroup of $M(\mathbb{A})$, and let X_M be the group of continuous homomorphisms of $M(\mathbb{A})$ into $\mathbb{C}^*$ which are trivial on M^1.

Given the importance of the group X_M in what follows, we give a further description of it here.

(2) Let $\lambda \in X_M$. Then there exist $\chi_1,\ldots,\chi_R 2 \in \mathrm{Rat}(M)$ and $s_1,\ldots,s_R \in \mathbb{C}$ such that

$$\forall m \in M(\mathbb{A}),\ m^{\lambda} = (m^{|\chi_1|})^{s_1}\ldots(m^{|\chi_R|})^{s_R}.$$

Proof The group $\mathrm{Rat}(M)$ is a free $\mathbb{Z}$-module since M is connected. Fix a basis $\chi_1,\ldots,\chi_R$ of this $\mathbb{Z}$-module and denote by V the subgroup of $\mathbb{R}_+^*$ which is the image of $\mathbb{A}^*$ under the absolute value map (if k is a number field then $V = \mathbb{R}_+^*$ and if k is a function field then $V = q^{\mathbb{Z}}$). Define a map

$$j : M(\mathbb{A}) \to V^R$$

by $j(m) = (m^{|\chi_1|},\ldots,m^{|\chi_R|})$. Its kernel is M^1 since

$$M^1 = \bigcap_{i=1}^{R} \mathrm{Ker}\,|\chi_i|.$$

Its image is all of V^R if k is a number field, a subgroup of V^R of finite index if k is a function field. For suppose first that M is a split torus. Then the map

$$M(\mathbb{A}) \longrightarrow \mathbb{A}^{*R}$$

$$m \longmapsto (m^{\chi_1},\ldots,m^{\chi_R})$$

is surjective and the assertion is clear. In the general case, it suffices to prove the same assertion for the subgroup $j(T_M(\mathbb{A}))$, where T_M is the maximal split torus of Z_M. The set of restrictions $\{\chi_{i|T_M}; i = 1,\ldots,R\}$ does not generate $\mathrm{Rat}(T_M)$ but remains linearly independent and generates a subgroup of finite index of $\mathrm{Rat}(T_M)$, which concludes the argument.

We check that j defines a topological group isomorphism of $M(\mathbb{A})/M^1$ onto the image $j(M(\mathbb{A}))$. A continuous homomorphism of $M(\mathbb{A})/M^1$ into $\mathbb{C}^*$ thus comes from a continuous homomorphism of V^R into $\mathbb{C}^*$ and the latter is always of the form

$$(x_1,\ldots,x_R) \longmapsto x_1^{s_1}\ldots x_R^{s_R}$$

for some suitable $s_1,\ldots,s_R \in \mathbb{C}$. □

In other words, X_M can be realised as a quotient of $\mathfrak{a}_M^*$, i.e. there exists a surjective morphism of groups

(3) $$\kappa : \mathfrak{a}_M^* \to X_M.$$

The above argument proves that κ is bijective if k is a number field; if k is a function field, the kernel of κ is of the form $(2\pi i/\log q)L$, where L is a lattice of $\mathrm{Rat}(M)\otimes_{\mathbb{Z}}\mathbb{Q}$ containing the image of $\mathrm{Rat}(M)$ in this vector space. We obtain the same result in an equivalent way by the following considerations. There exists a natural map of $M(A)$ into $\mathfrak{a}_M$, denoted by $\log_M$, defined by

(4) $$\forall m \in M(\mathbb{A}),\ \forall \chi \in \mathrm{Rat}(M),\ \langle \chi, \log_M(m)\rangle = \log(m^{|\chi|}).$$

The kernel of $\log_M$ is precisely M^1 and $M(\mathbb{A})/M^1$ can be identified via $\log_M$ with a subgroup of $\mathfrak{a}_M$ and even of $\mathrm{Re}\,\mathfrak{a}_M$. The kernel of κ is the set of $\lambda \in \mathfrak{a}_M^*$ such that $e^{\langle \lambda, H\rangle} = 1$ for all H in the image of $\log_M$. If k is a number field, then $\log_M(M(\mathbb{A})/M^1) \simeq \mathrm{Re}\,\mathfrak{a}_M$, and κ is bijective. If k is a function field, $\log_M(M(\mathbb{A})/M^1)$ is a lattice of $\mathrm{Re}\,\mathfrak{a}_M$ contained in $\mathrm{Hom}_{\mathbb{Z}}(\mathrm{Rat}(M), (\log q)\mathbb{Z})$.

Let us return to the situation where k is arbitrary. Set

$$\mathrm{Re}\,X_M = \kappa(\mathrm{Re}\,\mathfrak{a}_M^*),\ \mathrm{Im}\,X_M = \kappa(\mathrm{i}\,\mathrm{Re}\,\mathfrak{a}_M^*).$$

It is a consequence of the above arguments that κ induces an isomorphism

(5) $$\mathrm{Re}\,\mathfrak{a}_M^* \simeq \mathrm{Re}\,X_M.$$

We check that $\mathrm{Re}\,X_M$ is the group of characters of $M(\mathbb{A})/M^1$ (i.e. of continuous homomorphisms of $M(\mathbb{A})/M^1$ into $\mathbb{C}^*$) with values in $\mathbb{R}_+^*$. The group X_M can be identified with the group of characters of $\mathbf{M}$ trivial on $\mathbf{M}^1$, where $\mathbf{M}^1$ is the inverse image of M^1. The group $\mathrm{Re}\,X_M$ can then be identified with the group of characters of $\mathbf{M}/\mathbf{M}^1$ with values in $\mathbb{R}_+^*$. As we will see later on, if we equip $\mathbf{M}^1$ with a Haar measure, the quotient $M(k)\backslash\mathbf{M}^1$ is of finite volume. Thus a character of $\mathbf{M}^1$ with values in $\mathbb{R}_+^*$ which is trivial on $M(k)$ is necessarily trivial. Then $\mathrm{Re}\,X_M$ can also be identified with the group of characters of $\mathbf{M}$ with values in $\mathbb{R}_+^*$ which are trivial on $M(k)$. By Lemma I.1.6 below, it can similarly be identified with the group of characters of $Z_{\mathbf{M}}$ with values in $\mathbb{R}_+^*$ which are trivial on $Z_M(k) \cap Z_{\mathbf{M}}$.

(6) We write $X_M^{\mathbf{G}}$ for the subgroup of X_M consisting of the characters of $\mathbf{M}/\mathbf{M}^1$ trivial on $Z_{\mathbf{G}}$.

This group will play an essential role in our constructions. We will return to it later on (see I.1.6 (6)). Let us still write $\log_M$ for $\log_M$ composed with pr. Let p be the standard parabolic subgroup of Levi subgroup M and let U be its unipotent radical. Via the equality $\mathbf{G} = \mathbf{PK}$ (which is a consequence of I.1.4 (i)), we define a map

$$m_P : \mathbf{G} \longrightarrow \mathbf{M}^1\backslash\mathbf{M}$$

by $m_P(g) = \mathbf{M}^1 m$ if $g = umk$, where $u \in U(\mathbb{A})$, $m \in \mathbf{M}$, $k \in \mathbf{K}$. This is well-defined, for $\mathbf{M} \cap \mathbf{K}$ is contained in $\mathbf{M}^1$ and $\mathbf{M}^1$ is a normal subgroup of $\mathbf{M}$.

Let $M \subset M'$ be the standard Levis of parabolic subgroups $P \subset P'$;

the restriction map identifies $X_{M'}$ with a subgroup of X_M, so $\mathrm{Re}\, X_{M'}$ is naturally a vector subspace of $\mathrm{Re}\, X_M$ and there are maps $\mathfrak{a}^*_{M'} \hookrightarrow \mathfrak{a}^*_M$ and $\mathfrak{a}_M \to \mathfrak{a}_{M'}$. We also have a commutative diagram

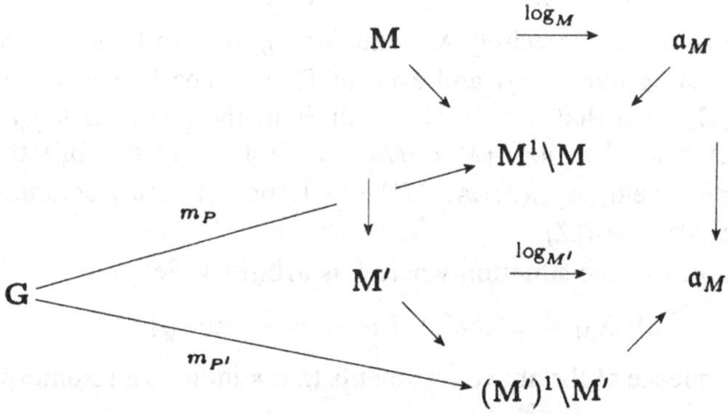

I.1.5. Description of G continued

Set $Z^1_G = Z_G(\mathbb{A}) \cap G^1$, $Z^1_{\mathbf{G}} = Z_{\mathbf{G}} \cap \mathbf{G}^1 = Z_{\mathbf{G}} \cap pr^{-1}(Z^1_G)$.

Lemma *The subgroups $Z_G(k)Z_{\mathbf{G}}$ and $G(k)Z_{\mathbf{G}}$ of* **G** *are closed. The quotient $Z_G(k) \cap Z_{\mathbf{G}} \backslash Z^1_{\mathbf{G}}$ is compact.*

Proof We begin by showing that, for every place v of k, there exists a normal open subgroup of finite index $\mathbf{G}^\circ_v$ of $\mathbf{G}_v$ which commutes with $\mathrm{pr}^{-1}(Z_G(k_v))$. For $g \in \mathbf{G}_v$ and $z \in \mathrm{pr}^{-1}(Z_G(k_v))$, we have an equality

$$z^{-1}gz = \chi(g, z)g,$$

where χ is a bicharacter of $\mathbf{G}_v \times \mathrm{pr}^{-1}(Z_G(k_v)) \to \mathbf{N}$. As pr is a covering, there exist open neighborhoods U, V of 1 in $\mathbf{G}_v$, $\mathrm{pr}^{-1}(Z_G(k_v))$ respectively, such that χ is trivial on $U \times V$. Let $\mathbf{G}^N_v$, $\mathrm{pr}^{-1}(Z_G(k_v))^N$ be the set of Nth powers in $\mathbf{G}_v$, $\mathrm{pr}^{-1}(Z_G(k_v))$ respectively, where $N = |\mathbf{N}|$. The bicharacter χ is also trivial on $\mathbf{G}^N_v \times pr^{-1}(Z_G(k_v))$ and $\mathbf{G}_v \times \mathrm{pr}^{-1}(Z_G(k_v))^N$. Let U_0, V_0 be the normal subgroups generated by U and $\mathbf{G}^N_v$, V and $\mathrm{pr}^{-1}(Z_G(k_v))^N$ respectively. Then χ is trivial on $U_0 \times V_0$. We show easily that U_0, V_0, is of finite index in $\mathbf{G}_v$, $\mathrm{pr}^{-1}(Z_G(k_v))$ respectively. Let $z_1, \ldots, z_n$ be a set of representatives of $\mathrm{pr}^{-1}(Z_G(k_v))/V_0$. For $i = 1, \ldots, n$, let U_i be the kernel of the character $g \longmapsto \chi(g, z_i)$. Then U_i is open and of finite index in $\mathbf{G}_v$. Set $\mathbf{G}^\circ_v = \bigcap_{i=0}^n U_i$. This group is thus the desired normal open subgroup of $\mathbf{G}_v$ of finite index.

Choose a finite set S of places of k, containing the archimedean places, and large enough so that

(i) $\left(\prod_{v\in S} G(k_v) \prod_{v\notin S} K_v\right) G(k) = G(\mathbb{A})$;

(ii) for $v \notin S$, K_v lifts to a subgroup denoted by $\tilde{K}_v$ of $\mathbf{G}_v$.

Let $H = \prod_{v\in S} G(k_v) \prod_{v\notin S} K_v$ and let $\mathbf{H}$ be its inverse image in $\mathbf{G}$. We will show that

(*) $$(Z_G(k) \cap Z_{\mathbf{G}})(\mathbf{H} \cap Z^1_{\mathbf{G}})$$

is of finite index in $Z^1_{\mathbf{G}}$.

Let Z^0_G be the connected component of the identity of Z_G. It is a torus. Set $Z^{01}_G = Z^0_G(\mathbb{A}) \cap G^1$. This is actually the group which is deduced from $Z^0_G(\mathbb{A})$ as G^1 is deduced from $G(\mathbb{A})$. By the finiteness of the class number of $k, Z^0_G(k)(H \cap Z^{01}_G)$ is of finite index in Z^{01}_G. By Lemma 1 of Appendix IV, $Z^{01}_G(H \cap Z^1_G)$ is of finite index in Z^1_G. Thus $Z_G(k)(H \cap Z^1_G)$ is of finite index in Z^1_G. It thus suffices to show that $(Z_G(k) \cap Z_{\mathbf{G}})(\mathbf{H} \cap Z^1_{\mathbf{G}})$ is of finite index in $[Z_G(k)(\mathbf{H} \cap \mathrm{pr}^{-1}(Z^1_G))] \cap Z^1_{\mathbf{G}}$. Indeed, it suffices to show that $\mathbf{H} \cap Z^1_{\mathbf{G}}$ is of finite index in $\mathbf{H} \cap Z_G(k)Z^1_{\mathbf{G}}$. Let us consider an element h of this set. For $v \in S$, choose a group $\mathbf{G}^\circ_v$ as above. As $h \in \mathrm{pr}^{-1}(Z_G(\mathbb{A}))$, h commutes with $\mathbf{G}^\circ_v$. Let $v \notin S$. As $\tilde{K}_v \to K_v$ is an isomorphism and $h \in \mathbf{H} \cap \mathrm{pr}^{-1}(Z_G(\mathbb{A}))$, h commutes with $\tilde{K}_v$. As $Z_G(k)$ and $Z^1_{\mathbf{G}}$ commute with $G(k)$, h does as well. Thus h commutes with the normal subgroup generated by $G(k)$ and $\prod_{v\in S} \mathbf{G}^\circ_v \prod_{v\notin S} \tilde{K}_v$. Denote this subgroup by C. By (i), we see that C is of finite index in $\mathbf{G}$. For $g \in \mathbf{G}$, we write $h^{-1}gh = \chi(g,h)g$ as above. The map $h \mapsto \chi(.,h)$ is an injective map of

$$\mathbf{H} \cap Z^1_{\mathbf{G}} \backslash \mathbf{H} \cap Z_G(k) Z^1_{\mathbf{G}}$$

into the finite group of characters of $C\backslash\mathbf{G}$. Thus the above group is finite, which is what we wanted to prove.

Let $E = Z^0_G(k) \cap H$. By the Dirichlet unit theorem we see that

(iii) $E\backslash H \cap Z^1_G$ is compact;

(iv) E is the product of a finite group by a finitely generated free group.

Let E^N be the subgroup of N-th powers of E. It can be identified with a subgroup of $\mathbf{H} \cap Z^1_{\mathbf{G}}$. The natural map

$$E^N\backslash\mathbf{H} \cap Z^1_{\mathbf{G}} \longrightarrow E^N\backslash H \cap Z^1_G$$

is proper. By (iii) and (iv), the right-hand space is compact, so the left-hand one must be as well. Thus, we can find a compact subset Γ of $\mathbf{H} \cap Z^1_{\mathbf{G}}$ such that $\mathbf{H} \cap Z^1_{\mathbf{G}} = E^N\Gamma$. Then

$$(Z_G(k) \cap Z_{\mathbf{G}})(\mathbf{H} \cap Z^1_{\mathbf{G}}) = (Z_G(k) \cap Z_{\mathbf{G}})\Gamma$$

and by (*), we see that $Z_G(k) \cap Z_{\mathbf{G}}\backslash Z^1_{\mathbf{G}}$ is compact. The groups

$Z_G(k)Z^1_{\mathbf{G}}$, $G(k)Z^1_{\mathbf{G}}$ contain

$$Z_G(k)(Z_G(k)\cap Z_{\mathbf{G}})(\mathbf{H}\cap Z^1_{\mathbf{G}}) = Z_G(k)\Gamma,$$

$$G(k)(Z_G(k)\cap Z_{\mathbf{G}})(\mathbf{H}\cap Z^1_{\mathbf{G}}) = G(k)\Gamma,$$

respectively, as subgroups of finite index. The last groups are closed since they are products of a discrete set by a compact set. Thus, $Z_G(k)Z^1_{\mathbf{G}}$ and $G(k)Z^1_{\mathbf{G}}$ are closed. Let us use the construction of the proof of I.1.4 (2) for $M = G$. We have a continuous map

$$j\circ \mathrm{pr} : \mathbf{G} \longrightarrow V^R$$

whose kernel is $\mathbf{G}^1$. We thus have the following diagram of continuous maps

$$\begin{array}{ccccccccc} 1 & \longrightarrow & \mathbf{G}^1 & \longrightarrow & \mathbf{G} & \longrightarrow & j\circ\mathrm{pr}(\mathbf{G}) & \longrightarrow & 1 \\ & & \uparrow & & \uparrow & & \uparrow & & \\ 1 & \longrightarrow & G(k)Z^1_{\mathbf{G}} & \longrightarrow & G(k)Z_{\mathbf{G}} & \longrightarrow & j\circ\mathrm{pr}(Z_{\mathbf{G}}) & \longrightarrow & 1 \end{array}$$

We check that $j\circ\mathrm{pr}(Z_{\mathbf{G}})$ is closed in $j\circ\mathrm{pr}(\mathbf{G})$ and that the map $Z_{\mathbf{G}} \to j\circ\mathrm{pr}(Z_{\mathbf{G}})$ admits a continuous section. Since $G(k)Z^1_{\mathbf{G}}$ is closed in $\mathbf{G}^1$, we see that $G(k)Z_{\mathbf{G}}$ is closed in $\mathbf{G}$. We argue similarly for $Z_G(k)Z_{\mathbf{G}}$.

□

I.1.6. Roots and coroots

We have already fixed a minimal parabolic subgroup P_0 of G and a Levi subgroup M_0 of P_0; let T_0 be the maximal split torus of the centre of M_0 (T_0 does not depend on the choices made and in particular on the choice of P_0 except up to conjugation by an element of $G(k)$). We denote by $R(T_0, G)$ the set of roots of G relative to T_0. It forms a root system in the sense of [B] Chapter 6. Recall that with every root one can associate a coroot which is a one-parameter subgroup of T_0 ([BT] §7 reduces this to the split case).

There is a canonical duality with values in $\mathbb{Z}$ between the subgroup of rational characters of a split torus and that of its one-parameter subgroups, which we denote by $\langle , \rangle$. For every $\chi \in \mathrm{Rat}(M_0)$ and every coroot $\check\beta$, this allows us to define $\langle \chi, \check\beta\rangle := \langle \mathrm{res}_0\chi, \check\beta\rangle$, where res_0 is the restriction of M_0 to T_0. This extends $\mathbb{R}$-linearly to define $\langle\lambda,\check\beta\rangle$ for all $\lambda \in \mathrm{Re}\,X_{M_0} \simeq \mathrm{Re}\,\mathfrak{a}^*_{M_0}$, and $\mathbb{C}$-linearly to define $\langle\lambda,\check\beta\rangle$ for all $\lambda\in\mathfrak{a}^*_{M_0}$. In particular, every coroot can be identified with an element of $\mathrm{Re}\,\mathfrak{a}_{M_0}$. If k is a number field, this defines $\langle\lambda,\check\beta\rangle$ for all $\lambda\in X_{M_0}$. If k is a function field, $\langle\lambda,\check\beta\rangle$ cannot be defined for $\lambda\in X_{M_0}$. See, however, I.1.8 below.

Remarks

(1) We write as above res_0 for the restriction map of $\mathrm{Rat}(M_0)$ into $\mathrm{Rat}(T_0)$. The kernel and the cokernel of this map are finite, so there is an isomorphism

$$\mathrm{Re}\,\mathfrak{a}^*_{M_0} = \mathrm{Rat}(M_0)\bigotimes_{\mathbb{Z}}\mathbb{R} \simeq \mathrm{Rat}(T_0)\bigotimes_{\mathbb{Z}}\mathbb{R}.$$

This allows one to consider the roots as elements of $\mathrm{Re}\,\mathfrak{a}^*_{M_0} \simeq \mathrm{Re}\,X_{M_0}$. From now on, unless we explicitly state the contrary, we will consider the roots as elements of $\mathrm{Re}\,\mathfrak{a}^*_{M_0} \simeq \mathrm{Re}\,X_{M_0}$ and the coroots as elements of $\mathrm{Re}\,\mathfrak{a}_{M_0}$, i.e. as linear forms on $\mathrm{Re}\,X_{M_0}$.

(2) In fact it is clear that roots and coroots are rational elements of $\mathrm{Re}\,\mathfrak{a}^*_{M_0}$, $\mathrm{Re}\,\mathfrak{a}_{M_0}$, i.e. they belong to $\mathrm{Rat}(M_0)\otimes_{\mathbb{Z}}\mathbb{Q}$, $\mathrm{Hom}_{\mathbb{Z}}(\mathrm{Rat}(M_0),\mathbb{Q})$ respectively, see I.1.4 (1).

(3) The choice of P_0 determines a set of positive roots of $R(T_0, G)$ denoted by $R^+(T_0, G)$ and a subset of simple roots denoted by Δ_0.

Now let $P = MU$ be a standard parabolic subgroup of G. The group $P_0 \cap M$ is a minimal parabolic subgroup of M; using what was done above, we define $R(T_0, M)$ to be the set of roots of T_0 in M as well as the set of coroots of M, which are identified with subsets of the sets of roots, coroots, respectively, of G. We write $\Delta_0^M = \Delta_0 \cap R(T_0, M)$.

Let T_M be the maximal split torus in the centre of M. It is contained in T_0. We denote by $R(T_M, G)$ the set of 'roots' of G relative to T_M. In general $R(T_M, G)$ is not a root system. As above, thanks to the restriction map of M to T_M, we identify $R(T_M, G)$ with a subset of $\mathrm{Re}\,\mathfrak{a}^*_M \simeq \mathrm{Re}\,X_M$ generating this space. We have a restriction map

(4) $$R(T_0, G) \to R(T_M, G) \cup \{0\},$$

which is trivial on $R(T_0, M)$. We denote by Δ_M the subset of $R(T_M, G)$ which consists of the non-trivial restrictions of elements of Δ_0. This, again, is a generating set for $\mathrm{Re}\,\mathfrak{a}^*_M$.

We denote by $\mathrm{Re}(\mathfrak{a}^M_{M_0})^*$ the real vector subspace $\mathrm{Re}\,\mathfrak{a}^*_{M_0}$ generated by the elements of $R(T_0, M)$. By the restriction map of M to M_0, $\mathrm{Re}\,\mathfrak{a}^*_M$ is identified with a real vector subspace of $\mathrm{Re}\,\mathfrak{a}^*_{M_0}$ and we have

$$\mathrm{Re}\,\mathfrak{a}^*_{M_0} \simeq \mathrm{Re}\,\mathfrak{a}^*_M \oplus \mathrm{Re}(\mathfrak{a}^M_{M_0})^*,$$

the restriction map (4) coming from the projection of $\mathrm{Re}\,\mathfrak{a}^*_{M_0}$ onto $\mathrm{Re}\,\mathfrak{a}^*_M$. The elements of $\mathrm{Re}(\mathfrak{a}^M_{M_0})^*$ are identified with those of $\mathrm{Re}\,X_{M_0}$ trivial on the centre of $M(\mathbb{A})$. We denote by $\mathrm{Re}\,X^M_{M_0}$ the set consisting of these elements. More generally, let $M \subset M'$ be standard Levis. We define $R(T_M, M')$ in an analogous way to $R(T_M, G)$; it is a subset of $R(T_M, G)$. We denote by $\mathrm{Re}(\mathfrak{a}^{M'}_M)^*$ the real vector subspace of $\mathrm{Re}\,\mathfrak{a}^*_M$ generated by

$R(T_M, M')$ and again, we have as above

(5) $$\operatorname{Re}\mathfrak{a}_M^* \simeq \operatorname{Re}\mathfrak{a}_{M'}^* \oplus \operatorname{Re}(\mathfrak{a}_M^{M'})^*.$$

The elements of $\operatorname{Re}(\mathfrak{a}_M^{M'})^*$ are identified with the elements of $\operatorname{Re}X_M$ which are trivial on the centre of $M'(\mathbb{A})$. We denote by $\operatorname{Re}X_M^{M'}$ the set of these elements. Recall that $\mathbf{M}'$ is the inverse image of $M'(\mathbb{A})$ in $\mathbf{G}$ and $Z_{\mathbf{M}'}$ is its centre; we saw that the quotient $\operatorname{pr}^{-1}(Z_{M'}(\mathbb{A}))/Z_{\mathbf{M}'}$ is $|\mathbf{N}|$-torsion. We denote by $X_M^{\mathbf{M}'}$ the set of elements of X_M trivial on the image under pr of the centre of $\mathbf{M}'$ and we have

$$\operatorname{Re} X_M^{\mathbf{M}'} = \operatorname{Re}X_M^{M'}.$$

Thanks to (5) we obtain a decomposition

(6) $$\mathfrak{a}_M^* = \mathfrak{a}_{M'}^* \oplus (\mathfrak{a}_M^{M'})^*, \text{ where } (\mathfrak{a}_M^{M'})^* = \operatorname{Re}(\mathfrak{a}_M^{M'})^* \bigotimes_{\mathbb{R}} \mathbb{C}.$$

Note that the map $\mathfrak{a}_M^* \to X_M$ restricts to a map

(7) $$(\mathfrak{a}_M^{M'})^* \to X_M^{\mathbf{M}'}.$$

If k is a number field, this map is an isomorphism. If k is a function field, it is in general neither injective nor surjective; its kernel is of the form $(2\pi i/\log q)L$, where L is a lattice of $(\operatorname{Rat}(M) \otimes_{\mathbb{Z}} \mathbb{Q}) \cap (\mathfrak{a}_M^{M'})^*$; its image is the connected component of the identity of $X_M^{\mathbf{M}'}$ and is of finite index in this group.

As in (6), we have a surjective map

(8) $$X_M \leftarrow X_{M'} \oplus X_M^{\mathbf{M}'},$$

the morphism being obtained by adding the restriction map $X_{M'} \to X_M$ to the inclusion $X_M^{\mathbf{M}'} \hookrightarrow X_M$. The kernel is $X_{M'}^{\mathbf{M}'}$ embedded antidiagonally.

Restricted to the real parts, the above map becomes an isomorphism

(9) $$\operatorname{Re}X_M \simeq \operatorname{Re}X_{M'} \oplus \operatorname{Re}X_M^{\mathbf{M}'}.$$

Let us return to M_0 and M and proceed dually. Let $\operatorname{Re}\mathfrak{a}_{M_0}^M$ be the real vector subspace of $\operatorname{Re}\mathfrak{a}_{M_0}$ generated by the coroots associated with the elements of $R(T_0, M)$ and $\mathfrak{a}_{M_0}^M = \operatorname{Re}\mathfrak{a}_{M_0}^M \otimes_{\mathbb{R}} \mathbb{C}$. Then $\mathfrak{a}_{M_0}^M$ is the orthogonal complement of $\mathfrak{a}_M^*$ and we have a decomposition (defined over $\mathbb{R}$ and even over $\mathbb{Q}$):

(10) $$\mathfrak{a}_{M_0} = \mathfrak{a}_{M_0}^M \oplus \mathfrak{a}_M.$$

I.1.7. The Weyl group

We denote by W the Weyl group of G, i.e. the quotient

$$W = \operatorname{Norm}_{G(k)} T_0(k)/\operatorname{Cent}_{G(k)} T_0(k)$$

(actually, $\operatorname{Cent}_{G(k)} T_0(k) = M_0(k)$). Every element w of W is represented by an element of $G(k)$: we fix such a representative $\dot{w}$. We have the

Bruhat decomposition

$$G(k) = \bigcup_{w \in W} P_0(k)\dot{w}P_0(k).$$

The group W acts naturally on the algebraic group T_0, since an element which normalises or centralises $T_0(k)$ also normalises or centralises, respectively, the whole group T_0. The group W thus also acts on $\mathrm{Re}\,\mathfrak{a}^*_{M_0}$ and is identified with the Weyl group of the root system $R(T_0, G)$. It is known that this is a Coxeter group generated by the symmetries relative to the simple roots. We will need a generalisation of this property in the following situation.

Let M be a standard Levi subgroup of G. Let W_M be the Weyl group of M (the analogue of W for the group M) and $W(M)$ the set of elements $w \in W$, of minimal length in their class wW_M, such that wMw^{-1} is again a standard Levi. This is not a group in general. However if $w \in W(M)$ and $w' \in W(wMw^{-1})$, then $w'w \in W(M)$.

Let $w \in W(M)$. It conjugates T_M to $T_{wMw^{-1}}$ and induces a map still denoted by $w : R(T_M, G) \to R(T_{wMw^{-1}}, G)$. Let $R_{\mathrm{ind}}(T_M, G)$ be the set of indivisible roots in $R(T_M, G)$. Set

$$\ell(w) = \#\{\alpha \in R_{\mathrm{ind}}(T_M, G); \alpha > 0,\ w\alpha < 0\}.$$

This defines a length function $\ell : W(M) \to \mathbb{N}$.

Let $\alpha \in \Delta_M$ and α_0 be the unique element of Δ_0 which projects onto α. Let M_α be the standard Levi subgroup of G such that $\Delta_0^{M_\alpha} = \{\alpha_0\} \cup \Delta_0^M$. Then M is a maximal proper Levi subgroup of M_α. Writing $W_{M_\alpha}(M)$ for the analogue of $W(M)$ when M_α replaces G, we easily show that $W_{M_\alpha}(M)$ has two elements: the identity and the element $s_\alpha = w_2 w_1$, where w_1, w_2, are the elements of greatest length of W_M, W_{M_α}, repsectively. Note that $W_{M_\alpha}(M)$ embeds naturally into $W(M)$. This defines an element $s_\alpha \in W(M)$ which is called an elementary symmetry.

Lemma

(a) $\{\beta \in R_{\mathrm{ind}}(T_M, G); \beta > 0,\ s_\alpha\beta < 0\} = \{\alpha\}$.

(b) *s_α is characterised by (a).*

Proof By construction, s_α preserves $R^+(T_0, G) - R^+(T_0, M_\alpha)$. However it sends $R^+(T_0, M_\alpha) - R^+(T_0, M)$ onto the opposite set, i.e. $\{-\beta; \beta \in R^+(T_0, M_\alpha) - R^+(T_0, M)\}$. Now, writing D_α for the set of elements of $R^+(T_M, G)$ proportional to α, $R^+(T_M, G) - D_\alpha$ is the projection into $\mathrm{Re}\,\mathfrak{a}^*_M$ of $R^+(T_0, G) - R^+(T_0, M_\alpha)$ and D_α is the projection of $R^+(T_0, M_\alpha) - R^+(T_0, M)$. This proves assertion (a).

Let $w \in W(M)$ be such that $\{\beta \in R_{\text{ind}}(T_M, G); \beta > 0, w\beta < 0\} = \{\alpha\}$. Set $M' = wMw^{-1}$ and $w' = s_\alpha w^{-1}$. Then $w' \in W(M')$ and $w'\beta > 0$ for all $\beta \in R^+(T_{M'}, G)$. By the definition of $W(M')$, i.e. the minimality of w' in $w'W_{M'}$, we see that $w'\beta > 0$ for all $\beta \in R^+(T_0, G)$. But then $w' = 1$, which proves (b). □

I.1.8. Decomposition into elementary symmetries

Lemma *Let $w \in W(M)$ and $M' = wMw^{-1}$.*

(a) *For all $w' \in W(M')$ we have $\ell(w'w) \leq \ell(w') + \ell(w)$.*

(b) *For all $\alpha \in \Delta_{M'}$, we have*

$$\ell(s_\alpha w) = \begin{cases} \ell(w) + 1 & \text{if } w^{-1}\alpha > 0, \\ \ell(w) - 1 & \text{if } w^{-1}\alpha < 0. \end{cases}$$

(c) *There exists a sequence of standard Levis $M_0 = M, M_1, \ldots, M_\ell = M'$, where $\ell = \ell(w)$, and for all $i \in \{1, \ldots, \ell\}$ a root $\alpha_i \in \Delta_{M_{i-1}}$, such that $M_i = s_{\alpha_i} M_{i-1} s_{\alpha_i}^{-1}$ for all i and*

$$w = s_{\alpha_\ell} \ldots s_{\alpha_1}.$$

For any such decomposition, we have the equality

$$\begin{aligned} &\{\alpha \in R_{\text{ind}}(T_M, G); \alpha > 0, w\alpha < 0\} \\ =&\{s_{\alpha_1}^{-1} s_{\alpha_2}^{-1} \ldots s_{\alpha_{i-1}}^{-1}(\alpha_i);\ i = 1, \ldots, \ell\}. \end{aligned}$$

Proof The set $\{\beta \in R_{\text{ind}}(T_M, G); \beta > 0, w'w\beta < 0\}$ is the disjoint union of

$$A = \{\beta \in R_{\text{ind}}(T_M, G); \beta > 0, w\beta > 0, w'w\beta < 0\}$$

and

$$B = \{\beta \in R_{\text{ind}}(T_M, G); \beta > 0, w\beta < 0, w'w\beta < 0\}.$$

The latter set has fewer than $\ell(w)$ elements. By the map $\beta \mapsto w\beta$, A is in bijection with

$$C = \{\beta \in R_{\text{ind}}(T_{M'}, G); \beta > 0, w'\beta < 0, w^{-1}\beta > 0\}$$

which has fewer than $\ell(w')$ elements. Thus we obtain (a). For $w' = s_\alpha$,

$$B = \begin{cases} \{\beta \in R_{\text{ind}}(T_M, G); \beta > 0, w\beta < 0\}, & \text{if } w^{-1}\alpha > 0, \\ \{\beta \in R_{\text{ind}}(T_M, G); \beta > 0, w\beta < 0\} - \{-w^{-1}\alpha\}, & \text{if } w^{-1}\alpha < 0, \end{cases}$$

$$C = \begin{cases} \{\alpha\}, & \text{if } w^{-1}\alpha > 0, \\ \emptyset, & \text{if } w^{-1}\alpha < 0. \end{cases}$$

Thus we obtain (b). Suppose $w \neq 1$. We see as in the proof of Lemma I.1.7 that there exists at least one root $\alpha \in \Delta_{M'}$ such that $w^{-1}\alpha < 0$. Choose such a root and set $M_{\ell-1} = s_\alpha M' s_\alpha^{-1}, w' = s_\alpha w, \alpha_\ell = -s_\alpha(\alpha)$. Then

$s_{\alpha_\ell} = s_\alpha^{-1}$. By (b) we see that $\ell(w') = \ell(w) - 1$. By induction on $\ell(w)$, we can choose $M_1, \ldots, M_{\ell-2}, \alpha_1, \ldots, \alpha_{\ell-1}$ in such a way that

$$w' = s_{\alpha_{\ell-1}} \ldots s_{\alpha_1}.$$

Then $w = s_{\alpha_\ell} s_{\alpha_{\ell-1}} \ldots s_{\alpha_1}$, and this decomposition satisfies the conditions of the statement. It is immediate that all possible decompositions are obtained by this procedure. With the above notation,

$$\{\beta \in R_{\text{ind}}(T_M, G); \beta > 0, w\beta < 0\}$$

is the union of

$$\{\beta \in R_{\text{ind}}(T_M, G); \beta > 0, w'\beta < 0\}$$

and

$$\{w'^{-1}(\alpha_\ell)\}.$$

We obtain the final assertion of the statement by induction. □

I.1.9. A composition lemma

Lemma *Let $M \subset L$ be two standard Levis and $w \in W$. Decompose w into $w = w_1 w_2$, where $w_2 \in W_L$ and w_1 is of minimal length in wW_L. Suppose $w \in W(M)$. Then $w_2 \in W_L(M)$ and $w_1 \in W(w_2 M w_2^{-1})$.*

Proof It is clear that w_2 is of minimal length in $w_2 W_M$. Let $\alpha \in \Delta_0^M$. As $w\alpha > 0$ and $\ell(w) = \ell(w_1) + \ell(w_2)$ (we are speaking of lengths in W), we also have $w_2\alpha > 0$. Write

$$w_2\alpha = \sum_{\beta \in \Delta_0^L} n_\beta \beta,$$

where the $n_\beta \geq 0$. As w_1 is of minimal length in $w_1 W_L$,

$$w_1\beta > 0 \text{ for all } \beta \in \Delta_0^L.$$

Since $w \in W(M)$, $w\alpha$ is simple. Then the equality

$$w\alpha = \sum_{\beta \in \Delta_0} n_\beta w_1 \beta$$

implies that n_β is zero for all β except one, for which $n_\beta = 1$. This means $w_2\alpha \in \Delta_0^L$. Then $M' = w_2 M w_2^{-1}$ is the standard Levi such that $\Delta_0^{M'} = \{w_2\alpha; \alpha \in \Delta_0^M\}$ and $w_2 \in W_L(M)$. We immediately deduce that $w_1 \in W(M')$. □

I.1.10. Decomposition of $\operatorname{Re}\mathfrak{a}_M$

Let $C_M = \{H \in \operatorname{Re}\mathfrak{a}_M; \forall\alpha \in \Delta_M, \langle\alpha, H\rangle > 0\}$. For $\alpha \in R^+(T_M, G)$, set $P_\alpha = \{H \in \operatorname{Re}\mathfrak{a}_M; \langle\alpha, H\rangle = 0\}$ and

$$(\operatorname{Re}\mathfrak{a}_M)^{\text{reg}} = \operatorname{Re}\mathfrak{a}_M - \bigcup_{\alpha \in R^+(T_M, G)} P_\alpha.$$

Lemma *The map which associates* $w^{-1}(C_{wMw^{-1}})$ *with* $w \in W(M)$ *is a bijection of* $W(M)$ *onto the set of connected components of* $(\mathrm{Re}\,\mathfrak{a}_M)^{\mathrm{reg}}$.

Proof The injectivity is easy. Let C_1 be a connected component of $(\mathrm{Re}\,\mathfrak{a}_M)^{\mathrm{reg}}$. Let $\ell(C_1)$ be the number of hyperplanes P_α separating C_1 from C_M. Suppose $C_1 \neq C_M$. There is at least one wall C_1 (i.e. a P_α such that $P_\alpha \cap \overline{C}_1$ contains an open set of P_α) which separates C_1 from C_M. Choose such a wall P, and let C_2 be the unique connected component which is separated from C_1 only by P. Then $\ell(C_2) = \ell(C_1) - 1$. By induction on the lengths, we can suppose that there exists $w_2 \in W(M)$ such that $C_2 = w_2^{-1}(C_{w_2Mw_2^{-1}})$. There exists $\alpha \in \Delta_{w_2Mw_2^{-1}}$ such that $P = w_2^{-1}P_\alpha$. Set $w_1 = s_\alpha w_2$. We then check that $C_1 = w_1^{-1}(C_{w_1Mw_1^{-1}})$. □

I.1.11. Return to the coroots

Let M be a standard Levi subgroup of G and $\alpha \in R(T_M, G)$. We want to define a coroot $\check{\alpha} \in \mathrm{Re}\,\mathfrak{a}_M$. There is a unique indivisible root $\alpha_0 \in R^+(T_M, G)$ and a unique $n \in \mathbf{Z}$ such that $\alpha = n\alpha_0$. Supposing that $\check{\alpha}_0$ is defined, set $\check{\alpha} = n^{-1}\check{\alpha}_0$. This reduces us to the case of a positive and indivisible root α. Let us show that with such a root we can uniquely associate a root $\beta \in R(T_0, G)$. There exists $w \in W(M)$ such that $w\alpha \in \Delta_{wMw^{-1}}$. Indeed, let w_1 and w_2 be the elements of greatest length in W_M and W. Then $w_2w_1 \in W(M)$ and $w_2w_1(\alpha) < 0$. Write $w_2w_1 = s_{\alpha_\ell} \dots s_{\alpha_1}$ as in I.1.8(c). The last assertion of that lemma ensures that some element $w = s_{\alpha_i} \dots s_{\alpha_1}$ satisfies the required condition. Let w be such an element. Let β_1 be the unique element of Δ_0 which projects onto $w\alpha$. Set $\beta = w^{-1}\beta_1$. We must show that this definition does not depend on the choice of w. Let $w' \in W(M)$ be such that $w'\alpha \in \Delta_{w'Mw'^{-1}}$ and β_2 be the unique element of Δ_0 which projects onto $w'\alpha$. Let M_1 and M_2 be the standard Levi subgroups of G such that $\Delta_0^{M_1} = \{\beta_1\} \cup \Delta_0^{wMw^{-1}}, \Delta_0^{M_2} = \{\beta_2\} \cup \Delta_0^{w'Mw'^{-1}}$. Consider the element $w'w^{-1} \in W(wMw^{-1})$. It sends wMw^{-1} onto $w'Mw'^{-1}$ and $w\alpha$ onto $w'\alpha$, thus M_1 onto M_2 and even $R^+(T_{M_1}, G)$ onto $R^+(T_{M_2}, G)$, thus also $\Delta_0^{M_1}$ onto $\Delta_0^{M_2}$. As it sends $\Delta_0^{wMw^{-1}}$ onto $\Delta_0^{w'Mw'^{-1}}$, it sends β_1 onto β_2, so we obtain the equality $w^{-1}\beta_1 = w'^{-1}\beta_2$ which we wanted to prove.

We define $\check{\alpha}$ to be the projection of $\check{\beta}$ onto $\mathrm{Re}\,\mathfrak{a}_M$, just as we defined $\check{\beta}$ in I.1.6.

Suppose that k is a function field. Rather than the coroot $\check{\alpha}$, we will sometimes use an element α^* of $\mathbf{M}/\mathbf{M}^1$ which we now define. Let M_α be the Levi subgroup of G, containing M as maximal proper Levi subgroup, associated with α (with the preceding notation, $M_\alpha = w^{-1}M_1w$). In general this is not a standard Levi subgroup, however it doesn't matter. The sequence

$$1 \to \mathbf{M} \cap \mathbf{M}_\alpha^1/\mathbf{M}^1 \to \mathbf{M}/\mathbf{M}^1 \to \mathbf{M}_\alpha/\mathbf{M}_\alpha^1 \to 1$$

is exact. The first group is thus a free $\mathbb{Z}$-module of rank 1. We naturally associate with α an element, say h_α, of $\mathbf{M}/\mathbf{M}^1$. Indeed, denote by $X_*(T_0)$ the group of one-parameter subgroups of T_0. We have a canonical identification

$$T_0(\mathbb{A}) \simeq \mathbb{A}^* \bigotimes_{\mathbb{Z}} X_*(T_0),$$

thus

$$\mathbf{T}_0/\mathbf{T}_0^1 \simeq \mathbb{A}^*/\mathbb{A}^1 \bigotimes_{\mathbb{Z}} X_*(T_0),$$

where $\mathbb{A}^1$ is the subgroup of $\mathbb{A}^*$ which consists of ideles of absolute value 1. Now $\mathbb{A}^*/\mathbb{A}^1 \simeq \mathbb{Z}$ by $x \mapsto \log|x|/\log q$. So

$$\mathbf{T}_0/\mathbf{T}_0^1 \simeq X_*(T_0).$$

Now let $\beta \in R(T_0, G)$ be the element associated with α. We have $\check\beta \in X_*(T_0)$ and we define h_α as the image of $\check\beta$ under the map

$$X_*(T_0) \simeq \mathbf{T}_0/\mathbf{T}_0^1 \to \mathbf{M}/\mathbf{M}^1.$$

It is easy to see that for all $\lambda \in \operatorname{Re}\mathfrak{a}_M \simeq \operatorname{Re}X_M$, we have

$$h_\alpha^\lambda = q^{\langle\lambda,\check\alpha\rangle}.$$

We deduce that $h_\alpha \in \mathbf{M} \cap \mathbf{M}_\alpha^1/\mathbf{M}^1$. We then define α^* to be the unique generator of $\mathbf{M} \cap \mathbf{M}_\alpha^1/\mathbf{M}^1$ of which h_α is a positive multiple.

I.1.12. The scalar product

Fix a positive definite scalar product on $\operatorname{Re}\mathfrak{a}_{M_0}^*$, invariant under W; by restriction for every standard Levi M, we obtain a scalar product on $\operatorname{Re}\mathfrak{a}_M^* \simeq \operatorname{Re}X_M$. By extension of scalars, we obtain a symmetric bilinear form $(,)$ on $\mathfrak{a}_M^* \times \mathfrak{a}_M^*$ and by Hermitian extension a positive definite Hermitian product $\langle,\rangle$ on $\mathfrak{a}_M^*$. By duality, we have analogous objects on $\mathfrak{a}_M$. We denote by $\|.\|$ the norms associated with the Hermitian products on $\mathfrak{a}_M$ or $\mathfrak{a}_M^*$. If k is a number field we have a positive definite Hermitian product on X_M, while if k is a function field, this is not the case and we only have a positive definite scalar product on $\operatorname{Re}X_M$. The choice of scalar product allows us to identify $\mathfrak{a}_M$ with $\mathfrak{a}_M^*$ (which we will do only occasionally in what follows); but note that for $\alpha \in R(T_0, G)$, the coroot $\check\alpha$ is proportional to α under such an identification, which allows us to check that the decompositions (5), (6) and (9) in I.1.6 are orthogonal decompositions.

I.1.13. Measures

For every unipotent group U, we equip $U(\mathbb{A})$ with the Haar measure such that, $U(k)$ being equipped with the counting measure and $U(k)\backslash U(\mathbb{A})$ with the quotient measure, we have $\text{meas}(U(k)\backslash U(\mathbb{A})) = 1$. We equip $\mathbf{K}$ with the Haar measure such that $\text{meas}(\mathbf{K}) = 1$ and we fix a measure on $\mathbf{M}_0$. Let U_0 be the unipotent radical of P_0 and ρ_0 the half-sum of positive roots of T_0. Then the map

$$f \longmapsto \int_{U_0(\mathbb{A})\times\mathbf{M}_0\times\mathbf{K}} f(umk)m^{-2\rho_0}\, dk\, dm\, du$$

defined for a continuous function with compact support on $\mathbf{G}$ is a Haar measure, denoted by dg, on $\mathbf{G}$. More generally, let $P = MU$ be a standard parabolic subgroup of G. Define a measure dm on $\mathbf{M}$ by replacing G by M in the above construction and denote by ρ_P the half-sum of positive roots of T_M. For a continuous function f with compact support on $\mathbf{G}$, we have the compatibility formula

$$\int_{\mathbf{G}} f(g)\, dg = \int_{U(\mathbb{A})\times\mathbf{M}\times\mathbf{K}} f(umk)m^{-2\rho_P}\, dk\, dm\, du.$$

Fix a Haar measure on $Z_{\mathbf{G}}$. Thanks to Lemma I.1.5, $G(k)Z_{\mathbf{G}}$ is also equipped with such a measure.

The real vector spaces $\text{Re}\,\mathfrak{a}_{M_0}$ and $\text{Re}\,\mathfrak{a}^*_{M_0}$ having been equipped with positive definite scalar products, are thus equipped with Haar measures (for which the measure of a cube with sides of length 1 is 1). The same remark is true for all their vector subspaces.

Let M be a standard Levi subgroup of G. We equip $\text{Im}\,X^{\mathbf{G}}_M$ with the Haar measure for which the map $\text{Re}\,\mathfrak{a}^G_M \to \text{Im}\,X^{\mathbf{G}}_M$, $\lambda \mapsto \kappa(i\lambda)$ locally preserves the measures. We can then deduce a measure on all sets of the form

$$\{\lambda \in X^{\mathbf{G}}_M;\ \text{Re}\,\lambda = \lambda_0\}.$$

I.2. Automorphic forms: growth, constant terms

I.2.1. Siegel domains

Fix an isomorphism $T_0 \to \mathbb{G}^R_m$ defined over k, where $\mathbb{G}_m$ is the multiplicative group. If k is a number field, embed $\mathbb{R}^*_+$ in $\mathbb{A}^*$, identifying $t \in \mathbb{R}^*_+$ with the element $(t_v) \in \mathbb{A}^*$ such that $t_v = t$ if v is archimedean, $t_v = 1$ if v is finite. Then $\mathbb{R}^{*R}_+$ is identified with a subgroup of $T_0(\mathbb{A})$ which is split in the extension $\mathbf{T}_0$ since $\mathbb{R}^*_+$ is simply connected. We denote by

$A_{\mathbf{M}_0}$ the unique connected subgroup of $\mathbf{T}_0$ which projects onto $\mathbb{R}_+^{*R}$. If k is a function field, fix once and for all a place v_0 of k and a uniformising parameter ϖ at v_0. The group $\varpi^{\mathbf{Z}}$ generated by ϖ can be identified with a subgroup of $\mathbb{A}^*$ and $(\varpi^{\mathbf{Z}})^R$ with a subgroup of $T_0(\mathbb{A})$. By the fact that $\mathbf{T}_0^{|\mathbf{N}|}$ is contained in $Z_{\mathbf{M}_0}$ (see I.1.3 (2)), we see that there exists a subgroup of $Z_{\mathbf{M}_0}$ such that the projection pr : $\mathbf{G} \to G(\mathbb{A})$ establishes an isomorphism between this group and a subgroup of finite index of $(\varpi^{\mathbf{Z}})^R$. Fix such a subgroup once and for all, and denote it by $A_{\mathbf{M}_0}$. Up to replacing it by

$$\bigcap_{w \in W} w A_{\mathbf{M}_0} w^{-1},$$

which satisfies the same conditions, we can suppose that $A_{\mathbf{M}_0}$ is invariant under the action of W.

More generally, if $P = MU$ is a standard parabolic subgroup of G, we set $A_{\mathbf{M}} = A_{\mathbf{M}_0} \cap Z_{\mathbf{M}}$. Let $\log^M$ be the composition of the map $\log : \mathbf{M}_0 \to \mathfrak{a}_{M_0}$ with the projection $\mathfrak{a}_{M_0} \to \mathfrak{a}_{M_0}^M$. We have the equality

$$A_{\mathbf{M}} = \{a \in A_{\mathbf{M}_0}; \log^M a = 0\}.$$

Proof We have $Z_{\mathbf{M}} \subset \operatorname{Ker}(\log^M)$ since $z^\alpha = 1$ for all $z \in Z_{\mathbf{M}}$, $\alpha \in \Delta_0^M$, so

$$A_{\mathbf{M}} \subset \{a \in A_{\mathbf{M}_0}; \log^M a = 0\}.$$

We must show the inclusion in the other direction. Let $a \in A_{\mathbf{M}_0}$ be such that $\log^M a = 0$. We want to show that $a \in Z_{\mathbf{M}}$, i.e. that it commutes with every element of $\mathbf{M}$. Now, $\mathbf{M}$ is generated by $\mathbf{M}_0$ and some unipotent subgroups on which $\mathbf{T}_0$ acts via elements of $R(T_0, M)$ (which we consider in this proof as algebraic characters). As a commutes with every element of $\mathbf{M}_0$, in order to show that it commutes with every element of $\mathbf{M}$, it will suffice to show that for all $\alpha \in R(T_0, M)$, we have $a^\alpha = 1$. But by the definition of $A_{\mathbf{M}_0}$, we have $a^\alpha \in \mathbb{R}_+^*$ if k is a number field, and $a^\alpha \in \varpi^Z$ if k is a function field. Thus the equalities $a^\alpha = 1$ and $a^{|\alpha|} = 1$ (i.e. $|a^\alpha| = 1$) are equivalent. Now the equality $a^{|\alpha|} = 1$ is true for $\alpha \in R(T_0, M)$ since $\log^M a = 0$. This concludes the proof. □

We define the group $A_{\mathbf{M}}^G = \{a \in A_{\mathbf{M}}; \log_G a = 0\}$. If k is a number field, we have the equality $A_{\mathbf{M}} = A_{\mathbf{G}} \oplus A_{\mathbf{M}}^G$. If k is a function field, $A_{\mathbf{G}}$ and $A_{\mathbf{M}}^G$ form a direct sum which is of finite index in $A_{\mathbf{M}}$.

The set $A_{\mathbf{M}} \backslash \mathbf{M}/\mathbf{M}^1$ is finite. Fix a set $\mathbf{M}^c$, finite union of classes modulo $\mathbf{M}^1$, in such a way that $\mathbf{M} = A_{\mathbf{M}}\mathbf{M}^c$. If k is a number field, choose $\mathbf{M}^c = \mathbf{M}^1$.

Let ω be a compact subset of $\mathbf{P}_0$, t_0 an element of $\mathbf{M}_0$. Set

$$A_{\mathbf{M}_0}(t_0) = \{a \in A_{\mathbf{M}_0}; \forall \alpha \in \Delta_0,\ a^\alpha > t_0^\alpha\},$$
$$S = \{pak;\ p \in \omega,\ a \in A_{\mathbf{M}_0}(t_0),\ k \in \mathbf{K}\}.$$

If ω is big enough and t_0 small enough, i.e. t_0^α close enough to 0 for all $\alpha \in \Delta_0$, we have the equalities

$$G(k)S = \mathbf{G},\ U_0(k)S = U_0(\mathbb{A})S,$$

where U_0 is the unipotent radical of P_0. More generally, for a standard parabolic subgroup P of G, we define S^P by replacing Δ_0 by Δ_0^M in the definition of $A_{\mathbf{M}_0}(t_0)$. For example if $P = P_0$, $S^P = \omega A_{\mathbf{M}_0}\mathbf{K}$. For ω big enough and t_0 small enough, we have the equality $P(k)S^P = \mathbf{G}$. We suppose from now on that ω and t_0 are fixed such that all the above equalities hold.

I.2.2. Heights

We equip $G(\mathbb{A})$ with a height in the following way. In I.1.1 we fixed an embedding $i_G : G \to SL_{2n}$. For $g \in G(\mathbb{A})$, write $i_G(g) = (g_{rs})_{r,s=1,\dots,2n}$. Set

$$\|g\| = \prod_v \sup\{|g_{rs}|_v; r, s = 1, \dots, 2n\}$$

where the product is over all places of k. We equip $\mathbf{G}$ with a height by composing the height of $G(\mathbb{A})$ with the projection $\mathrm{pr} : \mathbf{G} \to G(\mathbb{A})$. We remark certain properties of this height:

(i) there exists $c > 0$ such that $\|g\| > c$ for all $g \in \mathbf{G}$;

(ii) there exists $c > 0$ such that $\|g_1 g_2\| < c\|g_1\|\|g_2\|$ for every $g_1, g_2 \in \mathbf{G}$;

(iii) there exist $c, r > 0$ such that $\|g^{-1}\| < c\|g\|^r$ for all $g \in \mathbf{G}$;

(iv) the height is related to the norm on $\mathfrak{a}_{M_0}$: there exist $c, c'' > 0$ and $c' \in \mathbb{R}$ such that for all $a \in A_{\mathbf{M}_0}$, we have
$$c\|\log_{M_0} a\| \leq \log\|a\| + c' \leq c''\|\log_{M_0} a\|;$$

(v) let $\lambda \in \mathrm{Re}\, X_{M_0}$; there exist $c, r > 0$ such that
$$m_{P_0}(g)^\lambda \leq c\|g\|^r$$
for all $g \in \mathbf{G}$;

(vi) there exists $\lambda \in \mathrm{Re}\, X_{M_0}$ and $c > 0$ such that
$$\|g\| \leq c m_{P_0}(g)^\lambda$$
for all $g \in \mathbf{G}^c \cap S$;

(vii) there exists $c > 0$ such that for all $\gamma \in G(k)$, $g \in S$, we have
$$\|g\| \leq c\|\gamma g\|;$$

(viii) there exist $c, t, t' > 0$ such that for all $a \in A_\mathbf{G}$ and $g \in \mathbf{G}^c$, we have
$$c\|a\|^{t'}\|g\|^t \leq \|ag\|.$$

Let us prove (vii) and (viii) which are the only properties which are not obvious. For all $N \in \mathbb{N}$, $N \geq 1$, $g \in GL_N(\mathbb{A})$, $g = (g_{rs})_{r,s=1,\ldots,N}$, and for every place v of k, set

$$|g|_v = \sup\{|g_{rs}|_v;\ r,s = 1,\ldots,N\},$$
$$\|g\|_v = \sup(|g|_v, |g^{-1}|_v),$$
$$\|g\|'_v = \sup(|g|_v|\det g|_v^{-1/N}, |g^{-1}|_v|\det g|_v^{1/N}).$$

Let us prove (vii). As there exists a compact set Ω of $\mathbf{G}$ such that $S \subset A_{\mathbf{M}_0}(t_0)\Omega$, we can suppose that $g \in A_{\mathbf{M}_0}(t_0)$. Up to composing i'_G with an interior automorphism of GL_n, which transforms the norm into an equivalent norm, we can suppose that $i'_G(T_0)$ is diagonal. Then (vii) results from the following assertion:

(*) suppose $G = GL_n$, $i'_G = id$, T_0 =the diagonal torus, $\mathbf{G} = GL_n(A)$: then for all $\gamma \in G(k)$, $a \in A_{\mathbf{M}_0}$, we have $\|a\| \leq \|\gamma a\|$.

An element

$$a \in A_{\mathbf{M}_0}$$

is a diagonal matrix with coefficients $a_1,\ldots,a_n \in \mathbb{R}_+^*$ if k is a number field, $\in \mathfrak{w}^{\mathbb{Z}}$ if k is a function field. Let us identify W with the group of matrices of permutations. For $w \in W$, we have $\|waw^{-1}\| = \|a\|$, $\|\gamma waw^{-1}\| = \|\gamma wa\|$. We may replace a by a conjugate and then we can suppose

$$|a_1| \geq |a_2| \geq \cdots \geq |a_n|.$$

Then $\|a\| = \sup(|a_1|, |a_n|^{-1})^d$, where $d = [k : \mathbb{Q}]$ if k is a number field, $d = 1$ if k is a function field. Let $\gamma \in G(k)$. By the Bruhat–Tits decomposition, there exist $u, u' \in U_0(k)$ (where U_0 is the 'superior' unipotent group), $w \in W$ and $t \in T_0(k)$ such that $\gamma = uwtu'$. Let $t_1,\ldots,t_n$ be the diagonal coefficients of t and identify w with a permutation of $\{1,\ldots,n\}$. We check that

$$(\gamma a)_{w1,1} = a_1 t_1,\ (a^{-1}\gamma^{-1})_{n,wn} = a_n^{-1} t_n^{-1}.$$

By construction, for every place v,

$$\sup(|(\gamma a)_{w1,1}|_v, |(a^{-1}\gamma^{-1})_{n,wn}|_v) \leq \|\gamma a\|_v.$$

Thus,

$$\sup(|a_1 t_1|_v, |a_n^{-1} t_n^{-1}|_v) \leq \|\gamma a\|_v$$

and

$$\sup\left(\prod_v |a_1 t_1|_v, \prod_v |a_n^{-1} t_n^{-1}|_v\right) \leq \|\gamma a\|.$$

As $t_1, t_n \in k^*$, $\prod_v |t_1|_v = \prod_v |t_n|_v = 1$ and we obtain

$$\sup(|a_1|^d, |a_n|^{-d}) \leq \|\gamma a\|,$$

i.e. $\|a\| \leq \|\gamma a\|$, which concludes the proof of (*) and of (vii).

Let us prove (viii). First, we have the following elementary assertion, whose proof we leave to the reader:

(1) Let $S \subset V$ be two finite sets; let $s > 0$ and for all $v \in V$, let $x_v > 0$; suppose $S \neq \emptyset$ and $\prod_{v \in V} x_v = 1$. Then

$$\prod_{v \in S} \sup(sx_v, s^{-1}x_v^{-1}) \prod_{v \in V-S} \sup(x_v, x_v^{-1})$$
$$\geq \sup(s, s^{-1})^{1/2} \left(\prod_{v \in V} \sup(x_v, x_v^{-1}) \right)^{1/4}.$$

For $N \in \mathbb{N}$, $N \geq 1$, $g \in GL_N(\mathbb{A})$, we set $\|g\|' = \prod_v \|g\|'_v$. We have

(2) $\|g\| \leq \| \det g \|^{1/N} \|g\|'$.

Let us show that if $N \geq 2$,

(3) there exists $c > 0$ such that for all $g \in GL_N(\mathbb{A})$, we have the inequality

$$c \| \det g \|^{1/N} \|g\|'^{1/(N-1)} \leq \|g\|.$$

Fix a place v of k. By the formulae for calculating the determinant and the inverse of a matrix, there exist constants c'_v and $c''_v > 0$, $c'_v = c''_v = 1$ if v is finite, such that for all $g \in GL_N(\mathbb{A})$, we have

$$|g^{-1}|_v \leq c'_v |g|_v^{N-1} | \det g|_v^{-1},$$
$$| \det g|_v \leq c''_v |g|_v^N.$$

This last inequality implies that

$$\left(|g|_v | \det g|_v^{-1/N} \right)^{1/(N-1)} \leq c''^{1/(N-1)}_v |g|_v | \det g|_v^{-1/N}.$$

From the first, we see that

$$\left(|g^{-1}|_v | \det g|_v^{1/N} \right)^{1/(N-1)} \leq c'^{1/(N-1)}_v |g|_v | \det g|_v^{-1/N}.$$

These last two inequalities imply that

$$c_v \|g\|'^{1/(N-1)}_v \leq |g|_v | \det g|_v^{-1/N}$$

where $c_v = \inf(c'^{-1/(N-1)}_v, c''^{-1/(N-1)}_v)$. Suppose $| \det g|_v \geq 1$. Then $\| \det g\|_v = | \det g|_v$. As $|g|_v \leq \|g\|_v$, we obtain

$$c_v \| \det g\|_v^{1/N} \|g\|'^{1/(N-1)}_v \leq \|g\|_v.$$

This inequality remains true if $| \det g|_v \leq 1$: it suffices to replace g by g^{-1}. Taking the product of these inequalities, we obtain the inequality (3), where $c = \prod_v c_v$.

Suppose $G = GL_n$, $i'_G = id$ (see I.1.1), $\mathbf{G} = GL_n(\mathbb{A})$. Let $a \in A_{\mathbf{G}}$, $g \in \mathbf{G}^1$. First suppose $n \geq 2$. By (3),

$$c \| \det ag \|^{1/n} \|ag\|'^{1/(n-1)} \leq \|ag\|.$$

For every place v we have $|a|_v = | \det a|_v^{1/n}$, which shows that $\|ag\|' = \|g\|'$.

Apply (1), where for S we take the set of archimedean places if k is a number field and $S = \{v_0\}$ if k is a function field, and for V we take the union of S and of the set of places v for which $|\det g|_v \neq 1$, $s = |\det a|_v$ for $v \in S$, $x_v = |\det g|_v$. Then

$$\|\det a\|^{1/2} \|\det g\|^{1/4} \leq \|\det ag\|.$$

We obtain

$$c\|\det a\|^{1/2n} \|\det g\|^{1/4n} \|g\|'^{1/(n-1)} \leq \|ag\|.$$

Set $t = \inf(1/4, 1/(n-1))$. As $\|\det g\|$ and $\|g\|'$ are bounded below, there exists $c' > 0$ independent of g such that

$$\|\det g\|^{1/4n} \geq c'\|\det g\|^{t/n},$$

$$\|g\|'^{1/(n-1)} \geq c'\|g\|'^{t}.$$

By (2), we obtain

$$\|\det g\|^{1/4n} \|g\|'^{(1/(n-1))} \geq c'^{2} \|g\|$$

and finally

$$cc'^{2} \|\det a\|^{1/2n} \|g\|^{t} \leq \|ag\|,$$

which is the equality we wanted. If $n = 1$, this reasoning becomes simpler: we no longer need to consider the terms $\| \ \|'$.

Let us turn to the general case. Set $V = k^n$. The group G acts on V by i'_G. In particular T_G acts on V. This action being defined over k, we can decompose V into eigensubspaces for the action of T_G:

$$V = \bigoplus_{m=1}^{M} V_m.$$

Each V_m is stable under G. For each m, fix a basis of V_m; we deduce from it a homomorphism $i'_m : G \to GL_{n(m)}$ and an embedding

$$j'_G : G \hookrightarrow GL_n$$

$$g \longmapsto \begin{pmatrix} i'_1(g) & & \\ & \ddots & \\ & & i'_M(g) \end{pmatrix}$$

The induced norm of j'_G is equivalent to the one we fixed before. We can thus suppose that $j'_G = i'_G$. Let us show that

(4) there exists $c > 0$ such that for all $g \in G(\mathbb{A})$, we have

$$\prod_{m=1}^{M} \|i'_m(g)\|^{1/M} \leq \|g\| \leq c \prod_{m=1}^{M} \|i'_m(g)\|.$$

Fix a place v of k. There exists $c_v > 0$, $c_v = 1$ if v is finite, such that

$$\|i'_m(g)\|_v \geq c_v$$

for all $g \in G(\mathbb{A})$ and all $m \in \{1, \dots, M\}$. By definition, for $g \in G(\mathbb{A})$,

$$\|g\|_v = \sup\{\|i'_m(g)\|_v ; m = 1, \dots, M\}.$$

Now if $x_1, \dots, x_M$ are real numbers $> c_v$, we have the inequalities

$$\prod_{m=1}^{M} x_m^{1/M} \leq \sup\{x_m ; m = 1, \dots, M\} \leq c_v^{1-M} \prod_{m=1}^{M} x_m.$$

Then

$$\prod_{m=1}^{M} \|i'_m(g)\|_v^{1/M} \leq \|g\|_v \leq c_v^{1-M} \prod_{m=1}^{M} \|i'_m(g)\|_v,$$

which implies (4) with $c = \prod_v c_v^{1-M}$.

Let $g \in G(\mathbb{A})^1$ and $a \in A_{G(\mathbb{A})}$. Then for all m,

$$i'_m(g) \in GL_{n(m)}(\mathbb{A})^1 \quad \text{and} \quad i'_m(a) \in A_{GL_{n(m)}}(\mathbb{A}).$$

We thus have the inequalities

$$c(m)\|i'_m(a)\|^{t'(m)}\|i'_m(g)\|^{t(m)} \leq \|i'_m(ag)\|$$

where $c(m)$, $t(m)$, $t'(m) > 0$ are independent of a and g. Set

$$t = \inf\{t(m)/M;\ m = 1, \dots, M\},\ t' = \inf\{t'(m)/M;\ m = 1, \dots, M\}.$$

As the $\|i'_m(a)\|$ and $\|i'_m(g)\|$ are bounded below, we deduce the following inequalities from the preceding ones:

$$c'\|i'_m(a)\|^{t'}\|i'_m(g)\|^{t} \leq \|i'_m(ag)\|^{1/M}$$

where c' is independent of m, a and g. Applying (4) twice, we have the sequence of inequalities

$$\begin{aligned}\|ag\| &\geq \prod_{m=1}^{M} \|i'_m(ag)\|^{1/M} \\ &\geq c'^M \left(\prod_{m=1}^{M} \|i'_m(a)\|\right)^{t'} \left(\prod_{m=1}^{M} \|i'_m(g)\|\right)^{t} \\ &\geq c'^M c^{-t-t'} \|a\|^{t'} \|g\|^{t},\end{aligned}$$

which proves the desired assertion. □

The generalisation to the covering $\mathbf{G}$ and to two elements $a \in A_{\mathbf{G}}$, $g \in \mathbf{G}^c$ is immediate.

I.2.3. Functions with moderate growth

When V is a space of functions on $\mathbf{G}$ stable under right translation, we write δ for the action of $\mathbf{G}$ on V by right translation. More generally right translation sometimes makes it possible to define, in a space of functions V, the action of an algebra of functions on $\mathbf{G}$, or of an algebra of differential operators (for example an enveloping algebra). We still denote such an action by δ.

Let $\phi : \mathbf{G} \to \mathbb{C}$ be a function. We say that ϕ has moderate growth if there exist $c, r \in \mathbb{R}$ such that for all $g \in \mathbf{G}$, we have

$$|\phi(g)| \leq c\|g\|^r.$$

Suppose that k is a number field. Let $G_\infty = \prod_v G(k_v)$, where the product is over the archimedean places of k, and let $\mathbf{G}_\infty$ be its inverse image in $\mathbf{G}$. Let $\mathscr{U}$ be the enveloping algebra of the (complexified) Lie algebra of $\mathbf{G}$. This algebra acts by right translation on the space of C^∞ functions on $\mathbf{G}$. Let $\phi : \mathbf{G} \to \mathbb{C}$ be a C^∞ function. We say that ϕ and its derivatives have uniformly moderate growth if

$$\exists r \in \mathbb{R},\ \forall X \in \mathscr{U},\ \exists c_X \in \mathbb{R},\ \forall g \in \mathbf{G},\ |\delta(X)\phi(g)| \leq c_X\|g\|^r.$$

Let $P = MU$ be a standard parabolic subgroup of G and $\phi : U(\mathbb{A})M(k)\backslash\mathbf{G} \to \mathbb{C}$. Then ϕ has moderate growth if and only if there exist $c, r \in \mathbb{R}$ and $\lambda \in \mathrm{Re}\, X_{M_0}$ such that for all $a \in A_{\mathbf{M}}$, $k \in A_{\mathbf{M}}$, $m \in \mathbf{M}^c \cap S^P$, we have

$$|\phi(amk)| \leq c\|a\|^r m_{P_0}(m)^\lambda.$$

When k is a number field, there is an analogous interpretation of the condition: ϕ and its derivatives have uniformly moderate growth.

I.2.4. An upper bound lemma

For every subset $C \subset \mathbf{G}$, denote by 1_C the characteristic function of C.

Lemma (a) *There exists $r > 0$, and for every compact subset $C \subset \mathbf{G}$, there exists $c > 0$ such that*

(i) *for all $x, y \in \mathbf{G}$, we have*

$$\sum_{\gamma \in G(k)} 1_C(x^{-1}\gamma y) \leq c\|x\|^r;$$

(ii) *for all $x, y \in \mathbf{G}^c$, we have*

$$\int_{G(k)Z_G} 1_C(x^{-1}\gamma y)\, d\gamma \leq c\|x\|^r.$$

(b) *Let $r > 0$. There exists $r' > 0$, and for every compact subset $C \subset \mathbf{G}$, there exists $c > 0$ such that*

(i) *for all $x \in \mathbf{G}$, $y \in S$, we have*

$$\sum_{\gamma \in G(k)} 1_C(x^{-1}\gamma y) \leq c\|x\|^{r'}\|y\|^{-r};$$

(ii) *for all $x \in \mathbf{G}^c$, $y \in S \cap \mathbf{G}^c$, we have*

$$\int_{G(k)Z_G} 1_C(x^{-1}\gamma y)\, d\gamma \leq c\|x\|^{r'}\|y\|^{-r}.$$

Proof Let $\gamma_1, \gamma_2 \in G(k)$ be such that $1_C(x^{-1}\gamma_i y) = 1$ for $i = 1, 2$. Then $\gamma_i \in xCy^{-1}$ for $i = 1, 2$, so $\gamma_1\gamma_2^{-1} \in xCC^{-1}x^{-1}$ and

$$\|\gamma_1\gamma_2^{-1}\| \leq c'\|x\|^R,$$

where $c' > 0$ depends only on C and $R > 0$ is independent of x, y and C. The sum in (i) of Part (a) is bounded by

$$\#\{\gamma \in G(k); \|\gamma\| \leq c'\|x\|^R\}.$$

We show that there exist c and $r > 0$ such that this number is bounded by $c\|x\|^r$ (the definition of the norm reduces us to the case $G = GL_n$ for which it is a simple exercise).

Let $y \in S$ and $\gamma \in G(k)$, and suppose $1_C(x^{-1}\gamma y) = 1$. Then $\gamma y \in xC$, so $\|\gamma y\| \leq c'\|x\|$, where c' depends only on C. Thanks to I.2.2 (vii), we obtain $\|y\| \leq c''\|x\|$, where c'' depends only on C. Then (b)(i) can be deduced from (a)(i).

Let Γ be the image in $\mathbf{G}/\mathbf{G}^1$ of $\mathbf{G}^c \times C \times \mathbf{G}^c$ under the map

$$(x, g, y) \mapsto \text{ the projection of } xgy^{-1} \text{ into } \mathbf{G}/\mathbf{G}^1.$$

Clearly Γ is compact. In (a)(ii) and (b)(ii) we can restrict to integrating over the $\gamma \in G(k)Z_{\mathbf{G}}$ whose image in $\mathbf{G}/\mathbf{G}^1$ belongs to Γ. By Lemma I.1.5 we deduce that there exists a compact subset $\Gamma' \subset Z_{\mathbf{G}}$ such that the preceding set is contained in $G(k)\Gamma'$. Then there exists $c > 0$ such that for all $x, y \in \mathbf{G}^c$,

$$\int_{G(k)Z_{\mathbf{G}}} 1_C(x^{-1}\gamma y)d\gamma \leq c \sum_{\gamma \in G(k)} 1_{C\Gamma'}(x^{-1}\gamma y).$$

We deduce (a)(ii) from (a)(i) and (b)(ii) from (b)(i). □

I.2.5. Construction of functions with moderate growth

If ξ is a unitary character of $Z_{\mathbf{G}}$, trivial on $Z_G(k) \cap Z_{\mathbf{G}}$, we denote by $L^2(G(k)\backslash\mathbf{G})_\xi$ the space of functions ϕ on $G(k)\backslash\mathbf{G}$ of 'central character' ξ (i.e. $\phi(zg) = \xi(z)\phi(g)$ for all $z \in Z_{\mathbf{G}}$, $g \in \mathbf{G}$) and square integrable modulo $Z_{\mathbf{G}}$. This space is equipped with the L^2 norm. The group $\mathbf{G}$ acts on it by right translation.

Let $f : \mathbf{G} \to \mathbb{C}$ be a function. If k is a function field, we say that f is smooth if it is locally constant. If k is a number field, we say that f is smooth if the following condition is satisfied. Let $\mathbf{G}_f$ be the inverse image of $G(\mathbb{A}_f)$ in $\mathbf{G}$. Let $g \in \mathbf{G}$, $g = g_\infty g_f$, with $g_\infty \in \mathbf{G}_\infty$, $g_f \in \mathbf{G}_f$. We require that there exist a neighborhood V_∞ of g_∞ in $\mathbf{G}_\infty$, a neighborhood V_f of g_f in $\mathbf{G}_f$ and a C^∞ function $f' : V_\infty \to \mathbb{C}$, such that

$$f(g'_\infty g'_f) = f'(g'_\infty)$$

for all $g'_\infty \in V_\infty$, $g'_f \in V_f$.

Lemma *Let $f : \mathbf{G} \to C$ be a smooth function with compact support.*

(a) *Let $\phi : G(k)\backslash\mathbf{G} \to \mathbb{C}$ be a function having moderate growth. Then $\delta(f)\phi$ has moderate growth. If k is a number field, $\delta(f)\phi$ and its derivatives have uniformly moderate growth.*
(b) *Let ξ be a unitary character of $Z_{\mathbf{G}}$, trivial on $Z_G(k) \cap Z_{\mathbf{G}}$, and let $\phi \in L^2(G(k)\backslash\mathbf{G})_\xi$. Then $\delta(f)\phi$ satisfies the same properties as in (a).*

Proof For (a), we write

$$\delta(f)\phi(x) = \int_{G(k)\backslash\mathbf{G}} \sum_{\gamma\in G(k)} f(x^{-1}\gamma y)\phi(y)\, dy.$$

and we use I.2.4 (b)(i). If k is a number field and $X \in \mathcal{U}$, we similarly have

$$(\delta(X)\delta(f)\phi)(x) = \int_{G(k)\backslash\mathbf{G}} \sum_{\gamma\in G(k)} Xf(x^{-1}\gamma y)\phi(y)\, dy,$$

where X acts on f via left translation, so again we can conclude. The proof of (b) is analogous. □

Remark In the rest of this chapter, we only study functions with moderate growth. Certain results also have consequences for square integrable functions (which in general do not have moderate growth), because of the following fact. Let ξ be a unitary character of $Z_{\mathbf{G}}$, trivial on $Z_G(k) \cap Z_{\mathbf{G}}$, and let $\phi \in L^2(G(k)\backslash\mathbf{G})_\xi$. For all $\varepsilon > 0$, there exists a function $f : \mathbf{G} \to \mathbb{C}$, smooth with compact support, left and right $\mathbf{K}$-finite, such that $\|\delta(f)\phi - \phi\| < \varepsilon$. The function $\delta(f)\phi$ is smooth $\mathbf{K}$-finite and, as we just saw, has moderate growth.

I.2.6. Constant terms

Let $P = MU$ be a standard Levi subgroup of G, ϕ a measurable and locally L^1 function on $U(k)\backslash\mathbf{G}$. We define a measurable, locally L^1 function ϕ_P on $U(\mathbb{A})\backslash\mathbf{G}$ by

$$\phi_P(g) = \int_{U(k)\backslash U(\mathbb{A})} \phi(ug)\, du.$$

If ϕ is left invariant under $G(k)$, has moderate growth, is smooth, then ϕ_P is left invariant under $M(k)$, has moderate growth, is smooth, respectively.

I.2.7. Approximation of a function by its constant term, function field case

Lemma *Suppose k is a function field. Let $\mathbf{K}'$ be a compact open subgroup of $\mathbf{G}$. Then there exists $c > 0$ such that the following property holds. Let*

ϕ be a function on $G(k)\backslash\mathbf{G}$, right invariant under $\mathbf{K}'$, and let $g \in S$ be such that

$$m_{P_0}(g)^\alpha > c \text{ for all } \alpha \in R^+(T_0, G) - R^+(T_0, M).$$

Then we have the equality $\phi(g) = \phi_P(g)$.

Proof Choose an open subgroup $V \subset U(\mathbb{A})$ such that for all $k \in \mathbf{K}$, we have $k^{-1}Vk \subset \mathbf{K}'$. Let V' be a compact open subset of $U(\mathbb{A})$ such that the projection $V' \to U(k)\backslash U(\mathbb{A})$ is bijective. Let $g \in S$ and write $g = pak$, with $p \in \omega$, $k \in \mathbf{K}$, $a \in A_{\mathbf{M}_0}$. We have

$$\begin{aligned}\phi_P(g) &= \int_{U(k)\backslash U(\mathbb{A})} \phi(upak)\, du \\ &= \int_{V'} (\delta(k)\phi)(upa)\, du \\ &= (\text{meas } V'')^{-1} \int_{V''} (\delta(k)\phi)(pau)\, du\end{aligned}$$

where $V'' = (pa)^{-1}V'pa$. Suppose $V'' \subset V$. Then $\delta(k)\phi$ is invariant under V'' and we obtain $\phi_P(g) = (\delta(k)\phi)(pa) = \phi(g)$. As p remains in a compact set, clearly there exists $c > 0$ such that the condition $V'' \subset V$ is realised whenever $a^\alpha > c$ for all $\alpha \in R^+(T_0, G) - R^+(T_0, M)$. Up to modifying c, this is equivalent to the condition of the statement. □

Following Langlands, let us introduce a partition of $\mathbf{M}_0$. Let $t \in \mathbf{M}_0$. For a standard parabolic subgroup $P = MU$ of G, denote by $\hat{\Delta}^M$ the basis of $(\mathfrak{a}^M_{M_0})^*$ dual to the basis of coroots $\{\check{\alpha}; \alpha \in \Delta^M\}$ of $\mathfrak{a}^M_{M_0}$, and define the set $\mathbf{M}_0(P,t)$ to be the set of elements $m \in \mathbf{M}_0$ such that

$$\begin{aligned} m^\alpha &> t^\alpha \text{ for all } \alpha \in \Delta_M, \\ m^\omega &\le t^\omega \text{ for all } \omega \in \hat{\Delta}^M. \end{aligned}$$

We have the equality (see [A1] Lemma 6.3):

$$\mathbf{M}_0 = \bigcup_{P;P_0 \subset P \subset G} \mathbf{M}_0(P,t)$$

the union being disjoint.

Remark From the above partition we immediately deduce a partition of $\mathbf{M}_0^1\backslash\mathbf{M}_0$.

I.2.8. Corollary

Suppose k is a function field. Let $\mathbf{K}'$ be a compact open subgroup of $\mathbf{G}$. Then there exists $c > 0$ such that the following property is satisfied. Let ϕ be a function on $G(k)\backslash\mathbf{G}$, right invariant under $\mathbf{K}'$, let $t \in \mathbf{M}_0$ be such *that $t^\alpha > c$ for all $\alpha \in \Delta_0$, let P be a standard parabolic subgroup of G and let $g \in S$ be such that $m_{P_0}(g) \in \mathbf{M}_0(P,t)$. Then we have the equality $\phi(g) = \phi_P(g)$.*

Proof Write $P = MU$ and let $\alpha \in R^+(T_0,G) - R^+(T_0,M)$. There exist $\beta \in \Delta_0 - \Delta_0^M$ and $\gamma \in R^+(T_0,G) \cup \{0\}$ such that $\alpha = \beta + \gamma$. We can write

$$\beta = \beta_M + \sum_{\varpi \in \hat{\Delta}^M} c_\varpi \omega, \text{ with } \beta_M \in \Delta_M.$$

If $\omega \in \hat{\Delta}^M$ is associated with the coroot $\check{\delta}$, with $\delta \in \Delta^M$, we have

$$c_\omega = \langle \beta, \check{\delta} \rangle \leq 0.$$

As $m_{P_0}(g) \in \mathbf{M}_0(P,t)$, we deduce that

$$m_{P_0}(g)^\beta > t^\beta.$$

There exists c' such that for all $g' \in S$, we have $m_{P_0}(g')^\gamma > c'$. Thus,

$$m_{P_0}(g)^\alpha > c't^\beta.$$

Let c be the constant in the preceding lemma associated with P and $\mathbf{K}'$. If t is large enough, we obtain $m_{P_0}(g)^\alpha > c$ and the lemma allows us to conclude the proof. □

I.2.9. Corollary

Suppose k is a function field. Let $\mathbf{K}'$ be a compact open subgroup of $\mathbf{G}$. Then there exists a compact subset $C \subset A_\mathbf{G}\backslash S$ such that the following property is satisfied. Let ϕ be a function on $G(k)\backslash\mathbf{G}$, right invariant under $\mathbf{K}'$. Define a function $s\phi$ on S by

$$s\phi(g) = \sum_{P=MU;P_0 \subset P \subset G} (-1)^{\mathrm{rg}(G)-\mathrm{rg}(M)} \phi_P(g).$$

Then the image in $A_\mathbf{G}\backslash S$ of the support of $s\phi$ is contained in C.

(We have written $\mathrm{rg}(G)$ and $\mathrm{rg}(M)$ for the semi-simple ranks of G and M, i.e. for example $\mathrm{rg}(G) = |\Delta_0|$).

Proof Lemma I.2.7 can be generalised to functions on $U'(\mathbb{A})M'(k)\backslash\mathbf{G}$ for a parabolic subgroup $P' \subset P$. It suffices to replace $R(T_0,G)$ by $R(T_0,M')$. Then choose a c large enough to ensure the conclusion of Lemma I.2.7 for all pairs P, P' such that $P_0 \subset P \subset P'$. Let $g \in S$, and set

$$\Delta_0(g) = \{\alpha \in \Delta_0 ; m_{P_0}(g)^\alpha \leq c\}.$$

The set of standard parabolic subgroups is in bijection with the set of pairs (Γ,Γ') with $\Gamma \subset \Delta_0(g), \Gamma' \subset \Delta_0 - \Delta_0(g)$, via

$$(\Gamma,\Gamma') \longmapsto P(\Gamma,\Gamma') = M(\Gamma,\Gamma')U(\Gamma,\Gamma'),$$

with $\Delta_0^{M(\Gamma,\Gamma')} = \Gamma \cup \Gamma'$. The function of the statement can be written

$$\sum_{\Gamma\subset\Delta_0(g)} \sum_{\Gamma'\subset\Delta_0-\Delta_0(g)} (-1)^{|\Gamma|+|\Gamma'|}\phi_{P(\Gamma,\Gamma')}(g).$$

By Lemma I.2.7, generalised as explained above, we have the equality $\phi_{P(\Gamma,\Gamma')}(g) = \phi_{P(\Gamma,\emptyset)}(g)$. But then in the above sum we can factor out the sum

$$\sum_{\Gamma'\subset\Delta_0-\Delta_0(g)} (-1)^{|\Gamma'|}.$$

This sum is zero unless $\Delta_0 = \Delta_0(g)$. But the set of $g \in S$ for which $\Delta_0(g) = \Delta_0$ is compact modulo $A_{\mathbf{G}}$. □

I.2.10. The number field case

Over a number field, the situation is much more complicated.

Lemma *Let $P = MU$ be a proper maximal standard parabolic subgroup of G and let α be the root such that $\{\alpha\} = \Delta_0 - \Delta_0^M$. Let $\mathbf{K}'_f$ be a compact open subgroup of $\mathbf{G}_f$, $X \in \mathscr{U}$, $c > 0$ and $t > 0$. Then there exist two finite subsets $\{X_i; i = 1,\dots,N\} \subset \mathscr{U}$ and $\{c_i; i = 1,\dots,N\} \subset \mathbb{R}_+^*$ such that the following property is satisfied. Let ϕ be a smooth function on $U_0(k)\backslash\mathbf{G}$, right invariant under $\mathbf{K}'_f$. Let $r > 0$ and $\lambda \in ReX_{M_0}$. Suppose that for all $a \in A_{\mathbf{G}}$, $g \in \mathbf{G}^1 \cap S$, $i \in \{1,\dots,N\}$, we have the inequality*

$$|\delta(X_i)\phi(ag)| \le c_i\|a\|^r m_{P_0}(g)^\lambda.$$

Then for all $a \in A_{\mathbf{G}}$, $g \in \mathbf{G}^1 \cap S$, we have the inequality

$$|\delta(X)(\phi - \phi_P)(ag)| \le c\|a\|^r m_{P_0}(g)^{\lambda - t\alpha}.$$

Proof (see [HC] Lemma 10). We may fix a sequence

$$V_0 = \{1\} \subset V_1 \subset V_2 \subset \cdots \subset V_N = U$$

of subgroups defined over k, such that for all i, V_i is normal in U and $V_{i-1}\backslash V_i$ is isomorphic to the additive group $\mathbb{G}_a$. Let $j_i : V_i \to \mathbb{G}$ be the morphism given by the composition $V_i \to V_{i-1}\backslash V_i \to G_a$. Set

$$\Gamma_i = V_i(\mathbb{A}_f) \cap (\bigcap_{g\in S} g\mathbf{K}'_f g^{-1}).$$

Then $j_i(\Gamma_i)$ is a compact open subgroup of $\mathbb{A}_f$. The structure of $k\backslash\mathbb{A}/j_i(\Gamma_i)$ is well-known: there exists a lattice $L_i \subset \mathbb{A}_\infty$ such that

$$L_i\backslash\mathbb{A}_\infty \simeq k\backslash\mathbb{A}/j_i(\Gamma_i),$$

where $\mathbb{A}_\infty$ is sent into $\mathbb{A}$ by the natural map. Fix a supplementary subspace ϑ_i of $\mathrm{Lie}(V_{i-1}(\mathbb{A}_\infty))$ in $\mathrm{Lie}(V_i(\mathbb{A}_\infty))$; we are dealing here with real Lie algebras. The map $j_i \circ \exp : \vartheta_i \to \mathbb{A}_\infty$ is an isomorphism. We choose a basis $(X_{i\ell})_{\ell=1,\dots,d}$ (where $d = [k : \mathbb{Q}]$) of ϑ_i such that $(j_i \circ \exp(X_{i\ell}))_{\ell=1,\dots,d}$ is a basis of L_i. We finally obtain that the map

$$(\mathbb{Z}\backslash\mathbb{R})^d \longrightarrow V_{i-1}(\mathbb{A})V_i(k)\backslash V_i(\mathbb{A})/\Gamma_i$$

$$(x_\ell)_{\ell=1,\dots,d} \longmapsto \exp\left(\sum_{\ell=1}^{d} x_\ell X_{i\ell}\right)$$

is an isomorphism.

We will need the following assertion. Fix a basis $\{X_j; j \in J\}$ of $\mathrm{Lie}(\mathbf{G}_\infty)$, and denote by $\{X_j^*; j \in J\}$ the dual basis of

$$\mathrm{Hom}_{\mathbb{R}}(\mathrm{Lie}(\mathbf{G}_\infty), \mathbb{R}).$$

(1) There exists $c_1 > 0$ such that for all $i \in \{1,\dots,N\}$, $\ell \in \{1,\dots,d\}$, $j \in J$, $g \in S$, we have

$$|\langle \mathrm{Ad}(g_\infty^{-1})X_{i\ell}, X_j^*\rangle| \leq c_1 m_{P_0}(g)^{-\alpha},$$

where g_∞ is the component of g in $\mathbf{G}_\infty$.

Write $g = pak$, $p \in \omega$, $k \in \mathbf{K}$, $a \in A_{\mathbf{M}_0}(t_0)$ (see I.2.1). As $p \in \mathbf{P}_0$, $\mathrm{Ad}(p_\infty^{-1})$ preserves $\mathrm{Lie}(U(\mathbb{A}_\infty))$. As moreover ω is compact, we can write

$$\mathrm{Ad}(p_\infty^{-1})X_{i\ell} = \sum_{i',\ell'} c_{i'\ell'} X_{i'\ell'},$$

with bounded coefficients $c_{i'\ell'}$. The group $A_{\mathbf{M}_0}$ preserves $\mathrm{Lie}(U(\mathbb{A}_\infty))$ as well. It acts on it semi-simply, the eigencharacters being roots of the form $\alpha + \beta, \beta \geq 0$. We can thus write, for $a \in A_{\mathbf{M}_0}(t_0)$

$$\mathrm{Ad}(a^{-1})X_{i'\ell'} = \sum_{i'',\ell''} a^{-\alpha} c_{i',\ell';i''\ell''} X_{i'',\ell''},$$

with bounded coefficients $c_{i',\ell';i'',\ell''}$. Finally, since $\mathbf{K}$ is compact, it also acts boundedly. Composing these three actions and noting that $a^{-\alpha} < m_{P_0}(g)^{-\alpha}$, we obtain (1).

Let ϕ be as in the statement. For $i = 0,\dots,N$, define a function $\phi_i : \mathbf{G} \to \mathbb{C}$ by

$$\phi_i(g) = \int_{V_i(k)\backslash V_i(\mathbb{A})} \phi(ug)\, du.$$

We have $\phi_0 = \phi$, $\phi_P = \phi_N$, so

$$\phi - \phi_P = \sum_{i=1}^{N} (\phi_{i-1} - \phi_i). \tag{2}$$

Fix i. We will establish an upper bound for the function $\phi_{i-1} - \phi_i$. Note that

$$\phi_i(g) = \int_{V_{i-1}(\mathbb{A})V_i(k)\backslash V_i(\mathbb{A})} \phi_{i-1}(ug)\, du.$$

For $a \in A_G$ and $g \in \mathbf{G}^1 \cap S$, the function

$$u \longmapsto \phi_{i-1}(uag)$$

is right invariant under Γ_i. By the above discussion, we thus have

(i) the function

$$\mathbb{R}^d \longrightarrow \mathbb{C}$$

$$(x_\ell)_{\ell=1,\dots,d} \longmapsto \phi_{i-1}(\exp\left(\sum_{\ell=1}^{d} x_\ell X_{i\ell}\right) ag)$$

is invariant under $\mathbb{Z}^d$;

(ii)

$$\phi_i(ag) = \int_{\mathbb{Z}^d\backslash\mathbb{R}^d} \phi_{i-1}(\exp\left(\sum_{\ell=1}^{d} x_\ell X_{i\ell}\right) ag) \prod dx_\ell.$$

By Fourier inversion, we thus have

(3)
$$(\phi_{i-1} - \phi_i)(ag) = \sum_{\xi\in\mathbb{Z}^d-\{0\}} \phi_{i-1}^{\xi}(ag),$$

where, for $\xi = (\xi_\ell)_{\ell=1,\dots,d} \in \mathbb{Z}^d$, we set

$$\phi_{i-1}^{\xi}(ag) = \int_{\mathbb{Z}^d\backslash\mathbb{R}^d} \phi_{i-1}\left(\exp\left(\sum_{\ell=1}^{d} x_\ell X_{i\ell}\right) ag\right) \exp\left(2\pi i \sum_{\ell=1}^{d} x_\ell \xi_\ell\right) \prod dx_\ell.$$

Let $\xi = (\xi_\ell)_{\ell=1,\dots,d} \in \mathbb{Z}^d - \{0\}$. Choose $\ell(\xi) \in \{1,\dots,d\}$ such that $|\xi_{\ell(\xi)}|$ is maximal. We have

$$\phi_{i-1}^{\xi}(ag) = \int_{\mathbb{Z}^d\backslash\mathbb{R}^d} \phi_{i-1}\left(\exp\left(\sum_{\ell\neq\ell(\xi)} x_\ell X_{i\ell}\right) ag \exp\left(x_{\ell(\xi)} X'_{i\ell(\xi)}\right)\right)$$
$$\exp\left(2\pi i \sum_{\ell=1}^{d} x_\ell \xi_\ell\right) \prod dx_\ell,$$

where $X'_{i\ell(\xi)} = \mathrm{Ad}(g_\alpha^{-1}) X_{i\ell(\xi)}$. For $h \in \mathbb{N}$, we also have, via integration by parts

$$\phi_{i-1}^{\xi}(ag) = (2\pi i \xi_{\ell(\xi)})^{-h} \int_{\mathbb{Z}^d \backslash \mathbb{R}^d} \frac{d^h}{dx_{\ell(\xi)}} \phi_{i-1}\Bigg(\exp\Big(\sum_{\ell \neq \ell(\xi)} x_\ell X_{i\ell} \Big) ag \times$$

$$\exp\Big(x_{\ell(\xi)} X'_{i\ell(\xi)} \Big) \Bigg) \exp\Big(2\pi i \sum_{\ell=1}^{d} x_\ell \xi_\ell \Big) \prod dx_\ell$$

$$= (2\pi i \xi_{\ell(\xi)})^{-h} \int_{\mathbb{Z}^d \backslash \mathbb{R}^d} \delta(X'^{h}_{i\ell(\xi)}) \phi_{i-1}\Bigg(\exp\Big(\sum_{\ell \neq \ell(\xi)} x_\ell X_{i\ell} \Big) \times$$

$$ag \exp\Big(x_{\ell(\xi)} X'_{i\ell(\xi)} \Big) \Bigg) \exp\Big(2\pi i \sum_{\ell=1}^{d} x_\ell \xi_\ell \Big) \prod dx_\ell$$

$$= (2\pi i \xi_{\ell(\xi)})^{-h} \int_{\mathbb{Z}^d \backslash \mathbb{R}^d} \delta(X'^{h}_{i\ell(\xi)}) \phi_{i-1}\Bigg(\exp\Big(\sum_{\ell=1}^{d} x_\ell X_{i\ell} \Big) ag \Bigg) \times$$

$$\exp\Big(2\pi i \sum_{\ell=1}^{d} x_\ell \xi_\ell \Big) \prod dx_\ell .$$

Thus we obtain the upper bound

$$|\phi_{i-1}^{\xi}(ag)| \leq$$
$$(2\pi \sup\{|\xi_\ell|; \ell = 1, \ldots, d\})^{-h} \sup\{|\delta(X'^{h}_{i\ell(\xi)}) \phi_{i-1}(uag)|; u \in U(\mathbb{A})\}.$$

By (1), there exists $c_h > 0$, depending only on h, such that

$$|\phi_{i-1}^{\xi}(ag)| \leq c_h \sup\{|\xi_\ell|; \ell = 1, \ldots, d\}^{-h} m_{P_0}(g)^{-h\alpha} \times$$
$$\sup\{|\delta(X_{j_1} \ldots X_{j_h})\phi(uag)|; u \in U(\mathbb{A}), j_1, \ldots, j_h \in J\}.$$

Fix h such that

(4) $h \geq t$,

(5) the series $\sum_{\xi \in \mathbb{Z}^d - \{0\}} \sup\{|\xi_\ell|; \ell = 1, \ldots, d\}^{-h}$ is convergent.

Let $c' > 0$ and suppose that we have the upper bound

(6) $$|\delta(X_{j_1} \ldots X_{j_h})\phi(a'g')| \leq c' \|a'\|^r m_{P_0}(g')^{\lambda}$$

for all $a' \in A_{\mathbf{G}}$, $g' \in \mathbf{G}^1 \cap S$, $j_1, \ldots, j_h \in J$. As ϕ is invariant under $U_0(k)$, $U_0(\mathbb{A})S = U_0(k)S$ and m_{P_0} is invariant under $U_0(\mathbb{A})$, we deduce that

$$\sup\{|\delta(X_{j_1} \ldots X_{j_h})\phi(uag)|; u \in U(\mathbb{A}), j_1, \ldots, j_h \in J\} \leq c' \|a\|^r m_{P_0}(g)^{\lambda}.$$

Thus

$$|\phi_{i-1}^{\xi}(ag)| \leq c' c_h \sup\{|\xi_\ell|; \ell = 1, \ldots, d\}^{-h} m_{P_0}(g)^{\lambda - h\alpha} \|a\|^r.$$

From (2), (3), (4), (5) and (6) and the fact that $m_{P_0}(g)^{\alpha}$ is bounded below since $g \in S$, we deduce that there exists $c'_h > 0$ depending only on h such that

$$|(\phi - \phi_P)(ag)| \leq c' c'_h \|a\|^r m_{P_0}(g)^{\lambda - t\alpha}.$$

Taking $c' = c/c'_h$, we obtain the conclusion of the lemma for $X = 1$. We obtain the conclusion for all X by replacing ϕ by $\delta(X)\phi$. □

I.2.11. Corollary

Suppose that k is a number field. Let $\mathbf{K}'_f$ be a compact open subgroup of $\mathbf{G}_f$. Let $r > 0$, $\lambda \in \mathrm{Re}\, X_{M_0}$, $X \in \mathscr{U}$ and $c > 0$. Then there exist two finite subsets

$$\{X_i;\ i = 1, \dots, N\} \subset \mathscr{U},\ \{c_i; i = 1, \dots, N\} \subset \mathbb{R}^*_+$$

such that the following property is satisfied. Let ϕ be a smooth function on $G(k)\backslash\mathbf{G}$, right invariant under $\mathbf{K}'_f$. Suppose that for all $i \in \{1, \dots, N\}$ and all $g \in \mathbf{G}$, we have the inequality

$$|\delta(X_i)\phi(g)| \leq c_i \|g\|^r.$$

Define a function $s\phi$ on S by

$$s\phi(g) = \sum_{P=MU;P_0\subset P\subset G} (-1)^{rg(G)-rg(M)}\phi_P(g).$$

Then for all $a \in A_G$, $g \in \mathbf{G}^1 \cap S$, we have the inequality

$$|\delta(X)s\phi(ag)| \leq c\|a\|^r m_{P_0}(g)^\lambda.$$

Proof Let us number the simple roots and denote them by $\alpha_1, \dots, \alpha_n$. For $i = 1, \dots, n$, let $P_i = M_iU_i$, $P'_i = M'_iU'_i$, be the standard parabolic subgroups of G such that $\{\alpha_i\} = \Delta_0 - \Delta_0^{M_i}$, $\{\alpha_1, \dots, \alpha_i\} = \Delta_0 - \Delta_0^{M'_i}$, respectively. Define a function $s_i\phi$ on $U_0(k)\backslash\mathbf{G}$ by

$$s_i\phi(g) = \sum_{P=MU;P'_i\subset P\subset G} (-1)^{rg(G)-rg(M)}\phi_P(g).$$

Set $s_0\phi = \phi$. We check that

$$s_i\phi = s_{i-1}\phi - (s_{i-1}\phi)_{P_i}.$$

By induction, using the preceding lemma, we can obtain upper bounds for $s_i\phi$. But $s_n\phi$ is the function $s\phi$ of the statement. □

I.2.12. Rapidly decreasing functions

Let $\phi : S \to \mathbb{C}$. The field k being a number field, we say that ϕ is rapidly decreasing if there exists $r > 0$ and if for all $\lambda \in \mathrm{Re}\, X_{M_0}$, there exists $c > 0$ such that for all $a \in A_G$, $g \in \mathbf{G}^1 \cap S$, we have the inequality

$$|\phi(ag)| \leq c\|a\|^r m_{P_0}(g)^\lambda.$$

Let $\phi : G(k)\backslash\mathbf{G} \to \mathbb{C}$. We say that ϕ is rapidly decreasing if the restriction of ϕ to S is rapidly decreasing.

We can essentially summarize Corollaries I.2.9 and I.2.11 as follows.

Corollary *Let $\phi : G(k)\backslash \mathbf{G} \to \mathbb{C}$ be a smooth* **K***-finite function. Define a function $s\phi$ on S by*

$$s\phi(g) = \sum_{P=MU;P_0\subset P\subset G} (-1)^{rg(G)-rg(M)}\phi_P(g).$$

If k is a function field, the image in $A_{\mathbf{G}}\backslash S$ of the support of $s\phi$ is compact. If k is a number field, suppose in addition that ϕ and its derivatives have uniformly moderate growth. Then $s\phi$ is rapidly decreasing. □

I.2.13. Truncation

Arthur introduced another way of approximating a function by its constant terms. We recall below the definition and the principal properties of Arthur's truncation operator, citing [A2] for the proofs. For a standard parabolic subgroup $P = MU$ of G, denote by $\hat{\tau}_P$ the characteristic function of the sum of Re $\mathfrak{a}_G$ and of the interior of the cone of Re $\mathfrak{a}_M^G$ generated by the elements of Δ_M. Fix $T \in \mathrm{Re}\,\mathfrak{a}_{M_0}$. We suppose T sufficiently positive, i.e. $\langle \alpha, T\rangle$ large enough for all $\alpha \in \Delta_0$. We denote by T_M the projection of T onto $\mathrm{Re}\,\mathfrak{a}_M$, hoping that this will not create confusion with the subtorus T_M of M.

Lemma *Let $r > 0$, $c > 0$. There exist $r' > 0$ and $c' > 0$ such that the following property is satisfied. Let $\phi : P(k)\backslash \mathbf{G} \to \mathbb{C}$ be a function. Suppose that*

$$|\phi(g)| \leq c\|g\|^r$$

for all $g \in \mathbf{G}$. Let $g \in \mathbf{G}$. Then

$$\sum_{\gamma\in P(k)\backslash G(k)} |\phi(\gamma g)|\hat{\tau}_P(\log_M(m_P(\gamma g)) - T_M) \leq c'\|g\|^{r'}.$$

In particular, the sum is finite.

For the proof see [A1], Corollary 5.2. □

Let $\phi : G(k)\backslash \mathbf{G} \to \mathbb{C}$ be a locally L^1 function. We define a locally L^1 function $\wedge^T\phi$ over $G(k)\backslash \mathbf{G}$ by

$$\wedge^T\phi(g) = \sum_{P=MU;P_0\subset P\subset G} (-1)^{rg(G)-rg(M)} \times$$
$$\sum_{\gamma\in P(k)\backslash G(k)} \hat{\tau}_P(\log_M(m_P(\gamma g)) - T_M)\phi_P(\gamma g).$$

Remark If ϕ is a locally L^1 function, then $\wedge^T\phi(g)$ is only defined almost everywhere; if we suppose however that ϕ is locally bounded, *a fortiori* if we suppose ϕ continuous, then $\wedge^T\phi(g)$ is defined for all g and $\wedge^T\phi$ is locally bounded.

It is a consequence of the above lemma that if ϕ has moderate growth, then so does $\wedge^T\phi$.

I.2.14. Lemma

Let $\phi : G(k)\backslash\mathbf{G} \to \mathbb{C}$ be a locally L^1 function. Then we have the equality

$$\wedge^T \wedge^T \phi(g) = \wedge^T\phi(g)$$

for almost all g. If ϕ is also locally bounded then the above is true for all g).

For the proof see [A2] Corollary 1.2. □

I.2.15. Lemma

Let ξ be a unitary character of $Z_{\mathbf{G}}$, ϕ_1 and ϕ_2 two locally L^1 functions on $G(k)\backslash\mathbf{G}$ of 'central character' ξ. We suppose that ϕ_1 has moderate growth and ϕ_2 is either rapidly decreasing if k is a number field or bounded with compact support in $Z_{\mathbf{G}}G(k)\backslash\mathbf{G}$ if k is a function field. Then

$$\int_{Z_{\mathbf{G}}G(k)\backslash\mathbf{G}} \overline{\wedge^T\phi_1(g)}\phi_2(g)\, dg = \int_{Z_{\mathbf{G}}G(k)\backslash\mathbf{G}} \overline{\phi}_1(g) \wedge^T \phi_2(g)\, dg.$$

For the proofs see [A2] Lemma 1.3. □

I.2.16. Lemma

(1) *Suppose that k is a number field. Let $\mathbf{K}'_f$ be an open subgroup of $\mathbf{G}_f$,with r, r' two positive real numbers. Then there exists a subfinite set $\{X_i; i = 1,\ldots,N\} \subset \mathcal{U}$ such that the following property is satisfied. Let ϕ be a smooth function on $G(k)\backslash\mathbf{G}$, right invariant under $\mathbf{K}'_f$, and let $a \in A_{\mathbf{G}}$, $g \in \mathbf{G}^1 \cap S$. Then*

$$|\wedge^T \phi(ag)| \leq \|g\|^{-r} \sum_{i=1}^{N} \sup\{|\delta(X_i)\phi(ag')|\, \|g'\|^{-r'}; g' \in \mathbf{G}^1\}.$$

(2) *Suppose that k is a function field. Let $\mathbf{K}'$ be an open subgroup of $\mathbf{G}$. There exists a subset Γ of $G(k)\backslash\mathbf{G}$, of compact image in $G(k)Z_{\mathbf{G}}\backslash\mathbf{G}$, such that for any smooth function ϕ on $G(k)\backslash\mathbf{G}$, right invariant under $\mathbf{K}'$, the support of $\wedge^T\phi$ is contained in Γ.*

Proof See [A2] Lemma 1.4 in the number field case. For a function field, we note that Lemma 1.1 of [A2] remains valid. This lemma implies that for all standard parabolic subgroups P and all $t \in \mathbf{M}_0$, the intersection of the support of $(\wedge^T\phi)_P$ with $\{g \in S; m_{P_0}(g) \in \mathbf{M}_0(P,t)\}$ is contained in a set independent of ϕ and compact modulo $A_{\mathbf{G}}$. Applying corollary I.2.8 to the function $\wedge^T\phi$, we see that its support is contained in a set independent of ϕ and of compact image in $G(k)A_{\mathbf{G}}\backslash\mathbf{G}$. But $(G(k)\cap Z_{\mathbf{G}})A_{\mathbf{G}}\backslash Z_{\mathbf{G}}$ is compact and we thus deduce the lemma. □

I.2.17. Automorphic forms

Recall that if k is a number field, we write $\mathscr{U}$ for the enveloping algebra of the Lie algebra of $\mathbf{G}_\infty$ (see I.2.3). Let us write $\mathfrak{z}$ for the centre of $\mathscr{U}$. If k is a function field, we fix a place v_0 of k (see I.2.1). We denote by $\mathbf{G}_{v_0}$ the inverse image of $G(k_{v_0})$ in $\mathbf{G}$ (see I.1.2) and by $\mathfrak{z}$ the Bernstein centre of $\mathbf{G}_{v_0}$ (see [BDKV]).

Let $P = MU$ be a standard parabolic subgroup of G and ϕ : $U(\mathbb{A})M(k)\backslash\mathbf{G} \to \mathbb{C}$ a function. We say that ϕ is automorphic if it satisfies the following conditions:

(i) ϕ has moderate growth;
(ii) ϕ is smooth (see I.2.5);
(iii) ϕ is $\mathbf{K}$-finite;
(iv) ϕ is $\mathfrak{z}$-finite.

We can reduce this situation to the usual case of automorphic forms on $M(k)\backslash\mathbf{M}$ as follows. For $\phi : U(\mathbb{A})M(k)\backslash\mathbf{G} \to \mathbb{C}$ and $k \in \mathbf{K}$, define $\phi_k : M(k)\backslash\mathbf{M} \to \mathbb{C}$ by

$$\phi_k(m) = m^{-\rho_P}\phi(mk),$$

where ρ_P is the half-sum of roots of M in Lie U (ρ_P will be useful to us later). Then ϕ is automorphic if and only if it is smooth, $\mathbf{K}$-finite and for all $k \in \mathbf{K}$, ϕ_k is automorphic over $M(k)\backslash\mathbf{M}$.

Proof The only difficulty is to show the equivalence of the finiteness conditions (iv). Let $\mathfrak{z}^{\mathbf{M}}$ be the analogue of $\mathfrak{z}$ for the group $\mathbf{M}$. The algebras $\mathfrak{z}$ and $\mathfrak{z}^{\mathbf{M}}$ act by right translation and left translation respectively on the space V of smooth functions on $U(\mathbb{A})\backslash\mathbf{G}$. There exists a homomorphism of Harish-Chandra algebras (or Bernstein algebras in the function field case) $i : \mathfrak{z} \to \mathfrak{z}^{\mathbf{M}}$, such that if $\varphi \in V$ and $z \in \mathfrak{z}$, we have

$$z.\varphi = i(z).\varphi',$$

where φ' is defined by $\varphi'(g) = m_P(g)^{-\rho_P}\varphi(g)$. Moreover $\mathfrak{z}^{\mathbf{M}}$ is a finitely generated $i(\mathfrak{z})$-module (if k is a function field, this results from the description of $\mathfrak{z}$ and $\mathfrak{z}^{\mathbf{M}}$, see [BDKV] Theorem 2.13; if k is a number field, see [HC2]).

Suppose that ϕ is automorphic. By (iv) and the above properties, ϕ' is $\mathfrak{z}^{\mathbf{M}}$-finite for the action of $\mathfrak{z}^{\mathbf{M}}$ by left translation. Thus for all $k \in \mathbf{K}$, ϕ_k is $\mathfrak{z}^{\mathbf{M}}$-finite. But ϕ_k is a function on $\mathbf{M}$ and the action of $\mathfrak{z}^{\mathbf{M}}$ by left translation on the functions on **M** can be identified with its action by right translation, up to an automorphism of $\mathfrak{z}^{\mathbf{M}}$. Thus ϕ_k is $\mathfrak{z}^{\mathbf{M}}$-finite, $\mathfrak{z}^{\mathbf{M}}$ acting this time by right translation, i.e. ϕ_k satisfies the analogue of (iv) for the group **M**. Conversely, suppose that ϕ_k is automorphic for all $k \in \mathbf{K}$. By **K**-finiteness, the space generated by the ϕ_k is finite dimensional. Thus there exists an ideal $I \subset \mathfrak{z}^{\mathbf{M}}$ of finite codimension which kills all the ϕ_k. Then I kills ϕ (for its action by left translation) and the inverse image $i^{-1}(I) \subset \mathfrak{z}$ kills ϕ, thus ϕ satisfies (iv). □

We denote by $A(U(\mathbb{A})M(k)\backslash\mathbf{G})$ the space of automorphic forms on $U(\mathbb{A})M(k)\backslash\mathbf{G}$.

Remark If k is a function field, this space depends *a priori* on the choice of the place v_0; we show in I.3.6 that in fact it is independent of this choice.

Let ξ be a character of $Z_{\mathbf{M}}$, a continuous homomorphism of $Z_{\mathbf{M}}$ in $\mathbb{C}^*$. We denote by $A(U(\mathbb{A})M(k)\backslash\mathbf{G})_\xi$ the subspace of elements $\phi \in A(U(\mathbb{A})M(k)\backslash\mathbf{G})$ satisfying

(v) for all $g \in \mathbf{G}$, $z \in Z_{\mathbf{M}}$, $\phi(zg) = z^{\rho_P}\xi(z)\phi(g)$;
obviously we can suppose $\xi|_{Z_M(k)\cap Z_{\mathbf{M}}} = 1$, otherwise

$$A(U(\mathbb{A})M(k)\backslash\mathbf{G})_\xi = \{0\}.$$

If k is a function field, **G** acts by right translation on $A(U(\mathbb{A})M(k)\backslash\mathbf{G})$ and $A(U(\mathbb{A})M(k)\backslash\mathbf{G})_\xi$. If k is a number field, this is not the case, for condition (iii) is not stable by translation. But we can make the triple Lie $(\mathbf{G}_\infty) \times \mathbf{K}_\infty \times \mathbf{G}_f$ act on these spaces. This action replaces the missing action of **G**.

A certain number of classical theorems concerning automorphic forms on $M(k)\backslash\mathbf{M}$ can be immediately generalised to automorphic forms on $U(\mathbb{A})M(k)\backslash\mathbf{G}$. For example the finiteness theorem (see [HC] Theorem 1) and the following lemma:

Lemma *Let $\{P_i = M_iU_i; i = 1,\ldots,n\}$ be a finite set of standard parabolic subgroups of G and, for all i, V_i a finite dimensional subspace of $A(U_i(\mathbb{A})M_i(k)\backslash\mathbf{G})$. Then there exists a function $f : \mathbf{G} \to \mathbb{C}$, of compact support, smooth and right and left* **K***-finite, such that for all $i = 1,\ldots,n$, and all $\phi \in V_i$, we have the equality $\delta(f)\phi = \phi$.*

See [HC3] §8, Theorem 1. □

Suppose that k is a number field. The preceding lemma and Lemma

I.2.5 show that for all P and all $\phi \in A(U(\mathbb{A})M(k)\backslash \mathbf{G})$, ϕ and its derivatives have moderate growth.

Remark The proof of the finiteness theorem uses the theory of cuspidal components developed above. Thus we do not use this theorem until the end of Section I.3 in order to avoid any logical loop. The proof of the above lemma is entirely independent.

I.2.18. Cuspidal forms

Let $P = MU$ be a standard parabolic subgroup of G, ϕ an automorphic form on $U(\mathbb{A})M(k)\backslash \mathbf{G}$. We say that ϕ is cuspidal if for all parabolic subgroups P' such that $P_0 \subset P' \subsetneq P$, we have $\phi_{P'} = 0$. We denote by $A_0(U(\mathbb{A})M(k)\backslash \mathbf{G})$ the space of cuspidal automorphic forms on $U(\mathbb{A})M(k)\backslash \mathbf{G}$. Let ξ be a character of $Z_{\mathbf{M}}$. We write $A_0(U(\mathbb{A})M(k)\backslash \mathbf{G})_\xi$ for the subspace of elements of $A_0(U(\mathbb{A})M(k)\backslash \mathbf{G})$ which satisfy the condition (v) of I.2.17. It is a consequence of Corollary I.2.12, generalised to functions on $U(\mathbb{A})M(k)\backslash \mathbf{G}$, that if ϕ is cuspidal it is rapidly decreasing in a reasonable sense if k is a number field; its support is compact modulo $Z_{\mathbf{M}}$ if k is a function field.

Fix a character ξ of $Z_{\mathbf{M}}$. The representation of $\mathbf{G}$ in $A(U(\mathbb{A})M(k)\backslash \mathbf{G})$ – more precisely of $\mathrm{Lie}(\mathbf{G}_\infty) \times \mathbf{K}_\infty \times \mathbf{G}_f$ if k is a number field – preserves $A_0(U(\mathbb{A})M(k)\backslash \mathbf{G})$ and $A_0(U(\mathbb{A})M(k)\backslash \mathbf{G})_\xi$. In the case where $P = G$, the representation thus defined in $A_0(G(k)\backslash \mathbf{G})_\xi$ is semi-simple. For any character μ of $\mathfrak{z}$ (see I.2.17) and any irreducible representation σ of $\mathbf{K}$, denote by $A_0(\mu, \sigma)_\xi$ the eigensubspace of elements of $A_0(U(\mathbb{A})M(k)\backslash \mathbf{G})_\xi$ for $\mathfrak{z}$ with eigenvalue μ, on which $\mathbf{K}$ acts via σ (obviously, for this space to be non-trivial, μ and σ must satisfy a compatibility condition with ξ). Then $A_0(\mu, \sigma)_\xi$ is finite dimensional and $A_0(U(\mathbb{A})M(k)\backslash \mathbf{G})_\xi$ is the algebraic direct sum of these subspaces. These assertions can be deduced from the same assertions in the special case where $P = G$, for which we refer to [HC] Lemma 18 and Theorem 3. For $\phi \in A(U(\mathbb{A})M(k)\backslash \mathbf{G})_\xi$ and $\phi_0 \in A_0(U(\mathbb{A})M(k)\backslash \mathbf{G})_{\xi^*}$, where $\xi^* = \overline{\xi}^{-1}$, we can define

$$\langle \phi_0, \phi \rangle = \int_{Z_{\mathbf{M}} U(\mathbb{A})M(k)\backslash \mathbf{G}} \overline{\phi}_0(g)\phi(g)\, dg.$$

In particular, if ξ is unitary, $A_0(U(\mathbb{A})M(k)\backslash \mathbf{G})_\xi$ is a pre-Hilbertian space whose completion can be easily described (if $P = G$, it is $L_0^2(G(k)\backslash \mathbf{G})_\xi$). If $\phi \in A(U(\mathbb{A})M(k)\backslash \mathbf{G})_\xi$, there exists a unique $\phi^{\mathrm{cusp}} \in A_0(U(\mathbb{A})M(k)\backslash \mathbf{G})_\xi$ such that

$$\langle \phi_0, \phi^{\mathrm{cusp}} \rangle = \langle \phi_0, \phi \rangle \tag{1}$$

for all $\phi_0 \in A_0(U(\mathbb{A})M(k)\backslash \mathbf{G})_\xi$. Indeed for any character μ of $\mathfrak{z}$ and any irreducible representation σ of $\mathbf{K}$, we can define $\phi^{\text{cusp}}(\mu,\sigma) \in A_0(\mu,\sigma)_\xi$ such that (1) is true for $\phi_0 \in A_0(\mu^*,\sigma)_{\xi^*}$, where μ^* comes from μ via the automorphism coming from $g \longmapsto g^{-1}$ composed with complex conjugation. This is because the spaces $A_0(\mu,\sigma)_\xi$ and $A_0(\mu^*,\sigma)_{\xi^*}$ are finite dimensional and dual. The finiteness properties imposed on ϕ imply that $\phi^{\text{cusp}}(\mu,\sigma) = 0$ for almost all (μ,σ). Now set $\phi^{\text{cusp}} = \sum_{\mu,\sigma} \phi^{\text{cusp}}(\mu,\sigma)$.

I.3. Cuspidal components

I.3.1. $A_{\mathbf{M}}$-finite functions

Let M be a standard Levi subgroup of G.

Remark Let $\text{Hom}(A_{\mathbf{M}},\mathbb{C}^*)$ be the set of characters of $A_{\mathbf{M}}$, i.e. of continuous homomorphisms of $A_{\mathbf{M}}$ into $\mathbb{C}^*$. The restriction map of $\mathbf{M}$ to $A_{\mathbf{M}}$ defines a map

(1) $$X_M \to \text{Hom}(A_{\mathbf{M}},\mathbb{C}^*).$$

We can use this to prove surjectivity by giving a description of $\text{Hom}(A_{\mathbf{M}},\mathbb{C}^*)$ analogous to the description of X_M (see I.1.4). As $A_{\mathbf{M}}\mathbf{M}^1$ is of finite index in $\mathbf{M}$, the kernel of (1) is finite. It contains $X_M^{\mathbf{M}}$ since $A_{\mathbf{M}} \subset Z_{\mathbf{M}}$. We will call it $X_M(A_{\mathbf{M}})$. We obtain an isomorphism

$$\text{Hom}(A_{\mathbf{M}},\mathbb{C}^*) \simeq X_M/X_M(A_{\mathbf{M}}).$$

Suppose first that k is a number field. Let $\mathfrak{z}(A_{\mathbf{M}})$ be the enveloping algebra of the (complex) Lie algebra of the real group $A_{\mathbf{M}}$. We have

$$\mathfrak{z}(A_{\mathbf{M}}) \simeq \text{Sym}_{\mathbb{R}}(\mathfrak{a}_M) \bigotimes_{\mathbb{R}} \mathbb{C}.$$

Thus $\mathfrak{z}(A_{\mathbf{M}})$ is identified, by a map which we will denote by $z \longmapsto \hat{z}$, with the polynomial algebra over the complex space $\mathfrak{a}_M^*$, itself isomorphic to X_M and even to $X_M/X_M(A_{\mathbf{M}})$, since $X_M(A_{\mathbf{M}}) = \{0\}$. Suppose now that k is a function field. Let $\mathfrak{z}(A_{\mathbf{M}})$ be the convolution algebra of functions with compact support on $A_{\mathbf{M}}$. We associate with $z \in \mathfrak{z}(A_{\mathbf{M}})$ its Fourier-Mellin transform $\hat{z}$, which is the function on $X_M/X_M(A_{\mathbf{M}})$ defined by

$$\hat{z}(\lambda) = \int_{A_{\mathbf{M}}} z(a)a^{-\lambda}\, da.$$

Fix a basis $(a_i)_{i=1,\dots,n}$ of the $\mathbb{Z}$-module $A_{\mathbf{M}}$. Then $\mathfrak{z}(A_{\mathbf{M}})$ can be identified via $z \longmapsto \hat{z}$ with the space of polynomials in the variables $a_i^\lambda, a_i^{-\lambda}$ for $i = 1,\dots,n$.

Let us consider a finite set $\wedge \subset X_M/X_M(A_{\mathbf{M}})$ and a family $N = (N_\lambda)_{\lambda\in\wedge}$

of integers ≥ 0. Let $\mathfrak{z}(A_{\mathbf{M}}; \wedge, N)$ be the space of $z \in \mathfrak{z}(A_{\mathbf{M}})$ such that for all $\lambda \in \wedge$, $\hat{z}$ vanishes to the order at least N_λ in λ. This is an ideal of finite codimension of $\mathfrak{z}(A_{\mathbf{M}})$. The above description shows that conversely, if $I \subset \mathfrak{z}(A_{\mathbf{M}})$ is an ideal of finite codimension, there exists a pair $\wedge, N$ such that $I \supset \mathfrak{z}(A_{\mathbf{M}}; \wedge, N)$.

From the action of $A_{\mathbf{M}}$ on itself by translation, we deduce an action of the algebra $\mathfrak{z}(A_{\mathbf{M}})$ on the space of smooth functions on $A_{\mathbf{M}}$.

Fix once and for all a basis $\mathfrak{q}_M$ of the space $\mathbb{C}[\mathrm{Re}\, \mathfrak{a}_M]$ of polynomials over $\mathrm{Re}\, \mathfrak{a}_M$, consisting of homogeneous elements. We check that for $\wedge, N$ as above, the set of functions

$$(2) \qquad a \longmapsto a^\lambda Q(\log_M a),$$

for $\lambda \in \wedge$, $Q \in \mathfrak{q}_M$, $\deg(Q) < N_\lambda$, forms a basis of the space of smooth functions over $A_{\mathbf{M}}$ killed by $\mathfrak{z}(A_{\mathbf{M}}; \wedge, N)$. It is a consequence of this that every smooth and $\mathfrak{z}(A_{\mathbf{M}})$-finite function on $A_{\mathbf{M}}$ is a linear combination of functions of the form (2). This property will be useful to us later.

I.3.2. Decomposition of an automorphic form

Let $P = MU$ be a standard parabolic subgroup of G. Let $\mathrm{Hom}(Z_{\mathbf{M}}, \mathbb{C}^*)$ be the set of characters of $Z_{\mathbf{M}}$ (i.e. of continuous homomorphisms of $Z_{\mathbf{M}}$ into $\mathbb{C}^*$), and

$$A(U(\mathbb{A})M(k)\backslash\mathbf{G})_Z = \sum_{\xi \in \mathrm{Hom}(Z_{\mathbf{M}}, \mathbb{C}^*)} A(U(\mathbb{A})M(k)\backslash\mathbf{G})_\xi.$$

Let $Q \in \mathbb{C}[\mathrm{Re}\, \mathfrak{a}_M]$ and $\psi \in A(U(\mathbb{A})M(k)\backslash\mathbf{G})_Z$. The function

$$g \longmapsto Q(\log_M m_P(g))\psi(g)$$

on $\mathbf{G}$ is left invariant under $U(\mathbb{A})M(k)$. We check that it is an automorphic form on $U(\mathbb{A})M(k)\backslash\mathbf{G}$. By linearly continuing this construction, we obtain a map

$$(1) \qquad \mathbb{C}[\mathrm{Re}\, \mathfrak{a}_M] \bigotimes A(U(\mathbb{A})M(k)\backslash\mathbf{G})_Z \longmapsto A(U(\mathbb{A})M(k)\backslash\mathbf{G}).$$

Lemma *The above map is an isomorphism.*

Proof In I.2.17, we introduced the algebras $\mathfrak{z}$ and $\mathfrak{z}^{\mathbf{M}}$ and showed that every element of $A(U(\mathbb{A})M(k)\backslash\mathbf{G})$ is $\mathfrak{z}^{\mathbf{M}}$-finite, $\mathfrak{z}^{\mathbf{M}}$ acting by left translations (the translation by ρ_P introduced in I.2.17 is inessential). An analogous proof (this time using the **K**-finiteness of automorphic forms) shows that every element of $A(U(\mathbb{A})M(k)\backslash\mathbf{G})$ is $Z_{\mathbf{M}}$-finite, $Z_{\mathbf{M}}$ also acting by left translations. Let $\phi \in A(U(\mathbb{A})M(k)\backslash\mathbf{G})$. As the algebra $\mathfrak{z}(A_{\mathbf{M}})$ can be identified with a subalgebra of $\mathfrak{z}^{\mathbf{M}}$, ϕ is $\mathfrak{z}(A_{\mathbf{M}})$-finite. Thus there exists a finite set $\wedge \subset X_M/X_M(A_{\mathbf{M}})$ (see I.3.1) and a family $N = (N_\lambda)_{\lambda \in \wedge}$ of integers ≥ 0 such that ϕ is killed by $\mathfrak{z}(A_{\mathbf{M}};\wedge, N)$. Choose a set which we continue to call $\wedge$, contained in X_M, which is in bijection with the preceding $\wedge$. Recall that $\mathfrak{q}_M$ is a fixed basis of $\mathbb{C}[\mathrm{Re}\,\mathfrak{a}_M]$ (see I.3.1). Let $q \in \mathbf{G}$. The functions

$$a \longmapsto f_{\lambda,Q,g}(a) = (am_P(g))^\lambda Q(\log_M(am_P(g)),$$

on $A_{\mathbf{M}}$ for $\lambda \in \wedge$, $Q \in q_M$, $\deg(Q) < N_\lambda$, still form a basis of the space of smooth functions on $A_{\mathbf{M}}$ killed by $\mathfrak{z}(A_{\mathbf{M}};\wedge, N)$. We deduce from this the existence and uniqueness of elements $\psi_{\lambda,Q}(g) \in \mathbb{C}$ such that for all $a \in A_{\mathbf{M}}$, we have

$$\phi(ag) = \sum_{\lambda\in\wedge} \sum_{Q\in\mathfrak{q}_M;\deg(Q)<N_\lambda} f_{\lambda,Q,g}(a)\psi_{\lambda,Q}(g). \tag{2}$$

This defines functions $\psi_{\lambda,Q} : \mathbf{G} \to \mathbb{C}$. The uniqueness of the above expression allows us to show that $\psi_{\lambda,Q}$ is left invariant under $A_{\mathbf{M}}U(\mathbb{A})M(k)$ and is **K**-finite. Let us show that $\psi_{\lambda,Q}$ is automorphic. Fix λ and Q. Suppose first that $\deg(Q) = N_\lambda - 1$. Thanks to the description of $\mathfrak{z}(A_{\mathbf{M}})$ given in I.3.1, we show that there exists $z \in \mathfrak{z}(A_{\mathbf{M}})$ such that for all $a \in A_{\mathbf{M}}$, $g \in \mathbf{G}$, we have

$$(z.f_{\lambda,Q,g})(a) = (am_P(g))^\lambda,$$
$$(z.f_{\lambda',Q',g})(a) = 0$$

for all $\lambda' \in \wedge$, $Q' \in \mathfrak{q}_M$, $\deg(Q') < N_{\lambda'}$ and $(\lambda', Q') \neq (\lambda, Q)$. Fix $g \in \mathbf{G}$ and denote by $\phi_g : A_{\mathbf{M}} \to \mathbb{C}$ the function such that $\phi_g(a) = \phi(ag)$. From (2) we see that

$$(z \cdot \phi_g)(a) = (am_P(g))^\lambda \psi_{\lambda,Q}(g).$$

Clearly $(z.\phi_g)(a) = (z\phi)(ag)$, where z acts on ϕ via left translation. We deduce from this that

$$\psi_{\lambda,Q}(g) = m_P(g)^{-\lambda}(z\phi)(g).$$

It follows from I.2.17 that $z\phi$ is automorphic (for all $k \in \mathbf{K}$, $(z\phi)_k = z\phi_k$ is automorphic). Thus $\psi_{\lambda,Q}$ is also automorphic. Replacing ϕ by

$$\phi - \sum_{\lambda\in\wedge} \sum_{Q\in\mathfrak{q}_M;\deg(Q)=N_\lambda-1} m_P(.)^\lambda Q(\log_M m_P(.))\psi_{\lambda,Q},$$

we show that $\psi_{\lambda,Q}$ is automorphic if $\deg(Q) = N_\lambda - 2$, etc. ... Thus each function $\psi_{\lambda,Q}$ is automorphic, and thus (left) $Z_{\mathbf{M}}$-finite. Now, each one is invariant under $A_{\mathbf{M}}$ and $Z_{\mathbf{M}}/A_{\mathbf{M}}$ is compact. By 'Fourier inversion' on the compact abelian group $Z_{\mathbf{M}}/A_{\mathbf{M}}$, it can thus be uniquely decomposed into a finite sum

$$\psi_{\lambda,Q} = \sum_{\xi} \psi_{\lambda,Q,\xi} \tag{3}$$

where ξ runs through a subfinite set of $\mathrm{Hom}(Z_{\mathbf{M}}, \mathbb{C}^*)$ and $\psi_{\lambda,Q,\xi}$ satisfies

$$\psi_{\lambda,Q,\xi}(zg) = \xi(z)\psi_{\lambda,Q,\xi}(g)$$

for all $z \in Z_{\mathbf{M}}$, $g \in \mathbf{G}$. Now plug the equality (3) into (2) and set $a = 1$. We obtain

$$\phi(g) = \sum_{\lambda,Q,\xi} m_P(g)^\lambda Q(\log_M m_P(g))\psi_{\lambda,Q,\xi}(g)$$

where (λ, Q, ξ) runs through a certain finite set. For any such triple (λ, Q, ξ), define ξ^λ by $\xi^\lambda(z) = z^{\lambda-\rho_P}\xi(z)$ and $\psi'_{\lambda,Q,\xi}$ by

$$\psi'_{\lambda,Q,\xi}(g) = m_P(g)^\lambda \psi_{\lambda,Q,\xi}(g).$$

Then $\psi'_{\lambda,Q,\xi} \in A(U(\mathbb{A})M(k)\backslash\mathbf{G})_{\xi^\lambda}$ and we have

$$\phi(g) = \sum_{\lambda,Q,\xi} Q(\log_M m_P(g))\psi'_{\lambda,Q,\xi}(g).$$

Thus ϕ is in the image of the map (1), which proves the surjectivity of this map. Its injectivity can easily be deduced from the uniqueness of the decompositions (2) and (3). □

Remark Fix a character ξ of $Z_{\mathbf{G}}$ and denote by $A(U(\mathbb{A})M(k)\backslash\mathbf{G})_\xi$ the subspace of $\phi \in A(U(\mathbb{A})M(k)\backslash\mathbf{G})$ such that $\phi(zg) = \xi(z)\phi(g)$ for all $z \in Z_{\mathbf{G}}$, $g \in \mathbf{G}$ (there is no conflict of notation: we already defined a space denoted in the same way for a character ξ of $Z_{\mathbf{M}}$; if $M = G$, then this is the same space). Set

$$A(U(\mathbb{A})M(k)\backslash\mathbf{G})_{Z,\xi} = \bigoplus_\eta A(U(\mathbb{A})M(k)\backslash\mathbf{G})_\eta$$

where the sum is over the $\eta \in \mathrm{Hom}(Z_{\mathbf{M}}, \mathbb{C}^*)$ such that $\eta|_{Z_{\mathbf{G}}} = \xi$. The above constructions and lemma can be reformulated to define a map

$$\mathbb{C}[\mathrm{Re}\,\mathfrak{a}_M^G] \otimes A(U(\mathbb{A})M(k)\backslash\mathbf{G})_{Z,\xi} \longrightarrow A(U(\mathbb{A})M(k)\backslash\mathbf{G})_\xi;$$

and this map is an isomorphism.

I.3.3. Exponents of cuspidal automorphic forms

The constructions of I.3.2 possess obvious analogues for cuspidal automorphic forms. Defining the space $A_0(U(\mathbb{A})M(k)\backslash\mathbf{G})_Z$ in the obvious way, we have an isomorphism

(1) $$\mathbb{C}[\mathrm{Re}\,\mathfrak{a}_M] \otimes A_0(U(\mathbb{A})M(k)\backslash\mathbf{G})_Z \longrightarrow A_0(U(\mathbb{A})M(k)\backslash\mathbf{G}).$$

We can actually go further here. For a character ξ of $Z_{\mathbf{M}}$, let $\Pi_0(\mathbf{M})_\xi$ be the set of isomorphism classes of irreducible representations of $\mathbf{M}$ occurring as submodules of $A_0(M(k)\backslash\mathbf{M})_\xi$. For $\pi \in \Pi_0(\mathbf{M})_\xi$, let $A_0(M(k)\backslash\mathbf{M})_\pi$ be the isotypic component of type π of $A_0(M(k)\backslash\mathbf{M})_\xi$. Finally, let $A_0(U(\mathbb{A})M(k)\backslash\mathbf{G})_\pi$ be the subspace of

$$\phi \in A_0(U(\mathbb{A})M(k)\backslash\mathbf{G})_\xi$$

such that $\phi_k \in A_0(M(k)\backslash\mathbf{M})_\pi$ for all $k \in \mathbf{K}$. From the decomposition

$$A_0(M(k)\backslash\mathbf{M})_\xi = \bigoplus_{\pi\in\Pi_0(\mathbf{M})_\xi} A_0(U(\mathbb{A})M(k)\backslash\mathbf{M})_\pi,$$

(see I.2.18), we deduce the decomposition

$$A_0(U(\mathbb{A})M(k)\backslash\mathbf{G})_\xi = \bigoplus_{\pi\in\Pi_0(\mathbf{M})_\xi} A_0(U(\mathbb{A})M(k)\backslash\mathbf{G})_\pi$$

If we set

$$\Pi_0(\mathbf{M}) = \bigcup_{\xi\in\mathrm{Hom}(Z_\mathbf{M},\mathbb{C}^\bullet)} \Pi_0(\mathbf{M})_\xi,$$

we obtain

$$A_0(U(\mathbb{A})M(k)\backslash\mathbf{G})_Z = \bigoplus_{\pi\in\Pi_0(\mathbf{M})} A_0(U(\mathbb{A})M(k)\backslash\mathbf{G})_\pi.$$

The surjectivity of the map (1) signifies that for all

$$\phi \in A_0(U(\mathbb{A})M(k)\backslash\mathbf{G}),$$

there exists a finite set

$$D(M,\phi) \subset \mathbb{C}[\mathrm{Re}\,\mathfrak{a}_M] \times \Pi_0(\mathbf{M}) \times A_0(U(\mathbb{A})M(k)\backslash\mathbf{G})$$

such that

(i) if $(Q,\pi,\psi) \in D(M,\phi)$, then $\psi \in A_0(U(\mathbb{A})M(k)\backslash\mathbf{G})_\pi$;
(ii) for all $g \in \mathbf{G}$, we have the equality

$$\phi(g) = \sum_{(Q,\pi,\psi)\in D(M,\phi)} Q(\log_M m_P(g))\psi(g).$$

We call $D(M,\phi)$ a set of cuspidal data for ϕ. If we suppose moreover that for $(Q,\pi,\psi) \in D(M,\phi)$, we have $\phi \in \mathfrak{q}_M$ (see I.3.1) and $\psi \neq 0$, then $D(M,\phi)$ is unique. We then denote by $\Pi_0(\mathbf{M},\phi)$ the projection of $D(M,\phi)$ over $\Pi_0(\mathbf{M})$, i.e. the set of π such that there exist Q,ψ satisfying $(Q,\pi,\psi) \in D(M,\phi)$. We call $\Pi_0(\mathbf{M},\phi)$ the cuspidal support of ϕ.

Remark We can define $\Pi_0(\mathbf{M},\phi)$ without having recourse to a basis $\mathfrak{q}_M$: it is the set of $\pi \in \Pi_0(\mathbf{M})$ such that $\sum Q\otimes\psi \neq 0$ in

$$\mathbb{C}[\mathrm{Re}\,\mathfrak{a}_M] \bigotimes_{\mathbb{C}} A_0(U(\mathbb{A})M(k)\backslash\mathbf{G}),$$

where the sum is over the (Q,ψ) such that $(Q,\pi,\psi) \in D(M,\phi)$.

Let us introduce the following notation concerning the representations of $\mathbf{M}$. Let π be an irreducible representation of $\mathbf{M}$. We denote by χ_π the central character of π. It is a character of $Z_\mathbf{M}$. Suppose χ_π is trivial on $Z_M(k)$ (this is the case if π is automorphic, *a fortiori* if π is cuspidal). Recall that with such a character χ we can associate an element denoted by $\operatorname{Re}\chi$ of $\operatorname{Re}X_M$: it suffices to identify the character $|\chi|$ with an element of $\operatorname{Re}X_M$, see I.1.4. Simply set $\operatorname{Re}\pi = \operatorname{Re}\chi_\pi$. We define

$$\operatorname{Im}\pi = \pi \otimes (-\operatorname{Re}\pi).$$

If π is cuspidal, the representation $\operatorname{Im}\pi$ is unitary. Finally, we let $-\pi$ be the contragredient representation of π and $-\overline{\pi}$ the conjugate of the contragredient.

For ϕ as above, the set $\{\chi_\pi ; \pi \in \Pi_0(\mathbf{M}, \phi)\}$ is called the set of cuspidal exponents of ϕ.

Remark As in I.3.2, the above definitions and results have obvious analogues when we fix a character ξ of $Z_{\mathbf{G}}$ and consider the subspace $A_0(U(\mathbb{A})M(k)\backslash\mathbf{G})_\xi$ of elements $\phi \in A_0(U(\mathbb{A})M(k)\backslash\mathbf{G})$ such that $\phi(zg) = \xi(z)\phi(g)$ for all $z \in Z_{\mathbf{G}}$, $g \in \mathbf{G}$.

I.3.4. Cuspidal components

Let $P = MU$ be a standard parabolic subgroup of G. Let $C_0(U(\mathbb{A})M(k)\backslash\mathbf{G})$ be the space of functions on $U(\mathbb{A})M(k)\backslash\mathbf{G}$ which are linear combinations of functions of the form

$$g \longmapsto b(m_P(g))\phi(g),$$

where $\phi \in A_0(U(\mathbb{A})M(k)\backslash\mathbf{G})$ and b is a smooth function with compact support on $\mathbf{M}^1\backslash\mathbf{M}$. Now let ϕ be a measurable function on $G(k)\backslash\mathbf{G}$ having moderate growth. We define the cuspidal component of ϕ along P: it is the linear form on $C_0(U(\mathbb{A})M(k)\backslash\mathbf{G})$ given by:

$$\varphi \longmapsto \int_{U(\mathbb{A})M(k)\backslash\mathbf{G}} \overline{\phi}_P(g)\varphi(g)\,dg.$$

Proposition *Let $\phi : G(k)\backslash\mathbf{G} \to \mathbb{C}$ be a measurable function of moderate growth. Suppose that for every standard parabolic subgroup P of G the cuspidal component of ϕ along P is zero. Then $\phi = 0$ almost everywhere.*

Proof Reasoning by induction on the semi-simple rank of G, we can suppose that the lemma holds for all proper Levi subgroups of G. If $P = MU$ is a proper standard parabolic subgroup of G, we then deduce an analogous statement for the functions on $U(A)M(k)\backslash\mathbf{G}$. Thus let ϕ be as in the statement and P a proper standard parabolic subgroup of G. The function ϕ_P satisfies the same hypotheses as ϕ, for if P' is such that $P_0 \subset P' \subset P$, the cuspidal components of ϕ and of ϕ_P along P' coincide. By the induction hypothesis, $\phi_P = 0$. Let h be a smooth $\mathbf{K}$-finite function on $\mathbf{G}$ with compact support. Set $\phi' = \delta(h)\phi$. For P such that $P_0 \subset P \subsetneqq G$, we have $\phi'_P = \delta(h)\phi_P = 0$. The function ϕ' satisfies the hypotheses of Corollary I.2.12 and we have the equality $s\phi' = \phi'$ in a Siegel domain S. Corollary I.2.12 thus implies that ϕ' is rapidly decreasing if k is a number field, and its support has compact image in $Z_{\mathbf{G}}G(k)\backslash\mathbf{G}$ if k is a function field. Suppose $\phi' \neq 0$. As ϕ' is smooth, we can find a function b on $\mathbf{G}^1\backslash\mathbf{G}$, smooth and of compact support, such that the function ϕ'' defined by

$$\phi''(g) = \int_{A_G} \overline{b}(am_G(g))\phi'(ag)\,da$$

is non-zero. Now, ϕ'' is a function on $A_G G(k)\backslash\mathbf{G}$ satisfying $\phi''_P = 0$ for $P_0 \subset P \subsetneqq G$ and the decreasing properties of ϕ' imply that $\phi'' \in L^2_0(A_G G(k)\backslash\mathbf{G})$. Let $A_0(A_G G(k)\backslash\mathbf{G})$ denote the subspace of elements of $A_0(G(k)\backslash\mathbf{G})$ which are left invariant under A_G. The space $A_0(A_G(k)\backslash\mathbf{G})$ is dense in $L^2_0(A_G G(k)\backslash\mathbf{G})$ (this is an immediate generalisation of the situation of I.2.18). In particular there exists

$$\varphi \in A_0(A_G G(k)\backslash\mathbf{G})$$

such that

$$\int_{A_G G(k)\backslash\mathbf{G}} \overline{\phi''}(g)\varphi(g)\,dg \neq 0.$$

Set $\varphi_1(g) = b(m_G(g))\varphi(g)$. Then $\varphi_1 \in C_0(G(k)\backslash\mathbf{G})$ and we have

$$\int_{G(k)\backslash\mathbf{G}} \overline{\phi'}(g)\varphi_1(g)\,dg \neq 0.$$

Set $\varphi_2 = \delta(h^*)\varphi_1$, where $h^*(g) = \overline{h}(g^{-1})$. We can show that $\varphi_2 \in C_0(G(k)\backslash\mathbf{G})$. The preceding relation can be written

$$\int_{G(k)\backslash\mathbf{G}} \overline{\phi}(g)\varphi_2(g)\,dg \neq 0.$$

This contradicts the hypothesis that the cuspidal component of ϕ along G is zero. The supposition that $\phi' \neq 0$ leads to a contradiction, thus $\phi' = 0$, i.e. $\delta(h)\phi = 0$ for all h. But then $\phi = 0$ almost everywhere. □

Remarks (1) Here again, if we fix a character ξ of Z_G, the proposition has an analogue applicable to functions ϕ on $G(k)\backslash\mathbf{G}$ such that $\phi(zg) = \xi(z)\phi(g)$ for all $z \in Z_G$, $g \in \mathbf{G}$; for this one needs to use the spaces $C_0(U(\mathbb{A})M(k)\backslash\mathbf{G})_{\xi}$, generated by functions of the form

$$g \longmapsto b(m_P(g))\phi(g),$$

where $\phi \in A_0(U(\mathbb{A})M(k)\backslash\mathbf{G})_{\xi}$, (see I.3.3) and b is a smooth function of compact support on $Z_{\mathbf{G}}\mathbf{M}^1\backslash\mathbf{M}$.

(2) Because of Remark I.2.5 and the remark above, a reformulation of the lemma is also valid for the elements of $L^2(G(k)\backslash\mathbf{G})_\xi$. More generally, a variation of the lemma is valid for functions of the form $\phi_1+\phi_2$ where $\phi_1 \in L^2(G(k)\backslash\mathbf{G})_\xi$ and ϕ_2 has moderate growth, such that $\phi_2(zg)=\xi(z)\phi_2(g)$ for all $z \in Z_{\mathbf{G}}$, $g \in \mathbf{G}$.

I.3.5. Cuspidal components and automorphic functions

Let ϕ be a function over $G(k)\backslash\mathbf{G}$, measurable and of moderate growth, and $P = MU$ a standard parabolic subgroup of G. We have defined the cuspidal component of ϕ along P. We say that this cuspidal component is automorphic if there exists an automorphic form ϕ'_P on $U(\mathbb{A})M(k)\backslash\mathbf{G}$ such that the cuspidal component of ϕ along P is equal to the linear form (on $C_0(U(\mathbb{A})M(k)\backslash\mathbf{G})$):

$$\varphi \longmapsto \int_{U(\mathbb{A})M(k)\backslash\mathbf{G}} \overline{\phi'}_P(g)\varphi(g)\, dg.$$

If there exists such an automorphic form ϕ'_P, we can choose it to be cuspidal: if we write

$$\phi'_P(g) = \sum_{Q,\xi} Q(\log_M m_P(g))\psi_{Q,\xi}(g),$$

where (Q,ξ) runs through a certain subfinite set of

$$\mathbb{C}[\operatorname{Re}\mathfrak{a}_M] \times \operatorname{Hom}(Z_{\mathbf{M}}, \mathbb{C}^*),$$

and $\psi_{Q,\xi} \in A(U(\mathbb{A})M(k)\backslash\mathbf{G})_\xi$, see I.3.2, it suffices to replace ϕ'_P by the function

$$g \longmapsto \sum_{Q,\xi} Q(\log_M m_P(g))\psi_{Q,\xi}^{\mathrm{cusp}}(g),$$

see I.2.18. Such a cuspidal form ϕ'_P is unique; we will call it ϕ_P^{cusp}.

Remark

We hope that this notation will not create confusion; ϕ^{cusp} is defined and therefore so is $(\phi^{\mathrm{cusp}})_P$, but this term is not related to the term which we just defined ($(\phi^{\mathrm{cusp}})_P = 0$ if $P \neq G$).

Lemma *Let $\phi : G(k)\backslash\mathbf{G} \to \mathbb{C}$ be a measurable function of moderate growth. Then ϕ is automorphic (more precisely there exists an automorphic ϕ' such that $\phi' = \phi$ almost everywhere) if and only if for every standard parabolic subgroup P of G, the cuspidal component of ϕ along P is automorphic.*

Proof If ϕ is automorphic, ϕ_P is as well and the result is clear. Conversely, suppose that for all P the cuspidal component of ϕ along P is automorphic. Let us introduce the function ϕ_P^{cusp}. Choose a smooth and left-$\mathbf{K}$-finite function f on $\mathbf{G}$ with compact support, such that $\delta(f)\phi_P^{\text{cusp}} = \phi_P^{\text{cusp}}$ for all P (see Lemma I.2.17). We have $\delta(f)\phi_P^{\text{cusp}} = (\delta(f)\phi)_P^{\text{cusp}}$. Thus the functions ϕ and $\delta(f)\phi$ have the same cuspidal components along P for all P, thus $\phi = \delta(f)\phi$ almost everywhere. Replacing ϕ by $\delta(f)\phi$, we now have a smooth and $\mathbf{K}$-finite function with moderate growth. There exists an ideal I of $\mathfrak{z}$, of finite codimension, such that for all $X \in I$ and all P, $\delta(X)\phi_P^{\text{cusp}} = 0$. We have $\delta(X)\phi_P^{\text{cusp}} = (\delta(X)\phi)_P^{\text{cusp}}$. We deduce from this as above that $\delta(X)\phi = 0$. This gives the $\mathfrak{z}$-finiteness of ϕ, which is then automorphic. □

If ϕ is an automorphic form on $G(k)\backslash\mathbf{G}$, we simplify the notation by writing $D(M,\phi) = D(M,\phi_P^{\text{cusp}})$, $\Pi_0(\mathbf{M},\phi) = \Pi_0(\mathbf{M},\phi_P^{\text{cusp}})$ (see I.3.3). We call $\Pi_0(\mathbf{M},\phi)$ the cuspidal support of ϕ along P, and the union

$$\bigcup_M \Pi_0(\mathbf{M},\phi),$$

taken over the standard Levis M, the cuspidal support of ϕ.

I.3.6. The function field case: independence of the choice of v_0

Suppose k is a function field. We will briefly indicate that the notion of automorphic form is independent of the choice of the place v_0. Let $P = MU$ be a standard parabolic subgroup of G, ξ a character of $Z_{\mathbf{M}}$, $\mathbf{K}'$ an open subgroup of $\mathbf{K}$. Let $L_0(U(\mathbb{A})M(k)\backslash\mathbf{G})_\xi^{\mathbf{K}'}$ be the space of functions $\phi : U(\mathbb{A})M(k)\backslash\mathbf{G} \to \mathbb{C}$ of central character ξ, invariant under $\mathbf{K}'$, such that $\phi_{P'} = 0$ for all P' such that $P_0 \subset P' \subsetneq P$. Corollary I.2.8 shows that there exists a subset $\Gamma \subset U(\mathbb{A})M(k)\backslash\mathbf{G}$, of compact image in $Z_{\mathbf{M}}U(\mathbb{A})M(k)\backslash\mathbf{G}$, such that if $\phi \in L_0(U(\mathbb{A})M(k)\backslash\mathbf{G})_\xi^{\mathbf{K}'}$, then the support of ϕ is in Γ. Thus $L_0(U(\mathbb{A})M(k)\backslash\mathbf{G})_\xi^{\mathbf{K}'}$ is finite-dimensional. As this space is stable under $\mathfrak{z}$, every element of $L_0(U(\mathbb{A})M(k)\backslash\mathbf{G})_\xi^{\mathbf{K}'}$ is $\mathfrak{z}$-finite. Thus

$$L_0(U(\mathbb{A})M(k)\backslash\mathbf{G})_\xi^{\mathbf{K}'} \subset A_0(U(\mathbb{A})M(k)\backslash\mathbf{G})_\xi^{\mathbf{K}'},$$

where the exponent $\mathbf{K}'$ on the right signifies that we have taken the elements fixed by $\mathbf{K}'$. As the opposite inclusion is obvious, we see that these spaces are equal. Thus $A_0(U(\mathbb{A})M(k)\backslash\mathbf{G})_\xi^{\mathbf{K}'}$ is independent of v_0. As

$$A_0(U(\mathbb{A})M(k)\backslash\mathbf{G})_\xi$$

is the union of these subspaces as $\mathbf{K}'$ varies, it is itself independent of v_0. The description I.3.3 (1) shows that $A_0(U(\mathbb{A})M(k)\backslash\mathbf{G})$ also is. The space $C_0(U(\mathbb{A})M(k)\backslash\mathbf{G})$ used in I.3.5 is as well. But then Lemma I.3.5 shows that $A(G(k)\backslash\mathbf{G})$ is independent of v_0. This result can be generalised to the spaces $A(U(\mathbb{A})M(k)\backslash\mathbf{G})$.

I.4. Upper bounds as functions of the constant term

I.4.1. Lemma
Let ϕ be an automorphic form on $G(k)\backslash\mathbf{G}$. For every standard parabolic subgroup $P = MU$ of G, let us take a set

$$D(M,\phi) \subset \mathbb{C}[\mathrm{Re}\,\mathfrak{a}_M] \times \Pi_0(\mathbf{M}) \times A_0(M(k)U(\mathbb{A})\backslash\mathbf{G})$$

of cuspidal data for ϕ_P^{cusp} (see I.3.3). Then there exists $c > 0$ such that for all $g \in S$, we have the upper bound

$$|\phi(g)| \leq c \sum_P \sum_{(Q,\pi,\psi)\in D(M,\phi)} m_{P_0}(g)^{\mathrm{Re}\,\pi+\rho_P}(1 + \|\log_M m_P(g)\|)^{\deg(Q)};$$

see I.3.3 for the definition of $\mathrm{Re}\,\pi$; $\deg(Q)$ *is the total degree of Q.*

More generally for all $\mu \in \mathrm{Re}\,X_{M_0}$, there exists $c_\mu > 0$ such that for all $g \in S$, we have the upper bound

$$|\phi(g)| \leq c_\mu \sum_P \sum_{(Q,\pi,\psi)\in D(M,\phi)} m_{P_0}(g)^{\mathrm{Re}\,\pi+\mu^M+\rho_P}(1 + \|\log_M m_P(g)\|)^{\deg(Q)},$$

where μ^M is the projection of $\mu \in \mathrm{Re}\,X_{M_0}$ onto $\mathrm{Re}\,X_{M_0}^M$ (see I.1.6 (9)).

Proof (a) We proceed by induction on the semi-simple rank of G. Suppose the lemma is proved for every proper standard Levi subgroup M of G. We immediately deduce a similar lemma concerning automorphic forms on $M(k)U(\mathbb{A})\backslash\mathbf{G}$ for every proper standard parabolic $P = MU$ of G. Note that if $P' \subset P$ are two such subgroups, we have the equality of cuspidal components

$$(\phi_P)_{P'}^{\mathrm{cusp}} = \phi_{P'}^{\mathrm{cusp}}.$$

We deduce from this that for all $P \subsetneq G$ and all $G \in S^P$, we have an upper bound

$$|\phi_P(g)| \leq c \sum_{P'\subset P} \sum_{(Q,\pi,\psi)\in D(M',\phi)} m_{P_0}(g)^{\mathrm{Re}\,\pi+\mu^{M'}+\rho_{P'}}(1 + \|\log_{M'} m_{P'}(g)\|)^{\deg(Q)}.$$

A fortiori, we can replace the sum over P' by the sum over all $P' \subset G$ and restrict ourselves to $g \in S$.

(b) As in I.3.1, let us introduce a basis $\mathfrak{q}_G$ of $\mathbb{C}[\mathrm{Re}\,\mathfrak{a}_G]$ consisting of homogeneous polynomials. Choose a set $D \subset \mathfrak{q}_G \times \mathrm{Hom}(Z_G, \mathbb{C}^*)$ and for all $(Q,\xi) \in D$ an element $\psi_{Q,\xi} \in A(G(k)\backslash\mathbf{G})_\xi$ such that we have the equality

$$\phi(g) = \sum_{(Q,\xi)\in D} Q(\log_G g)\psi_{Q,\xi}(g)$$

for all $g \in \mathbf{G}$. We suppose moreover that $\psi_{Q,\xi} \neq 0$ for all $(Q, \xi) \in D$. Then the above data are unique (see I.3.2). Let us show that

(1) for all $(Q, \xi) \in D$, there exists a standard Levi subgroup M and $(Q', \pi', \psi') \in D(M, \phi)$ such that $\chi_{\pi'}|_{Z_G} = \xi$, $\deg(Q') \geq \deg(Q)$.

Indeed let $(Q, \xi) \in D$. Choose $P = MU$ such that $\psi^{\text{cusp}}_{Q,\xi;P} \neq 0$. We obviously have

$$\phi_P^{\text{cusp}}(g) = \sum_{(Q'',\xi'')\in D} Q''(\log_G g)\psi^{\text{cusp}}_{Q'',\xi'';P}(g).$$

Let $Z_{\mathbf{G}}$ act by translation. The partial sum

$$\sum_{(Q'',\xi)\in D} Q''(\log_G g)\psi^{\text{cusp}}_{Q'',\xi;P}(g) \tag{2}$$

is the component of ϕ_P^{cusp} over the generalised eigenspace (for the action of $Z_{\mathbf{G}}$) of eigenvalue ξ. Using the equality of the statement, we obtain another expression for this component, namely

(3) $\sum Q'(\log_M m_P(g))\psi'(g)$, sum over the $(Q', \pi', \psi') \in D(M, \phi)$ such that $\chi_{\pi'}|Z_{\mathbf{G}} = \xi$.

Thus we obtain the equality of (2) and (3). Fix g_0 such that $\psi^{\text{cusp}}_{Q,\xi;P}(g_0) \neq 0$, replace g by zg_0 with $z \in Z_{\mathbf{G}}$ in formulae (2) and (3) and multiply them by $\xi(z)^{-1}$. The expression (2) is then a polynomial (in $\log_G z$) of degree $\geq \deg(Q)$. The expression (3) is a polynomial of degree $\leq \sup\{\deg(Q')\colon (Q', \pi', \psi') \in D(M, \phi),\ \chi_{\pi'}|_{Z_G} = \xi\}$. Thus $\deg(Q)$ is less than or equal to this term. But that is precisely the assertion of (1).

(c) Let us introduce the function on S

$$s\phi = \sum_{P=MU} (-1)^{\dim T_M - \dim T_G}\phi_P. \tag{4}$$

Clearly,

$$s\phi(g) = \sum_{(Q,\xi)\in D} Q(\log_G g)s\psi_{Q,\xi}(g).$$

Fix $(Q, \xi) \in D$. The function $s\psi_{Q,\xi}$ is an eigenfunction for the action of $Z_{\mathbf{G}}$, rapidly decreasing if k is a number field, of support having compact image in $Z_{\mathbf{G}}G(k)\backslash\mathbf{G}$ if k is a function field (see I.2.12). We deduce that for all $\nu \in \operatorname{Re} X_{M_0}$, there exists $c > 0$ such that for all $g \in S$, we have the upper bound

$$|s\psi_{Q,\xi}(g)| \leq c m_{P_0}(g)^{\operatorname{Re}\xi + \nu^G},$$

where ν^G is the projection of $\nu \in \operatorname{Re} X_{M_0}$ onto $\operatorname{Re} X^{\mathbf{G}}_{M_0}$, see I.1.6 (9). In particular, fix $P = MU$ and (Q', π', ψ') satisfying (1). Choose $\nu = \nu^G = (\operatorname{Re}\pi')^G + \mu^M + \rho_P$. We have $\operatorname{Re}\xi + \nu^G = \operatorname{Re}\pi' + \mu^M + \rho_P$, which shows that

$$|s\psi_{Q,\xi}(g)| \leq c m_{P_0}(g)^{\operatorname{Re}\pi' + \mu^M + \rho_P}.$$

On the other hand, by (1), we have an upper bound (with another c):

$$|Q(\log_G g)| \leq c(1 + \| \log_G g \|)^{\deg(Q')} \leq c(1 + \| \log_M m_P(g) \|)^{\deg(Q')}$$

which gives

$$|Q(\log_G g) s\psi_{Q,\xi}(g)| \leq c m_{P_0}(g)^{\mathrm{Re}\,\pi' + \mu^M + \rho_P}(1 + \| \log_M m_P(g) \|)^{\deg(Q')}.$$

We can replace the right-hand side by the sum over $P = MU$ and $(Q', \pi', \psi') \in D(M, \phi)$ of the same expression, then sum the left-hand terms over $(Q, \xi) \in D$. We obtain

$$|s\phi(g)| \leq c \sum_P \sum_{(Q',\pi',\psi') \in D(M,\phi)} m_{P_0}(g)^{\mathrm{Re}\,\pi' + \mu^M + \rho_P}(1 + \| \log_M m_P(g) \|)^{\deg(Q')}.$$

Thanks to (4), we can express ϕ as a linear combination of $s\phi$ and of ϕ_P for $P \subsetneq G$. The upper bound above and those of (a) imply those of the statement. □

I.4.2. Exponential polynomial functions

Let $n, d \in \mathbb{N}$, $R \in \mathbb{R}$, $n, d \geq 1$, $R > 0$. We consider the set Y, equal either to $\mathbb{R}^n$ or to $\mathbb{Z}^n$. Let $A(d, R, Y)$ be the set of complex-valued functions on Y, of the form

$$y \longmapsto \sum_{i=1}^{N} P_i(y) e^{y\lambda_i},$$

where

(i) for all i, P_i is a polynomial on Y with complex coefficients;
(ii) $\sum_{i=1}^{N}(1 + \deg(P_i)) \leq d$;
(iii) for all i, $\lambda_i \in \mathbb{C}^n$ and $\|\lambda_i\| < R$;

for $y = (y_1, \ldots, y_n)$, $\lambda = (\lambda_1, \ldots, \lambda_n) \in \mathbb{C}^n$, we write in the obvious way

$$y\lambda = \sum_{j=1}^{n} y_j \lambda_j, \; \|\lambda\| = \left(\sum_{j=1}^{n} |\lambda_j|^2 \right)^{1/2}.$$

The set $A(d, R, Y)$ is not a vector space but it doesn't matter. For $r \in \mathbb{R}$, $r > R$, $f \in A(d, R, Y)$, the function

$$y \longmapsto |f(y)| e^{-r\|y\|}$$

is bounded on Y. We can set

$$s(f, r) = \sup\{|f(y)| e^{-r\|y\|}; y \in Y\}.$$

Lemma *Let $n, d \in \mathbb{N}$, $R \in \mathbb{R}$, $n, d \geq 1$, $R > 0$. Set $Y = \mathbb{R}^n$ or $\mathbb{Z}^n$. There exists a subfinite set $Y_0 \subset Y$, and for all $r \in \mathbb{R}$, $r \geq 2^{(n-1)/2}(R + 1)$, there exists $c(r) > 0$ such that, for all $f \in A(d, R, Y)$, we have the inequality*

$$s(f, r) \leq c(r) \sup\{|f(y)|; y \in Y_0\}.$$

Proof The proof is an adaptation of [W] Lemma 6.3.1. First suppose $n = 1$ and $Y = \mathbb{R}$. We consider the function $f : \mathbb{R} \longmapsto \mathbb{C}$ given by

$$f(y) = \sum_{i=1}^{N} \sum_{j=0}^{d_i - 1} c_{ij} y^j e^{y\lambda_i}$$

where the λ_i are distinct complex numbers such that $|\lambda_i| < R$, the c_{ij} are complex numbers and the integer

$$d' = \sum_{i=1}^{N} d_i$$

satisfies $d' \leq d$. Take an integer $m \geq 4R/\pi$. For $z \in \mathbb{C}$, we set $e(z) = e^{z/m}$. Let $(\alpha_1, \ldots, \alpha_{d'})$ be the sequence consisting of the λ_i, each λ_i being repeated d_i times. Let $y \in \mathbb{R}$. For all $\lambda \in \mathbb{C}$ such that $|\lambda| < R$, all $z \in \mathbb{C}$ and all $k = 1, \ldots, d'$, we have the identity

$$\frac{1}{e(z) - e(\lambda)} = \frac{1}{e(z) - e(\alpha_k)} + \frac{e(\lambda) - e(\alpha_k)}{e(z) - e(\alpha_k)} \frac{1}{e(z) - e(\lambda)}$$

Thus by induction,

$$\frac{1}{e(z) - e(\lambda)} = \left[\sum_{k=1}^{d'} \left(\prod_{\ell=1}^{k-1} (e(\lambda) - e(\alpha_\ell)) \right) \left(\prod_{\ell=1}^{k} (e(z) - e(\alpha_\ell)) \right)^{-1} \right]$$

$$+ \left(\prod_{\ell=1}^{d'} (e(\lambda) - e(\alpha_\ell)) \right) \left[(e(z) - e(\lambda)) \prod_{\ell=1}^{d'} (e(z) - e(\alpha_\ell)) \right]^{-1} .$$

Multiply this identity by $e^{zy} dz$ and integrate over the contour Γ:

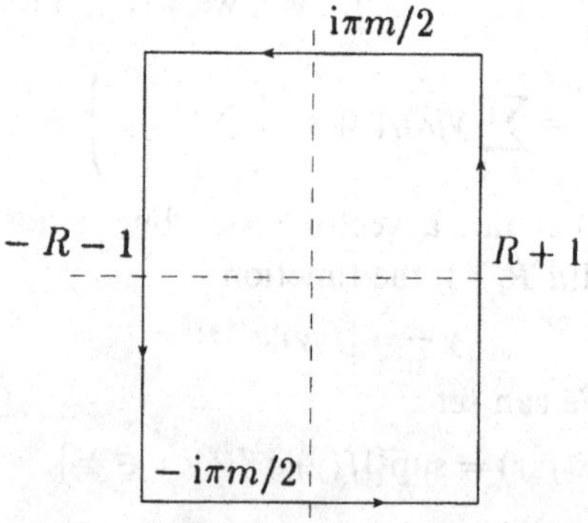

Note that when z runs along Γ, $e(z)$ runs along the following contour Γ', while $e(\lambda)$ is always in the striped region.

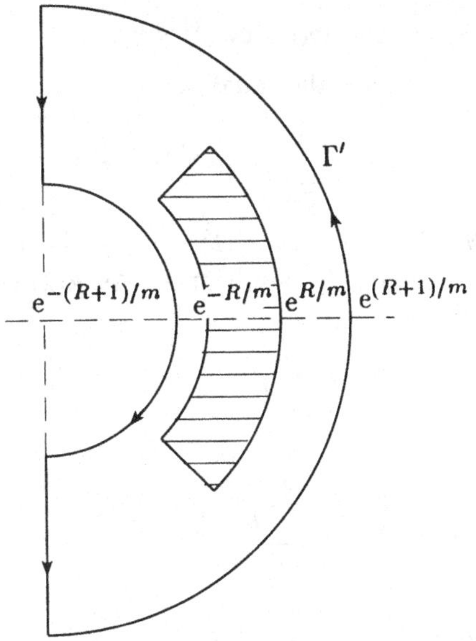

The functions that we integrate are thus holomorphic on Γ. The only pole of the left-hand term in the interior of Γ is at $z = \lambda$. By the residue formula, the integral of the left-hand term is equal to $2\pi i m e^{\lambda y - \lambda/m}$. On the right we obtain

$$\sum_{k=1}^{d'} a_k \prod_{\ell=1}^{k-1}(e(\lambda) - e(\alpha_\ell)) + \varphi(\lambda) \prod_{\ell=1}^{d'}(e(\lambda) - e(\alpha_\ell)) \tag{2}$$

where $\varphi(\lambda)$ is a holomorphic function for $|\lambda| < R$ and

$$a_k = \int_\Gamma e^{zy} \left(\prod_{\ell=1}^{k}(e(z) - e(\alpha_\ell)) \right)^{-1} dz$$

for $k = 1, \ldots, d'$. It can be seen from the figure that there exists $c_1 > 0$ depending only on R and m such that $|e(z) - e(\alpha_\ell)| \geq c_1$ for all $z \in \Gamma$, all $\ell = 1, \ldots, d'$. Thus

$$|a_k| \leq c_2 e^{(R+1)|y|}$$

where $c_2 > 0$ is independent of y and of λ_i. Formula (2) can also be written

$$2\pi i m \left(\sum_{k=0}^{d'-1} b_k e(\lambda)^k + R(\lambda) \right)$$

where

$$R(\lambda) = (2\pi i m)^{-1} \varphi(\lambda) \prod_{\ell=1}^{d'}(e(\lambda) - e(\alpha_\ell))$$

and the b_k satisfy an inequality

(3) $$|b_k| \leq c_3 e^{(R+1)|y|}.$$

With this notation, we have the equality

$$e^{\lambda y} = \sum_{k=1}^{d'} b_{k-1} e^{\lambda k/m} + R(\lambda).$$

Let $i \in \{1, \dots, N\}$, $j \in \{0, \dots, d_i - 1\}$: apply the operator $\mathrm{d}^j/\mathrm{d}\lambda^j$ to this equality and evaluate the result at λ_i. As $R(\lambda)$ contains the factor $(e(\lambda) - e(\lambda_i))^{d_i}$, it is clear that

$$\left(\frac{\mathrm{d}^j}{\mathrm{d}\lambda^j} R\right)(\lambda_i) = 0.$$

We then obtain

$$y^j e^{\lambda_i y} = \sum_{k=1}^{d'} b_{k-1} (k/m)^j e^{\lambda_i k/m}.$$

Then

$$f(y) = \sum_{i=1}^{N} \sum_{j=0}^{d_i - 1} c_{ij} y^j e^{y\lambda_i} = \sum_{i=1}^{N} \sum_{j=0}^{d_i - 1} \sum_{k=1}^{d'} c_{ij} b_{k-1} (k/m)^j e^{\lambda_i k/m}$$
$$= \sum_{k=1}^{d'} b_{k-1} f(k/m).$$

For $r \geq R + 1$, the following inequality is a consequence of (3):

$$|f(y)| e^{-r|y|} \leq c_3 d' \sup\{|f(k/m)|;\ k = 1, \dots, d'\}.$$

But this is the inequality we needed.

Continue to suppose that $n = 1$, but let $Y = \mathbb{Z}$. We can find $\beta \in \mathbb{R}$ such that

$$\inf\{|\mathrm{Im}(\lambda_i) + 2nm - \beta|;\ i = 1, \dots, N,\ m \in \mathbb{Z}\} \geq \pi/(d+1).$$

Fix such a β. Up to replacing λ_i by $\lambda_i + 2\pi i m_i$ for a certain $m_i \in \mathbb{Z}$, we can suppose

$$\beta \leq \mathrm{Im}(\lambda_i) \leq \beta + 2\pi$$

and thus in fact

$$\beta + \pi/(d+1) \leq \mathrm{Im}(\lambda_i) \leq \beta - \pi/(d+1) + 2\pi.$$

Consider formula (1) once again, but this time with $e(z) = e^z$ and now supposing that

(4) $$\beta + \pi/(d+1) \leq \mathrm{Im}(\lambda) \leq \beta - \pi/(d+1) + 2\pi,\ |\lambda| < R.$$

We multiply it by $e^{zy} dz$; we now integrate over the new contour

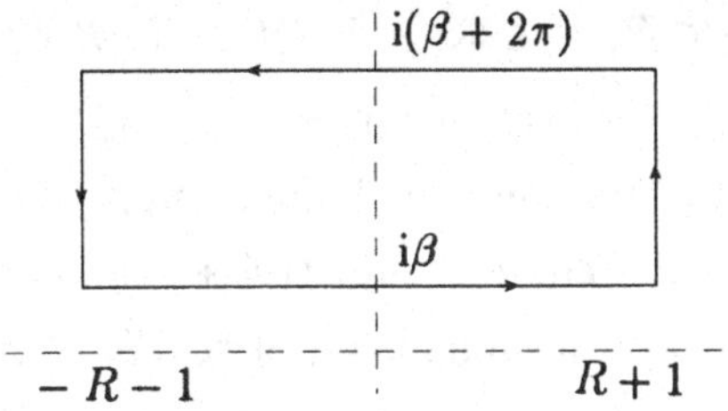

When z runs along Γ, e^z runs along the contour Γ' given in the diagram below, whereas for λ satisfying (4), $e(\lambda)$ remains in the striped region.

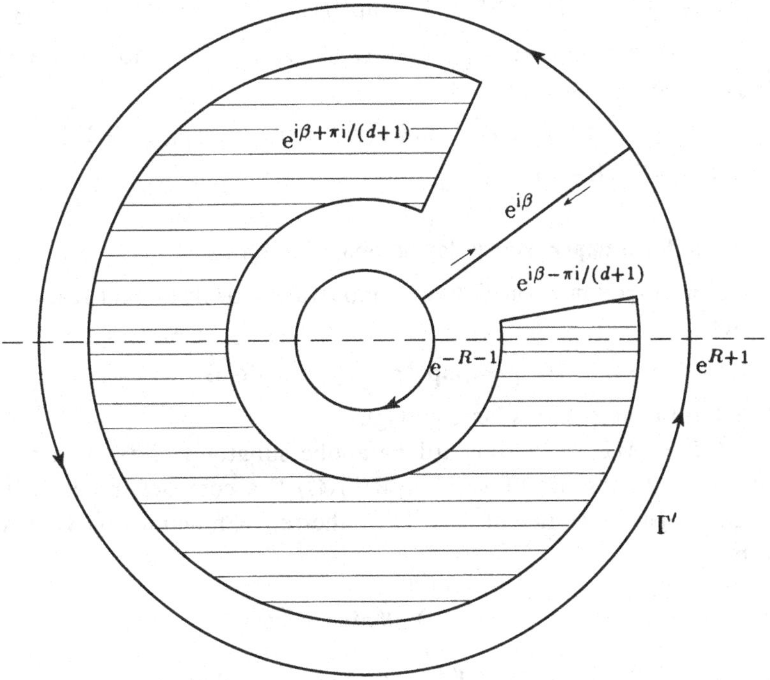

The same calculations as in the case $Y = \mathbb{R}$ lead to an equality

$$f(y) = \sum_{k=1}^{d'} b_{k-1} f(k),$$

where the b_k satisfy an inequality $|b_k| \leq ce^{|y|(R+1)}$, c being independent of y and λ_i. We conclude as in the case $Y = \mathbb{R}$.

In order to pass to the general case, we reason by induction on n. We thus suppose the lemma proved for all integers $n' < n$. We write

$\mathbb{R}^n = \mathbb{R} \oplus \mathbb{R}^{n-1}$ (an analogous reasoning is valid for $\mathbb{Z}^n$). For $y \in \mathbb{R}^n$, set $y = y_1 + y_2$, with $y_1 \in \mathbb{R}$, $y_2 \in \mathbb{R}^{n-1}$. For $f \in A(d, R, \mathbb{R}^n)$, the function

$$y_1 \longmapsto f(y_1 + y_2)$$

belongs to $A(d, R, \mathbb{R})$. From this we obtain an upper bound

$$|f(y_1 + y_2)| \leq c_1(r) e^{r|y_1|} \sup\{|f(y_1' + y_2)|;\ y_1' \in Y_1\}$$

for $r \geq R + 1$, where Y_1 is a subfinite set of $\mathbb{R}$. For $y_1' \in \mathbb{R}$, the function

$$y_2 \longmapsto f(y_1' + y_2)$$

belongs to $A(d, R, \mathbb{R}^{n-1})$. From this we obtain an upper bound

$$|f(y_1' + y_2)| \leq c_2(r) e^{r\|y_2\|} \sup\{|f(y_1' + y_2')|;\ y_2' \in Y_2\},$$

for $r \geq 2^{(n-2)/2}(R + 1)$, where Y_2 is a subfinite set of $\mathbb{R}^{n-1}$. Finally,

$$|f(y_1 + y_2)| \leq c_1(r) c_2(r) e^{r|y_1| + r\|y_2\|} \sup\{|f(y_1' + y_2')|;\ y_1' \in Y_1;\ y_2' \in Y_2\}.$$

It remains to note that $|y_1| + \|y_2\| \leq \sqrt{2}\|y_1 + y_2\|$ and to replace r by $r/\sqrt{2}$ to obtain the upper bound

$$|f(y)| \leq c_1(r) c_2(r) e^{r\|y\|} \sup\{|f(y_1' + y_2')|;\ y_1' \in Y_1,\ y_2' \in Y_2\}$$

for $r \geq 2^{(n-1)/2}(R + 1)$. □

I.4.3. Uniform upper bounds for automorphic forms

For every function φ on $\mathbf{G}$ having moderate growth and all real $r > 0$, we set

$$s(\varphi, r) = \sup\{\|g\|^{-r} |\varphi(g)|;\ g \in \mathbf{G}\}.$$

This term is finite if r is large enough.

Let $P = MU$ be a standard parabolic subgroup of G, V a finite-dimensional subspace of $A_0(U(\mathbb{A})M(k)\backslash\mathbf{G})$, Γ a compact subset of X_M, N an integer ≥ 1. Let $A(V, \Gamma, N)$ be the set of functions on $\mathbf{G}$ of the form

$$g \longmapsto \sum_{i=1}^{N} m_P(g)^{\lambda_i} \varphi_i(g),$$

where, for all i, $\varphi_i \in V$ and $\lambda_i \in \Gamma$. It is clear that there exists a real number $R > 0$ such that for all $\varphi \in A(V, \Gamma, N)$, $s(\varphi, R)$ is finite.

Lemma *Let P, V, Γ, N and R be as above. There exists $R' \geq R$ and a subfinite set $C \subset \mathbf{G}$ such that, for all $r \geq R'$, there exists $c(r) > 0$ such that, for all $\varphi \in A(V, \Gamma, N)$, we have the inequality*

$$s(\varphi, r) \leq c(r) \sup\{|\varphi(g)|;\ g \in C\}.$$

Remark *It is clear that we also have an inequality in the other direction:*

$$\sup\{|\varphi(g)|;\ g \in C\} \leq c'(r)s(\varphi, r)$$

where $c'(r) = \sup\{\|g\|^r;\ g \in C\}$.

Proof Let $A(\mathbf{M}^1\backslash\mathbf{M})$, $A(A_{\mathbf{M}})$, be the spaces of functions on $\mathbf{M}^1\backslash\mathbf{M}$, $A_{\mathbf{M}}$, respectively, generated by the functions

$$m \longmapsto m^\lambda P(\log_M m),$$

where $\lambda \in X_M$, $P \in \mathbb{C}[\mathrm{Re}\,\mathfrak{a}_M]$. Let $A_0(A_{\mathbf{M}}U(\mathbb{A})M(k)\backslash\mathbf{G})$ be the subspace of elements of $A_0(U(\mathbb{A})M(k)\backslash\mathbf{G})$ which are left invariant under $A_{\mathbf{M}}$. From I.3.3 we see that the map

$$A(\mathbf{M}^1\backslash\mathbf{M}) \bigotimes_{\mathbb{C}} A_0(A_{\mathbf{M}}U(\mathbb{A})M(k)\backslash\mathbf{G}) \to A_0(U(\mathbb{A})M(k)\backslash\mathbf{G}),$$

which associates with $f \otimes \varphi$ the function $g \longmapsto f(m_P(g))\varphi(g)$, is surjective. In order to obtain an isomorphism, fix a set of representatives $\mathfrak{m}$ of the (finite) quotient $\mathbf{M}^1A_{\mathbf{M}}\backslash\mathbf{M}$. For $m \in \mathfrak{m}$, denote by $A_0(A_{\mathbf{M}}U(\mathbb{A})M(k)\backslash\mathbf{G})_m$ the subspace of elements of $A_0(A_{\mathbf{M}}U(\mathbb{A})M_k\backslash\mathbf{G})$ whose support is in $\{g \in \mathbf{G}; m_P(g) \in \mathbf{M}^1A_{\mathbf{M}}m\}$. We define a map

$$A(A_{\mathbf{M}}) \bigotimes_{\mathbb{C}} A_0(A_{\mathbf{M}}U(\mathbb{A})M(k)\backslash\mathbf{G})_m \to A_0(U(\mathbb{A})M(k)\backslash\mathbf{G})$$

which associates with $f \otimes \varphi$ the function $g \longmapsto f(m_P(g)m^{-1})\varphi(g)$, where we consider $m_P(g)m^{-1}$ as an element of $A_{\mathbf{M}}$. Then the sum map, denoted by ζ:

$$\zeta : \bigoplus_{m\in\mathfrak{m}}[A(A_{\mathbf{M}}) \bigotimes_{\mathbb{C}} A_0(A_{\mathbf{M}}U(\mathbb{A})M(k)\backslash\mathbf{G})_m] \to A_0(U(\mathbb{A})M(k)\backslash\mathbf{G})$$

is an isomorphism. As V is finite-dimensional, we can find

(i) a subfinite set $(\varphi_i)_{j\in J}$ of linearly independent elements of $A_0(A_{\mathbf{M}}U(\mathbb{A})M(k)\backslash\mathbf{G})$, such that for all $j \in J$, there exists $m(j) \in \mathfrak{m}$ such that $\varphi_j \in A_0(A_{\mathbf{M}}U(\mathbb{A})M(k)\backslash\mathbf{G})_{m(j)}$;

(ii) a finite-dimensional subspace $V' \subset A(A_{\mathbf{M}})$ such that every element $\varphi \in V$ can be written in the form

$$(*) \qquad \varphi = \zeta\left[\sum_{m\in\mathfrak{m}}\left(\sum_{j\in J;m(j)=m} f_j \otimes \varphi_j\right)\right]$$

where the $f_j \in V'$.

For an integer $d \geq 1$ and a compact subset $\Gamma' \subset X_M$, denote by $A(d, \Gamma', A_{\mathbf{M}})$ the set of functions on $A_{\mathbf{M}}$ of the form

$$m \longmapsto \sum_{i=1}^{N'} m^{\lambda_i}P_i(\log_M m)$$

where

(iii) for all i, $P_i \in \mathbb{C}[\mathrm{Re}\, \mathfrak{a}_M]$;

(iv) $\sum_{i=1}^{N'}(1+\deg(P_i)) \leq d$;

(v) for all i, $\lambda_i \in \Gamma'$.

It is clear that one can find d and Γ' such that all $\varphi \in A(V,\Gamma,N)$ can be written in the form (*) where the $f_j \in A(d,\Gamma',A_{\mathbf{M}})$. There exists $R_1 > 0$ such that for $f \in A(d,\Gamma',A_{\mathbf{M}})$, the function $m \longmapsto |f(m)|\,\|m\|^{-R_1}$ is bounded on $A_{\mathbf{M}}$. For $r \geq R_1$, we set

$$s(f,r) = \sup\{|f(m)|\,\|m\|^{-r};\ m \in A_{\mathbf{M}}\}.$$

There exists $R_2 > 0$, $c_2 > 0$ such that for all $g \in \mathbf{G}$, $m \in \mathfrak{m}$ satisfying $m_P(g)m^{-1} \in \mathbf{M}^1 A_{\mathbf{M}}$, we have

$$\|m_P(g)m^{-1}\| \leq c_2\|g\|^{R_2},$$

where we consider $m_P(g)m^{-1}$ as an element of $A_{\mathbf{M}}$. Finally, there exists $R_3 > 0$, $c_3 > 0$ such that for all $j \in J$ and all $g \in \mathbf{G}$,

$$|\varphi_j(g)| \leq c_3\|g\|^{R_3}.$$

Set $R' = R_1R_2 + R_3$. We can suppose $R' \geq R$. Let $r \geq R'$, $g \in \mathbf{G}$ and $\varphi \in A(V,\Gamma,N)$, which we write in the form (*). We have

$$\varphi(g) = \sum_{j \in J(g)} f_j(m_P(g)m(j)^{-1})\varphi_j(g),$$

where $J(g) = \{j \in J;\ m_P(g)m(j)^{-1} \in \mathbf{M}^1 A_{\mathbf{M}}\}$. Thus,

$$\begin{aligned}
&|\varphi(g)|\,\|g\|^{-r}\\
&\quad\leq \sum_{j\in J(g)} |f_j(m_P(g)m(j)^{-1})|\,\|g\|^{-r+R_3}|\varphi_j(g)|\,\|g\|^{-R_3}\\
&\quad\leq c_3 \sum_{j\in J(g)} s(f_j,(r-R_3)/R_2)\|m_P(g)m(j)^{-1}\|^{(r-R_3)/R_2}\|g\|^{-r+R_3}\\
&\quad\leq c_3c_2^{(r-R_3)/R_2} \sum_{j\in J(g)} s(f_j,(r-R_3)/R_2).
\end{aligned}$$

As $A_{\mathbf{M}}$ is isomorphic to $\mathbb{R}^n$ or $\mathbb{Z}^n$, where $n = \dim X_M$, we can apply Lemma I.4.2 to $A(d,\Gamma',A_{\mathbf{M}})$: up to increasing R_1, there exists a subfinite set $Y_0 \subset A_{\mathbf{M}}$ and for all $r' \geq R_1$ a real number $c_1(r') > 0$ such that

$$s(f,r') \leq c_1(r')\sup\{|f(y)|;\ y \in Y_0\}.$$

From this we obtain

$$|\varphi(g)|\,\|g\|^{-r} \leq c_4(r)\sum_{j\in J(g)} \sup\{|f_j(y)|;\ y \in Y_0\},$$

where $c_4(r) = c_1((r-R_3)/R_2)c_2^{(r-R_3)/R_2}c_3$.

Fix a set $(g_j)_{j\in J}$ of elements of $\mathbf{G}$ such that

$$\det(\varphi_k(g_j))_{j,k\in J} \neq 0.$$

We can suppose $m_P(g_j)m(j)^{-1} \in \mathbf{M}^1$ for all j. Let Ξ be the matrix $(\varphi_k(g_j))_{j,k\in J}$. For $m \in A_\mathbf{M}$, we introduce the column matrices

$$\varphi(mg.) = (\varphi(mg_j))_{j\in J}, \ f(m) = (f_j(m))_{j\in J}.$$

We have the equality

$$\varphi(mg.) = \Xi f(m)$$

which gives

$$f(m) = \Xi^{-1}\varphi(mg.).$$

Letting c_5 be the product of $|J|$ and of the sup of the absolute values of the coefficients of Ξ^{-1}, we obtain for all j

$$\sup\{|f_j(y)|;\ y \in Y_0\} \leq c_5 \sup\{|\varphi(g')|;\ g' \in C\},$$

where we have put

$$C = \{yg_k;\ y \in Y_0,\ k \in J\}.$$

Setting $c(r) = |J|c_4(r)c_5$, we then obtain

$$|\varphi(g)|\ \|g\|^{-r} \leq c(r) \sup\{|\varphi(g')|;\ g' \in C\}$$

and this holds for all g. It is the desired upper bound. □

I.4.4. Sequences of automorphic forms

For every standard parabolic subgroup $P = MU$ of G, consider a finite-dimensional space V_P of the space of cuspidal automorphic forms on $U(\mathbb{A})M(k)\backslash\mathbf{G}$, a compact subset $\Gamma_P \subset X_M$ and an integer N_P. We denote by $A((V_P, \Gamma_P,\ N_P)_{P_0\subset P\subset G})$ the set of automorphic forms ϕ on $G(k)\backslash\mathbf{G}$ such that for all P, there exists a family $(\lambda_{P,i}, \varphi_{P,i})_{i=1,\dots,N_P}$ of elements of $\Gamma_P \times V_P$ such that we have the equality

$$\phi_P^{\text{cusp}}(g) = \sum_{i=1}^{N_P} m_P(g)^{\lambda_{P,i}}\varphi_{P,i}(g)$$

for all $g \in \mathbf{G}$. This is not a vector space. By Lemma I.4.1, there exists $R > 0$ such that $s(\phi, R)$ is finite for all $\phi \in A((V_P, \Gamma_P, N_P)_{P_0\subset P\subset G})$.

Lemma *For every standard parabolic subgroup $P = MU$ of G, let V_P be a finite-dimensional subspace of the space of cuspidal automorphic forms over $U(\mathbb{A})M(k)\backslash\mathbf{G}$, let Γ_P be a compact subset of X_M and let N_P be an integer. Then*

(a) *there exists a smooth, right and left $\mathbf{K}$-finite function $f : \mathbf{G} \to \mathbb{C}$, with compact support, and a map*

$$A((V_P, \Gamma_P, N_P)_{P_0\subset P\subset G}) \longrightarrow A(G(k)\backslash\mathbf{G})$$

$$\phi \longmapsto \phi',$$

and for all $r > 0$, *there exist* $c_1(r)$, $c_2(r) > 0$, *such that for all* $\phi \in A((V_P,\Gamma_P,N_P)_{P_0\subset P\subset G})$, *we have*

(i) $\phi = \delta(f)\phi'$;

(ii) $c_1(r)s(\phi',r) \leq s(\phi,r) \leq c_2(r)s(\phi',r)$;

(b) *there exists* $R > 0$ *and for all* $r \geq R$, *there exists* $c(r) > 0$ *such that for all* $\phi \in A((V_P,\Gamma_P,N_P)_{P_0\subset P\subset G})$, *we have*

$$s(\phi,r) \leq c(r)\sup\{s(\phi_P^{\text{cusp}},r);\ P_0 \subset P \subset G\};$$

(c) *let* $(\phi_n)_{n\in\mathbb{N}}$ *be a sequence of elements of* $A((V_P,\Gamma_P,N_P)_{P_0\subset P\subset G})$ *and for all* P, *let* $\varphi_P \in V_P$. *Suppose that for all* P, *the sequence* $(\phi_{n,P}^{\text{cusp}})_{n\in\mathbb{N}}$ *converges uniformly to* φ_P *on every compact subset of* $\mathbf{G}$. *Then there exists a unique automorphic form* ϕ *on* $G(k)\backslash\mathbf{G}$ *such that the sequence* $(\phi_n)_{n\in\mathbb{N}}$ *converges uniformly to* ϕ *on all compact sets. For all* P, *we have the equality* $\phi_P^{\text{cusp}} = \varphi_P$.

I.4.5. Proof of I.4.4(a)

As we can always make V_P bigger, we can suppose that there exists a finite set $\mathfrak{F}$ of $\mathbf{K}$-types and for all $P = MU$ an ideal $\mathfrak{I}_M \subset \mathfrak{z}_M$ of finite codimension such that V_P is the space of cuspidal automorphic forms φ on $U(\mathbb{A})M(k)\backslash\mathbf{G}$ such that

(1) the space generated by $\{\delta(k)\varphi;\ k \in \mathbf{K}\}$ decomposes into a sum of irreducible representations of $\mathbf{K}$ whose class belongs to $\mathfrak{F}$;

(2) for all $k \in \mathbf{K}$, φ_k is killed by $\mathfrak{I}_M$.

We can suppose $\mathfrak{F}$ stable under passage to the contragredient. Let $\mathscr{H}^{\mathfrak{F}}$ be the algebra of smooth functions $f : \mathbf{G} \to \mathbb{C}$ with compact support, such that the representation of $\mathbf{K}\times\mathbf{K}$ in the space generated by the (right and left) translates of f decomposes into a sum of irreducible representations whose class belongs to $\mathfrak{F}\times\mathfrak{F}$.

For all $P = MU$ and $\lambda \in X_M$, write $V_P[\lambda] = \{\lambda\varphi;\ \varphi \in V_P\}$, where, for $\varphi \in V_P$, we denote by $\lambda\varphi$ the function $g \longmapsto m_P(g)^\lambda\varphi(g)$. Then if $f \in \mathscr{H}^{\mathfrak{F}}$, $\delta(f)$ leaves $V_P[\lambda]$ stable. Let $R_P(\lambda,f;X)$ be the characteristic polynomial of $\delta(f)$ in $V_P[\lambda]$ and d_P the dimension of V_P. Let us show that:

(3) for all $\varepsilon > 0$, there exists $f \in \mathscr{H}^{\mathfrak{F}}$ such that for all P and all $\lambda \in \Gamma_P$, (the absolute value of) each coefficient of $R_P(\lambda,f;X)-(X-1)^{d_P}$ is bounded by ε.

Note that $R_P(\lambda,f)$ is the characteristic polynomial of the endomorphism $\varphi \longmapsto (-\lambda)\delta(f)\lambda\varphi$ of V_P. Fix a basis $(\varphi_{P,i})_{i=1,\dots,N_P}$ of V_P and a subset $(g_{P,i})_{i=1,\dots,N_P}$ of $\mathbf{G}$ such that

$$\det(\varphi_{P,i}(g_{P,j}))_{i,j=1,\dots,N_P} \neq 0.$$

For all $u \in \mathrm{End}(V_P)$, the coefficients of the characteristic polynomial of u are polynomials in the variables $u(\varphi_{P,i})(g_{P,j})$. Then (3) is a consequence of

(4) for all $\varepsilon > 0$, there exists $f \in \mathscr{H}^{\mathfrak{F}}$ such that for all P, all $\lambda \in \Gamma_P$, all $i, j = 1, \ldots, N_P$, we have

$$|((-\lambda)\delta(f)\lambda\varphi_{P,i})(g_{P,j}) - \varphi_{P,i}(g_{P,j})| < \varepsilon.$$

Without losing too much we can replace the above inequality by

$$|(\delta(f)\lambda\varphi_{P,i})(g_{P,j}) - (\lambda\varphi_{P,i})(g_{P,j})| < \varepsilon.$$

As Γ_P is compact, the set of functions $\{\lambda\varphi_{P,i}; P_0 \subset P \subset G,\ \lambda \in \Gamma_P,\ i = 1, \ldots, N_P\}$ is uniformly equicontinuous on all compact sets of $\mathbf{G}$. Letting a smooth function with support in a small neighborhood of the origin tend towards the Dirac measure, we show that for every compact set C of $\mathbf{G}$, there exists a smooth function f' with compact support in $\mathbf{G}$ such that for all $P, \lambda \in \Gamma_P,\ i = 1, \ldots, N_P,\ g \in C$, we have

$$|(\delta(f')\lambda\varphi_{P,i})(g) - (\lambda\varphi_{P,i})(g)| < \varepsilon.$$

Fix such an f' for a compact set C containing all the $g_{P,j}$ and stable on the left under $\mathbf{K}$. Let χ be the distribution on $\mathbf{G}$ with support in $\mathbf{K}$ and which, on $\mathbf{K}$, is the sum of the characters of elements of $\mathfrak{F}$. Set $f = \chi * f' * \chi$. Then $f \in \mathscr{H}^{\mathfrak{F}}$ and f satisfies the desired inequalities. This terminates the proof of (3).

Now, for every P and every $i = 1, \ldots, N_P$, choose an element $\lambda_{P,i} \in \Gamma_P$. Consider the space

$$V((\lambda_{P,i})) = \bigoplus_P \bigoplus_{i=1}^{N_P} V_P[\lambda_{P,i}].$$

Its dimension is independent of $\lambda_{P,i}$, let us call it d. The algebra $\mathscr{H}^{\mathfrak{F}}$ acts on this space via the representation δ. For $f \in \mathscr{H}^{\mathfrak{F}}$, denote by $R((\lambda_{P,i}), f; X)$ the characteristic polynomial of $\delta(f)$ acting on $V((\lambda_{P,i}))$. It is a consequence of (3) that for all $\varepsilon > 0$, there exists $f \in \mathscr{H}^{\mathfrak{F}}$ such that, for every family $(\lambda_{P,i})$ as above, (the absolute values of) the coefficients of

$$R((\lambda_{P,i}), f; X) - (X-1)^d$$

are bounded by ε. Fix such an f associated with $\varepsilon = 1/2$. Let $\phi \in A((V_P, \Gamma_P, N_P)_{P_0 \subset P \subset G})$. For all P, let $(\lambda_{P,i})_{i=1,\ldots,N_P}$ be a family of elements of Γ_P such that $\phi_P^{\mathrm{cusp}} \in \sum_{i=1}^{N_P} V_P[\lambda_{P,i}]$. Simply set $R(X) = R((\lambda_{P,i}),\ f; X)$, let r_d be its constant term, and set $Q(X) = 1 - R(X)/r_d$. Then

(5) the coefficients of $Q(X)$ are bounded, for example by $2(d! + 1)$;

(6) the constant term of $Q(X)$ is zero;

(7) the operator $\delta(Q(f))$ of $V((\lambda_{P,i}))$ is equal to the identity.

It is a consequence of (7) that ϕ and $\delta(Q(f))\phi$ have the same cuspidal components. By Proposition I.3.3, we thus have $\phi = \delta(Q(f))\phi$. Set $Q'(X) = X^{-1}Q(X)$ (see (6)) and $\phi' = \delta(Q'(f))\phi$. The preceding relation can be written $\phi = \delta(f)\phi'$. From this relation we obtain an inequality

$$s(\phi,r) \leq c_2(r)s(\phi',r)$$

for all $r > 0$, where $c_2(r)$ is independent of ϕ. Similarly, from the relation $\phi' = \delta(Q'(f))\phi$ and from (5) we obtain an inequality

$$c_1(r)s(\phi',r) \leq s(\phi,r).$$

This proves (a) of Lemma I.4.4.

I.4.6. Partial proof of I.4.4(c)

Let us prove (c) under the following supplementary condition on $(\phi_n)_{n\in\mathbb{N}}$:

(1) there exist $r > 0$ and $c > 0$ such that for all $n \in \mathbb{N}$, we have $s(\phi_n,r) \leq c$.

Fix f and the map $\phi \longmapsto \phi'$ satisfying (a) of Lemma I.4.4. From the sequence $(\phi_n)_{n\in\mathbb{N}}$ we deduce a sequence $(\phi'_n)_{n\in\mathbb{N}}$. The inequalities

$$c_1(r)s(\phi'_n,r) \leq s(\phi_n,r)$$

and the relation (1) show that this sequence is uniformly bounded on all compact sets of $\mathbf{G}$. As $\phi_n = \delta(f)\phi'_n$, the sequence $(\phi_n)_{n\in\mathbb{N}}$ is uniformly continuous on all compact sets of $\mathbf{G}$. It is also uniformly bounded over all compact sets. We can apply Ascoli's theorem to it: from every subsequence, we can extract a subsequence which converges uniformly on all compact sets. Let ϕ be the limit of such a subsequence $(\phi_{n(i)})_{i\in\mathbb{N}}$. The function ϕ is necessarily continuous and has moderate growth. Let us show that:

(2) for every standard parabolic subgroup P, we have the equality

$$\phi_P^{\text{cusp}} = \varphi_P.$$

Fix $P = MU$, let $\varphi \in C_0(U(\mathbb{A})M(k)\backslash\mathbf{G})$. Set

$$\langle\phi_P^{\text{cusp}},\varphi\rangle = \int_{U(\mathbb{A})M(k)\backslash\mathbf{G}} \overline{\phi_P^{\text{cusp}}}(g)\varphi(g)\,dg.$$

We have the equalities

$$\begin{aligned}\langle\phi_P^{\text{cusp}},\varphi\rangle &= \int_{U(\mathbb{A})M(k)\backslash\mathbf{G}} \overline{\phi_P}(g)\varphi(g)\,dg\\ &= \int_{U(\mathbb{A})M(k)\backslash\mathbf{G}} \lim_{i\to\infty}\overline{\phi_{n(i),P}}(g)\varphi(g)\,dg.\end{aligned}$$

Hypothesis (1) allows us to exchange integration and passage to the limit.

We obtain

$$(3)\qquad \begin{aligned}\langle \phi_P^{\text{cusp}}, \varphi\rangle &= \lim_{i\to\infty} \int_{U(\mathbb{A})M(k)\backslash \mathbf{G}} \overline{\phi_{n(i),P}}(g)\varphi(g)\, dg\\ &= \lim_{i\to\infty} \int_{U(\mathbb{A})M(k)\backslash \mathbf{G}} \overline{\phi^{\text{cusp}}_{n(i),P}}(g)\varphi(g)\, dg.\end{aligned}$$

Lemma I.4.3 and the hypothesis show that, for r big enough,

$$\lim_{n\to\infty} s(\phi_{n,P}^{\text{cusp}} - \varphi_P, r) = 0.$$

Remark In I.4.3, we only considered terms $s(\varphi, r)$ and not $s(\varphi_1 - \varphi_2, r)$; but with the notation of I.4.3, if $\varphi_1, \varphi_2 \in A(V, \Gamma, N)$, we have $\varphi_1 - \varphi_2 \in A(V, \Gamma, 2N)$ and it suffices to apply I.4.3 for the integer $2N$ to obtain upper bounds for $s(\varphi_1 - \varphi_2, r)$.

As the function $g \longmapsto \|g\|^r |\varphi(g)|$ is integrable, say over S^P, the theorem of dominated convergence shows that one can exchange limit and integration in the expression (3). Thus

$$\begin{aligned}\langle \phi_P^{\text{cusp}}, \varphi\rangle &= \int_{U(\mathbb{A})M(k)\backslash \mathbf{G}} \lim_{i\to\infty} \overline{\phi^{\text{cusp}}_{n(i),P}}(g)\varphi(g)\, dg\\ &= \int_{U(\mathbb{A})M(k)\backslash \mathbf{G}} \overline{\varphi_P}(g)\varphi(g)\, dg\end{aligned}$$

and the desired equality $\phi_P^{\text{cusp}} = \varphi_P$.

By (2), we see that the function ϕ is well-determined (Proposition I.3.3). But then the sequence $(\phi_n)_{n\in\mathbf{N}}$ itself converges to a function ϕ which, as we already remarked, is continuous, has moderate growth and satisfies (2). These properties imply that ϕ is automorphic (Lemma I.3.4). This proves (c) of Lemma I.4.4 under hypothesis (1).

I.4.7. Proof of I.4.4 (b)

Reasoning by induction on the semi-simple rank of G, we can suppose the lemma proved for every proper Levi subgroup M of G. We deduce from this a result similar to the statement of the lemma for the automorphic forms on $U(\mathbb{A})M(k)\backslash\mathbf{G}$. In particular there exists $R_1 > 0$ and for all $r \geq R_1$, there exists $c_1(r) > 0$ such that for all $\phi \in A((V_P, \Gamma_P, N_P)_{P_0\subset P\subset G})$ and every standard parabolic subgroup P of G, we have

$$(1)\qquad s(\phi_P, r) \leq c_1(r) \sup\{s(\phi_{P'}^{\text{cusp}}, r);\ P_0 \subset P' \subset P\}.$$

On the other hand, we have the property:

(2) there exists $R_2 > 0$, $t > 0$, a compact subset $C \subset \mathbf{G}^1\backslash\mathbf{G}$ and,

for all $r \geq R_2$, a real number $c_2(r) > 0$ such that for all $\phi \in A((V_P, \Gamma_P, N_P)_{P_0 \subset P \subset G})$, we have

$$s(\phi, r) \leq c_2(r) \sup\{|\phi(g)| \, \|g\|^{-rt}; \; g \in \mathbf{G}, \; m_G(g) \in C\}.$$

Indeed, the same reasoning as in I.4.1 (b) shows that there exists an integer $d \geq 1$ and a compact subset $\Gamma \subset X_M$ such that for all $\phi \in A((V_P, \Gamma_P, N_P)_{P_0 \subset P \subset G})$ and all $g \in \mathbf{G}$, the function on $A_G : a \longmapsto \phi(ag)$ is of the form

$$a \longmapsto \sum_{i=1}^{N} a^{\lambda_i} P_i(\log_G a)$$

where

(i) for all i, $P_i \in \mathbb{C}[\mathrm{Re}\, \mathfrak{a}_G]$;

(ii) $\sum_{i=1}^{N} (1 + \deg(P_i)) \leq d$;

(iii) for all i, $\lambda_i \in \Gamma$.

Let us apply I.4.2. There exists $R_2' > 0$, a subfinite set $Y_0 \subset A_{\mathbf{G}}$ and, for all $r \geq R_2'$, a real number $c_2'(r) > 0$ such that for all ϕ, g as above, we have

$$\sup\{|\phi(ag)| \, \|a\|^{-r}; a \in A_{\mathbf{G}}\} \leq c_2'(r) \sup\{|\phi(ag)|; \; a \in Y_0\}.$$

Fix c, t, t' as in I.2.2 (viii), set $R_2 = R_2'/t'$. For $r \geq R_2$, $g \in \mathbf{G}^c$, $a \in A_{\mathbf{G}}$ and $\phi \in A((V_p, \Gamma_P, N_P)_{P_0 \subset P \subset G})$, we have

$$\begin{aligned} |\phi(ag)| \, \|ag\|^{-r} &\leq c^{-r} |\phi(ag)| \, \|a\|^{-rt'} \, \|g\|^{-rt} \\ &\leq c^{-r} c_2'(rt') \|g\|^{-rt} \sup\{|\phi(ag)|; \; a \in Y_0\} \\ &\leq c_2(r) \sup\{|\phi(ag)| \, \|ag\|^{-rt}; \; a \in Y_0\}, \end{aligned}$$

for a certain real number $c_2(r)$. Take for C a compact subset of $\mathbf{G}^1 \backslash \mathbf{G}$ containing $m_G(a'g')$ for all $a \in Y_0$, $g' \in \mathbf{G}^c$. Then

$$\begin{aligned} &\sup\{|\phi(ag)| \, \|ag\|^{-rt}; \; a \in Y_0\} \\ &\qquad \leq \sup\{|\phi(g')| \, \|g'\|^{-rt}; \; g' \in \mathbf{G}, \; m_G(g') \in C\}. \end{aligned}$$

This gives assertion (2).

Let $R_3 > 0$ be such that for all $\phi \in A((V_P, \Gamma_P, N_P)_{P_0 \subset P \subset G})$, $s(\phi, R_3)$ and $s(\phi_P^{\text{cusp}}, R_3)$ are finite for all P. Let $r \geq R_3$. If assertion (b) of Lemma I.4.4 is false, we can find a sequence $(\psi_n)_{n \in \mathbb{N}}$ of elements of $A((V_P, \Gamma_P, N_P)_{P_0 \subset P \subset G})$ such that for all n,

$$n \sup\{s(\psi_{n,P}^{\text{cusp}}, r); \; P_0 \subset P \subset G\} \leq s(\psi_n, r) \neq 0.$$

Set

$$\phi_n = s(\psi_n, r)^{-1} \psi_n.$$

Then

$$s(\phi_n, r) = 1. \tag{3}$$

For all P,

(4) $$s(\phi_{n,P}^{\text{cusp}}, r) \leq 1/n :$$

the sequence $(\phi_{n,P}^{\text{cusp}})_{n\in\mathbb{N}}$ thus converges uniformly to 0 on all compact sets. The sequence $(\phi_n)_{n\in\mathbb{N}}$ thus satisfies the hypotheses of Lemma I.4.4 (c) and condition (1) of I.4.6. It converges uniformly on all compact sets to a function whose cuspidal components are zero. Thus it converges to zero.

We introduce the following functions on S:

(5) $$s\phi_n = \sum_{P=MU} (-1)^{rg(G)-rg(M)} \phi_{n,P}.$$

By I.4.4 (a) and (3) above, there exists a smooth, right and left $\mathbf{K}$-finite function f on $\mathbf{G}$ with compact support, and a sequence $(\phi'_n)_{n\in\mathbb{N}}$ of functions on $G(k)\backslash\mathbf{G}$ such that

$$\phi_n = \delta(f)\phi'_n.$$

Moreover there exists $c > 0$ such that for all $n \in \mathbb{N}$, $s(\phi'_n, r) \leq c$. If k is a function field, it is a consequence of this and of Corollary I.2.7 that there exists a subset C' of S, compact modulo $A_{\mathbf{G}}$, such that the support of $s\phi_n$ is in C' for all n. If k is a number field, we use Corollary I.2.9: for all $\lambda \in \operatorname{Re} X_{M_0}$, there exists $c' > 0$ such that for all $n \in \mathbb{N}$, $g \in \mathbf{G}^1 \cap S$, $a \in A_{\mathbf{G}}$, we have the inequality

$$|s\phi_n(g)| \leq c' \|a\|^r m_{P_0}(g)^\lambda.$$

Suppose $r \geq R_2$, and introduce the compact set C and the real number t of (2). Fix $\varepsilon > 0$. According to what was said above, there exists a compact subset $C' \subset \mathbf{G}^c \cap S$ such that for all $n \in \mathbb{N}$, $g \in \mathbf{G}^c \cap S - C'$, $a \in A_{\mathbf{G}}$, we have the inequality

$$|s\phi_n(ag)| \leq \varepsilon \|a\|^r \|g\|^{rt}.$$

Let us limit ourselves to pairs (a, g) such that $m_G(ag) \in C$. As the set

$$\{\|a\|^r \|g\|^{rt} \|ag\|^{-rt};\ a \in A_{\mathbf{G}},\ g \in \mathbf{G}^c,\ m_G(ag) \in C\}$$

is bounded above, say by $c_4(r)$, we obtain

(6) $$|s\phi_n(ag)| \leq \varepsilon c_4(r) \|ag\|^{rt}$$

for all $n \in \mathbb{N}$, $a \in A_{\mathbf{G}}$, $g \in \mathbf{G}^c \cap S - C'$ such that $m_G(ag) \in C$.

Suppose $r \geq R_1/t$. By (1), under the same hypotheses as above, we have

$$\sum_{P_0 \subset P \subsetneq G} |\phi_{n,P}(ag)| \leq c'_1(r) \|ag\|^{rt} \sup\{s(\phi_{n,P}^{\text{cusp}}, rt); P_0 \subset P \subset G\}$$

for a certain real number $c'_1(r) > 0$. By (4), Lemma I.4.3 and the remark which follows it, we have

$$\lim_{n\to\infty} s(\phi_{n,P}^{\text{cusp}}, rt) = 0$$

for all P, at least when r is big enough, say $r \geq R_4$. Thus there exists N such that, for $n \geq N$ and a, g as above,

$$\sum_{P_0 \subset P \subsetneq G} |\phi_{n,P}(ag)| \leq \varepsilon \|ag\|^{-rt}.$$

This inequality, together with (5) and (6), implies

$$|\phi_n(ag)| \leq \varepsilon(1 + c_4(r))\|ag\|^{-rt}$$

for $n \geq N$, $a \in A_{\mathbf{G}}$, $g \in \mathbf{G}^c \cap S - C'$, $m_G(ag) \in C$.

On the other hand, the set $\{ag;\ a \in A_{\mathbf{G}},\ g \in C',\ m_G(ag) \in C\}$ is compact. As the sequence $(\phi_n)_{n \in \mathbb{N}}$ tends to 0 on all compact sets, there exists N' such that for $n \geq N'$ and a, g in the set in question, we have

$$|\phi_n(ag)| \leq \varepsilon(1 + c_4(r))\|ag\|^{-rt}.$$

Finally, for $n \geq N, N'$, we have

$$|\phi_n(g)| \leq \varepsilon(1 + c_4(r))\|g\|^{-rt}$$

for all $g \in S$ such that $m_G(g) \in C$. From I.2.2 (vii), we then deduce the existence of a real number $c_5(r) > 0$ such that

$$|\phi_n(g)| \leq \varepsilon c_5(r)\|g\|^{-rt}$$

for all $g \in \mathbf{G}$ such that $m_G(g) \in C$ and all $n \geq N, N'$. By (2), we obtain

$$s(\phi_n, r) \leq \varepsilon c_2(r) c_5(r)$$

for $n \geq N, N'$. For ε small enough, this contradicts (3). This contradiction proves assertion (b) of Lemma I.4.4.

I.4.8. End of the proof of I.4.4

Let us complete the proof of (c). The hypothesis and Lemma I.4.3 show that for r large enough, the set

$$\{s(\phi_{n,P}^{\text{cusp}}, r);\ n \in \mathbb{N},\ P_0 \subset P \subset G\}$$

is bounded above. By I.4.4 (b), the set

$$\{s(\phi_n, r);\ n \in \mathbb{N}\}$$

is thus also bounded above, i.e. the sequence $(\phi_n)_{n \in \mathbb{N}}$ satisfies hypothesis (1) of I.4.6. This gives (c). □

I.4.9. Fréchet spaces and holomorphic functions

We recall some properties of maps with values in Fréchet spaces. Let H be such a space, i.e. a vector space over $\mathbb{C}$ equipped with a topology defined by a countable family of semi-norms, for which it is separated and complete. Let U be an open set of $\mathbb{C}$ and $f : U \to H$ a function. We say that f is holomorphic if for all $z \in U$, the function $z' \longmapsto (z' - z)^{-1}(f(z') - f(z))$ has a limit in H when z' tends to z. It is equivalent

([Ru] 3.31) to ask that $\ell \circ f$ be holomorphic for every continuous linear form $\ell : H \to \mathbb{C}$. Now if U is an open set of $\mathbb{C}^N$ and $f : U \to H$ is a function, we say that f is holomorphic if it is holomorphic with respect to each variable $z_1, \ldots, z_N$ of $\mathbb{C}^N$. Finally, if X is a complex analytic manifold, and $f : X \to H$ is a function, we say that f is holomorphic if for every local chart

$$\mathbb{C}^N \supset U \hookrightarrow^i X$$

the function $f \circ i$ is holomorphic. Now if f is only defined almost everywhere on X, we say that f is meromorphic if for every $x \in X$ and every sufficiently small neighborhood V of x, there exist two holomorphic functions

$$f_1 : V \to H, \; d : V \to \mathbb{C},$$

$d \neq 0$, such that $d(y)f(y) = f_1(y)$ at every point $y \in V$ where $f(y)$ is defined.

Several classical theorems on complex-valued holomorphic functions can be generalised to this situation: expansion into a convergent series of a holomorphic function in a polydisk, Cauchy's formula and, if X is connected, the uniqueness of the meromorphic continuation of a function defined on a non-empty open set.

Let Y be a locally compact topological space which is countable at infinity, equipped with a measure. Let $L^2_{\text{loc}}(Y)$ be the space of complex-valued functions on Y which are locally square integrable. We of course identify functions which are equal almost everywhere. For every compact subset $C \subset Y$, define the semi-norm $\varphi \mapsto \|\varphi_{|C}\|$ on $L^2_{\text{loc}}(Y)$ by

$$\varphi \longmapsto \|\varphi|_C\|^2 = \int_C |\varphi(y)|^2 \, dy.$$

Then this family of semi-norms equips $L^2_{\text{loc}}(Y)$ with the structure of a Fréchet space. For every compact subset $C \subset Y$, let us equip $L^2(C)$ with the topology defined by the L^2 norm. Then it is also a Fréchet space. The restriction map

$$\varphi \longmapsto \varphi|_C$$

$$L^2_{\text{loc}}(Y) \longrightarrow L^2(C)$$

is continuous. Let U be an open set of $\mathbb{C}^N$ and $f : U \to L^2_{\text{loc}}(Y)$ a function. We check that f is holomorphic if and only if, for all C, the function

$$U \longrightarrow L^2(C)$$

$$x \longmapsto f(z)|_C$$

is holomorphic. It is equivalent to ask that for all $\varphi \in L^2(C)$, the function

$$U \longrightarrow \mathbb{C}$$

$$z \longmapsto \int_C f(z)(y)\overline{\varphi(y)}\, dy$$

be holomorphic.

I.4.10. Automorphic forms depending holomorphically on a parameter

Lemma *For every standard parabolic subgroup $P = MU$ of G, let V_P be a finite-dimensional subspace of the space of cuspidal automorphic forms on $U(\mathbb{A})M(k)\backslash\mathbf{G}$, Γ_P a compact subset of X_M and N_P an integer. Let n be an integer, D an open connected subset of $\mathbb{C}^n$ and $f : D \to \mathbb{C}$ a holomorphic function not identically zero. Set $D' = \{z \in D;\ f(z) \neq 0\}$. Let*

$$\phi : D' \longrightarrow L^2_{\rm loc}(\mathbf{G})$$

$$z \longmapsto \phi_z$$

be a function. Suppose that

(i) *for all $z \in D'$, $\phi_z \in A((V_P, \Gamma_P, N_P)_{P_0 \subset P \subset G})$.*

For all P, let $\psi_P : D' \to L^2_{\rm loc}(\mathbf{G})$ be the function defined by

$$z \longmapsto \psi_{P,z} = \phi^{\rm cusp}_{z,P}.$$

(ii) *Also suppose that for all P, ψ_P is holomorphic on D'.*

Then the function ϕ can be continued to a holomorphic function on D in $L^2_{\rm loc}(\mathbf{G})$ if and only if for all P, the function ψ_P satisfies the same property. Suppose these conditions are satisfied: the continuations of these functions are still denoted by ϕ and ψ_P. They satisfy the properties:

(a) *for all $z \in D$, ϕ_z is automorphic;*

(b) *for all $z \in D$ and all P, $\phi^{\rm cusp}_{z,P} = \psi_{P,z}$;*

(c) *the function*

$$\mathbf{G} \times D \longrightarrow \mathbb{C}$$

$$(g, z) \longmapsto \phi_z(g)$$

is smooth;

(d) *if $g \in \mathbf{G}$, the function $z \longmapsto \phi_z(g)$ is holomorphic on D.*

Proof Let us first prove the following assertion:

(1) let C be a compact subset of $\mathbf{G}$; then there exists a compact subset C' of $\mathbf{G}$ and a real number $c > 0$ such that for all standard parabolic subgroups P of G, all $z, z' \in D'$ and all $g \in C$, we have

$$|\psi_{P,z}(g)| \leq c\|\psi_{P,z}|_{C'}\|,$$

$$|\psi_{P,z}(g) - \psi_{P,z'}(g)| \leq c\|(\psi_{P,z} - \psi_{P,z'})|_{C'}\|.$$

It was proved in I.4.5 that there exists a smooth, right and left **K**-finite function $\varphi : \mathbf{G} \to \mathbb{C}$ with compact support, an integer d, a real number $c_1 > 0$ and, for all $z \in D'$, a polynomial with no constant term of degree d, say $Q_z(X) = \sum_{j=1}^{d} q_{z,j} X^j$, such that

(i) $|q_{z,j}| \le c_1$ for all z, j;

(ii) $\psi_{P,z} = \delta(Q_z(\varphi))\psi_{P,z}$ for all P, z.

For $g \in \mathbf{G}$, we thus have

$$\psi_{P,z}(g) = \sum_{j=1}^{d} q_{z,j} \int_{\mathbf{G}} \varphi^j(g^{-1}h)\psi_{P,z}(h)\,\mathrm{d}h,$$

where of course $\varphi^j = \varphi * \ldots * \varphi$ (j times). Set

$$c_2 = \sup\{|\varphi^j(h)|;\ h \in \mathbf{G},\ j = 1,\ldots,d\},$$

and let $C' \subset \mathbf{G}$ be a compact set containing $g\mathrm{Supp}(\varphi^j)$ for all $g \in C$, $j = 1,\ldots,d$. Then for $g \in C$, we have

$$|\psi_{P,z}(g)| \le dc_1c_2\,\mathrm{meas}(C')\|\psi_{P,z}|_{C'}\|.$$

This is the first of the desired inequalities. As we noted in I.4.5, the same proof can be applied to the functions $\psi_{P,z} - \psi_{P,z'}$. This proves (1).

It is a consequence of (1) and of Lemma I.4.3 that there exists $R > 0$ such that for all $r \ge R$, $s(\psi_{P,z}, r)$ and $s(\psi_{P,z} - \psi_{P,z'}, r)$ are finite for all P and all $z, z' \in D'$. Let $z_0 \in D'$. As the map $z \longmapsto \psi_{P,z}$ is holomorphic on D', the map $z \longmapsto \|(\psi_{P,z} - \psi_{P,z_0})|_{C'}\|$ is continuous there. It is a consequence of (1) and of Lemma I.4.3 that for $r \ge R$, $\lim_{z\to z_0} s(\psi_{P,z} - \psi_{P,z_0}, r) = 0$. We deduce from these properties that the map

$$\mathbf{G} \times D' \longrightarrow \mathbb{C}$$

$$(g, z) \longmapsto \psi_{P,z}(g)$$

is continuous.

We deduce from I.4.4 (b) and from what was said above the existence of a real number $r \ge R$ such that $s(\phi_z, r)$ and $s(\phi_z - \phi_{z'}, r)$ are finite for all $z, z' \in D'$. For all $z_0 \in D'$, we have $\lim_{z\to z_0} s(\phi_z - \phi_{z_0}, r) = 0$. We also deduce from this that the map

$$\mathbf{G} \times D' \longrightarrow \mathbb{C}$$

$$(g, z) \longmapsto \phi_z(g)$$

is continuous.

Let $z_0 \in D$. To simplify the notation, suppose that $z_0 = 0$. Up to changing coordinates by a linear transformation, we can suppose that the function

$$\{u \in \mathbb{C};\ (u, 0, \ldots, 0) \in D\} \longrightarrow \mathbb{C}$$

$$u \longmapsto f(u, 0, \ldots, 0)$$

is not identically zero. For $\varepsilon > 0$, set

$$D(\varepsilon) = \{u \in \mathbb{C};\ |u| < \varepsilon\},\ \overline{D}(\varepsilon) = \{u \in \mathbb{C};\ |u| \leq \varepsilon\},$$

$$C(\varepsilon) = \{u \in \mathbb{C};\ |u| = \varepsilon\}.$$

There exists $\varepsilon_1 > 0$ such that $\{(u,0,\ldots,0);\ u \in \overline{D}(\varepsilon_1)\} \subset D$ and $f(u,0,\ldots,0) \neq 0$ for all $u \in C(\varepsilon_1)$. Fix such an ε_1. We can then find $\varepsilon_2,\ldots,\varepsilon_n > 0$ such that

$$\overline{D}(\varepsilon_1) \times \overline{D}(\varepsilon_2) \times \ldots \times \overline{D}(\varepsilon_n) \subset D$$

and $f(u_1,\ldots,u_n) \neq 0$ for $(u_1,\ldots,u_n) \in C(\varepsilon_1) \times \ldots \times C(\varepsilon_n)$.

For $z = (z_1,\ldots,z_n) \in D(\varepsilon_1) \times \ldots \times D(\varepsilon_n)$ and $g \in \mathbf{G}$, set

$$\text{(2)} \qquad \phi'_z(g) = (2\pi \mathrm{i})^{-n} \int_{C(\varepsilon_1)\times\ldots\times C(\varepsilon_n)} \phi_u(g) \prod_{i=1}^{n} (u_i - z_i)^{-1} \prod_{i=1}^{n} du_i,$$

and, for all P,

$$\text{(3)} \qquad \psi'_{P,z}(g) = (2\pi \mathrm{i})^{-n} \int_{C(\varepsilon_1)\times\ldots\times C(\varepsilon_n)} \psi_{P,u}(g) \prod_{i=1}^{n} (u_i - z_i)^{-1} \prod_{i=1}^{n} du_i;$$

all functions being integrated are continuous.

The compactness of $C(\varepsilon_1)\times\ldots\times C(\varepsilon_n)$ allows us to extend the properties of ϕ_u for $u \in D'$ to the function ϕ'_z. Indeed, $s(\phi'_z, r)$ and $s(\phi'_z - \phi'_{z'}, r)$ are finite for $z, z' \in D(\varepsilon_1) \times \ldots \times D(\varepsilon_n)$. In particular ϕ'_z has moderate growth. For $z' \in D(\varepsilon_1) \times \ldots \times D(\varepsilon_n)$, $\lim_{z \to z'} s(\phi'_z - \phi'_{z'}, r) = 0$. The map

$$\mathbf{G} \times D(\varepsilon_1) \times \ldots \times D(\varepsilon_n) \longrightarrow \mathbb{C}$$

$$(g, z) \longmapsto \phi'_z(g)$$

is continuous. The same properties are true for the functions $\psi'_{P,z}$.

Let us show that for all $z \in D(\varepsilon_1) \times \ldots \times D(\varepsilon_n)$ and all $P = MU$, the cuspidal component of $\phi'_{z,P} - \psi'_{P,z}$ (along P) is zero.

Since calculating a constant term consists of integrating over a compact set, continuity of the map $(g,u) \longmapsto \phi_u(g)$ implies the equality

$$\phi'_{z,P}(g) = (2\pi \mathrm{i})^{-n} \int\limits_{C(\varepsilon_1)\times\ldots\times C(\varepsilon_n)} \phi_{u,P}(g) \prod_{i=1}^{n} (u_i - z_i)^{-1} \prod_{i=1}^{n} du_i$$

for all $g \in \mathbf{G}$. Let $\psi \in C_0(U(\mathbb{A})M(k)\backslash \mathbf{G})$. Set

$$\langle \psi, \phi'_{z,P} \rangle = \int\limits_{U(\mathbb{A})M(k)\backslash \mathbf{G}} \overline{\psi}(g) \phi'_{z,P}(g)\, dg.$$

Since the function $u \longmapsto s(\phi_{u,P}, r)$ is bounded on $C(\varepsilon_1) \times \ldots \times C(\varepsilon_n)$, we can exchange these integrations. We obtain

$$\langle \psi, \phi'_{z,P} \rangle = (2\pi i)^{-n} \int_{C(\varepsilon_1)\times\ldots\times C(\varepsilon_n)} \langle \psi, \phi_{u,P} \rangle \prod_{i=1}^{n} (u_i - z_i)^{-1} \prod_{i=1}^{n} du_i$$

$$= (2\pi i)^{-n} \int_{C(\varepsilon_1)\times\ldots\times C(\varepsilon_n)} \langle \psi, \psi_{P,u} \rangle \prod_{i=1}^{n} (u_i - z_i)^{-1} \prod_{i=1}^{n} du_i.$$

We can again exchange the integrations. We obtain

$$\langle \psi, \phi'_{z,P} \rangle = \langle \psi, \psi'_{P,z} \rangle$$

and this proves the assertion.

Suppose ϕ can be continued to a holomorphic function of D into $L^2_{\mathrm{loc}}(\mathbf{G})$. Let us show that $\phi_z = \phi'_z$ for all $z \in D(\varepsilon_1) \times \ldots \times D(\varepsilon_n)$. It suffices to show that for all compact subsets C of $\mathbf{G}$ and all $\psi \in L^2(C)$, we have

$$\langle \psi, \phi'_z \rangle = \langle \psi, \phi_z \rangle,$$

where we set for example

$$\langle \psi, \phi'_z \rangle = \int_C \overline{\psi}(g) \phi'_z(g)\, dg.$$

The function $(g, u) \longmapsto \phi_u(g)$ is bounded on $C \times C(\varepsilon_1) \times \ldots \times C(\varepsilon_n)$. We can exchange the integrations:

$$\langle \psi, \phi'_z \rangle = (2\pi i)^{-n} \int_{C(\varepsilon_1)\times\ldots\times C(\varepsilon_n)} \langle \psi, \phi_u \rangle \prod_{i=1}^{n} (u_i - z_i)^{-1} \prod_{i=1}^{n} du_i.$$

As the function $u \longmapsto \langle \psi, \phi_u \rangle$ is holomorphic, Cauchy's formula shows that the right-hand side above is equal to $\langle \psi, \phi_z \rangle$, which gives our assertion. Note that, with no hypothesis on ϕ, the above calculation shows that $z \longmapsto \langle \psi, \phi'_z \rangle$ is holomorphic, and thus that the map

$$D(\varepsilon_1) \times \ldots \times D(\varepsilon_n) \longrightarrow L^2_{\mathrm{loc}}(\mathbf{G})$$

$$z \longmapsto \phi'_z$$

is holomorphic.

Let $z \in D' \cap (D(\varepsilon_1) \times \ldots \times D(\varepsilon_n))$. Under the hypothesis that ϕ can be holomorphically continued, we have $\phi'_z = \phi_z$. As $z \in D'$, ϕ_z is automorphic and for all P, $\phi^{\mathrm{cusp}}_{z,P}$ exists. The cuspidal component along P of $\phi^{\mathrm{cusp}}_{z,P} - \psi'_{P,z}$ is zero. As it is clear that $(\phi^{\mathrm{cusp}}_{z,P} - \psi'_{P,z})_{P'} = 0$ for all $P' \subsetneq P$, all the cuspidal components of $\phi^{\mathrm{cusp}}_{z,P} - \psi'_{P,z}$ are zero. By I.3.4, $\phi^{\mathrm{cusp}}_{z,P} = \psi'_{P,z}$. As $\phi^{\mathrm{cusp}}_{z,P} = \psi_{P,z}$ by hypothesis, we obtain $\psi_{P,z} = \psi'_{P,z}$. We show as above that the map

$$D(\varepsilon_1) \times \ldots \times D(\varepsilon_n) \longrightarrow L^2_{\mathrm{loc}}(\mathbf{G})$$

$$z \longmapsto \psi'_{P,z}$$

is holomorphic. Thus ψ_P can be continued to a holomorphic function on $D(\varepsilon_1) \times \ldots \times D(\varepsilon_n)$, then on D by letting the point z_0 vary.

Conversely, suppose that all the functions ψ_P can be holomorphically continued. We show as above that $\psi'_{P,z} = \psi_{P,z}$ for all $z \in D(\varepsilon_1)\times\ldots\times D(\varepsilon_n)$ and all P. For all $z \in D' \cap (D(\varepsilon_1)\times\ldots\times D(\varepsilon_n))$, the cuspidal component of $\phi'_z - \phi_z$ along P is the same as that of $\phi'_{z,P} - \phi^{\text{cusp}}_{z,P}$, i.e. that of $\phi'_{P,z} - \psi'_{P,z}$, which is zero. By I.3.4 $\phi'_z = \phi_z$. As $z \longmapsto \phi'_z$ is holomorphic, we see that ϕ can be continued to a holomorphic function on $D(\varepsilon_1) \times \ldots \times D(\varepsilon_n)$, then on D.

Suppose from now on that the functions ϕ and ψ_P can be holomorphically continued to D. In order to prove properties (a) to (d), we can suppose that $z \in D(\varepsilon_1) \times \ldots \times D(\varepsilon_n)$ as above. The functions ϕ and ψ_P are then given by formulae (2) and (3). Assertion (d) is immediate. It is a consequence of I.4.4(a) that there exists an open subgroup $\mathbf{K}'_f$ of $\mathbf{G}_f$ which leaves ϕ_u invariant for all $u \in D'$. If k is a function field, assertion (c) results from this property and from (d). Suppose k is a number field. Let us use the objects introduced at the beginning of the proof. For all $u \in D'$, we have the equality

$$\phi_u = \sum_{j=1}^{d} q_{u,j}\delta(\varphi^j)\phi_u.$$

Let $X \in \mathcal{U}$. We then have

$$\delta(X)\phi_u = \sum_{j=1}^{d} q_{u,j}\delta(X\varphi^j)\phi_u$$

where X acts on φ^j by left translation. There exists $c > 0$ such that for every function ψ over $\mathbf{G}$ and all $j = 1, \ldots, d$, we have

$$s(\delta(X\varphi^j)\psi, r) \leq c\, s(\psi, r).$$

The properties of the coefficients $q_{u,j}$ and the fact that $s(\phi_u, r)$ is locally bounded on D' then implies that $s(\delta(X)\phi_u, r)$ is locally bounded on D'. We prove similarly that for all $u' \in D'$,

$$\lim_{u \to u'} s(\delta(X)\phi_u - \delta(X)\phi_{u'}, r) = 0.$$

Then the function

$$\mathbf{G} \times D' \longrightarrow \mathbb{C}$$

$$(g, u) \longmapsto (\delta(X)\phi_u)(g)$$

is continuous. Let Δ be a differential operator on D, with constant coefficients. Using the above properties, we show by induction on the degrees of X and Δ that for $z \in D(\varepsilon_1)\times\ldots\times D(\varepsilon_n)$ and $g \in \mathbf{G}$, $\Delta\delta(X)\phi_z(g)$

and $\delta(X)\Delta\phi_z(g)$ exist and are both equal to

$$(2\pi i)^{-n} \int_{C(\varepsilon_1)\times\ldots\times C(\varepsilon_n)} \delta(X)\phi_u(g)\Delta\left(\prod_{i=1}^{n}(u_i - z_i)^{-1}\right) \Pi_{i=1}^{n}\, du_i.$$

Thus they are equal and continuous in (z, g). This means that the function $(g, z) \longmapsto \phi_z(g)$ is C^∞ as a function of 'archimedean' variables, which concludes the proof of (c).

Fix $P = MU$ and let us show that $\psi_{P,z}$ is automorphic for $z \in D(\varepsilon_1) \times \ldots \times D(\varepsilon_n)$. We have already noted that $\psi_{P,z}$ has moderate growth. We show as above that it is smooth. It is a consequence of the beginning of the proof (existence of the function φ) that there exists a finite set $\mathfrak{F}$ of **K**-types such that for all $z \in D'$, the space generated by $\{\delta(k)\psi_{P,z}; k \in \mathbf{K}\}$ can be decomposed into irreducible representations of **K** whose class belongs to $\mathfrak{F}$. Thanks to formula (3), this property can be extended to all $z \in D(\varepsilon_1) \times \ldots \times D(\varepsilon_n)$ and $\psi_{P,z}$ is **K**-finite. We saw in the course of the proof of I.4.3 that, with the notation introduced in I.4.3, there exists:

(i) a finite set $(\varphi_j)_{j\in J}$ of linearly independent elements of $A_0(A_{\mathbf{M}}U(\mathbb{A})M(k)\backslash\mathbf{G})$ such that for all $j \in J$, there exists $m(j) \in \mathfrak{m}$ such that $\varphi_j \in A_0(A_{\mathbf{M}}U(\mathbb{A})M(k)\backslash\mathbf{G})_{m(j)}$

(ii) an integer $d \geq 1$ and a compact subset $\Gamma' \subset X_M$;

(iii) for all $z \in D'$, a subset $\{f_{z,j};\ j \in J\} \subset A(d, \Gamma', A_{\mathbf{M}})$ such that
$$\psi_{P,z}(g) = \sum_{j\in J(g)} f_{z,j}(m_P(g)m(j)^{-1})\varphi_j(g)$$
for all $g \in \mathbf{G}$;

(iv) a subset $\{g_j;\ j \in J\} \subset \mathbf{G}$ and a matrix $\Xi = (\xi_{jk})_{j,k\in J}$ such that for all $z \in D'$, all $m \in A_{\mathbf{M}}$ and all $j \in J$, we have the equality
$$f_{z,j}(m) = \sum_{k\in J} \xi_{jk}\psi_{P,z}(mg_k).$$

This last property allows us to define $f_{z,j}$ for all $z \in D$. As the function

$$\mathbf{G} \times D \longrightarrow \mathbb{C}$$

$$(g, z) \longmapsto \psi_{P,z}(g)$$

is smooth (same proof as for the function ϕ), for all $j \in J$, the function

$$A_{\mathbf{M}} \times D \longrightarrow \mathbb{C}$$

$$(m, z) \longmapsto f_{z,j}(m)$$

is also and formula (iii) can be extended by continuity to all $z \in D$. In order to show that $\psi_{P,z}$ is $\mathfrak{z}$-finite, it then suffices to show that for $j \in J$, the function $f_{z,j}$ is $(A_{\mathbf{M}})$-finite (see I.3.1). Fix $j \in J$ and $z \in D$. Let $(z(\ell))_{\ell\in\mathbb{N}}$

be a sequence of elements of D' such that $\lim_{\ell\to\infty} z(\ell) = z$. For all $\ell \in \mathbb{N}$, choose $\lambda_1(\ell),\ldots,\lambda_d(\ell) \in \Gamma'$ and polynomials $P_1(\ell),\ldots,P_d(\ell) \in \mathbb{C}[\mathrm{Re}\,\mathfrak{a}_M]$ such that, for all $m \in A_{\mathbf{M}}$,

$$f_{z(\ell),j}(m) = \sum_{i=1}^{d} m^{\lambda_i(\ell)} P_i(\ell; \log_M m)$$

and $\sum_{i=1}^{d}(1+\deg(P_i(\ell))) \leq d$; see I.4.3. By compactness of Γ', we can suppose, up to extracting a subsequence of $(z(\ell))_{\ell\in\mathbb{N}'}$, that for all i, the sequence $(\lambda_i(\ell))_{\ell\in\mathbb{N}}$ converges to a limit denoted by $\lambda_i \in \Gamma'$. Recall that $\mathfrak{z}(A_{\mathbf{M}})$ can be identified with a polynomial algebra over X_M (see I.3.1). Let I be the ideal of $\mathfrak{z}(A_{\mathbf{M}})$ generated by the products $Q_1,\ldots,Q_d$, where $Q_1 \in \mathfrak{z}(A_{\mathbf{M}})$ vanishes to the order d in λ_i. We easily show that I is of finite codimension in $\mathfrak{z}(A_{\mathbf{M}})$. Let us show that for $Q \in I$, $\delta(Q)f_{z,j} = 0$. We can suppose $Q = Q_1 \ldots Q_d$ as above. For $\ell \in N$, define an element $Q(\ell) \in \mathfrak{z}(A_{\mathbf{M}})$ by

$$Q(\ell;\lambda) = Q_1(\lambda+\lambda_1-\lambda_1(\ell))\ldots Q_d(\lambda+\lambda_d-\lambda_d(\ell))$$

for all $\lambda \in X_M$. These polynomials are of bounded degrees and their coefficients converge to those of Q when ℓ tend towards ∞. From the smoothness of the function $(m,z') \longmapsto f_{z',j}(m)$, we then deduce that for all $m \in A_{\mathbf{M}}$,

$$(\delta(Q)f_{z,j})(m) = \lim_{\ell\to\infty}(\delta(Q(\ell))f_{z(\ell),j})(m).$$

Fix $\ell \in \mathbb{N}$. For all $i = 1,\ldots,d$, the polynomial $\lambda \longmapsto Q_i(\lambda+\lambda_i-\lambda_i(\ell))$ vanishes to the order d in $\lambda_i(\ell)$, and thus kills the function $m \longmapsto m^{\lambda_i(\ell)}P_i(\ell;\log_M m)$ on $A_{\mathbf{M}}$. Thus $\delta(Q(\ell))f_{z(\ell),j} = 0$, so $\delta(Q)f_{z,j} = 0$. This proves that $f_{z,j}$ is $\mathfrak{z}(A_{\mathbf{M}})$-finite, which concludes the proof that $\psi_{P,z}$ is automorphic for all $z \in D(\varepsilon_1)\times\ldots\times D(\varepsilon_n)$.

But then the fact that the cuspidal component of $\phi_{z,P} - \psi_{P,z}$ is zero means that the cuspidal component of ϕ_z along P is automorphic and equal to $\psi_{P,z}$. This proves (b) and then (a) by applying I.3.5. □

I.4.11. Exponents of square integrable automorphic forms

Lemma *Let ξ be a unitary character of $Z_{\mathbf{G}}$ and $\phi \in A(G(k)\backslash\mathbf{G})_\xi$* (see I.2.17). *For every standard parabolic subgroup $P = MU$ of G, let $\Pi_0(\mathbf{M},\phi)$ be the cuspidal support of ϕ along P* (see I.3.5). *Then for ϕ to be square integrable, it is necessary and sufficient that for all $P = MU$ and all $\pi \in \Pi_0(\mathbf{M},\phi)$, the character $\mathrm{Re}\,\pi$ can be written in the form*

$$\mathrm{Re}\,\pi = \sum_{\alpha\in\Delta_M} x_\alpha \alpha,$$

with coefficients $x_\alpha \in \mathbb{R}$, $x_\alpha < 0$.

Remarks (a) As ξ is unitary, we have in any case $\operatorname{Re}\pi \in \operatorname{Re}X_M^{\mathbf{G}}$ for all $\pi \in \Pi_0(\mathbf{M}, \phi)$;

(b) Introduce the basis $\{\varpi_{\check\alpha};\ \alpha \in \Delta_0\}$ of $\operatorname{Re}\mathfrak{a}_{M_0}^G$ which is the dual of the basis Δ_0 of $\operatorname{Re}(\mathfrak{a}_{M_0}^G)^*$. For $\alpha \in \Delta_M$, let $\beta \in \Delta_0$ be the unique simple root which projects onto α; define $\varpi_{\check\alpha}$ by $\varpi_{\check\alpha} = \varpi_{\check\beta}$. We know that $\varpi_{\check\alpha} \in \operatorname{Re}\mathfrak{a}_M^G$ and that $\{\varpi_{\check\alpha};\ \alpha \in \Delta_M\}$ is the basis of $\operatorname{Re}\mathfrak{a}_M^G$ which is the dual of the basis Δ_M of $\operatorname{Re}(\mathfrak{a}_M^G)^*$. Here and in the proof below we identify $\operatorname{Re}X_M^{\mathbf{G}}$ with $\operatorname{Re}(\mathfrak{a}_M^G)^*$. By remark (a), the conclusion of the statement can also be written

$$\langle \varpi_{\check\alpha}, \operatorname{Re}\pi\rangle < 0 \text{ for all } \alpha \in \Delta_M.$$

Proof Let us begin with the sufficiency. By Lemma I.4.1, it suffices to show that there exists $\mu \in \operatorname{Re}X_{M_0}$ such that for all $P = MU$, all $\pi \in \Pi_0(\mathbf{M}, \phi)$ and all integers N, the function on $A_{\mathbf{G}}\backslash S$ given by

$$g \longmapsto m_{P_0}(g)^{\operatorname{Re}\pi+\mu^M+\rho_P}(1 + \|\log_M m_P(g)\|)^N$$

is square integrable. Taking into consideration the form of the measure on S, it suffices to show the existence of μ such that the following integral is convergent:

$$\int a^{2\operatorname{Re}\pi+2\mu^M+2\rho_P}(1 + \|\log_M a\|)^{2N} a^{-2\rho_0}\, da$$

where we integrate over

$$\{a \in A_{\mathbf{G}}\backslash A_{\mathbf{M}_0};\ \forall\alpha \in \Delta_0,\ a^\alpha > t_0^\alpha\},$$

see I.2.1, and where $\rho_0 = \rho_{P_0}$. Let $\log_{M_0}^G$ be the composition of $\log_{M_0}$ with the projection $a_{M_0} \to \mathfrak{a}_{M_0}^G$. For $a \in A_{\mathbf{M}_0}$, let us write

$$\log_{M_0}^G a = \sum_{\alpha\in\Delta_0} y_\alpha \varpi_{\check\alpha}$$

and similarly,

$$\log_{M_0}^G t_0 = \sum_{\alpha\in\Delta_0} t_\alpha \varpi_{\check\alpha},$$

with $y_\alpha,\ t_\alpha \in \mathbb{R}$. The conditions $a^\alpha > t_0^\alpha$ can be interpreted as $y_\alpha > t_\alpha$. For the preceding integral to converge, it thus suffices to check that

$$\text{(1)} \qquad \langle \varpi_{\check\alpha}, \operatorname{Re}\pi + \mu^M + \rho_P - \rho_0\rangle < 0$$

for all $\alpha \in \Delta_0$. If $\alpha \in \Delta_0^M$, we can choose μ such that $\langle \varpi_{\check\alpha}, \mu^M\rangle$ is very negative and the above condition is satisfied. If $\alpha \in \Delta_0 - \Delta_0^M$, we have $\varpi_{\check\alpha} \in a_M^G$, which gives

$$\langle \varpi_{\check\alpha}, \mu^M + \rho_P - \rho_0\rangle = 0$$

and the inequality (1) is a result of the hypothesis on $\operatorname{Re}\pi$.

Let us prove the necessity. Let $P = MU \not\subseteq G$ (if $P = G$, there is nothing to prove) and $\alpha \in \Delta_M$. Let $P' = M'U'$ be the maximal proper parabolic subgroup of G such that $P \subset P'$, associated with the root α (i.e. α is the unique root of Δ_M which is not zero on $\mathfrak{a}_{M'}$). Let $\alpha_0 \in \Delta_0$ be the unique element which is projected onto $\alpha \in \Delta_M$. Set

$$S(\alpha_0) = \{g \in S; m_{P_0}(g)^{\alpha_0} > \sup\{m_{P_0}(g)^{\beta};\ \beta \in \Delta_0 - \{\alpha_0\}\}.$$

It is a consequence of Lemmas I.2.7 and I.2.10 that the function $\phi - \phi_{P'}$ is square integrable on $A_{\mathbf{G}}\backslash S(\alpha_0)$. The function ϕ being supposed square integrable, $\phi_{P'}$ is thus itself square integrable on $A_{\mathbf{G}}\backslash S(\alpha_0)$. We will deduce from this that:

(2) there exists a subset $S_n \subset S$ of measure zero, and for all $g \in S - S_n$, there exists $c_g > 0$ such that the function $a \longmapsto a^{-\rho_{P'}}\phi_{P'}(ag)$ is square integrable on $A_{\mathbf{G}}\backslash\{a \in A_{\mathbf{M}'};\ a^{\alpha} > c_g\}$.

Recall (see I.2.1) that $S = \{pak;\ p \in \omega,\ k \in \mathbf{K},\ a \in A_{\mathbf{M}_0}(t_0)\}$. The invariant measure is given on $A_{\mathbf{G}}\backslash S$ by the formula

$$\int_S f(g)\, dg = \int_{\omega \times A_{\mathbf{G}}\backslash A_{\mathbf{M}_0}(t_0) \times \mathbf{K}} f(pak) a^{-2\rho_0}\, dk\, da\, dp$$

for every function f which is, say, continuous with compact support on S, d dp being a left invariant measure. On the other hand, set $A_{\mathbf{M}_0}^{\mathbf{M}'} = A_{\mathbf{M}_0} \cap \mathbf{M}'^1$. Then $A_{\mathbf{M}'}A_{\mathbf{M}_0}^{\mathbf{M}'}$ is of finite index in $A_{\mathbf{M}_0}$ (and is equal to it if k is a number field). Fix a set B of representatives of the quotient. Finally, denote by ε_α and ε_{t_0} the characteristic functions of $A_{\mathbf{G}}\backslash S(\alpha_0)$ and $A_{\mathbf{G}}\backslash A_{\mathbf{M}_0}(t_0)$ respectively. Then the square integrability of $\phi_{P'}$ over $A_{\mathbf{G}}\backslash S(\alpha_0)$ signifies that the following integral is convergent:

$$\int_{\omega \times A_{\mathbf{G}}\backslash A_{\mathbf{M}'} \times A_{\mathbf{M}_0}^{\mathbf{M}'} \times \mathbf{K}} \sum_{b \in B} |\phi_{P'}(paa'bk)|^2 (aa'b)^{-2\rho_0} \times$$

$$\varepsilon_\alpha(paa'bk)\varepsilon_{t_0}(aa'b)\, dk\, da'\, da\, dp.$$

By Fubini's theorem, there exists a set $S_n' \subset \omega \times A_{\mathbf{M}_0}^{\mathbf{M}'} \times B \times \mathbf{K}$ of measure zero, such that if $(p, a', b, k) \notin S_n'$, the following integral is convergent:

$$\int_{A_{\mathbf{G}}\backslash A_{\mathbf{M}'}} |\phi_{P'}(paa'bk)|^2 (aa'b)^{-2\rho_0} \varepsilon_\alpha(paa'bk)\varepsilon_{t_0}(aa'b)\, da.$$

Fix $(p, a', b, k) \notin S_n'$ such that the function $a \longmapsto \varepsilon_{t_0}(aa'b)$ on $A_{\mathbf{M}}$ is not identically zero. Note that one can write $p = mu$ with $m \in \mathbf{M}'$, $u \in U'(\mathbb{A})$. As $\phi_{P'}$ is left invariant under $U'(\mathbb{A})$ and $A_{\mathbf{M}'}$ commutes with $\mathbf{M}'$ and normalises $U'(\mathbb{A})$, we obtain

$$\phi_{P'}(paa'bk) = \phi_{P'}(apa'bk).$$

On the other hand, we easily see that there exists $c > 0$ such that

$$A_{\mathbf{G}} \backslash \{paa'bk;\ a \in A_{\mathbf{M}'},\ a^{\alpha} > c\} \subset S(\alpha_0).$$

Finally for $a \in A_{\mathbf{G}} \backslash A_{\mathbf{M}'}$, we have $a^{\rho_0} = a^{\rho_{P'}}$. So the following integral is convergent:

$$\int_{A_{\mathbf{G}} \backslash \{a \in A_{\mathbf{M}'};\ a^{\alpha} > c\}} |a^{-\rho_{P'}} \phi_{P'}(apa'bk)|^2 \, da.$$

This proves (2) for $g = pa'bk$. The assertion can be immediately generalised to $g \in S - S_n$, where we set

$$S - S_n = S \cap \{paa'bk;\ a \in A_{\mathbf{M}'},\ (p, a', b, k) \notin S'_n\}.$$

Fix $g \in S - S_n$ and choose a finite set $\wedge'_g \subset \mathrm{Hom}(A_{\mathbf{M}'}, \mathbb{C}^*) \simeq X_{M'}/X_{M'}(A_{\mathbf{M}'})$ (see I.3.1) and for all $\lambda \in \wedge'_g$ a non-zero polynomial $Q_\lambda \in \mathbb{C}[\mathrm{Re}\, \mathfrak{a}_{M'}^G]$ such that we have

$$\phi_{P'}(ag) = \sum_{\lambda \in \wedge'_g} a^{\lambda + \rho_{P'}} Q_\lambda(\log_{M'}^G a)$$

for all $a \in A_{\mathbf{M}'}$ (see I.3.1; the condition $Q_\lambda \in \mathbb{C}[\mathrm{Re}\, \mathfrak{a}_{M'}^G]$ rather than $Q_\lambda \in \mathbb{C}[\mathrm{Re}\, \mathfrak{a}_{M'}]$ is possible since we supposed that ϕ had a central character; $\log_{M'}^G$ is of course the composition of $\log_{M'}$ with the projection $\mathfrak{a}_{M'} \to \mathfrak{a}_{M'}^G$). Note that $\log_{M'}^G a$ is colinear to $\varpi_{\check{\alpha}}$ for all $a \in A_{\mathbf{M}'}$. It is then an integration exercise over $A_{\mathbf{G}} \backslash A_{\mathbf{M}'}$ (which is isomorphic to $\mathbb{R}$ or to $\mathbb{Z} \times$a finite group, according to whether k is a number field or a function field) to show that the integrability of $|\phi_{P'}(ag)|^2$ implies $\langle \varpi_{\check{\alpha}}, \mathrm{Re}\, \lambda \rangle < 0$ for all $\lambda \in \wedge'_g$; recall $\mathrm{Re}\, \lambda := \mathrm{Re}\, \lambda'$ for any $\lambda' \in X_{M'}$ whose image in $X_{M'}/X_{M'}(A_{\mathbf{M}'})$ is λ).

An argument analogous to that of (b) in the proof of Lemma I.4.1 shows that if $\pi \in \Pi_0(\mathbf{M}, \phi)$, there exists $g \in S - S_n$ such that $\chi_{\pi|_{A'_{\mathbf{M}}}} \in \wedge'_g$. As $\mathrm{Re}(\chi_{\pi|_{A'_{\mathbf{M}}}})$ is the projection of $\mathrm{Re}\, \pi$ onto $\mathrm{Re}\, X_{M'}$, the preceding result implies that $\langle \varpi_{\check{\alpha}}, \mathrm{Re}\, \pi \rangle < 0$. This holds for all $\pi \in \Pi_0(\mathbf{M}, \phi)$. □

II

Decomposition According to Cuspidal Data

II.1. Definitions

II.1.1. Equivalence classes of irreducible subrepresentations of the space of automorphic forms and complex analytic structure
Let M be a standard Levi subgroup of G. We denote by $\mathbf{M}$ the inverse image of $M(\mathbb{A})$ in $\mathbf{G}$ (see I.1.4) and by P a parabolic standard subgroup of Levi subgroup M; we denote by U the unipotent radical of P. In I.2.17 we defined spaces $A(M(k)\backslash\mathbf{M})$ and $A(U(\mathbb{A})M(k)\backslash\mathbf{G})$ of automorphic forms; the first is a $\mathbf{M}_f \times ((\mathbf{M}_\infty \cap \mathbf{K}), \operatorname{Lie}\mathbf{M}_\infty \otimes_{\mathbb{R}} \mathbb{C})$-module. The second is a $\mathbf{G}_f \times (\mathbf{K}, \operatorname{Lie}\mathbf{G}_\infty \otimes_{\mathbb{R}} \mathbb{C})$-module. Let V be an irreducible submodule in $A(M(k)\backslash\mathbf{M})$. We denote by π_0 the representation of $\mathbf{M}_f \times ((\mathbf{M} \cap \mathbf{K}), \operatorname{Lie}\mathbf{M}_\infty \otimes_{\mathbb{R}} \mathbb{C})$ into V. We say that π_0 is an automorphic subrepresentation of $\mathbf{M}$. (In general an automorphic representation of $\mathbf{M}$ is the isomorphism class of an irreducible subquotient of $A(M(k)\backslash\mathbf{M})$.) We denote by $A(M(k)\backslash\mathbf{M})_{\pi_0}$ the isotypic submodule π_0 , i.e. the set of automorphic forms of $\mathbf{M}$ generating a semi-simple isotypic $\mathbf{M}_f \times ((\mathbf{M}\cap\mathbf{K}), \operatorname{Lie}\mathbf{M}_\infty \otimes_{\mathbb{R}} \mathbb{C})$-module of type π_0. There is a canonical isomorphism:

$$V \otimes \operatorname{Hom}_{\mathbf{M}_f \otimes ((\mathbf{M}\cap\mathbf{K}),\ \operatorname{Lie}\mathbf{M}_\infty \otimes_{\mathbb{R}} \mathbb{C})}(V, A(M(k)\backslash\mathbf{M})) \to A(M(k)\backslash\mathbf{M})_{\pi_0}.$$

In the notation ϕ_k of I.2.17, we set:

$$A(U(\mathbb{A})M(k)\backslash\mathbf{G})_{\pi_0} := \{\phi \in A(U(\mathbb{A})M(k)\backslash\mathbf{G}) | \forall k \in \mathbf{K}, \phi_k \in A(M(k)\backslash\mathbf{M})_{\pi_0}\}.$$

This space can be canonically identified with

$$\mathrm{ind}_{\mathbf{M}\cap\mathbf{K}}^{\mathbf{K}} A(M(k)\backslash\mathbf{M})_{\pi_0}.$$

Suppose that π_0 is cuspidal, i.e. that $A(M(k)\backslash\mathbf{M})_{\pi_0}$ contains cuspidal automorphic forms. In general, we do not know if $A(M(k)\backslash\mathbf{M})_{\pi_0}$ is then contained in the set of cuspidal automorphic forms, i.e. in $A_0(M(k)\backslash\mathbf{M})$; but it will be easier for us to use the following convention:

if π_0 is explicitly supposed to be cuspidal, then by convention, $A(M(k)\backslash\mathbf{M})_{\pi_0}$ is the isotypic subspace of $A_0(M(k)\backslash\mathbf{M})$ of type π_0. We will use the same convention for $A(U(\mathbb{A})M(k)\backslash\mathbf{G})_{\pi_0}$.

Let π be an irreducible automorphic representation of $\mathbf{M}$ and let π_0 be as above; we say that π_0 is equivalent to π if there exists $\lambda \in X_M^{\mathbf{G}}$ such that $\pi \simeq \pi_0 \otimes \lambda$. Note that if π satisfies this isomorphism then π can also be realised as a submodule of the set of automorphic forms of $\mathbf{M}$. More precisely, we have:

(1) $$A(M(k)\backslash\mathbf{M})_{\pi} = \lambda A(M(k)\backslash\mathbf{M})_{\pi_0},$$

where λ is considered as a function on $\mathbf{M}$, and

(2) $$A(U(\mathbb{A})M(k)\backslash\mathbf{G})_{\pi} = \lambda \circ m_P A(U(\mathbb{A})M(k)\backslash\mathbf{G})_{\pi_0}.$$

We denote by $\mathfrak{P}$ the equivalence class of π_0 and we note that the subgroup of $X_M^{\mathbf{G}}$ given by

(3) $$\{\nu \in X_M^{\mathbf{G}} | \nu \otimes \pi \simeq \pi\}$$

is independent of the chosen point π of $\mathfrak{P}$. We denote it by $\mathrm{Fix}_{X_M^{\mathbf{G}}}\mathfrak{P}$. Clearly $\mathrm{Fix}_{X_M^{\mathbf{G}}}\mathfrak{P}$ is contained in $X_M^{\mathbf{M}}$ and is a finite group. It is trivial if k is a number field. The group $X_M^{\mathbf{G}}/\mathrm{Fix}_{X_M^{\mathbf{G}}}\mathfrak{P}$ operates freely and transitively on $\mathfrak{P}$; in particular $\mathfrak{P}$ is a principal homogeneous space under $X_M^{\mathbf{G}}/\mathrm{Fix}_{X_M^{\mathbf{G}}}\mathfrak{P}$. In order to put a complex analytic manifold structure on $\mathfrak{P}$, it thus suffices to put one on $X_M^{\mathbf{G}}/\mathrm{Fix}_{X_M^{\mathbf{G}}}\mathfrak{P}$ for which the natural operation of $X_M^{\mathbf{G}}$ is analytic. It suffices in fact to put one on $X_M^{\mathbf{G}}$ having the same property. We use the realisation of X_M as a quotient of $\mathfrak{a}_M^*$ by a lattice of the imaginary part (see I.1.4); $\mathfrak{a}_M^*$ is a complex vector space and we use the analytic complex structure coming from it. We obtain the structure on X_M by passing to the quotient. Every element $m \in \mathbf{M}$ defines an analytic function on X_M by $\epsilon_m(\lambda) := m^\lambda - 1$; $X_M^{\mathbf{G}}$ is the set of zeros of these functions when m is in the set $Z_{\mathbf{G}}$. Thus $X_M^{\mathbf{G}}$ is a closed analytic set of X_M and all the desired properties are obvious. This gives the complex analytic manifold structure of $\mathfrak{P}$ which we use from now on.

Notation A pair $(M, \mathfrak{P})$, where $\mathfrak{P}$ consists of cuspidal representations as above, will be called a cuspidal datum. We will sometimes need to specify the character of $Z_{\mathbf{G}}$ corresponding to $\mathfrak{P}$. Namely, let ξ be the character of $Z_{\mathbf{G}}$ such that the restriction of the central character of every element of $\mathfrak{P}$ to $Z_{\mathbf{G}}$ is ξ; we say that $(M, \mathfrak{P})$ is a cuspidal datum of central character ξ.

Remark Let ξ be a unitary character of $Z_{\mathbf{G}}$ and let $(M, \mathfrak{P})$ be a cuspidal datum of central character ξ. For all $\pi \in \mathfrak{P}$, we defined $\operatorname{Re}\pi$ in I.3.3. Under our hypothesis, $\operatorname{Re}\pi \in \operatorname{Re}X_M^{\mathbf{G}}$, and $\operatorname{Im}\pi$, also defined in I.3.3, belongs to $\mathfrak{P}$, which means that $-\overline{\pi} \in \mathfrak{P}$, for all $\pi \in \mathfrak{P}$. On the other hand the set of elements of $\mathfrak{P}$ for which the central character is unitary is non-empty – it is a homogeneous space under $\operatorname{Im}X_M^{\mathbf{G}}$.

II.1.2. Paley-Wiener functions

Fix $M, \mathbf{M}$ and $\mathfrak{P}$ as in II.1.1. Let ϕ be a function on $\mathfrak{P}$ with the properties:

(1) $$\forall \pi \in \mathfrak{P}, \phi(\pi) \in A(U(\mathbb{A})M(k)\backslash \mathbf{G})_\pi,$$

(2) $$\phi \text{ is } \mathbf{K}-\text{finite}.$$

Fix $\pi_0 \in \mathfrak{P}$. We associate with a function ϕ, satisfying the above properties, a function $\tilde{\phi}$ on $X_M^{\mathbf{G}}$ with values in $A(U(\mathbb{A})M(k)\backslash \mathbf{G})_{\pi_0}$ defined by:

$$\tilde{\phi}(\lambda) = \lambda^{-1}\phi(\pi_0 \otimes \lambda).$$

This function satisfies:

(3) $$\forall \mu \in \operatorname{Fix}_{X_M^{\mathbf{G}}}\mathfrak{P}, \quad \phi(\lambda + \mu) = \mu^{-1}\phi(\lambda)$$

(4) the vector subspace of $\big(A(U(\mathbb{A})M(k)\backslash \mathbf{G})\big)_{\pi_0}$ generated by the set of elements $\phi(\lambda)$ for λ in the set $X_M^{\mathbf{G}}$ is finite-dimensional.

Clearly (1) is equivalent to (3) and (2) is equivalent to (4). Conversely, with every function $\tilde{\phi}$ on $X_M^{\mathbf{G}}$ with values in $A(U(\mathbb{A})M(k)\backslash \mathbf{G})_{\pi_0}$ satisfying (3) and (4), we associate a function ϕ on $\mathfrak{P}$ satisfying (1) and (2) by the formula:

$$\phi(\pi) = \mu_\pi \tilde{\phi}(\mu_\pi)$$

where μ_π is an element of $X_M^{\mathbf{G}}$ satisfying $\pi \simeq \pi_0 \otimes \mu_\pi$.

With this double point of view, it is easy to introduce functions which are continuous, differentiable, holomorphic and so on. We explain here the definition of a Paley-Wiener function as it will be important later on.

Denote by $P(X_M^{\mathbf{G}})$ the set of complex-valued functions on $X_M^{\mathbf{G}}$ which are Paley-Wiener, i.e. which are Fourier transforms of functions on $\mathbf{M}^1 Z_{\mathbf{G}} \backslash \mathbf{M}$, C^∞ with compact support, or else $f \in P(X_M^{\mathbf{G}})$ if for λ_0 fixed in $\operatorname{Re} X_M^{\mathbf{G}}$, the function:

$$\hat{f}(m) := \int_{\substack{\lambda \in X_M^{\mathbf{G}} \\ \operatorname{Re}\lambda = \lambda_0}} |\operatorname{Fix}_{X_M^{\mathbf{G}}} \mathfrak{P}|^{-1} f(\lambda) m^\lambda \, d\lambda$$

is a function on $\mathbf{M}^1 Z_{\mathbf{G}} \backslash \mathbf{M}$ with compact support (it does not depend on the choice of λ_0). We then have

$$f(\lambda) = * \int_{\mathbf{M}^1 Z_{\mathbf{G}} \backslash \mathbf{M}} \hat{f}(m) m^{-\lambda} \, dm$$

where $*$ is a scalar which depends on the choice of measure. Clearly f is then holomorphic on $X_M^{\mathbf{G}}$. Let us return to the ϕ considered above, satisfying (1) and (2). We say that ϕ is Paley-Wiener if $\tilde{\phi}$ can be identified with an element of

$$A(U(\mathbb{A})M(k)\backslash \mathbf{G})_{\pi_0} \otimes_{\mathbb{C}} P(X_M^{\mathbf{G}}).$$

We denote by $P_{(M,\mathfrak{P})}$ the set of Paley-Wiener functions. The group $\operatorname{Fix}_{X_M^{\mathbf{G}}} \mathfrak{P}$ acts on

$$A(U(\mathbb{A})M(k)\backslash \mathbf{G})_{\pi_0} \otimes_{\mathbb{C}} P(X_M^{\mathbf{G}})$$

by the tensor product of multiplication by the functions $\mu \circ m_P$ (for $\mu \in \operatorname{Fix}_{X_M^{\mathbf{G}}} \mathfrak{P}$) with the translation T_μ defined by the elements of $X_M^{\mathbf{G}}$, where $T_\mu f(\lambda) = f(\lambda + \mu)$. We use an exponent to denote the space of invariants under this action and we have a natural isomorphism:

$$(5) \qquad P_{(M,\mathfrak{P})} \xleftarrow{\sim} \left(A(U(\mathbb{A})M(k)\backslash \mathbf{G})_{\pi_0} \otimes_{\mathbb{C}} P(X_M^{\mathbf{G}}) \right)^{\operatorname{Fix}_{X_M^{\mathbf{G}}} \mathfrak{P}}.$$

We will characterise $P_{(M,\mathfrak{P})}$ via Fourier transforms.

II.1.3. Fourier transforms

Let $\phi \in P_{(M,\mathfrak{P})}$ and $\pi \in \mathfrak{P}$. We define the Fourier transform of ϕ; the convergence of the integral will be considered below, see just before (5).

$\forall g \in \mathbf{G}$, where $g = umk$ with $u \in U(\mathbb{A}), m \in \mathbf{M}, k \in \mathbf{K}$ and $\forall m' \in \mathbf{M}$,

$$F_\pi(\phi)(g)(m') := |\operatorname{Fix}_{X_M^{\mathbf{G}}} \mathfrak{P}|^{-1} \int_{\lambda \in \operatorname{Im} X_M^{\mathbf{G}}} (m')^{-\lambda - \rho_P} \phi(\pi \otimes \lambda)(\pi \otimes \lambda)(m' mk) \, d\lambda.$$

If k is a function field, then $F_\pi(\phi)(g) \in A(M(k)\backslash \mathbf{M}))_\pi$ for all $g \in \mathbf{G}$. If k is a number field, the property of $\mathbf{K} \cap \mathbf{M}$-finiteness is lost, for arbitrary g in $\mathbf{G}$, at the archimedean places. Denote by $\tilde{A}(M(k)\backslash \mathbf{M})_\pi$ the vector space generated by the translations under M of elements of $A(M(k)\backslash\mathbf{M})_\pi$ (we will use this notation very rarely). Then:

$$F_\pi(\phi)(g) \in \tilde{A}(M(k)\backslash\mathbf{M})_\pi, \forall g \in \mathbf{G}.$$

We use δ to denote right translation in $\tilde{A}(M(k)\backslash\mathbf{M})_\pi$ and we have:

(1) $\forall m \in Z_{\mathbf{G}}\mathbf{M}^1,\ \forall u \in U(\mathbb{A}),\ \forall g \in \mathbf{G},\ \ F_\pi(\phi)(umg) = \delta(m)F_\pi(\phi)(g).$

This depends only weakly on π, since we have:

(2) $$\forall \lambda \in X_M^{\mathbf{G}},\ \ F_{(\pi\otimes\lambda)}(\phi)(g) = \lambda F_\pi(\phi)(g).$$

In particular, for all $\mu \in \operatorname{Fix}_{X_M^{\mathbf{G}}}\mathfrak{P}$, we have $F_\pi(\phi)(g) = \mu F_\pi(\phi)(g)$ i.e.

(3) the support of $F_\pi(\phi)(g)$ is in the subgroup of $\mathbf{M}$ given by $\bigcap_{\mu\in\operatorname{Fix}_{X_M^{\mathbf{G}}}\mathfrak{P}} \operatorname{Ker}\mu$.

We denote by ϵ the evaluation of elements of $\tilde{A}(M(k)\backslash\mathbf{M})_\pi$ at the unit element of $\mathbf{M}$. We will mainly be interested in functions of the form $\epsilon F_\pi(\phi)$ which do not depend on the choice of π and are thus written $\epsilon F(\phi)$; we will characterize these functions below.

We define

$$\left(A(U(\mathbb{A})M(k)\backslash\mathbf{G})_\pi \otimes_{\mathbb{C}} P(X_M^{\mathbf{G}})\right)^{\operatorname{Fix}_{X_M^{\mathbf{G}}}\mathfrak{P}}$$

as in II.1.2(5); here π replaces π_0. We let $\operatorname{Fix}_{X_M^{\mathbf{G}}}\mathfrak{P}$ act on

$$A(U(\mathbb{A})M(k)\backslash\mathbf{G})_\pi \otimes C_c^\infty(M^1Z_{\mathbf{G}}\backslash\mathbf{M})$$

in such a way that for all

$$\mu \in \operatorname{Fix}_{X_M^{\mathbf{G}}}\mathfrak{P},\ \psi \in A(U(\mathbb{A})M(k)\backslash\mathbf{G})_\pi$$

and $f \in C_c^\infty(\mathbf{M}^1Z_{\mathbf{G}}\backslash\mathbf{M})$, we have

$$\mu(\psi \otimes f) = (\mu \circ m_P)\psi \otimes \mu^{-1}f.$$

We put $\operatorname{Fix}_{X_M^{\mathbf{G}}}\mathfrak{P}$ as an exponent to denote the space of elements fixed under this action and we also write $\hat{P}_{(M,\mathfrak{P})}$ for the image of $P_{(M,\mathfrak{P})}$ under the Fourier transform. We define a map v from:

$$\left(A(U(\mathbb{A})M(k)\backslash\mathbf{G})_\pi \otimes C_c^\infty(M^1Z_{\mathbf{G}}\backslash\mathbf{M})\right)^{\operatorname{Fix}_{X_M^{\mathbf{G}}}\mathfrak{P}}$$

into the set of functions on $\mathbf{G}$ with values in $\tilde{A}(M(k)\backslash\mathbf{M})_\pi$ by

(4) $$v\left(\sum \psi_i \otimes f_i\right)(g)(m') = \sum f_i(m_P(g))\psi_i(m'g)(m')^{-\rho_P}.$$

It is elementary to verify that we have a commutative diagram whose arrows are as follows: the first horizontal arrow is that of II.1.2: $\phi \leftrightarrow \tilde{\phi}$, the first vertical arrow is $id \otimes F$ where F is the Fourier transform, the second vertical arrow is the transformation F_π defined above (the convergence of the integral is a consequence of the diagram) and the second horizontal arrow is ν:

$$(5)\qquad \begin{array}{ccc} \left(A(U(\mathbb{A})M(k)\backslash\mathbf{G})_\pi \otimes_{\mathbb{C}} P(X^{\mathbf{G}}_M)\right)^{\mathrm{Fix}_{X^{\mathbf{G}}_M}\mathfrak{P}} & \rightleftarrows & P_{(M,\mathfrak{P})} \\ \downarrow\wr & & \downarrow\wr \\ \left(A(U(\mathbb{A})M(k)\backslash\mathbf{G})_\pi \otimes_{\mathbb{C}} C^\infty_c(\mathbf{M}^1 Z_{\mathbf{G}}\backslash\mathbf{M})\right)^{\mathrm{Fix}_{X^{\mathbf{G}}_M}\mathfrak{P}} & \rightleftarrows & \hat{P}_{(M,\mathfrak{P})}. \end{array}$$

Thanks to this diagram, we obtain the following characterisation of $\epsilon\hat{P}_{(M,\mathfrak{P})}$:

$$(6)\quad \epsilon\hat{P}_{(M,\mathfrak{P})} = \left\{\begin{array}{l} \mathbf{K}\text{-finite complex-valued functions } \Phi \text{ on} \\ U(\mathbb{A})M(k)\backslash\mathbf{G} \text{ such that } \forall\pi\in\mathfrak{P}, \text{there exists a set} \\ \text{of functions} \{f_i\} \subset C^\infty_c(\mathbf{M}^1 Z_{\mathbf{G}}\backslash\mathbf{M}) \text{ and a} \\ \text{set of functions } \{\psi_i\} \subset A(U(\mathbb{A})M(k)\backslash\mathbf{G})_\pi \\ \text{indexed by the same finite set of indices,} \\ \text{such that } \Phi = \sum_i (f_i \circ m_P)\psi_i. \end{array}\right\}$$

Indeed, the inclusion of the left-hand side in the right-hand side is a consequence of the diagram and it is clear that the right-hand side is given by

$$\epsilon \circ \nu(A(U(\mathbb{A})M(k)\backslash\mathbf{G})_\pi \otimes C^\infty_c(\mathbf{M}^1 Z_{\mathbf{G}}\backslash\mathbf{M}))$$

where π is any fixed element of $\mathfrak{P}$. We denote by proj the projection of

$$A(U(\mathbb{A})M(k)\backslash\mathbf{G})_\pi \otimes C^\infty_c(\mathbf{M}^1 Z_{\mathbf{G}}\backslash\mathbf{M})$$

onto the space of invariants under $\mathrm{Fix}_{X^{\mathbf{G}}_M}\mathfrak{P}$. The assertion is then a consequence of:

$$\nu \circ \mathrm{proj} = \nu.$$

Before concluding, we note that on the one hand it is very easy to write the inverse Fourier transform:

> let $\Phi \in \hat{P}_{(M,\mathfrak{P})}$; then Φ is the Fourier transform of the following function ϕ:
>
> $$\forall k \in \mathbf{K}, \forall\pi' \in \mathfrak{P},\quad \phi(\pi')_k = \mu_{\pi'} \int_{\mathbf{M}^1 Z_{\mathbf{G}}\backslash\mathbf{M}} (\delta(m)^{-1}\Phi(mk))m^{-(\rho_P+\mu_{\pi'})}\,dm$$
>
> where δ denotes translation in $\tilde{A}(M(k)\backslash\mathbf{M})_\pi$ as before and $\mu_{\pi'}$ is any element of $X^{\mathbf{G}}_M$ satisfying $\mu_{\pi'} \otimes \pi \simeq \pi'$.

On the other hand, for all ϕ as above, all $\pi_0 \in \mathfrak{P}$ and all $g \in \mathbf{G}$:

$$\epsilon F(\phi)(g) = \int_{\pi \in \mathfrak{P}, \mathrm{Re}\,\pi = \mathrm{Re}\,\pi_0} \phi(\pi)(g)\, d\pi.$$

II.1.4. The spaces $P^R_{(M,\mathfrak{P})}$

If k is a number field, $P(X^{\mathbf{G}}_M)$ is the space of complex-valued holomorphic functions f on $X^{\mathbf{G}}_M$ satisfying:

$$\exists r > 0 \text{ such that } \forall n \in \mathbf{N}, \quad \sup_{\lambda \in X^{\mathbf{G}}_M} |f(\lambda)| e^{-r\|\mathrm{Re}\,\lambda\|}(1 + \|\lambda\|)^n < \infty.$$

If k is a function field, $P(X^{\mathbf{G}}_M)$ is simply the space of linear combinations of functions of the form $\lambda \mapsto m^\lambda$ where m is an element of $\mathbf{M}$, or even more simply, it is the algebra of polynomial functions on $X^{\mathbf{G}}_M$.

For some of the proofs, we need to consider larger spaces. Let R be a positive real number; denote by $P^R(X^{\mathbf{G}}_M)$ the set of holomorphic functions f onto $D_R := \{\lambda \in X^{\mathbf{G}}_M \mid \|\mathrm{Re}\,\lambda\| < R\}$ satisfying:

if k is a number field, then for all $n \in \mathbf{N}$,

$$\sup_{\lambda \in D_R} |f(\lambda)|(1 + \|\lambda\|)^n < \infty.$$

if k is a function field, then f is bounded on D_R.

For $\pi \in \mathfrak{P}$, we denote by $(\mathrm{Re}\,\pi)^G$ the projection of $\mathrm{Re}\,\pi$ onto $\mathrm{Re}\,X^{\mathbf{G}}_M$; see I.1.6(9). Considering the isomorphism $\phi \mapsto \tilde{\phi}$ of II.1.2, we define $P^R_{(M,\mathfrak{P})}$ to be the set of holomorphic functions on

$$\mathfrak{P}_R := \{\pi \in \mathfrak{P} \mid \|(\mathrm{Re}\,\pi)^G\| < R\}$$

which can be identified with an element of

$$\left(A(U(\mathbb{A})M(k)\backslash\mathbf{G})_{\pi_0} \otimes P^R(X^{\mathbf{G}}_M)\right)^{\mathrm{Fix}_{X^{\mathbf{G}}_M}\mathfrak{P}}$$

where π_0 is fixed in $\mathfrak{P}$ and satisfies $(\mathrm{Re}\,\pi_0)^G = 0$.

For $\phi \in P^R_{(M,\mathfrak{P})}$ and for all $\pi \in \mathfrak{P}_R$, we define $F_\pi(\phi)$, as in II.1.3. Denote by $\tilde{P}^R(X^{\mathbf{G}}_M)$ the space of C^∞ functions $\tilde{f}$ on $\mathbf{M}^1 Z_{\mathbf{G}}\backslash\mathbf{M}$ satisfying:

$$\exists\, c > 0 \text{ such that } \forall m \in \mathbf{M}, |\tilde{f}(m)| \leq c \exp(-R)\|\log^{\mathbf{G}}_M m\|.$$

We define as in II.1.3 a map ν of

$$\left(A(U(\mathbb{A})M(k)\backslash\mathbf{G})_{\pi_0} \otimes \tilde{P}^R(X^{\mathbf{G}}_M)\right)^{\mathrm{Fix}_{X^{\mathbf{G}}_M}\mathfrak{P}}$$

into the set of functions $\Phi : \mathbf{G} \to \tilde{A}(M(k)\backslash\mathbf{M})_{\pi_0}$ (see II.1.3 for the notation).

Lemma *Denote by* $\hat{P}^R_{(M,\mathfrak{P})}$ *the space of Fourier transforms* $P^R_{(M,\mathfrak{P})}$. *Then* $\hat{P}^R_{(M,\mathfrak{P})}$ *is contained in the image of the map* ν *defined above.*

We have a commutative diagram analogous to II.1.3(5), by adding Rs as exponents and replacing $C_c^\infty(\mathbf{M}^1 Z_{\mathbf{G}}\backslash\mathbf{M})$ by $\hat{P}^R(X_M^{\mathbf{G}})$, which is, by definition, the image of $P^R(X_M^{\mathbf{G}})$ under the Fourier transform. It thus suffices to prove the inclusion $\hat{P}^R(X_M^{\mathbf{G}}) \subset \tilde{P}^R(X_M^{\mathbf{G}})$. Let $f \in P^R(X_M^{\mathbf{G}})$. Its Fourier transform $\hat{f}$ is defined by:

$$\hat{f}(m) = \int_{\lambda \in X_M^{\mathbf{G}} | \operatorname{Re}\lambda = \lambda_0} m^\lambda f(\lambda)\, \mathrm{d}\lambda,$$

where here $\lambda_0 \in \operatorname{Re} X_M^{\mathbf{G}}$ satisfies $\|\lambda_0\| < R$. The conditions on the decreasing of f imply that $\hat{f}$ is C^∞. Fix m. We have:

$$|\hat{f}(m)| \leq m^{\lambda_0} \int_{\lambda \in X_M^{\mathbf{G}} | \operatorname{Re}\lambda = \lambda_0} |f(\lambda)|\, \mathrm{d}\lambda.$$

This last integral is bounded above independently of λ_0. Thus:

$$|\hat{f}(m)| \leq c m^{\lambda_0},$$

for some constant c. Let us identify $\operatorname{Re}\mathfrak{a}_M^{\mathbf{G}}$ and $\operatorname{Re}\mathfrak{a}_M^{\mathbf{G}*}$ with $\operatorname{Re} X_M^{\mathbf{G}}$. Choose $\lambda_0 := -R(1-\epsilon)\|\log_M^{\mathbf{G}} m\|^{-1}$, with $\epsilon > 0$. We have the equality:

$$m^{\lambda_0} = \exp(\langle \lambda_0, \log_M^{\mathbf{G}} m\rangle) = \exp(-R(1-\epsilon)\|\log_M^{\mathbf{G}} m\|),$$

so:

$$|\hat{f}(m)| \leq c \exp(-R(1-\epsilon)\|\log_M^{\mathbf{G}} m\|).$$

Letting ϵ tend to 0, we obtain the upper bound which ensures that $\hat{f} \in \tilde{P}^R(X_M^{\mathbf{G}})$.

II.1.5. Eisenstein series

Let π be an irreducible automorphic representation of M; for example, with the above notation, $\pi \in \mathfrak{P}$. Let ϕ_π be an element of $A(U(\mathbb{A})M(k)\backslash\mathbf{G})_\pi$. Whenever the series below converges, we define a function on $G(k)\backslash\mathbf{G}$ by:

$$E(\phi_\pi, \pi)(g) = \sum_{\gamma \in P(k)\backslash G(k)} \phi_\pi(\gamma g).$$

For all $\lambda \in X_M^{\mathbf{G}}$, we denote by $\lambda\phi_\pi$ the element $\lambda \circ m_P \phi_\pi$ in

$$A(U(\mathbb{A})M(k)\backslash\mathbf{G})_{\pi\otimes\lambda}.$$

Proposition *There exists a (positive) open cone of* $\operatorname{Re} X_M^{\mathbf{G}}$ *such that for all* π *whose real part is in this cone, the series defining* $E(\lambda\phi_\pi, \pi \otimes \lambda)(g)$ *converges normally for* g *in a compact set and* λ *in a neighborhood of* 0 *in $X_M^{\mathbf{G}}$. It then defines an automorphic form on $G(k)\backslash\mathbf{G}$. Suppose that $\mathfrak{P}$ consists of cuspidal representations. Then the following cone works:*

$$\{\lambda \in \mathrm{Re}X_M^{\mathbf{G}} | \langle\lambda, \check{\alpha}\rangle > \langle\rho_P, \check{\alpha}\rangle \forall\alpha \in \Delta^+(T_M, G)\}.$$

If we suppose that $\mathfrak{P}$ consists of cuspidal representations, in particular for all $\mathfrak{P}$ if $M = M_0$, the proposition is proved in [G] §3: Godement considers the number field case, but the function field case is analogous. We do not make this cuspidality hypothesis, but nevertheless use the same arguments as Godement. Let us write ρ_0 instead of ρ_{P_0}. Fix $\mu \in \mathrm{Re}X_{M_0}^{\mathbf{M}}$ such that $\langle\mu, \check{\alpha}\rangle > \langle\rho_0, \check{\alpha}\rangle$ for any simple positive root α of M_0 in M. Let $y : \mathbf{G} \to \mathbb{R}_+^*$ be a function such that for all $u \in U_0(\mathbb{A})$, all $m \in \mathbf{M}_0$, all $g \in \mathbf{G}$ and all $k \in \mathbf{K}$,

$$y(umg) = m^{\mu+\rho_0} y(g) \text{ and } 1 < y(k).$$

For all $g \in \mathbf{G}$, set

$$E^M(g) := \sum_{\gamma \in M(k)\cap P_0(k)\backslash M(k)} y(\gamma g).$$

By the part of the proposition already proved by Godement, which we use for $\mathbf{G} = \mathbf{M}$, this series converges and defines an element of $A(U(\mathbb{A})M(k)\backslash\mathbf{G})$. Moreover $E^M(g)$ is a series with only positive terms, one of which is $y(g)$, so:

$$E^M(g) \geq y(g), \forall g \in \mathbf{G}.$$

For $\lambda \in X_M^{\mathbf{G}}$, the function $m \in \mathbf{M} \longmapsto \lambda\phi_\pi(m) = m^\lambda\phi_\pi(m)$ has moderate growth. Thus, if μ is an element of $\mathrm{Re}X_{M_0}^{\mathbf{M}}$ which is sufficiently positive, there exists c such that for all $k \in \mathbf{K}$ and for all $m \in \mathbf{M}^c \cap S^P$, in the notation of I.2.1:

$$|\lambda\phi_\pi(mk)| \leq c m_{P_0}(m)^{\mu+\rho_0}.$$

Now:

$$m_{P_0}(m)^{\mu+\rho_0} = cy(mk)y(k)^{-1} \leq cy(mk) \leq cE^M(mk).$$

Using the left invariance under $M(k)$ of $\lambda\phi_\pi$ and of E^M, we obtain, for all $m \in M(k)(\mathbf{M}^c \cap S^P)$ and for all $k \in \mathbf{K}$:

$$|\lambda\phi_\pi(mk)| \leq cE^M(mk).$$

Taking ω large enough in I.2.1, we have $M(k)(\mathbf{M}^c \cap S^P) = \mathbf{M}^c$, so the above inequality is true for all $m \in \mathbf{M}^c$ and all $k \in \mathbf{K}$. We also know that the centre of $\mathbf{M}$ acts on $\lambda\phi_\pi$ via the central character of $\pi \otimes \lambda \otimes \rho_P$. In particular for all $z \in A_{\mathbf{M}}$:

$$|\lambda\phi_\pi(zmk)| \leq cz^{\mathrm{Re}\pi+\mathrm{Re}\lambda+\rho_P}\lambda\phi_\pi(mk).$$

For z as above, we have $z^{\mu+\rho_P} = z^{\rho_P}$ and $E^M(zmk) = z^{\rho_0}E^M(mk)$. We deduce the inequalities:

$$\forall z \in A_{\mathbf{M}}, \forall m \in \mathbf{M}^c, \forall k \in \mathbf{K},\ |\lambda\phi_\pi(zmk)| \le cz^{\mathrm{Re}\,\pi+\mathrm{Re}\,\lambda}E^M(zmk),$$

and also

$$\forall m \in \mathbf{M}, \forall k \in \mathbf{K},\ |\lambda\phi_\pi(mk)| \le cm^{\mathrm{Re}\,\pi+\mathrm{Re}\,\lambda}E^M(mk).$$

Each term is left invariant under $U(\mathbb{A})$ which gives the inequality:

$$cm_P(g)^{\mathrm{Re}\,\pi+\mathrm{Re}\,\lambda}E^M(g) \ge |\lambda\phi_\pi(g)|, \forall g \in \mathbf{G}.$$

To prove the proposition, we can thus replace $\lambda\phi_\pi$ by the function:

$$g \longmapsto m_P(g)^{\mathrm{Re}\,\pi+\mathrm{Re}\,\lambda}E^M(g).$$

A step-by-step argument brings us back immediately to the minimal parabolic case. But in this case the result is already known (see [G]).

II.1.6. Intertwining operators

Let M, M' be standard Levis of G corresponding to standard parabolics P and P' of unipotent radicals U and U'. Let $w \in G(k)$. We suppose that:

$$wMw^{-1} = M'.$$

Fix an equivalence class $\mathfrak{P}$ of irreducible automorphic subrepresentations of $\mathbf{M}$; thus w sends $\mathfrak{P}$ to an equivalence class of irreducible automorphic subrepresentations $w\mathfrak{P}$ in $\mathbf{M}'$. Let $\pi \in \mathfrak{P}$ and ϕ_π be an element of $A(U(\mathbb{A})M(k)\backslash\mathbf{G})_\pi$. For $g \in \mathbf{G}$, whenever the integral below is convergent, we set:

$$(M(w,\pi)\phi_\pi)(g) = \int_{(U'(k)\cap wU(k)w^{-1})\backslash U'(\mathbb{A})} \phi_\pi(w^{-1}ug)\, du.$$

Remark When it exists, this integral is equal to the integral of the same function over

$$(U'(\mathbb{A}) \cap wU(\mathbb{A})w^{-1})\backslash U'(\mathbb{A}),$$

since the volume of $(U'(k) \cap wU(k)w^{-1})\backslash(U'(\mathbb{A}) \cap wU(\mathbb{A})w^{-1})$ is equal to 1.

Proposition *Let w, π be as above. Let $\mathfrak{P}$ denote the equivalence class of π.*

(i) *The integral defined above converges absolutely and uniformly when g varies in a compact set if $\langle \mathrm{Re}\,\pi, \check{\alpha}\rangle \gg 0$ for every positive simple* *root α of G relative to M. Under this hypothesis the integral depends only on the double class*

$$M'(k)wM(k)$$

and defines an element of $A(U'(\mathbb{A})M'(k)\backslash \mathbf{G})_{w\pi}$.

(ii) *Suppose that $\mathfrak{P}$ consists of cuspidal representations and let C be a compact set of $\{\mathrm{Re}\,\pi; \pi \in \mathfrak{P}\}$ contained in the domain of convergence described in (i). Let U be an open set of $\mathfrak{P}$ consisting of elements whose real part is in C. Let ϕ, ϕ' be holomorphic functions on U, $-w\overline{U}$ respectively, satisfying II.1.2 (1) and (2). Then the map:*

$$\pi \in U \longmapsto \langle \phi'(-w\overline{\pi}), M(w,\pi)\phi(\pi)\rangle,$$

where $\langle , \rangle$ is the natural pairing of $A(U'(\mathbb{A})M'(k)\backslash \mathbf{G})_{w\pi}$ and $A(U'(\mathbb{A})M'(k)\backslash \mathbf{G})_{-w\overline{\pi}}$, is holomorphic and bounded.

(iii) *Suppose moreover that π is unitary and cuspidal. Then the properties (i) are true under the more precise condition: $\langle \mathrm{Re}\,\pi, \check{\alpha}\rangle > \langle \rho_P, \check{\alpha}\rangle$ for all α as in (i).*

(iv) *More precisely, it suffices in (i) and (iii), to suppose that*

$$\langle \mathrm{Re}\,\pi, \check{\alpha}\rangle \gg 0, \text{ and } \langle \mathrm{Re}\,\pi, \check{\alpha}\rangle > \langle \rho_P, \check{\alpha}\rangle, \text{ respectively,}$$

for every root $\alpha \in R(T_M, G)$ such that $\alpha > 0$ and $w\alpha < 0$.

For (i) and (iii), for the integral defining $M(w,\pi)(\phi_\pi)(g)$ an upper bound is given by:

$$\int_{U'(k)\cap wU(k)w^{-1}\backslash U'(\mathbb{A})} |\phi_\pi(w^{-1}ug)|\, du.$$

We will see farther on (II.1.7 (4) and (6)) that this integral is uniformly bounded above on all compact sets of $\mathbf{G}$ by:

$$\int_{U'(k)\backslash U'(\mathbb{A})} \sum_{\gamma \in P(k)\backslash G(k)} |\phi_\pi(\gamma ug)|\, du.$$

As $U'(k)\backslash U'(\mathbb{A})$ is compact and the functions are continuous, this term is finite as long as the series converge uniformly on all compact sets. It then suffices to apply Proposition II.1.5. to obtain the result.

(ii) We consider $\lambda \in X_M^{\mathbf{G}}$ as a function on $\mathbf{G}$ by precomposing with m_P. Fix π. We have

$$(w\lambda)^{-1}M(w,\pi\otimes\lambda)(\lambda\phi(\pi)) \in A(U'(\mathbb{A})M'(k)\backslash \mathbf{G})_{w\pi}.$$

By $\mathbf{K}$-finiteness, this function remains in a finite-dimensional subspace denoted by $\mathfrak{M}$ of $A(U'(\mathbb{A})M'(k)\backslash \mathbf{G})_{w\pi}$. The product with $\phi'(-\overline{w\pi})$ defines a linear form on $\mathfrak{M}$. Now, the dual of $\mathfrak{M}$ is generated by the values at the different points of $\mathbf{G}$. It thus suffices, for all $g \in \mathbf{G}$, to prove the same properties as in (ii) for the map:

$$\lambda \longmapsto (w\lambda)^{-1}M(w,\pi\otimes\lambda)(\lambda\phi(\pi))(g).$$

It is obvious that it is holomorphic whenever $\lambda\otimes\pi\in U$. To show that this map is bounded for λ such that $\lambda\otimes\pi\in U$, it suffices as in (i) to prove it for

$$\sum_{\gamma\in P(k)\backslash G(k)} |\lambda\phi(\pi)(\gamma g)|,$$

which is a consequence, via devissage (see II.1.5), of [G] §3.

(iv) Let us make (i) more precise. We introduce the same function $E^M : U(\mathbb{A})M(k)\backslash\mathbf{G}\to\mathbb{R}_+^*$ as in the proof of II.1.5. There exists $c>0$ such that for all $g\in\mathbf{G}$,

(1) $$|\phi_\pi(g)|\leq c m_P(g)^{\mathrm{Re}\,\pi}E^M(g).$$

Let ϕ' be the function defined by $\phi'(g)=m_P(g)^{\mathrm{Re}\,\pi}E^M(g)$ and π' the automorphic representation of $\mathbf{M}$ generated by the functions ϕ'_k for $k\in\mathbf{K}$. This representation is not always irreducible but admits a central character, and we have $\mathrm{Re}\,\pi'=\mathrm{Re}\,\pi$. By relation (1), for all $w\in W$, the integral defining $M(w,\pi)\phi_\pi$ converges absolutely and uniformly when g remains in a compact set as long as this is also true of the integral defining $M(w,\pi')\phi'$. We study the latter integral. Denote by $W(M)$ the set of $w\in W$ such that wMw^{-1} is standard and w is minimal in its right class modulo W_M, where W_M is the Weyl group of M. We can suppose that $w\in W(M)$ thanks to (i). We know that with every simple root $\alpha\in R(T_M,G)$ we can associate an element $w_\alpha\in W(M)$ such that $w_\alpha(\alpha)<0$ but $w_\alpha(\beta)>0$ for all $\beta\in R(T_M,G)$ such that $\beta>0$ and β is not a multiple of α (see I.1.7). Let $\alpha\in R(T_M,G)$ be a simple root. Let us introduce the standard Levi subgroup M_α containing M such that $\mathfrak{a}_{M_\alpha}=\{H\in\mathfrak{a}_M;\langle H,\check{\alpha}\rangle = 0\}$, and define $W^{M_\alpha}(M)$ in the same way as $W(M)$. The element w_α can be identified with an element of $W^{M_\alpha}(M)$. We see that $M(w_\alpha,\pi')$ is induced by an operator defined on $A(U(\mathbb{A})M(k)\cap\mathbf{M}_\alpha\backslash\mathbf{M}_\alpha)_{\pi'}$. The results of (i) can be applied to the group M_α and to this operator (the proof is also valid if π' is not irreducible). The integral defining $M(w_\alpha,\pi')\phi'$ thus converges as long as $\langle\mathrm{Re}\,\pi',\check{\beta}\rangle\gg 0$ for a simple root β of $R(T_M,M_\alpha)$. The only simple root in $R(T_M,M_\alpha)$ is α. Moreover $\mathrm{Re}\,\pi'=\mathrm{Re}\,\pi$. The condition thus becomes $\langle\mathrm{Re}\,\pi,\check{\alpha}\rangle\gg 0$, i.e. $\langle\mathrm{Re}\,\pi,\check{\beta}\rangle\gg 0$ for every root $\beta\in R(T_M,G)$ such that $\beta>0$ and $w_\alpha\beta<0$. This concludes the proof for $w=w_\alpha$. In general, we know that we can find a decomposition:

$$w=w_{\alpha_n}\cdots w_{\alpha_1},$$

such that, if we set $w_i = w_{\alpha_i} \cdots w_{\alpha_1}, M_i = w_i M w_i^{-1}$, then α_{i+1} is a simple root of $R(T_{M_i}, G), w_i^{-1}(\alpha_{i+1}) > 0$ and:

(2) the set of positive multiple roots of an element of

$$\{w_i^{-1}(\alpha_{i+1}); i = 0, \ldots, n-1\}$$

coincides with $\{\alpha \in R(T_M, G) | \alpha > 0, w\alpha < 0\}$.

Denote by U_i the unipotent radical of the parabolic standard P_i of Levi M_i. We can decompose integration over $U_n(\mathbb{A}) \cap wU(\mathbb{A})w^{-1} \backslash U_n(\mathbb{A})$ into n integrals over

$$(U_n(\mathbb{A}) \cap ww_{i-1}^{-1} U_{i-1}(\mathbb{A}) w_{i-1} w^{-1}) \backslash (U_n(\mathbb{A}) \cap ww_i^{-1} U_i(\mathbb{A}) w_i w^{-1})$$

which by $u \longmapsto w_i w^{-1} u w w_i^{-1}$ is isomorphic to

$$(U_i(\mathbb{A}) \cap w_{\alpha_i} U_{i-1}(\mathbb{A}) w_{\alpha_i}^{-1}) \backslash U_i(\mathbb{A}).$$

This decomposition and some variable changes lead to the equality:

$$M(w, \pi')\phi' = M(w_{\alpha_n}, w_{n-1}\pi') M(w_{\alpha_{n-1}}, w_{n-2}\pi') \cdots M(w_{\alpha_1}, \pi')\phi'.$$

Note that all these integrals involve real positive-valued functions. By Fubini's theorem, in order for the left-hand integral to converge, it suffices that each partial integral on the right-hand side converge. But each one belongs to the case already considered: in order to treat the operator $M(w_{\alpha_i}, w_{i-1}\pi')$, we apply the above result to the representation $\pi'' = w_{i-1}\pi'$ and to the function

$$\phi'' = M(w_{\alpha_{i-1}}, w_{i-2}\pi') \cdots M(w_{\alpha_1}, \pi')\phi'.$$

The conditions of convergence thus become:

$$\langle \mathrm{Re}\, w_i \pi, \check{\alpha}_{i+1} \rangle \gg 0,$$

for $i = 0, \cdots, n-1$, i.e. $\langle \mathrm{Re}\, \pi, w_i^{-1}(\check{\alpha}_{i+1}) \rangle \gg 0$ for $i = 0, \cdots, n-1$ and we recover the condition of the statement by applying (2).

Remark With the above notation, we have the equality:

$$M(w, \pi')\phi_{\pi'} = M(w_{\alpha_n}, w_{n-1}\pi') M(w_{\alpha_{n-1}}, w_{n-2}\pi') \cdots M(w_{\alpha_1}, \pi')\phi_{\pi'}$$

if $\mathrm{Re}\, \pi$ satisfies the conditions of convergence of the statement. This is a result of the above calculation.

Let us now make (iii) more precise. Let ϕ' denote the function defined by $\phi'(g) = m_P(g)^{\mathrm{Re}\,\pi + \rho_P}$. As π is cuspidal, ϕ_π is bounded on $\mathbf{M}^c\mathbf{K}$ and there exists $c > 0$ such that:

$$|\phi_\pi(g)| \leq c\phi'(g)$$

for all $g \in \mathbf{G}$. We are reduced to considering the integral $M(w, \pi')\phi'$, where π' is the representation of $\mathbf{M}$ generated by the functions ϕ'_k for $k \in \mathbf{K}$, i.e π' is the character $\mathrm{Re}\,\pi$. Suppose first that w is an elementary symmetry w_α. We can then suppose that $G = M_\alpha$. But under this hypothesis Godement proves that the Eisenstein series:

$$\sum_{\gamma \in P(k)\backslash G(k)} m_P(\gamma g)^{\mathrm{Re}\,\pi + \rho_P}$$

converges for all g whenever $\langle \mathrm{Re}\,\pi, \check{\alpha} \rangle > \langle \rho_P, \check{\alpha} \rangle$, see [G], Theorem 3. As we noted at the beginning of the proof, this implies that under the same hypothesis the integral $M(w_\alpha, \pi')\phi'$ converges.

For $w = w_{\alpha_n} \cdots w_{\alpha_1}$, the above result and the decomposition of integrals explained earlier lead to the following convergence conditions:

(2) $$\langle \mathrm{Re}\, w_i \pi, \check{\alpha}_{i+1} \rangle > \langle \rho_{P_i}, \check{\alpha}_{i+1} \rangle,$$

for $i = 0, \cdots, n-1$, where we recall that P_i is the parabolic standard subgroup of Levi subgroup $w_i M w_i^{-1}$.

Fix i, set $\alpha = \alpha_{i+1}$ and $w = w_i$ (we may forget the initial w which will no longer play any role), $P' = P_i, P' = M'U'$. The root α is simple in $R(T_{M'}, G)$. Let β be a simple root of $R(T_0, G)$ which projects onto α by the natural map; see I.1.5. We have $\check{\beta} - \check{\alpha} \in \mathfrak{a}^{M'}$ so $\langle \rho_{P'}, \check{\beta} - \check{\alpha} \rangle = 0$. As $w \in W(M)$, we have $w^{-1}(\check{\beta} - \check{\alpha}) \in \mathfrak{a}^M$, so $\langle \rho_P, w^{-1}(\check{\beta} - \check{\alpha}) \rangle = 0$, i.e. $\langle w\rho_P, \check{\beta} - \check{\alpha} \rangle = 0$. We must thus check that we have

(3) $$\langle \rho_{P'}, \check{\beta} \rangle \leq \langle w\rho_P, \check{\beta} \rangle.$$

Since w is of minimal length in its class modulo the Weyl group of M, we have $w\rho_{P_0}^M = \rho_{P_0}^{M'}$, where we denote by $\rho_{P_0}^M$, $\rho_{P_0}^{M'}$ the half-sum of roots > 0 in M, M' respectively. As $\rho_{P_0}^M + \rho_P = \rho_{P_0}$ and $\rho_{P_0}^{M'} + \rho_{P'} = \rho_{P_0}$, we can replace (3) by the condition:

(4) $$\langle \rho_{P_0}, \check{\beta} \rangle \leq \langle w\rho_{P_0}, \check{\beta} \rangle.$$

This reduces us to the situation of a root system. By a devissage similar to the proof of II.1.10 below, we reduce to the case where G is simple and split over k. As β is simple, we have:

$$\langle \rho_{P_0}, \check{\beta} \rangle = 1.$$

By hypothesis, $w^{-1}\beta$ is a positive root, so $\langle \rho_{P_0}, w^{-1}\check{\beta} \rangle \geq 1$, which proves (4) and concludes the proof of the proposition.

II.1.7. Constant terms of Eisenstein series

Let M be a standard Levi of G. Let $W(M)$ be the set of elements w of the Weyl group of G of minimal length modulo the Weyl group of M, and such that wMw^{-1} is also a standard Levi of G. Let $(M, \mathfrak{P})$ be a cuspidal datum.

Proposition *Let $\pi \in \mathfrak{P}$ be such that $\langle \mathrm{Re}\,\pi, \check{\alpha}\rangle \gg 0\ \forall \alpha \in R^+(T_M, G)$, and let $\phi_\pi \in A(U(\mathbb{A})M(k)\backslash \mathbf{G})_\pi$. The Eisenstein series $E(\phi_\pi, \pi)$ is defined by a convergent series (see II.1.5). Let $P' = U'M'$ be a parabolic standard of Levi M' and of unipotent radical U'. Let $E_{P'}(\phi_\pi, \pi)$ denote the constant term of $E(\phi_\pi, \pi)$ along U'. Then*

(i) *if M' is conjugate to M, then*

$$E_{P'}(\phi_\pi, \pi) = \sum_{w \in W(M),\, wMw^{-1} = M'} M(w, \pi)\phi_\pi,$$

in particular the intertwining operators $M(w, \pi)$ are defined by absolutely convergent integrals which converge uniformly on all compact sets of G.

(ii) *More generally:*

$$E_{P'}(\phi_\pi, \pi) = \sum_{w \in W(M,M')} E^{M'}\big(M(w, \pi)\phi_\pi, w\pi\big),$$

where $W(M, M') = \{w \in W | w^{-1}(\alpha) > 0, \forall \alpha \in R^+(T_0, M')$ and wMw^{-1} is a standard Levi of $M'\}$. This constant term is zero if $W(M, M') = \emptyset$.

The proof is elementary: we will use the Bruhat decomposition (see I.1.6):

(1)
$$G(k) = \bigcup_{w \in W_{M'} \backslash W / W_M} P(k) w^{-1} P'(k),$$

where the union is disjoint, and the facts:

(2)
$$\begin{cases} W^{\bullet}_{M,M'} := \{w \in W | \forall \alpha \in R^+(T_0, M), w(\alpha) > 0, \\ \forall \alpha \in R^+(T_0, M'), w^{-1}(\alpha) > 0\} \\ \text{is a system of representatives for } W_{M'}\backslash W / W_M, \end{cases}$$

(3)
$$\{w \in W^{\bullet}_{M,M'} | wMw^{-1} \subset M'\} = W(M, M').$$

According to the definition, we see that

$$E_{P'}(\phi_\pi, \pi)(g) = \int_{U'(k)\backslash U'(\mathbb{A})} \sum_{\gamma \in P(k)\backslash G(k)} \phi_\pi(\gamma u g)\, du.$$

Thanks to the convergence properties of the series, we can exchange sum and integration (the integral being over a compact set) and we find, using (1) and (2):

$$E_{P'}(\phi_\pi, \pi)(g) = \sum_{w \in W^{\bullet}_{M,M'}} \sum_{m' \in (M'(k) \cap wP(k)w^{-1})\backslash M'(k)} \int_{U'(k)\backslash U'(\mathbb{A})} \sum_{u' \in (U'(k) \cap m'^{-1} wP(k) w^{-1} m')\backslash U'(k)} \phi_\pi(w^{-1} m' u' u)\, du.$$

We obtain an analogous formula by replacing ϕ_π by $|\phi_\pi|$. In particular, for all $w \in W^\bullet_{M,M'}$ and all $m' \in M'(k)$, we find that:

$$(4)\qquad \int_{U'(k)\backslash U'(\mathbb{A})} \sum_{u' \in U'(k)\cap m'^{-1}wP(k)w^{-1}m'\backslash U'(k)} |\phi_\pi|(w^{-1}m'u'ug)\, du$$
$$= \int_{U'(k)\cap wP(k)w^{-1}\backslash U'(\mathbb{A})} |\phi_\pi|(w^{-1}um'g)\, du$$

is uniformly convergent on all compact sets of $\mathbf{G}$. This integral and its analogue for ϕ_π are zero by the cuspidality property if $w^{-1}U'w\cap M \neq \{1\}$ i.e. if wMw^{-1} is not contained in M', or (by (3)), even if w is not an element of $W(M, M')$. Let $w \in W(M, M')$. Then we have:

(5) $M' \cap wPw^{-1}$ is the parabolic standard of M' of Levi wMw^{-1},

(6) $U'(k) \cap wP(k)w^{-1} = U'(k) \cap wU(k)w^{-1}$,

so:

$$E_{P'}(\phi_\pi, \pi)(g) = \sum_{w\in W(M,M')} \sum_{m'\in M'(k)\cap wP(k)w^{-1}\backslash M'(k)} M(w,\pi)\phi(\pi)(m'g)$$
$$= \sum_{w\in W(M,M')} E^{M'}(M(w,\pi)\phi(\pi), w\pi)(g).$$

II.1.8. Adjunction of intertwining operators

Lemma *Let M, M' be standard Levis and let $w \in W(M)$ be such that $wMw^{-1} = M'$, in the notation of the preceding paragraph. Also let π be a cuspidal subrepresentation of $\mathbf{M}$ and $\phi_\pi \in A(U(\mathbb{A})M(k)\backslash\mathbf{G})_\pi$, $\phi'_{-w\overline{\pi}} \in A(U'(\mathbb{A})M'(k)\backslash\mathbf{G})_{-w\overline{\pi}}$. Then, if $\mathrm{Re}\,\pi$ is in the positive cone of absolute convergence for the intertwining operator, we have:*

$$\langle \phi'_{-w\overline{\pi}}, M(w,\pi)\phi_\pi\rangle = \langle M(w^{-1}, -w\overline{\pi})\phi'_{-w\overline{\pi}}, \phi_\pi\rangle.$$

We have:

$$\langle \phi'_{-w\overline{\pi}}, M(w,\pi)\phi_\pi\rangle =$$
$$\int_{(U'(\mathbb{A})M'(k)Z_{\mathbf{M}'}\backslash\mathbf{G})} \overline{\phi}'_{-w\overline{\pi}}(g') \int_{(U'(k)\cap wU(k)w^{-1})\backslash U'(\mathbb{A})} \phi_\pi(w^{-1}u'g')d\, du'\, dg'.$$

Each integral converges absolutely and uniformly on all compact sets, which allows us to write:

$$\langle \phi'_{-w\overline{\pi}}, M(w,\pi)\phi_\pi\rangle = \int_{(U'(k)\cap wU(k)w^{-1})M'(k)Z_{M'}\backslash\mathbf{G}} \overline{\phi}'_{-w\overline{\pi}}(g')\phi_\pi(w^{-1}g')\, dg'.$$

Set $g' = wg$, so the preceding expression is equal to

$$\int_{(w^{-1}U'(k)w\cap U(k))M(k)Z_M\backslash\mathbf{G}} \overline{\phi}'_{-w\overline{\pi}}(wg)\phi_\pi(g)\,dg,$$

which we decompose into an integral over

$$U(\mathbb{A})M(k)Z_M\backslash\mathbf{G}\times(U(k)\cap w^{-1}U'(k)w)\backslash U(\mathbb{A}).$$

The result is a consequence of this.

II.1.9. Euler expansions of intertwining operators

In the extremely frequent case where for $\pi\in\mathfrak{P}$ the right regular representation in $A(M(k)\backslash\mathbf{M})_\pi$ is not irreducible, the operators $M(w^{-1},w\pi)$ defined in II.1.6 cannot have an Euler expansion although, taking the Remark into consideration, we can define one via an integral over a domain possessing an expansion as a product over all places of k. To obtain an Euler expansion, we need to adopt another point of view. Fix $\rho\in\mathfrak{P}$ and a model V_ρ of ρ. We canonically identify:

$$A(U(\mathbb{A})M(k)\backslash\mathbf{G})_\rho\simeq(\mathrm{ind}_{\mathbf{M}\cap\mathbf{K}}^{\mathbf{K}}V_\rho)\otimes\mathrm{Hom}_{\mathbf{M}}(V_\rho,A(M(k)\backslash\mathbf{M})).$$

We also fix a model $V_{w\rho}$ for the representation $w\rho$ of $\mathbf{M}'$ and an isomorphism $A: V_\rho\tilde{\rightarrow}V_{w\rho}$ such that:

(1) $$\forall m\in\mathbf{M},\forall y\in V_\rho, A(\rho(m)y)=(w\rho)(wmw^{-1})A(y),$$

if k is a function field; if k is a number field, m does not run through $\mathbf{M}$ but only through $\mathbf{M}_f\times(\mathrm{Lie}\,M_\infty,\mathbf{M}_\infty\cap\mathbf{K})$.

Let $\pi\in\mathfrak{P}$, and fix $\mu_\pi\in X_M^{\mathbf{G}}$ such that $\pi\simeq\rho\otimes\mu_\pi$. This allows us to identify V_ρ with a model, which we denote by V_π, of π. Since $w\pi\simeq w\rho\otimes w\mu_\pi$ we also have a (compatible) identification of $V_{w\rho}$ with $V_{w\pi}$. Thus A defines a isomorphism of V_π onto $V_{w\pi}$ satisfying the analogue of (1).

We define:

$$\tilde{A}:\mathrm{Hom}_{\mathbf{M}'}(V_{w\pi},A(M'(k)\backslash\mathbf{M}'))\simeq\mathrm{Hom}_{\mathbf{M}}(V_\pi,A(M(k)\backslash\mathbf{M}))$$

by $\tilde{A}(h)(y)(m)=hA(y)(wmw^{-1})$ for all $y\in V_\pi$ and all $m\in\mathbf{M}$. We define the operator $M_A(w^{-1},w\pi)$ (when the integral converges) of $\mathrm{ind}_{\mathbf{M}'\cap\mathbf{K}}^{\mathbf{K}}V_{w\pi}$ into $\mathrm{ind}_{\mathbf{M}\cap\mathbf{K}}^{\mathbf{K}}V_\pi$ by, $\forall\tilde{\phi}\in\mathrm{ind}_{\mathbf{M}'\cap\mathbf{K}}^{\mathbf{K}}V_{w\pi}$:

$$(M_A(w^{-1},w\pi)\tilde{\phi})(k)=\int_{(U(\mathbb{A})\cap w^{-1}U'(\mathbb{A})w\backslash U(\mathbb{A}))}A^{-1}(i_{P',G}\tilde{\phi})(wnk)\,dn$$

where $(i_{P'G}\tilde{\phi})(g):=(m'_g)^{\rho_{P'}}w\pi(m'_g)\tilde{\phi}(k_g)$ for all $g\in\mathbf{G}$ where $g=u'm'_gk_g$ with $u'\in U'(\mathbb{A}), m'_g\in\mathbf{M}'$ and $k_g\in\mathbf{K}$. We check that

$$M(w^{-1},w\pi)\simeq M_A(w^{-1},w\pi)\otimes\tilde{A}:$$

in particular the right-hand side does not depend on the choice of A; the operator $M_A(w^{-1}, w\pi)$ has an Euler expansion.

II.1.10. Pseudo-Eisenstein series

Let $\phi \in P_{(M,\mathfrak{P})}$, or, more generally, let $R > \|\rho_0\|$ and $\phi \in P^R_{(M,\mathfrak{P})}$. We denote by $\epsilon F(\phi)$ the function on $\mathbf{G}$ obtained by evaluating the Fourier transform of ϕ at the unit element of $\mathbf{M}$ (see II.1.3). If the series below converges, we define for $g \in \mathbf{G}$:

$$\theta_\phi(g) := \sum_{\gamma \in P(k)\backslash G(k)} \epsilon F(\phi)(\gamma g).$$

Proposition *Suppose that $\mathfrak{P}$ consists of cuspidal representations whose restriction to $\mathbf{Z_G}$ is a character, denoted by ξ. Let $\phi \in P^R_{(M,\mathfrak{P})}$. The series defining $\theta_\phi(g)$ is absolutely convergent, uniformly when g remains in a compact set. $\mathbf{Z_G}$ acts on it via the character ξ. If ξ is unitary, then θ_ϕ is square integrable modulo the centre. If $\phi \in P_{(M,\mathfrak{P})}$, then θ_ϕ is rapidly decreasing if k is a number field and has compact support modulo the centre if k is a function field.*

Let $\phi \in P^R_{(M,\mathfrak{P})}$, we make no hypotheses on ξ to begin with. Choose $\mu \in \mathrm{Re}\, X^{\mathbf{G}}_{M_0}$ such that $\langle \mu, \check{\alpha} \rangle > \langle \rho_0, \check{\alpha} \rangle$ for every positive root α of $R(T_0, G)$ and μ is very near ρ_0. Choose a continuous function $y : \mathbf{G} \to \mathbb{R}^*_+$ satisfying:

$$y(umg) = m^{\mu+\rho_0+\mathrm{Re}\,\xi} y(g),$$

for all $u \in U_0(\mathbb{A})$, $m \in \mathbf{M}_0$, $g \in \mathbf{G}$. As in the proof of II.1.5, we introduce the series:

$$E^M(g) = \sum_{\gamma \in P_0 \cap M(k)\backslash M(k)} y(\gamma g).$$

It is convergent and has a lower bound, since its terms are all positive and one of them is $y(g)$:

$$E^M(g) \geq y(g).$$

Denote by μ^M and μ_M the components of μ on $\mathrm{Re}\, X^{\mathbf{M}}_{M_0}$ and $\mathrm{Re}\, X^{\mathbf{G}}_M$ respectively. Fix $\pi_0 \in \mathfrak{P}$ such that $\mathrm{Re}\,\pi_0 = \mathrm{Re}\,\xi$ (identifying $\mathrm{Re}\, X_G$ with a subspace of $\mathrm{Re}\, X_M$). Every element ϕ of $A(U(\mathbb{A})M(k)\backslash\mathbf{G})_{\pi_0}$ (which by convention is a subset of $A_0(U(\mathbb{A})M(k)\backslash\mathbf{G})$, see II.1.1), possesses an upper bound of the form:

$$|\Phi(g)| \leq c m_{P_0}(g)^{\rho_0+\mu^M+\mathrm{Re}\,\xi}$$

for all $g \in S^P$ (see I.2.13 and I.2.10). As R is supposed strictly greater than $\|\rho_P\|$ and as μ is very near ρ_0 and thus μ_M very near ρ_P, every element f of $\hat{P}^R(X_M^{\mathbf{G}})$ (see II.1.4) possesses an upper bound of the form:

$$|f(m)| \leq cm^{\mu_M}, \forall m \in \mathbf{M}.$$

By Lemma II.1.4., the function $\epsilon F(\phi)(g)$ possesses an upper bound of the form:

$$|\epsilon F(\phi)(g)| \leq cm_{P_0}(g)^{\rho_P+\mu+\mathrm{Re}\,\xi}, \forall g \in S^P,$$

or even:

$$|\epsilon F(\phi)(g)| \leq cE^M(g), \ \forall g \in S^P,$$

for a function y as above. As in the proof of II.1.5, we deduce from this the upper bounds:

$$|\epsilon F(\phi)(g)| \leq cE^M(g), \ \forall g,$$

$$\sum_{\gamma \in P(k)\backslash G(k)} |\epsilon F(\phi)(\gamma g)| \leq c \sum_{\gamma \in P(k)\backslash G(k)} E^M(\gamma g) =: E(g).$$

The right-hand series is convergent by the hypothesis on μ: from this we deduce the convergence and even the uniformity of the convergence. It is clear that $\theta_\phi(zg) = \xi(z)\theta_\phi(g)$ for all $z \in Z_{\mathbf{G}}$ and for all $g \in \mathbf{G}$. Suppose now that ξ is unitary. According to the choice made, π_0 is unitary, i.e. $\mathrm{Re}\,\pi_0 = 0$. To show that θ_ϕ is square integrable modulo the centre, it suffices by the upper bound above to show that:

$$\int_{A_{\mathbf{G}}G(k)\backslash \mathbf{G}} |\theta_\phi(g)|E(g)\ dg < \infty.$$

By an easy calculation, the integral is bounded above by:

(1) $$\int_{A_{\mathbf{G}}U(\mathbb{A})M(k)\backslash \mathbf{G}} |\epsilon F(\phi)(g)|E_P(g)\ dg$$

$$= \int_{A_{\mathbf{G}}\backslash A_{\mathbf{M}}} \int_{M(k)\backslash \mathbf{M}^c} \int_{\mathbf{K}} |\epsilon F(\phi)(amk)|E_P(amk)\ dk(am)^{-2\rho_P}\ dm\ da.$$

We can decompose $E_P(amk)$ into a sum of functions of the form:

$$a\mu' + \rho_P E'(mk),$$

where E' has moderate growth on $\mathbf{M}^c\mathbf{K}$ and μ' is conjugate to μ by an element of W. As π_0 is cuspidal of unitary central character, every element Φ of $A_0(U(\mathbb{A})M(k)\backslash \mathbf{G})_{\pi_0}$ satisfies an equality:

$$|\Phi(amk)| = a^{\rho_P}\psi(mk),$$

for all $a \in A_{\mathbf{M}}$, $m \in \mathbf{M}^c$, $k \in \mathbf{K}$, where the function ψ defined on $\mathbf{M}^c\mathbf{K}$ is rapidly decreasing (in the obvious way) if k is a number field and has support with compact image in $M(k)\backslash\mathbf{M}^c\mathbf{K}$ if k is a function field. As $m_P(m)$ remains in a compact set when $m \in \mathbf{M}^c$, it is a consequence of the definition that every element f of $\hat{P}^R(X_M^{\mathbf{G}})$ possesses an upper bound of the form:

$$|f(am)| \leq c\exp(-R\|\log_M^G a\|), \forall a \in A_{\mathbf{M}},\ m \in \mathbf{M}^c.$$

By Lemma II.1.4, we see that we have a upper bound

$$|\epsilon F(\phi)(amk)| \leq a^{\rho_P}\exp(-R\|\log_M^G a\|)\psi(mk),$$

for $a \in A_{\mathbf{M}}$, $m \in \mathbf{M}^c$, $k \in \mathbf{K}$, where ψ satisfies the same conditions as above.

In (1) the integral on $M(k)\backslash\mathbf{M}^c \times \mathbf{K}$ is thus convergent. It remains to prove the convergence of

$$\int_{A_{\mathbf{G}}\backslash A_{\mathbf{M}}} a^{\mu'}\exp(-R\|\log_M^G a\|)\, da.$$

In what follows, we identify $(\mathrm{Re}\,\mathfrak{a}_{M_0}^G)^*$ and $\mathrm{Re}\,X_{M_0}^{\mathbf{G}}$. If $\lambda \in \mathrm{Re}\,X_{M_0}^{\mathbf{G}}$, we denote by λ^M and λ_M its components on $\mathrm{Re}\,X_{M_0}^{\mathbf{G}}$ and $\mathrm{Re}\,X_M^{\mathbf{G}}$ respectively (see I.1.6 (9)). We denote by $\{\omega_\alpha^M; \alpha \in \Delta_{M_0}^M\}$ the basis of $\mathrm{Re}\,X_{M_0}^{\mathbf{M}}$ dual to the basis $\{\check{\alpha}; \alpha \in \Delta_{M_0}^M\}$ of $\mathrm{Re}\,\mathfrak{a}_{M_0}^M$ identified, via the fixed scalar product, with $\mathrm{Re}\,X_{M_0}^{\mathbf{M}}$. Thanks to the hypotheses on μ and R, it suffices to show that if $w \in W$, we have $\|(w\rho_{P_0})_M\| \leq \|\rho_P\|$.

First suppose G split over k and simple. Consider the space $\mathfrak{a}^*$. Set $\rho_0 := \rho_{P_0}$. We have the equalities:

$$\rho_P = (\rho_0)_M,$$

$$\|(\rho_0)_M\|^2 + \|(\rho_0)^M\|^2 = \|\rho_0\|^2 = \|w\rho_0\|^2 = \|(w\rho_0)_M\|^2 + \|(w\rho_0)^M\|^2,$$

which reduce us to proving the inequality:

$$\|(\rho_0)^M\| \leq \|(w\rho_0)^M\|.$$

We can suppose that w is of minimal length in its left class modulo the Weyl group of M. It is well-known that

$$\langle\rho_0, \check{\alpha}\rangle = 1$$

for every simple root α. Applying this in M, we have:

(2) $$(\rho_0)^M = \sum_{\alpha\in\Delta_{M_0}^M} \omega_\alpha^M.$$

Let $\alpha \in \Delta_{M_0}^M$. We have:

$$\langle(w\rho_0)^M, \check{\alpha}\rangle = \langle w_{P_0}, \check{\alpha}\rangle = \langle\rho_0, w^{-1}\check{\alpha}\rangle.$$

By the hypothesis on w, the root $w^{-1}\alpha$ is positive. Thus

$$\langle\rho_0, w^{-1}\check{\alpha}\rangle \geq 1$$

and

(3) $$(w\rho_0)^M = \sum_{\alpha\in\Delta_{M_0}^M} c_\alpha\omega_\alpha^M,$$

with coefficients c_α greater than or equal to 1. As we have

$$(\omega_\alpha^M, \omega_\beta^M) \geq 0,$$

for all $\alpha, \beta \in \Delta_{M_0}^M$, it is a consequence of (2) and (3) that

$$\|(\rho_0)^M\| \leq \|(w\rho_0)^M\|,$$

which is what we wanted.

When G is not simple and split, we must introduce the separable closure k_s of k. Over k_s, choose a Borel subgroup B and a maximal subtorus T of G such that:

$$\begin{array}{ccc} T & \subset & M_0 \\ \cap & & \cap \\ B & \subset & P_0. \end{array}$$

Denote by $\mathfrak{a}_s^*$ and W_s the analogues of $\mathfrak{a}^*$ and W over k_s with respect to these choices. We have the equality

$$\mathfrak{a}^* = (\mathfrak{a}_s^*)^{\mathrm{Gal}(k_s/k)}.$$

For $w \in W$, there exists $w_s \in W_s$ preserving M_0 and $\mathfrak{a}^*$, whose action in $\mathfrak{a}^*$ coincides with that of w: it suffices to represent w by an element of $G(k)$ which we multiply by an element of $M_0(k_s)$ to ensure that it preserves T; the element obtained represents w_s. For such a w_s, we have:

$$(w\rho_0)_M = (w_s\rho_0)_M = (w_s\rho_B)_M.$$

On the other hand,

$$\rho_P = (\rho_B)_M.$$

We are thus reduced to the split case. By decomposing $\mathfrak{a}_s^*$ into simple components, we are reduced to the simple split case. This concludes the proof of the first part of the proposition.

The second part of the proposition is proved, for number fields, in [G]4. The function field case is very simple: we check with the help of II.1.4 (6) that there exists a compact set ω of $\mathbf{G}$ such that the support of $\epsilon \circ F\phi$ is contained in $P(k)Z_{\mathbf{G}}\omega$ and thus *a fortiori* in $G(k)Z_{\mathbf{G}}\omega$. This proves the assertion. □

By an argument similar to the one in this proof, we prove the following remark:

Remark *With the above notation, let $\phi \in P_{(M,\mathfrak{P})}$ and let ϕ' be an automorphic form on* $\mathbf{G}$ *of central character $-\overline{\xi}$. Then $\overline{\theta_\phi}\phi'$ is integrable on $G(k)Z_{\mathbf{G}}\backslash\mathbf{G}$: we denote the integral by $\langle\theta_\phi, \phi'\rangle$.*

II.1.11. Relation between pseudo-Eisenstein series and Eisenstein series

Remark *Keep the hypotheses on* $\mathfrak{P}$ *made in Proposition II.1.10. Fix* $\pi_0 \in \mathfrak{P}$ *in the domain of definition of* ϕ *satisfying* $\langle \mathrm{Re}\,\pi_0, \check{\alpha}\rangle \gg 0$, *for all positive roots of* $R(T_M, G)$. *Then, for all* $g \in \mathbf{G}$, *we have:*

$$\theta_\phi(g) = \int_{\pi\in\mathfrak{P}|\mathrm{Re}\,\pi=\mathrm{Re}\,\pi_0} E(\phi,\pi)(g)\,\mathrm{d}\pi,$$

where $\mathrm{d}\pi$ *can be obtained from the Lebesgue measure* $\mathrm{d}\lambda$ *on* $\mathrm{Im}\,X_M^{\mathbf{G}}$ *which serves to define the Fourier transform after the identification* $\lambda \in \mathrm{Im}\,X_M^{\mathbf{G}}/\mathrm{Fix}_{X_M^{\mathbf{G}}}\mathfrak{P} \mapsto \lambda \otimes \pi_0$.

We have:

$$(\epsilon F(\phi))(g) = \int_{\pi\in\mathfrak{P}|\mathrm{Re}\,\pi=\mathrm{Re}\,\pi_0} \phi(\pi)(g)\,\mathrm{d}\pi.$$

Thus

$$\theta_\phi(g) = \sum_{\gamma\in P(k)\backslash G(k)} \int_{\pi\in\mathfrak{P}|\mathrm{Re}\,\pi=\mathrm{Re}\,\pi_0} \phi(\pi)(\gamma g)\,\mathrm{d}\pi.$$

The hypotheses on $\mathrm{Re}\,\pi_0$ and Proposition II.1.5. ensure that we can exchange the sum and the integral. This gives the result.

II.1.12. Density of pseudo-Eisenstein series

Fix a unitary character ξ of the centre of $\mathbf{G}$ and denote by $L^2(G(k)\backslash\mathbf{G})_\xi$ the set of functions on $G(k)\backslash\mathbf{G}$ whose central character coincides with ξ and which are square integrable modulo the centre of $\mathbf{G}$.

Theorem *The closure of the space generated by the pseudo-Eisenstein series* θ_ϕ *as* ϕ *runs through* $P_{(M,\mathfrak{P})}$, *where* $(M,\mathfrak{P})$ *runs through the set of cuspidal data such that* $\mathfrak{P}$ *satisfies the hypotheses of II.1.10, is* $L^2(G(k)\backslash\mathbf{G})_\xi$ *itself.*

Let $f \in L^2(G(k)\backslash\mathbf{G})_\xi$ be orthogonal to all pseudo-Eisenstein series. We must show that f is zero. By hypothesis we have, for all $(M,\mathfrak{P})$ as in the statement, and for all $\phi \in P_{(M,\mathfrak{P})}$:

$$0 = \int_{G(k)Z_{\mathbf{G}}\backslash\mathbf{G}} \bar{f}(g)\theta_\phi(g)\mathrm{d}g = \int_{P(k)Z_{\mathbf{G}}\backslash\mathbf{G}} \bar{f}(g)\epsilon F(\phi)(g)\,\mathrm{d}g$$

$$= \int_{P(k)U(\mathbb{A})Z_{\mathbf{G}}\backslash\mathbf{G}} \bar{f}_P(g)\epsilon F(\phi)(g)\,\mathrm{d}g,$$

where U is the unipotent radical of the parabolic standard of Levi M and where:

$$f_P(g) := \int_{U(k)\backslash U(\mathbb{A})} f(ug)\, du.$$

Fix $(M, \mathfrak{P})$ and let ϕ vary in $P_{(M,\mathfrak{P})}$; these equalities prove that the cuspidal part of the constant term of f along U is zero. The proposition is thus an immediate consequence of Remark 2 of I.3.4.

Remarks (1) Let ϕ, ϕ' be automorphic forms on **G** having the same central character. The proof above shows that if $\phi - \phi'$ is orthogonal to all pseudo-Eisenstein series whose central character is the complex conjugate of that of $\phi - \phi'$, then $\phi - \phi' = 0$.

(2) This remains true if we suppose ϕ automorphic and ϕ' square integrable modulo the centre.

(3) Let ϕ be an automorphic form on **G** and let $\mathfrak{E} := \{(Q, \pi, \psi)\}$ be the set of cuspidal components of ϕ (see I.3.5 and I.3.3). For every standard Levi M of G, we set $\mathfrak{E}_M = \{(Q, \pi, \psi) \in \mathfrak{E} | \pi \in \Pi_0(M)\}$, with notation as in I.3.3. Then:

$$\forall \text{ cuspidal datum } (\mathfrak{P}', M'),\ \forall \phi' \in P_{(\mathfrak{P}', M')} :$$

$$\langle \theta_{\phi'}, \phi \rangle = \sum_{(Q,\pi,\psi) \in \mathfrak{E}_{M'},\ \pi \in -\overline{\mathfrak{P}'}} \langle \Delta_Q \phi'(-\overline{\pi}), \psi \rangle,$$

where Δ_Q is a differential operator on $P_{(\mathfrak{P}', M')}$ obtained as a Fourier transform of the contragredient polynomial of Q (a more precise calculation is given in the proof of III.2.2).

II.2. Calculation of the scalar product of two pseudo-Eisenstein series

II.2.1. Statement

Let M, M' be standard Levis, $P = MU$ and $P' = M'U'$ the corresponding standard parabolics, $\mathfrak{P}, \mathfrak{P}'$ equivalence classes of automorphic representations of **M** and **M'** respectively (see II.1.1). We say that $(M, \mathfrak{P}) \sim (M', \mathfrak{P}')$ if there exists $\gamma \in G(k)$ such that:

$$\gamma M \gamma^{-1} = M', \quad \gamma \mathfrak{P} = \mathfrak{P}'.$$

By the Bruhat decomposition, if there exists such a γ, we can suppose that γ is an element of the Weyl group W. Suppose that $(M, \mathfrak{P})$ and $(M', \mathfrak{P}')$ are equivalent: we set

$$W((M, \mathfrak{P}), (M', \mathfrak{P}')) := \{w \in W | w(M, \mathfrak{P}) = (M', \mathfrak{P}')\}.$$

Theorem *Fix $\lambda_0 \in \operatorname{Re} X_M^{\mathbf{G}}$ such that the conditions of positivity of II.1.6 are satisfied, then fix $R \in \mathbb{R}_+^*$ such that $\|\lambda_0\| < R$. We suppose that $\mathfrak{P}$ and $\mathfrak{P}'$ consist of cuspidal representations whose restriction to $Z_{\mathbf{G}}$ is a unitary character. Let $\phi \in P^R_{(M,\mathfrak{P})}$ and $\phi' \in P^R_{(M',\mathfrak{P}')}$. Form the pseudo-Eisenstein series θ_ϕ and $\theta_{\phi'}$, which are functions on $G(k)\backslash\mathbf{G}$. We have:*

$$\langle \theta_{\phi'}, \theta_\phi \rangle = \begin{cases} 0, \text{ if } (M,\mathfrak{P}) \not\sim (M',\mathfrak{P}') \\ \displaystyle\int_{\pi\in\mathfrak{P},\ \operatorname{Re}\pi=\lambda_0} \sum_{w\in W((M,\mathfrak{P}),(M',\mathfrak{P}'))} \\ \langle M(w^{-1}, -w\overline{\pi})\phi'(-w\overline{\pi}), \phi(\pi)\rangle \, d\pi, \text{ otherwise,} \end{cases}$$

where $\langle , \rangle$ is the Hermitian product on

$$A(U(\mathbb{A})M(k)\backslash\mathbf{G})_{-\overline{\pi}} \times A(U(\mathbb{A})M(k)\backslash\mathbf{G})_\pi$$

obtained in the usual way, namely

$$(\psi', \psi) \mapsto \int_{(M(k)Z_{\mathbf{M}}\backslash\mathbf{M})\times\mathbf{K}} m^{-2\rho_P}\overline{\psi'}(mk)\psi(mk)\, dm\, dk.$$

Remark *In the above theorem, we do not need to suppose that the restriction of representations of $\mathfrak{P}$ and $\mathfrak{P}'$ to $Z_{\mathbf{G}}$ is a unitary character. Without this hypothesis the result becomes:*

$$\langle \theta_{\phi'}, \theta_\phi \rangle = \begin{cases} 0, \text{ if } (M,\mathfrak{P}) \not\sim (M',-\overline{\mathfrak{P}'}) \\ \displaystyle\int_{\pi\in\mathfrak{P},\ (\operatorname{Re}\pi)^G=\lambda_0} \sum_{w\in W((M,\mathfrak{P}),(M',-\overline{\mathfrak{P}'}))} \\ \langle M(w^{-1}, -w\overline{\pi})\phi'(-w\overline{\pi}), \phi(\pi)\rangle \, d\pi, \text{ otherwise.} \end{cases}$$

We begin with the definitions:

$$\langle \theta_{\phi'}, \theta_\phi \rangle = \int_{G(k)Z_{\mathbf{G}}\backslash\mathbf{G}} \overline{\theta_{\phi'}}(g)\theta_\phi(g)\, dg = \int_{P(k)Z_{\mathbf{G}}\backslash\mathbf{G}} \overline{\theta_{\phi'}}(g)\epsilon F(\phi)(g)\, dg$$

$$= \int_{U(\mathbb{A})M(k)Z_{\mathbf{G}}\backslash\mathbf{G}} \epsilon F(\phi)(g)\overline{\theta_{\phi',P}}(g)\, dg,$$

where $\theta_{\phi',P} := \int_{U(k)\backslash U(\mathbb{A})} \theta_{\phi'}(ug)\, du$ is the constant term along U of $\theta_{\phi'}$. We must calculate this term.

II.2.2. Calculation of constant terms of pseudo-Eisenstein series

Since it will be useful in what follows, we do this calculation in a more general framework than that of II.2.1.

In II.1.4, we defined the spaces $P^R_{(M,\mathfrak{P})}$. Consider two standard Levis $M \subset M'$ and an equivalence class of cuspidal automorphic representations $\mathfrak{P}$ of $\mathbf{M}$ whose restriction to $Z_{\mathbf{G}}$ is a unitary character, ξ. Let $\xi_{M'}$

be a unitary character of the centre of $\mathbf{M}'$ which extends ξ and suppose that

$$\mathfrak{P}_{M'} := \{\pi \in \mathfrak{P} | \pi_{|Z_{\mathbf{M}'}} = \xi_{M'}\} \neq \emptyset.$$

Then $(M, \mathfrak{P}_{M'})$ is a cuspidal datum for M' in the sense of II.1.1. We define the space $P^{R,M'}_{(M,\mathfrak{P}_{M'})}$ in a similar way to $P^{R}_{(M,\mathfrak{P})}$ and we consider the induced space for $\mathbf{G}$, i.e.

$$P^{R}_{(M,\mathfrak{P}_{M'})} := \{\text{functions } \phi \text{ on } \mathfrak{P}_{M'} \text{ with values in the set of functions on } U'(\mathbb{A})M'(k)\backslash\mathbf{G} \text{ such that } \forall k \in \mathbf{K},\ \phi_k \in P^{R,M'}_{(M,\mathfrak{P}_{M'})}\}.$$

Lemma *Let $P = MU$, $P' = M'U'$ be two standard parabolics of G and $\mathfrak{P}$ as above. Also let R and R' be two positive real numbers, with $R' > \|\rho_P\|$ and R very large with respect to R'. Fix $\mu_0 \in \operatorname{Re} X^{\mathbf{G}}_{M'}$ satisfying $\|\mu_0\| + R' < R$ but $\langle -\mu_0, \check{\alpha}\rangle \gg 0$, for all $\alpha \in R^+(T_{M'}, G)$. Define $W(M, M')$ as in I.1.7.*

(i) *for all $w \in W(M, M')$, the operator $M(w, w^{-1}\pi)$ is defined by a convergent integral if*

$$\pi \in D_{w,\mu_0} := \{\pi' \in w\mathfrak{P} | \operatorname{Re}\pi' = \mu_0 + \nu, \text{ where } \nu \in \operatorname{Re} X^{\mathbf{M}'}_{wMw^{-1}} \text{ and } \|\nu\| < R'\}.$$

(ii) *Let $w \in W(M, M')$ and $\pi'_w \in w\mathfrak{P}$ such that $\operatorname{Re}\pi'_w = \mu_0$. Set:*

$$[\pi'_w]_{M'} := \{\pi'_w \otimes X^{\mathbf{M}'}_{wMw^{-1}}\},$$

$$[\pi'_w]_{M',R'} := \{\pi' \in [\pi'_w]_{M'} | \ \|\operatorname{Re}\pi' - \mu_0\| < R'\}.$$

We have the inclusion $[\pi'_w]_{M',R'} \subset D_{w,\mu_0}$ and the function $\pi' \in [\pi'_w]_{M',R'} \mapsto M(w, w^{-1}\pi')\phi(w^{-1}\pi') =: \phi(\pi'_w; \pi')$ belongs to $P^{R'}_{(wMw^{-1},[\pi'_w]_{M'})}$.

(iii) *Under the hypotheses of (ii), we can form the pseudo-Eisenstein series for M', which is the function on $\{\pi'_w \in w\mathfrak{P} | \operatorname{Re}\pi'_w = \mu_0\} \times \mathbf{G}$ defined by:*

$$\theta^{M'}_{\phi}(\pi'_w, g) = \int_{\substack{\pi' \in [\pi'_w]_{M'} \\ \operatorname{Re}\pi' = \mu_0 + \nu_0}} \sum_{\gamma \in (P^w \cap M')(k)\backslash M'(k)} \phi(\pi'_w; \pi')(\gamma g)\, d\pi',$$

where $\nu_0 \in \operatorname{Re} X^{M'}_{wMw^{-1}}$ such that

$$\langle \nu_0, \check{\alpha}\rangle > \langle \rho^{P'}_{P^w}, \check{\alpha}\rangle, \forall \alpha \in R(T_{wMw^{-1}}, M')$$

and $\|\nu_0\| < R'$, where we write $P^w = wMw^{-1}U^w$ for the standard parabolic of Levi wMw^{-1} of G and where $\rho^{P'}_{P^w}$ is the module function of wMw^{-1} acting on the unipotent radical of $M' \cap P^w$. If π'_w and π''_w differ only by an element of $\operatorname{Im} X^{\mathbf{M}'}_{wMw^{-1}}$, we have the *equality* $\theta_\phi^{M'}(\pi'_w, g) = \theta_\phi^{M'}(\pi''_w, g)$. *Set* $D^\bullet_{w,\mu_0} := \{\pi'_w \in w\mathfrak{P} | \mathrm{Re}\,\pi'_w = \mu_0\}/\mathrm{Im}\,X^{\mathbf{M}'}_{\mathbf{w}\mathbf{M}\mathbf{w}^{-1}}$. *For* $g \in \mathbf{G}$, *the function* $\pi'_w \mapsto \theta_\phi^{M'}(\pi'_w, g)$ *defined on* $D^\bullet_{w,\mu_0}$ *is rapidly decreasing (bounded if k is a function field) and uniformly rapidly decreasing when g remains in a compact set.*

(iv) *Let* $\theta_{\phi,P'}$ *denote the constant term of* θ_ϕ *along* P'*; then for all* $g \in \mathbf{G}$ *we have:*

$$\theta_{\phi,P'}(g) = \sum_{w \in W(M,M')} \int_{D^\bullet_{w,\mu_0}} \theta_\phi^{M'}(\pi'_w, g)\, d\pi'_w.$$

(i) By II.1.6 it suffices to prove that for every root $\alpha \in R(T_M, G)$ with $\alpha > 0$ and $w\alpha < 0$, and for all $\nu \in \mathrm{Re}\,X^{M'}_{wMw^{-1}} \|\nu\| < R'$, we have:

$$\langle \rho_P, \check{\alpha} \rangle < \langle w^{-1}(\mu_0 + \nu), \check{\alpha} \rangle,$$

i.e.

$$\langle \rho_P, \check{\alpha} \rangle < \langle \mu_0 + \nu, w\check{\alpha} \rangle.$$

By the hypotheses on w and α, we know that $-w\alpha > 0$ and $-w\alpha$ is not a root of wT_Mw^{-1} in M'. We thus have $\langle \mu_0, w\check{\alpha} \rangle \gg 0$, so $\langle \nu, w\check{\alpha} \rangle$ remains bounded by a term which depends only on R' and on the root system. This proves the assertion.

(ii) As for (i), we can apply Proposition II.1.6 (ii), when $\pi' \in [\pi'_w]_{M',R'}$ which shows that the operator $M(w, w^{-1}\pi')$ is bounded in a reasonable sense. As $\phi \in P^R_{(M,\mathfrak{P})}$, we have an upper bound of $\phi(w^{-1}\pi')$ from which we deduce an upper bound for $\phi(\pi'_w; \pi')$ and the assertion.

(iii) The first assertions are trivial. For the question of growth, we consider the number field case, the function field case being simpler. Denote by $\epsilon F^{\mathbf{M}'}(\phi)(\pi'_w; g)$ the partial Fourier transform of $\phi(\pi'_w; \pi')$ evaluated at the unit element of wMw^{-1}, i.e.

$$\epsilon F^{\mathbf{M}'}(\phi)(\pi'_w; g) := \int_{\lambda \in \mathrm{Im}\,X^{\mathbf{M}'}_{wMw^{-1}}} \phi(\pi'_w; \pi'_w \otimes \lambda)(g)\, d\lambda.$$

We see as in II.1.11 that we have the equality:

$$\theta_\phi^{M'}(\pi'_w; g) = \sum_{\gamma \in (P^w \cap M')(k) \backslash M'(k)} \epsilon F(\phi)(\pi'_w; \gamma g). \tag{1}$$

Denote by U^w the unipotent radical of P^w. Let us establish an upper bound for $|\epsilon F^{M'}(\phi)(\pi'_w; muk)|$ for $m \in w\mathbf{M}w^{-1}$, $u \in U^w(\mathbb{A})$, $k \in \mathbf{K}$. As in (ii), using the fact that $M(w, w^{-1}\pi')$ is bounded and $\phi \in P^R_{(M,\mathfrak{P})}$, we obtain the following upper bound: for all $n, n' \in \mathbf{N}$, there exists a constant $c > 0$ such that we have:

$$|\phi(\pi'_w; \pi'_w \otimes \lambda)(muk)| \leq c(1 + \|\mathrm{Im}\,\pi'_w\|)^{-n}(1 + \|\mathrm{Im}\,\lambda\|)^{-n'} m^{\mu_0 + \mathrm{Re}\,\lambda + \rho_{P^w}},$$

this inequality being valid for $\pi'_w \in w\mathfrak{P}$, with $\operatorname{Re}\pi'_w = \mu_0$ and $\lambda \in X^{\mathbf{M}'}_{wMw^{-1}}$, $\|\operatorname{Re}\lambda\| < R'$.

As in II.1.4, such an upper bound leads to the following upper bound for the Fourier transform:

$$(2)\qquad \begin{aligned}&|\epsilon F^{M'}(\phi)(\pi'_w; muk)| \leq \\ &c'(1+\|\operatorname{proj}_{X_{M'}}\operatorname{Im}\pi'_w\|)^{-n} m^{\mu_0+\rho_{P^w}} \exp(-R'\|\log^{M'}_{wMw^{-1}} m\|).\end{aligned}$$

Let us introduce the function:

$$\psi(muk) = m^{\mu_0+\rho_{P^w}} \exp(-R'\|\log^{\mathbf{M}'}_{wMw^{-1}} m\|).$$

We prove as in II.1.10 that the series:

$$E^{M'}_{\psi}(g) := \sum_{\gamma\in(P^w\cap M')(k)\backslash M'(k)} \psi(\gamma g)$$

converges absolutely for $g \in \mathbf{G}$ and defines a function of g. By (1) and (2), we have an upper bound:

$$|\theta^{M'}_{\phi}(\pi'_w; g)| \leq c'(1+\|\operatorname{proj}_{X_{M'}}\operatorname{Im}\pi'_w\|)^{-n} E^{M'}_{\psi}(g).$$

Such an upper bound being valid for all n, this signifies that $\theta^{M'}_{\phi}(\pi'_w; g)$ is rapidly decreasing in π'_w when $\operatorname{Re}\pi'_w = \mu_0$, and even uniformly when g varies in a compact set.

(iv) By II.1.11 and II.1.7, we have the equality:

$$\theta_{\phi,P'}(g) = \sum_{w\in W(M,M')} I_w(g),$$

where:

$$I_w(g) = \int_{\pi\in\mathfrak{P}|\operatorname{Re}\pi=\lambda_0} E^{M'}(M(w,\pi)\phi(\pi), w\pi, g)\, d\pi,$$

λ_0 being an element of $\operatorname{Re}X^{\mathbf{G}}_M$ with $\lambda_0 - \rho_P$ in the positive cone and $\|\lambda_0\| < R$. Fixing w and making the variable change $\pi \mapsto w^{-1}\pi'$ in the above integral, we obtain:

$$I_w(g) = \int_{\pi'\in w\mathfrak{P}|\operatorname{Re}\pi'=w\lambda_0} E^{M'}(M(w,w^{-1}\pi')\phi(w^{-1}\pi'), \pi', g)\, d\pi'.$$

Fix $\nu_0 \in \operatorname{Re}X^{\mathbf{M}'}_{wMw^{-1}}$ as in (iii) of the statement. Consider the segment S in $\operatorname{Re}X^{\mathbf{G}}_{wMw^{-1}}$ which joins $w\lambda_0$ to $\mu_0+\nu_0$. The operator $M(w,w^{-1}\pi')$ and the Eisenstein series $E^{M'}(M(w,w^{-1}\pi')\phi(w^{-1}\pi')\pi', g)$ are defined and holomorphic at every point π' such that $\operatorname{Re}\pi' \in S$. Indeed for $M(w,w^{-1}\pi')$, it suffices to see that:

$$(3)\qquad \langle w^{-1}\operatorname{Re}\pi', \check{\alpha}\rangle > \langle \rho_P, \check{\alpha}\rangle,$$

for all π' such that $\operatorname{Re}\pi' \in S$ and $\alpha \in R(T_M, G)$ with $\alpha > 0$ and $w\alpha < 0$. We already checked this in (i) for $\operatorname{Re}\pi' = \mu_0 + \nu_0$. It is immediate for

$\operatorname{Re}\pi' = w\lambda_0$. As the condition (3) defines a convex set, it is thus satisfied for $\operatorname{Re}\pi' \in S$. Similarly, for the Eisenstein series, it suffices to see that:

$$(4) \qquad \langle \operatorname{Re}\pi', \check{\alpha}\rangle > \langle \rho_{P^w}^{P'}, \check{\alpha}\rangle,$$

for all $\alpha \in R^+(T_{wMw^{-1}}, M')$ where $\rho_{P^w}^{P'}$ is the half-sum of roots of $T_{wMw^{-1}}$ in $\operatorname{Lie} M' \cap U^w$. It is immediate if $\operatorname{Re}\pi' = \mu_0 + \nu_0$ by hypothesis on ν_0. If $\operatorname{Re}\pi' = w\lambda_0$, note that we have:

$$\langle \rho_{P^w}^{P'}, \check{\alpha}\rangle = \langle \rho_{P^w}, \check{\alpha}\rangle \leq \langle w\rho_P, \check{\alpha}\rangle,$$

(see II.1.6., proof of (iv)). It thus suffices to see that:

$$\langle w\lambda_0, \check{\alpha}\rangle > \langle w\rho_P, \check{\alpha}\rangle,$$

or even $\langle \lambda_0 - \rho_P, w^{-1}\check{\alpha}\rangle > 0$, which is a result of the hypothesis on λ_0 and of the fact that $w^{-1}\alpha > 0$ since $w \in W(M, M')$. Here again the condition (4) defines a convex set and is thus satisfied for $\operatorname{Re}\pi' \in S$.

The function:

$$\pi' \mapsto E^{M'}(M(w, w^{-1}\pi')\phi(w^{-1}\pi'), \pi', g)$$

is holomorphic in $\{\pi' \in w\mathfrak{P} | \operatorname{Re}\pi' \in S\}$, bounded if k is a function field and rapidly decreasing if k is a number field. We can thus move the base of integration along S (see the residue theorem farther on in chapter 4) to obtain:

$$I_w(g) = \int_{\pi' \in w\mathfrak{P} | \operatorname{Re}\pi' = \mu_0 + \nu_0} E^{M'}(M(w, w^{-1}\pi')\phi(w^{-1}\pi'), \pi', g)\, d\pi',$$

or even

$$I_w(g) = \int_{\pi'_w \in D^\bullet_{w,\mu_0}} \int_{\substack{\pi' \in [\pi'_w]_{M'} \\ \operatorname{Re}\pi' = \mu_0 + \nu_0}} E^{M'}(M(w, w^{-1}\pi')\phi(w^{-1}\pi'), \pi')(g)\, d\pi'\, d\pi'_w.$$

The interior integral is just $\theta_\phi^{M'}(\pi'_w, g)$ so:

$$I_w(g) = \int_{\pi'_w \in D^\bullet_{w,\mu_0}} \theta_\phi^{M'}(\pi'_w, g)\, d\pi'_w,$$

which leads to the equality of the statement.

II.2.3. End of the calculation of the scalar product

We use II.2.2 (iii), in exchanging the roles of M and M'; by symmetry, we suppose that the semi-simple rank of M' is less than or equal to that of M. Under this hypothesis $W(M, M')$ is the set of elements w of W of minimal length in their class modulo the Weyl group of M satisfying

$wMw^{-1} = M'$. Thus:

$$\langle \theta_{\phi'}, \theta_{\phi} \rangle = \sum_{w \in W(M,M')} \int_{U(\mathbb{A})M(k)Z_{\mathbf{G}}\backslash \mathbf{G}} \epsilon F(\phi)(g) \times \int_{\pi | -w\overline{\pi} \in \mathfrak{P}', \mathrm{Re}\,\pi = \lambda_0} \overline{M(w^{-1}, -w\overline{\pi})\phi'(-w\overline{\pi})(g)}\, d\pi\, dg$$

where λ_0 is in the very positive cone.

We can exchange the integrals. Now we have:

$$\int_{U(\mathbb{A})M(k)Z_{\mathbf{G}}\backslash \mathbf{G}} \epsilon F(\phi)(g)\overline{M(w^{-1}, -w\overline{\pi})\phi'(-w\overline{\pi})(g)}\, dg = \int_{\mathbf{K}\times(M(k)Z_{\mathbf{G}}\backslash \mathbf{M})} m^{-2\rho_P} \epsilon F(\phi)(mk)\overline{M(w^{-1}, -w\overline{\pi})\phi'(-w\overline{\pi})(mk)}\, dm\, dk.$$

For all π such that $-w\overline{\pi} \in \mathfrak{P}'$, we must calculate the projection of the function $m \mapsto m^{-2\rho_P}\epsilon F(\phi)(mk)$ onto the π-isotypic component. We calculate this projection by first integrating over the centre of $\mathbf{M}$ against the central character of π, denoted by χ. With II.1.3 and the Fourier inversion formulae, we obtain for $\pi_0 \in \mathfrak{P}$ of real part λ_0:

$$\int_{Z_{\mathbf{M}} \cap \mathbf{M}^1 Z_{\mathbf{G}} \backslash Z_{\mathbf{M}}} \int_{\lambda \in \mathrm{Im}\, X_M^{\mathbf{G}} / \mathrm{Fix}_{X_M^{\mathbf{G}}} \mathfrak{P}} \phi(\rho \otimes \lambda_0 \otimes \lambda)(zmk)\chi(z)^{-1} dz\, d\lambda = \sum_{\pi' \in \mathfrak{E}} \phi(\pi')(mk),$$

where $\mathfrak{E}$ is the set of elements of $\mathfrak{P}$ of central character χ and of real part λ_0. This set is finite and it is even a homogeneous space under $X_M^{\mathbf{M}}$. Thus this projection is zero except if $\pi \in \mathfrak{E}$, in which case it is equal to $m \mapsto \phi(\pi)(mk)$. Now, we see that $\pi \in \mathfrak{E}$ if and only if $-w\overline{\pi} \in \mathfrak{P}'$ or even $w\pi \in \mathfrak{P}'$. This implies that we must have $w(M', \mathfrak{P}') = (M, \mathfrak{P})$ – recall that $\mathrm{Re}\,\pi = \lambda_0$. This gives the desired result.

II.2.4. Decomposition of $L^2(G(k)\backslash \mathbf{G})_\xi$ along the cuspidal support

Let ξ be a unitary character of $Z_{\mathbf{G}}$. We denote by $\mathfrak{E}$ the set of equivalence classes of pairs $(M, \mathfrak{P})$ formed by a standard Levi of G, denoted by M, and an equivalence class of cuspidal representations of $\mathbf{M}$ whose restriction to $Z_{\mathbf{G}}$ is ξ. For $\mathfrak{X} \in \mathfrak{E}$, we denote by $L^2(G(k)Z_{\mathbf{G}}\backslash \mathbf{G})_{\mathfrak{X}}$ the closed subspace of $L^2(G(k))\backslash \mathbf{G})_\xi$ generated by the pseudo-Eisenstein series θ_ϕ where ϕ runs through $P_{(M',\mathfrak{P}')}$ and where $(M', \mathfrak{P}')$ describes $\mathfrak{X}$. Then we have:

Proposition *With the preceding notation,*

$$L^2(G(k)\backslash \mathbf{G})_\xi = \hat{\oplus}_{\mathfrak{X}\in\mathfrak{E}} L^2(G(k)\backslash \mathbf{G})_{\mathfrak{X}},$$

the sum being orthogonal.

This is an immediate consequence of II.1.8 and II.2.1.

II.2.5. Lemma

Keeping the same hypotheses as in Lemma II.2.2, we have:

Lemma *Let f be a smooth function of moderate growth on*

$$M'(k)U'(\mathbb{A})\backslash \mathbf{G},$$

on which $Z_{\mathbf{G}}$ acts via a unitary character and let

$$T \in \mathfrak{a}_{M_0}, R > \|\rho_P\| \text{ and } \mu_0 \in \operatorname{Re} X_{M'}^{\mathbf{G}}.$$

Suppose that R is big enough, as well as $\langle \mu_0, \check{\alpha}\rangle$ for all $\alpha \in R^+(T_M, G)$; these conditions depend on f. Then, in the notation of II.2.2, the following integral is convergent:

$$\int_{Z_{\mathbf{G}}M'(k)U'(\mathbb{A})\backslash \mathbf{G}} \sum_{w\in W(M,M')} \int_{D^\bullet_{w,\mu_0}} |f(g)|\hat{\tau}_{P'}(\log_{M'} g - T) \times |\theta_\phi^{M'}(\pi'_w, g)| \, d\pi'_w \, dg;$$

see I.2.11 for the definition of $\hat{\tau}_{P'}$.

It suffices to consider the integral corresponding to a fixed $w \in W(M, M')$. Using the definition of $\theta_\phi^{M'}(\pi', g)$, we reduce by devissage to the integral:

$$(1)\qquad \int_{(Z_{\mathbf{G}}(P^w\cap M')(k)U'(\mathbb{A}))\backslash \mathbf{G}} |f(g)|\hat{\tau}_{P'}(\log_{M'} g - T) \times \int_{D^\bullet_{w,\mu_0}} |\epsilon F^{M'}(\phi)(\pi'_w; g)| \, d\pi'_w \, dg.$$

We can replace the domain of integration by $Z_{\mathbf{G}}P^w(k)\backslash \mathbf{G}$, then again by $S^{P^w} \cap \mathbf{G}^c$ (see I.2.1) and write $g = amuk$ with

$$a \in A_{wMw^{-1}}, m \in \mathbf{M}^c_w w^{-1},\ u \in U^w(\mathbb{A}), k \in \mathbf{K};$$

the measure dg becomes $(am)^{-2\rho_{P^w}}$ da dm du dk. When

$$g = amuk \in S^{P^w} \cap \mathbf{G}^c,$$

we can improve the upper bound of $\epsilon F^{M'}\phi(\pi'_w, g)$ established in the proof of II.2.2 as follows: for all $n \in \mathbb{N}$ and all $\nu \in \operatorname{Re} X_{M_0}^{wMw^{-1}}$, there exists $c > 0$ such that:

$$|\epsilon F^{M'}(\phi)(\pi'_w; amuk)| \le c(1 + \|\operatorname{proj}_{X_{M'}} \operatorname{Im} \pi'_w\|)^{-n} a^{\mu_0+\rho_{P^w}} \exp(-R'\|\log^{M'}_{wMw^{-1}} a\|) m^\nu,$$

for $amuk \in S^{P^w} \cap \mathbf{G}^c$. It suffices to employ the same reasoning as in II.2.2 and to recall that $\mathfrak{P}$ consists of cuspidal representations, which allows us to introduce the term m^v. As f has moderate growth, there exists N such that:

$$|f(amuk)| \leq \|a\|^N \|m\|^N, \forall\, amuk \in S^{P^w} \cap \mathbf{G}^c.$$

Choosing v sufficiently negative, we see that in the integral (1) the partial integral in (m, u, k, π'_w) is convergent and bounded above by:

$$ca^{\mu_0 + \rho_{P^w}} \|a\|^N \exp(-R' \| \log_{wMw^{-1}}^{M'} a \|).$$

Up to increasing N, we can put the terms $a^{\rho_{P^w}}$ into $\|a\|^N$ and it remains to see that the following integral is convergent:

$$J := \int_{A_G \backslash A_{wMw^{-1}} \cap \mathbf{G}^c} a^{\mu_0} \|a\|^N \hat{\tau}_{P'}(\log_{M'} a - T) \exp(-R \| \log_{wMw^{-1}}^{M'} a \|)\, da.$$

We see that the map:

$$A_{M'} \cap \mathbf{G}^c \times A_{wMw^{-1}} \cap \mathbf{M'}^1 \to A_G \backslash A_{wMw^{-1}} \cap \mathbf{G}^c$$

is injective and of finite cokernel. We can essentially limit ourselves to integrating over the left-hand set. Choose N', N'' and c such that:

$$\|a'a''\|^N \leq c \|a'\|^{N'} \|a''\|^{N''}.$$

Then J is bounded above by the product of two integrals:

$$J'' := \int_{A_{wMw^{-1}} \cap \mathbf{M'}^1} \|a''\|^{N''} \exp(-R \| \log_{wMw^{-1}}^{M'} a'' \|)\, da'',$$

$$J' := \int_{A_{M'} \cap \mathbf{G}^c} \hat{\tau}_{P'}(\log_{M'} a' - T) a'^{\mu_0} \|a'\|^{N'}\, da'.$$

The first converges for R large enough. In the second, set

$$\log_{M'}^G a' := \sum_{\alpha \in \Delta_{M'}} h_\alpha \check{\alpha}, \quad \log_{M'}^G T = \sum_{\alpha \in \Delta_{M'}} t_\alpha \check{\alpha},$$

where h_α belongs to $\mathbb{R}$ if k is a number field, or to a certain monogenous subgroup of $\mathbb{R}$ if k is a function field. Introduce numbers N'_α such that:

$$\|a'\|^{N'} \leq \exp \sum |h_\alpha| N'_\alpha.$$

Then J' is bounded by:

$$\prod_{\alpha \in \Delta_{M'}} \int_{h_\alpha > t_\alpha} \exp(h_\alpha (N'_\alpha + \langle \mu_0, \check{\alpha} \rangle))\, dh_\alpha.$$

These integrals are convergent whenever $\langle -\mu_0, \check{\alpha} \rangle$ is large enough.

III

Hilbertian Operators and Automorphic Forms

III.1. Hilbertian operators

III.1.1. A family of operators

Fix an equivalence class $\mathfrak{X}$ of pairs $(M, \mathfrak{P})$ where M is a standard Levi of G and $\mathfrak{P}$ an orbit under $X_M^{\mathbf{G}}$ of irreducible cuspidal automorphic representations of $\mathbf{M}$, for the equivalence relation defined in II.2.1. We denote by $\tilde{\mathfrak{X}}$ the set of pairs (M, π), where π is an irreducible cuspidal automorphic representation of $\mathbf{M}$ such that if $\mathfrak{P}$ is the orbit of π, then $(M, \mathfrak{P}) \in \mathfrak{X}$. Denote by ξ the character of $Z_{\mathbf{G}}$ which is the restriction to $Z_{\mathbf{G}}$ of the central character of π for any pair $(M, \pi) \in \tilde{\mathfrak{X}}$. Fix a real number R such that if $(M, \mathfrak{P}) \in \mathfrak{X}$ and if P is the standard parabolic subgroup of G of Levi M, then $R > \|\rho_P\|$; the norm $\|\rho_P\|$ is the same for all pairs $(M, \mathfrak{P}) \in \mathfrak{X}$. Denote by $\Theta_{\mathfrak{X}}^R$ the space generated by the functions θ_φ for $\varphi \in P_{(M,\mathfrak{P})}^R$ and $(M, \mathfrak{P}) \in \mathfrak{X}$. In the case where ξ is unitary, we write $L_{\mathfrak{X}}^2$ for its closure in $L^2(G(k)\backslash \mathbf{G})_\xi$ (the space $L_{\mathfrak{X}}^2$ is independent of R).

Let us introduce the space $H_{\mathfrak{X}}^R$ whose elements are the functions f defined on $\tilde{\mathfrak{X}}^R := \{(M, \pi) \in \tilde{\mathfrak{X}}; \|\mathrm{Re}\,\pi - \mathrm{Re}\,\xi\| < R\}$ that associate with (M, π) an element $f(M, \pi) \in \mathrm{End}_{\mathbf{M}}\left(A_0(M(k)\backslash \mathbf{M})_\pi\right)$ (see I.3.3; this last space is finite-dimensional), and that satisfy:

(i) f is holomorphic on $\tilde{\mathfrak{X}}^R$.

That is, fix $(M, \pi) \in \tilde{\mathfrak{X}}$ and consider the function $\lambda \mapsto f(M, \pi \otimes \lambda)$ defined on a certain subset of $X_M^{\mathbf{G}}$. We can identify $A_0(M(k)\backslash \mathbf{M})_{\pi\otimes\lambda}$ with the space

$A_0(M(k)\backslash\mathbf{M})_\pi$, which is independent of λ, by the map which associates with the function φ the function $m \mapsto m^{-\lambda}\varphi(m)$. We denote this operation by $(-\lambda)$. Then we want the function

$$\lambda \longmapsto f_{M,\pi}(\lambda) := (-\lambda) \circ f(M, \pi \otimes \lambda) \circ \lambda$$

to be holomorphic.

(ii) If k is a function field or a number field, then f is bounded, or slowly increasing, respectively.

That is, in the case of a function field, the function $f_{M,\pi}$ defined in (i) is bounded on $\{\lambda \in X_M^{\mathbf{G}}; \|\mathrm{Re}\,\pi \otimes \lambda - \mathrm{Re}\,\xi\| < R\}$; when k is a number field, there exist $c > 0$ and $n \in \mathbb{N}$ such that we have

$$|f_{M,\pi}(\lambda)| < c(1 + \|\lambda\|)^n$$

in this domain, where the absolute value of the left-hand side is a fixed norm on the finite-dimensional space $\mathrm{End}_{\mathbf{M}}(A_0(M(k)\backslash\mathbf{M})_\pi)$;

(iii) Let (M, π), $(M', \pi') \in \tilde{\mathfrak{X}}$, $w \in G(k)$ be such that

$$wMw^{-1} = M', w\pi \ = \pi';$$

conjugation by w induces a isomorphism, denoted by $\mathrm{ad}(w)$, as follows:

$$A_0(M(k)\backslash\mathbf{M})_\pi \xrightarrow{\mathrm{ad}(w)} A_0(M'(k)\backslash\mathbf{M})'_\pi.$$

Then we have the equality

$$f(M, \pi) = \mathrm{ad}(w)^{-1} \circ f(M', \pi') \circ \mathrm{ad}(w).$$

Remarks

(1) By (iii), the function f is determined by giving $f_{M,\pi}$ for just one pair $(M, \pi) \in \tilde{\mathfrak{X}}$;

(2) $H_{\mathfrak{X}}^R$ is an algebra for the product defined by

$$(f_1 f_2)(M, \pi) = f_1(M, \pi) \circ f_2(M, \pi);$$

(3) $H_{\mathfrak{X}}^R$ contains the subalgebra $H_{\mathfrak{X},\mathbb{C}}^R$ of elements with values in the homotheties, which can be identified with complex-valued functions.

III.1.2. Operations on pseudo-Eisenstein series

Note that if $P = MU$ is a standard parabolic and π an irreducible cuspidal representation of $\mathbf{M}$, $\mathrm{End}_{\mathbf{M}}(\mathbf{A_0}(\mathbf{M}(\mathbf{k})\backslash\mathbf{M})_\pi)$ can be identified with a subalgebra of $\mathrm{End}_{\mathbf{G}}(A_0(M(k)U(\mathbb{A})\backslash\mathbf{G})_\pi)$ via $(u(\varphi))_k = u(\varphi_k)$ for $u \in \mathrm{End}_{\mathbf{M}}(A_0(M(k)\backslash\mathbf{M})_\pi)$, $\varphi \in A_0(M(k)U(\mathbb{A})\backslash\mathbf{G})_\pi, k \in \mathbf{K}$, see I.2.12 for the definition of φ_k.

This being so, let $(M, \mathfrak{P}) \in \mathfrak{X}$; see III.1.1. Then $H_{\mathfrak{X}}^R$ acts on $P_{(M,\mathfrak{P})}^R$ by

$$(f\varphi)(\pi) = f(M, \pi)(\varphi(\pi)),$$

for $\pi \in \mathfrak{P}, \|\mathrm{Re}\,\pi - \mathrm{Re}\,\xi\| < R$, $f \in H^R_{\mathfrak{X}}, \varphi \in P^R_{(M,\mathfrak{P})}$. For a Levi M and an orbit $\mathfrak{P}$ of representations of $\mathbf{M}$, we write $-\overline{\mathfrak{P}} = \{-\overline{\pi}; \pi \in \mathfrak{P}\}$. Set $-\overline{\mathfrak{X}} = \{(M, -\overline{\mathfrak{P}}); (M, \mathfrak{P}) \in \mathfrak{X}\}$. We defined a map $f \mapsto f^*$ of $H^R_{\mathfrak{X}}$ into $H^R_{-\overline{\mathfrak{X}}}$ by

$$f^*(M, \pi) = f(M, -\overline{\pi})^*,$$

where the second $*$ indicates adjunction. We check directly on the scalar product that for $(M, \mathfrak{P}) \in \mathfrak{X}, (M', \mathfrak{P}') \in -\overline{\mathfrak{X}}, \varphi \in P^R_{(M,\mathfrak{P})}, \varphi' \in P^{R'}_{(M',\mathfrak{P}')}$, and $f \in H^R_{\mathfrak{X}}$, we have

$$\langle \theta_{\varphi}, \theta_{f\varphi} \rangle = \langle \theta_{f^*\varphi'}, \theta_{\varphi} \rangle.$$

We deduce from this that the map $\varphi \mapsto f\varphi$ 'passes to the quotient' by the map $\varphi \longmapsto \theta_{\varphi}$ and defines a map $\Delta_{\theta}(f) \in \mathrm{End}\,\Theta^R_{\mathfrak{X}}$. This map commutes with right translation by elements of $\mathbf{G}$ (or of Lie $\mathbf{G}_{\infty}$, etc.). In the case where ξ is unitary, it is a problem to determine whether this operation can be extended to $L^2_{\mathfrak{X}}$.

III.1.3. Transformations of Hilbert spaces

Recall some definitions concerning transformations in Hilbert spaces. Let H be a Hilbert space. A (linear) transformation of H is a pair (D, T) where D is a vector subspace of H and $T \in \mathrm{Hom}_{\mathbb{C}}(D, H)$. We say that D is the domain of T. We say that two transformations (D, T) and (D', T') are pre-adjoint if for all $h \in D$, $h' \in D'$, we have

$$\langle h', Th \rangle = \langle T'h', h \rangle.$$

Let (D, T) be a transformation. Suppose that D is dense in H. Let D^* be the set of $h' \in H$ such that there exists h'' satisfying

$$\langle h'', h \rangle = \langle h', Th \rangle$$

for every $h \in D$. For $h' \in D^*$, h'' is unique and we define $T^* : D^* \to H$ by $T^*h' = h''$. Then (D^*, T^*) is a transformation called the adjoint of (D, T). It is the largest transformation which is pre-adjoint to (D, T). We say that (D, T) is auto-adjoint if $D = D^*$ and $T = T^*$.

Let (D, T) be a transformation, and let $V \subset H \times H$ be the closure

$$\{(h, Th);\ h \in D\}.$$

Suppose that there exists $(\hat{D}, \hat{T})$ such that $V = \{(h, \hat{T}h);\ h \in \hat{D}\}$. We then say that $(\hat{D}, \hat{T})$ is the closure of (D, T). In particular if (D, T) admits a pre-adjoint transformation of dense domain, its closure exists. We say that a transformation is closed if it is equal to its closure. This is the case of an auto-adjoint transformation.

Let (D, T) be a transformation. We say that T is bounded if there exists c such that $\|Th\| \leq c\|h\|$ for every $h \in D$. Suppose that T is bounded and also that D is dense in H. Then the closure of (D, T) exists, its domain is all of H, and it is also bounded. The adjoint of (D, T) exists and has these properties as well; see [St] Theorems 2.23 and 2.30.

III.1.4. Bounded operators

Suppose ξ is unitary. It is a consequence of the adjunction formula that the transformation $(\Theta^R_{\mathfrak{X}}, \Delta_\theta(f))$ of $L^2_{\mathfrak{X}}$ (see III.1.2) admits a closure $(D_f, \Delta(f))$ which we denote more simply by $\Delta(f)$. It commutes with right translation by elements of $\mathbf{G}$.

Lemma *Let $f \in H^R_{\mathfrak{X}}$. Suppose that f is bounded. Then $D_f = L^2_{\mathfrak{X}}$ and $\Delta(f)$ is bounded.*

Proof Since f, and thus also f^*, is bounded, we can choose $c > 0$ such that the eigenvalues of the element

$$c - f^*(M, \pi) f(M, \pi)$$

of $\mathrm{End}_{\mathbf{M}}(A_0(M(k)\backslash \mathbf{M})_\pi)$ have real parts > 0, for every

$$(M, \pi) \in \mathfrak{X}^R.$$

We can extract the square root of such an endomorphism in a unique way if we require that the eigenvalues continue to have strictly positive real part. We can thus define a function g such that

$$g(M, \pi)^2 = c - f^*(M, \pi) f(M, \pi).$$

By uniqueness, g satisfies the condition (iii) of III.1.1, so $g \in H^R_{\mathfrak{X}}$. Still by uniqueness, we have the equality $g = g^*$. Then for $\psi \in \Theta^R_{\mathfrak{X}}$, we have

$$\begin{aligned}\|\Delta_\theta(f)\psi\|^2 &= \langle \Delta_\theta(f^*)\Delta_\theta(f)\psi, \psi\rangle \\ &= c\|\psi\|^2 - \langle \Delta_\theta(g^*)\Delta_\theta(g)\psi, \psi\rangle \\ &= c\|\psi\|^2 - \|\Delta_\theta(g)\psi\|^2 \\ &\leq c\|\psi\|^2,\end{aligned}$$

and thus $\Delta_\theta(f)$ is bounded, which implies the lemma. □

We denote by $H^R_{\mathfrak{X},b}$ the set of bounded $f \in H^R_{\mathfrak{X}}$. When $f \in H^R_{\mathfrak{X},b}$, $\Delta(f)$ has domain equal to L^2; we have $f^* \in H^R_{\mathfrak{X},b}$ and $\Delta(f^*)$ is the adjoint of $\Delta(f)$.

III.1.5. Adjunction of non-bounded operators

The preceding lemma concludes the function field case since in this case $H^R_{\mathfrak{X}} = H^R_{\mathfrak{X},b}$. In the number field case, it is necessary to work with unbounded transformations. It would be tempting to try to prove that for every $f \in H_{\mathfrak{X}}^R$, $\Delta(f^*)$ is the adjoint of $\Delta(f)$, but we do not know how to prove this statement. In any case we do have:

Lemma *Let $f \in H_{\mathfrak{X}}^R$, $z \in \mathbb{C}$ and $c > 0$. Suppose that for every $(M,\pi) \in \tilde{\mathfrak{X}}^R$, $f(M,\pi) - z$ is invertible and*

$$\|(f(M,\pi) - z)^{-1}\| < c.$$

Then $\Delta(f)$ and $\Delta(f^*)$ are adjoints of each other.

Proof Define a function g by $g(M,\pi) = (f(M,\pi) - z)^{-1}$. Then $g \in H_{\mathfrak{X}}^R$ and in fact $g \in H_{\mathfrak{X},b}^R$. The transformation $\Delta(g)$ thus has domain equal to $L_{\mathfrak{X}}^2$. Set $D = \Delta(g)(L_{\mathfrak{X}}^2)$. The map $\Delta(g) : L_{\mathfrak{X}}^2 \to D$ is bijective. Indeed, let $\psi \in L_{\mathfrak{X}}^2$ be such that $\Delta(g)\psi = 0$ and let $(\psi_n)_{n\in\mathbb{N}}$ be a sequence of elements of $\Theta_{\mathfrak{X}}^R$ such that $\lim_{n\to\infty} \psi_n = \psi$. Set $\psi_n' = \Delta_\theta(f^*)\psi_n - \bar{z}\psi_n$. We have $\Delta(g^*)\psi_n' = \psi_n$, and thus

$$0 = \langle \Delta(g)\psi, \psi_n' \rangle = \langle \psi, \Delta(g^*)\psi_n' \rangle = \langle \psi, \psi_n \rangle,$$

and

$$0 = \lim_{n\to\infty} \langle \psi, \psi_n \rangle = \|\psi\|^2,$$

which shows that $\psi = 0$. We can thus define $\Delta(g)^{-1} : D \to L_{\mathfrak{X}}^2$. We then check that $\Delta(f)$ has domain equal to D and is equal to $\Delta(g)^{-1} + z$. We define D^* in a similar way and $\Delta(g^*)^{-1} : D^* \to L_\chi^2$. Then $\Delta(f^*)$ has domain D^* and is equal to $\Delta(g^*)^{-1} + \bar{z}$. It remains to check that the transformations $(D, \Delta(g)^{-1} + z)$ and $(D^*, \Delta(g^*)^{-1} + \bar{z})$ are adjoint, which is easy. □

Suppose that k is a number field. Recall that we have equipped $\mathfrak{a}_{M_0}^*$ with a bilinear form (,) which linearly extends the scalar product on $\mathrm{Re}\,\mathfrak{a}_{M_0}^*$. On the other hand if M is a standard Levi subgroup of G and π an irreducible representation of $\mathbf{M}$, we define a term $\lambda_\pi \in \mathfrak{a}_M^*$ by restricting the central character of π to $A_{\mathbf{M}}$ and using the fact that $\mathfrak{a}_M^*$ is naturally isomorphic to the group of characters of $A_{\mathbf{M}}$. Define a function f on $\tilde{\mathfrak{X}}^R$ by

$$f(M,\pi) = (\lambda_\pi, \lambda_\pi).$$

Then $f \in H_{\mathfrak{X},\mathbb{C}}^R$. We denote the associated transformation simply by Δ. As $f = f^*$ and f satisfies the hypothesis of the preceding lemma (for $\mathrm{Re} f(M,\pi) < R^2$), the transformation Δ is auto-adjoint.

III.1.6. Spectral projections

Let us recall the properties of auto-adjoint transformations which will be useful to us. Let H be a Hilbert space and (D,Δ) an auto-adjoint transformation of H (D is supposed dense in H). Let $\mathscr{Q}$ be the set of $z \in \mathbb{C}$ such that $\Delta - z : D \to H$ is bijective, of bounded inverse. We show that $\mathscr{Q}$ is open and contains $\mathbb{C} - \mathbb{R}$. The set $\mathbb{C} - \mathscr{Q}$ is called the spectrum of Δ and for $z \in \mathscr{Q}$, $(\Delta - z)^{-1}$ is called the resolvent of Δ. For every $x \in \mathbb{R}$, we can define a closed subspace $H_x \subset H$ such that if we denote by $p_x : H \to H_x$ the orthogonal projection, the following properties are satisfied:

(i) $x \leq x' \Longrightarrow H_x \subset H_{x'}$;
(ii) $H_x = \bigcap_{x'>x} H_{x'}$;
(iii) $\bigcap_x H_x = \{0\}, \bigcup_x H_x$ is dense in H;
(iv) for all $h \in D$, $h' \in H$, $\langle h', \Delta h\rangle = \int_{-\infty}^{+\infty} x \, \mathrm{d}\langle h', p_x h\rangle$ the integral being taken in the sense of Stieltjes (see [St] §V.1).

We also have

(v) for $h \in D$, $\langle \Delta h, \Delta h\rangle = \int_{-\infty}^{+\infty} x^2 \, \mathrm{d}\langle h, p_x h\rangle$;
(vi) for $h \in H$, $\langle h, h\rangle = \int_{-\infty}^{+\infty} \mathrm{d}\langle h, p_x h\rangle$;
(vii) for $z \in \mathscr{Q}$, $h, h' \in H$, $\langle h', (\Delta - z)^{-1} h\rangle =$

$$\int_{-\infty}^{+\infty} (x - z)^{-1} \, \mathrm{d}\langle h', p_x h\rangle.$$

Denote by H_{x-0} the closure of $\bigcup_{x'<x} H_{x'}$ and p_{x-0} the orthogonal projection onto H_{x-0}. Then for $x < x'$, $h, h' \in H$, we have

(viii)
$$\langle h', p_{x'} h\rangle + \langle h', p_{x'-0} h\rangle - \langle h', p_x h\rangle - \langle h', p_{x-0} h\rangle$$
$$= \lim_{\varepsilon \to 0} -(\pi i)^{-1} \int_{\Gamma(x,x',\varepsilon)} \langle h', (\Delta - z)^{-1} h\rangle \, \mathrm{d}z,$$

where $\Gamma(x, x', \varepsilon)$ is the contour:

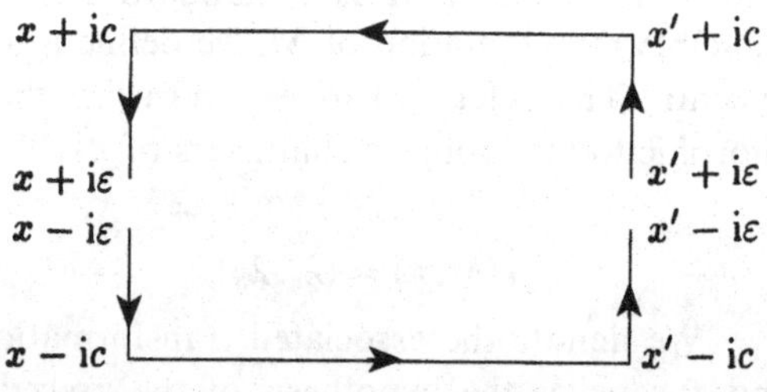

c being any fixed number > 0.

We define a point of continuity of (H_x) to be a point x such that $H_x = H_{x-0}$. The complement in $\mathbb{R}$ of this set is countable.

Remark We will apply formula (viii) in the following situation: we suppose that there exists a real number $R > 0$ such that the spectrum of Δ is contained in $(-\infty, R]$. Take a 'very large' real number T, suppose that $-T$ is a point of continuity of (H_x), set $q_T = 1 - p_{-T}$ and fix $R' > R$: we have $1 = p'_R = p_{R'-0}$, and formula (viii) gives

$$\langle h', q_T h\rangle = \lim_{\varepsilon\to 0} -(2\pi \mathrm{i})^{-1} \int_{\Gamma(-T,R',\varepsilon)} \langle h', (\Delta - z)^{-1} h\rangle \, \mathrm{d}z.$$

III.1.7. Transformations of $L^2(G(k)\backslash \mathbf{G})_\xi$

We can regroup the constructions of III.1.1 as follows. Let ξ be a character of $Z_{\mathbf{G}}$. Introduce the spaces

$$\Theta_\xi^R = \oplus \Theta_{\mathfrak{X}}^R, H_\xi^R = \oplus H_{\mathfrak{X}}^R,$$

where we sum over the classes $\mathfrak{X}$ whose associated character is ξ with an algebraic direct sum. For $f \in H_\xi^R$, we define an element $\Delta_\theta(f) \in \operatorname{End} \Theta_\xi^R$. We define the spaces $H_{\xi,b}^R$, $H_{\xi,\mathbb{C}}^R$ in the obvious way. We still define

$$\Theta_\xi = \oplus \Theta_{\mathfrak{X}}$$

where $\Theta_{\mathfrak{X}}$ is the analogue of $\Theta_{\mathfrak{X}}^R$ obtained by replacing the spaces $P_{(M,\mathfrak{P})}^R$ by the spaces $P_{(M,\mathfrak{P})}$ in the definitions.

Suppose ξ is unitary. We know that we have a decomposition into a direct Hilbertian sum

$$L^2(G(k)\backslash \mathbf{G})_\xi = \hat{\oplus} L^2_{\mathfrak{X}},$$

where we sum over the same set of classes as above; see II.2.4. For $f \in H_\xi^R$, we define a transformation $\Delta(f)$ of $L^2(G(k)\backslash \mathbf{G})_\xi$.

III.2. A decomposition of the space of automorphic forms

III.2.1. Operators on the space of automorphic forms

Let ξ be a character of $Z_{\mathbf{G}}$. Denote by ξ^* the character defined by $\xi^*(z) = \overline{\xi}(z^{-1})$. It is a consequence of Lemma II.2.5 applied to $M' = G$ that if $\phi \in A(G(k)\backslash \mathbf{G})_\xi$, if R is large enough with respect to ϕ, then the integral

$$\int_{Z_{\mathbf{G}} G(k)\backslash \mathbf{G}} \overline{\theta}(g)\phi(g) \, \mathrm{d}g$$

is absolutely convergent for every $\theta \in \Theta_{\xi^*}^R$. We denote by $\langle \theta, \phi\rangle$ the term defined in this way.

Proposition *Let $\phi \in A(G(k)\backslash \mathbf{G})_\xi$ and $R > 0$. Suppose R sufficiently large* *with respect to* ϕ. *Then for every* $f \in H_\xi^R$, *there exists a unique element* $\Delta(f)\phi \in A(G(k)\backslash\mathbf{G})_\xi$ *such that for every* $\theta \in \Theta_{\xi^*}$, *we have the equality*

$$\langle \Delta_\theta(f^*)\theta, \phi\rangle = \langle \theta, \Delta(f)\phi\rangle.$$

The proof will be given in Section III.2.5.

III.2.2. Cuspidal components

We first determine what the cuspidal components of $\Delta(f)\phi$ should be. Let $\phi \in A(G(k)\backslash\mathbf{G})_\xi$ and $P = MU$ be a standard parabolic subgroup of G. Let

$$D(M,\phi) \subset \mathbb{C}[\mathrm{Re}\,\mathfrak{a}_M^G] \times \Pi_0(\mathbf{M}) \times A_0(M(k)U(\mathbb{A})\backslash\mathbf{G})$$

be a set of cuspidal data for ϕ_P^{cusp}; see I.3.3. We thus have the equality

$$\phi_P^{\mathrm{cusp}}(g) = \sum_{(Q,\pi,\psi)\in D(M,\phi)} Q(\log_M^G m_P(g))\psi(g)$$

for every $g \in \mathbf{G}$. Denote by $\Pi_0(\mathbf{M},\phi)$ the cuspidal support of ϕ_P^{cusp}; see I.3.3. We suppose that $D(M,\phi)$ is such that $\Pi_0(\mathbf{M},\phi)$ is the projection of $D(M,\phi)$ onto $\Pi_0(\mathbf{M})$. Let R be such that $R > \|\mathrm{Re}\,\pi - \mathrm{Re}\,\xi\|$ for every $\pi \in \Pi_0(\mathbf{M},\phi)$, and let $f \in H_\xi^R$. We will define a function $\Delta_P(f)\phi_P^{\mathrm{cusp}}$. Denote by $\Pi_0(\mathbf{M})_\xi$ the set of $\pi \in \Pi_0(\mathbf{M})$ such that $\chi_{\pi|Z_G} = \xi$. Note that $\Pi_0(\mathbf{M},\phi) \subset \Pi_0(\mathbf{M})_\xi$. For $\pi \in \Pi_0(\mathbf{M})_\xi$, the term $f(M,\pi)$ is defined. We have already noted in III.1.2 that it can be interpreted as an element of

$$\mathrm{End}_{\mathbf{G}}(A_0(M(k)U(\mathbb{A})\backslash\mathbf{G})_\pi).$$

We can thus define $(f(M,\pi)\psi)(g)$ for

$$\psi \in A_0(M(k)U(\mathbb{A})\backslash\mathbf{G})_\pi$$

and $g \in \mathbf{G}$. More generally, for $\lambda \in X_M^{\mathbf{G}}$, define $\lambda \circ \psi$ by

$$(\lambda \circ \psi)(g) = m_P(g)^\lambda \psi(g).$$

Then

$$\lambda \circ \psi \in A_0(M(k)U(\mathbb{A})\backslash\mathbf{G})_{\pi\otimes\lambda}$$

and we can define $(f(M,\pi\otimes\lambda)(\lambda\circ\psi))(g)$. On the other hand the space of polynomials $\mathbb{C}[\mathrm{Re}\,\mathfrak{a}_M^G]$ is isomorphic to

$$\mathrm{Sym}\left(\mathrm{Re}(\mathfrak{a}_M^G)^*\right) \otimes_{\mathbb{R}} \mathbb{C}$$

and can be interpreted as a space of derivations on $X_M^{\mathbf{G}}$ as follows: with $\mu \in \mathrm{Re}(\mathfrak{a}_M^G)^*$ associate the derivation ∂_μ defined by

$$\partial_\mu F(\lambda) = \frac{\mathrm{d}}{\mathrm{d}t} F(\lambda + t\mu)_{|t=0},$$

for every function F on $X_M^{\mathbf{G}}$, where we identify $t\mu \in \mathrm{Re}(\mathfrak{a}_M^G)^*$ with an element of $\mathrm{Re}\, X_M^{\mathbf{G}}$. Thus with $Q \in \mathbb{C}[\mathrm{Re}\,\mathfrak{a}_M^G]$ we associate a derivation ∂_Q. We check that if we fix $g \in \mathbf{G}$ and write $m_P(g)^\bullet$ for the function on $X_M^{\mathbf{G}}$ which at λ equals $m_P(g)^\lambda$, we have the equality

(1) $$(\partial_Q m_P(g)^\bullet)(\lambda) = Q(\log_M^G m_P(g)) m_P(g)^\lambda.$$

This being so, we set for $g \in \mathbf{G}$:

$$(\Delta_P(f)\phi_P^{\mathrm{cusp}})(g) = \sum_{(Q,\pi,\psi)\in D(M,\phi)} \partial_Q(\lambda \longmapsto (f(M, \pi\otimes\lambda)(\lambda\circ\psi))(g))_{|\lambda=0}.$$

We can check that this term does not depend on the choice of the set $D(M,\phi)$. This is actually also a consequence of the proof of the following lemma.

Lemma *Let $\phi, \phi' \in A(G(k)\backslash \mathbf{G})_\xi$, $R > 0$, $f \in H_\xi^R$. We suppose R sufficiently large. Then the following conditions are equivalent:*

(i) *for every $\theta \in \Theta_{\xi^*}$, $\langle \theta, \phi' \rangle = \langle \Delta_\theta(f^*)\theta, \phi\rangle$;*
(ii) *for every $\theta \in \Theta_{\xi^*}^R$, $\langle \theta, \phi' \rangle = \langle \Delta_\theta(f^*)\theta, \phi\rangle$;*
(iii) *for every standard parabolic subgroup P of G*

$$\phi_P^{\mathrm{cusp}} = \Delta_P(f)\phi_P'^{\mathrm{cusp}}.$$

Proof Let $P = MU$ be a standard parabolic subgroup of G and $\mathfrak{P}$ an orbit under $X_M^{\mathbf{G}}$ of irreducible cuspidal representations of $\mathbf{M}$, the restrictions of whose central characters to $Z_{\mathbf{G}}$ are equal to ξ^*. Let $\varphi \in P_{(M,\mathfrak{P})}^R$. By an easy calculation, we have

$$\langle \theta_\varphi, \phi' \rangle = \int_{Z_{\mathbf{G}} M(k) U(\mathbb{A})\backslash \mathbf{G}} \overline{\varepsilon F\varphi(g)} \phi_P'(g)\, dg;$$

see II.1.3 for the definition of $\varepsilon F\varphi$. As we only have $\varphi \in P_{(M,\mathfrak{P})}^R$ and not $\varphi \in P_{(M,\mathfrak{P})}$, the function $\varepsilon F\varphi$ does not belong to the space $C_0(M(k)U(\mathbb{A})\backslash \mathbf{G})_{\xi^*}$ of I.3.4. But choose a sequence $(F_n)_{n\in\mathbb{N}}$ of functions on $Z_{\mathbf{G}}\mathbf{M}^1 U(\mathbb{A})\backslash \mathbf{G}$, all smooth with compact support, such that

(i) for every $g \in \mathbf{G}$, $0 \le F_0(g) \le F_1(g) \le \ldots \le 1$;
(ii) for every $g \in \mathbf{G}$, $\lim_{n\to\infty} F_n(g) = 1$.

By the theorem of dominated convergence, we have

$$\langle \theta_\varphi, \phi' \rangle = \lim_{n\to\infty} \int_{Z_{\mathbf{G}} M(k) U(\mathbb{A})\backslash \mathbf{G}} F_n(g)\overline{\varepsilon F\varphi(g)} \phi_P'(g)\, dg.$$

The function $F_n(\varepsilon F\varphi)$ belongs to $C_0(M(k)U(\mathbb{A})\backslash \mathbf{G})_{\xi^*}$. By the definition of $\phi_P'^{\mathrm{cusp}}$, we can thus replace ϕ_P', by $\phi_P'^{\mathrm{cusp}}$ in the preceding integral.

Passing to the limit, we obtain

$$\langle \theta_\varphi, \phi' \rangle = \int_{Z_G M(k)U(\mathbb{A})\backslash \mathbf{G}} \overline{\varepsilon F\varphi(g)} \phi'^{\text{cusp}}_P(g)\, dg,$$

as long as this integral is absolutely convergent, which is the case if R is sufficiently large, as can be seen from the upper bounds established in II.1.4.

Suppose we have proved the formula

$$(2) \qquad \langle \Delta_\theta(f^*)\theta_\varphi, \phi \rangle = \int_{Z_G M(k)U(\mathbb{A})\backslash \mathbf{G}} \overline{\varepsilon F\varphi(g)} (\Delta_P(f)\phi_P^{\text{cusp}})(g)\, dg.$$

When $\mathfrak{P}$ runs through the orbits in $\Pi_0(\mathbf{M})_{\xi^\bullet}$ and φ runs through $P_{(M,\mathfrak{P})}$, $\varepsilon F\varphi$ runs through a set which linearly generates $C_0(M(k)U(\mathbb{A})\backslash \mathbf{G})_{\xi^\bullet}$. The two formulae above then give the implications (i) $\Longrightarrow$ (iii) $\Longrightarrow$ (ii). As (ii) $\Longrightarrow$ (i) trivially, the lemma thus results from (2) which we now proceed to prove.

We have as above the equality

$$\langle \Delta_\theta(f^*)\theta_\varphi, \phi \rangle = \int_{Z_G M(k)U(\mathbb{A})\backslash \mathbf{G}} \overline{\varepsilon F(f^*\varphi)(g)} \phi_P^{\text{cusp}}(g)\, dg.$$

Choosing a set $D(M, \phi)$ of cuspidal data for ϕ_P^{cusp}, we obtain

$$\langle \Delta_\theta(f^*)\theta_\varphi, \phi \rangle = \sum_{(Q,\pi,\psi)\in D(M,\phi)} \int_{Z_G M(k)U(\mathbb{A})\backslash \mathbf{G}} \overline{\varepsilon F(f^*\varphi)(g)} Q(\log_M^G m_P(g))\psi(g)\, dg.$$

Fix $(Q, \pi, \psi) \in D(M, \phi)$. For $\lambda \in X_M^{\mathbf{G}}$, set

$$I(\varphi, \psi; \lambda) = \int_{Z_G M(k)U(\mathbb{A})\backslash \mathbf{G}} \overline{\varepsilon F\varphi(g)} m_P(g)^\lambda \psi(g)\, dg.$$

The following equality is a consequence of (1):

$$(3) \qquad \langle \Delta_\theta(f^*)\theta_\varphi, \phi \rangle = \sum_{(Q,\pi,\psi)\in D(M,\phi)} \partial_Q I(f^*\varphi, \psi; 0).$$

Let us calculate $I(\varphi, \psi; \lambda)$. It is zero if for $\pi' \in \mathfrak{P}$, the restrictions of χ_π and $-\overline{\chi}_{\pi'}$ to $Z_{\mathbf{M}}^1 (:= Z_{\mathbf{M}} \cap \mathbf{M}^1)$ are distinct (note that the restriction of $\chi_{\pi'}$ to $Z_{\mathbf{M}}^1$ does not depend on the choice of π' in $\mathfrak{P}$): it suffices to decompose the integral into a double integral over

$$(Z_G(Z_M(k) \cap Z_{\mathbf{M}})\backslash Z_G Z_{\mathbf{M}}^1) \times (Z_G Z_{\mathbf{M}}^1 M(k)U(\mathbb{A})\backslash \mathbf{G}),$$

the integral over the first factor being easy to calculate. Suppose the restrictions of χ_π and $-\overline{\chi}_{\pi'}$ to $Z^1_{\mathbf{M}}$ equal, for $\pi' \in \mathfrak{P}$. We can write

$$I(\varphi,\psi;\lambda) = m \int_{Z_{\mathbf{M}}M(k)U(\mathbb{A})\backslash \mathbf{G}} \int_{Z_{\mathbf{G}}Z^1_{\mathbf{M}}\backslash Z_{\mathbf{M}}} m_P(zg)^\lambda \psi(zg) \times$$

$$\int_{\pi'\in\mathfrak{P},\mathrm{Re}\,\pi'=\lambda_0+\mathrm{Re}\,\xi^*} \overline{\varphi(\pi')(zg)}\, d\pi'\, dz\, dg$$

where $m = \mathrm{meas}(Z_{\mathbf{G}}(Z_M(k)\cap Z_{\mathbf{M}})\backslash Z_{\mathbf{G}}Z^1_{\mathbf{M}})$ and $\lambda_0 \in \mathrm{Re}\,X^{\mathbf{G}}_M$, $\|\mathrm{Re}\,\lambda_0\| < R$. The interior double integral can be calculated by Fourier inversion whenever $R > \|\mathrm{Re}\,\pi - \mathrm{Re}\,\xi\|$. Supposing the measure on $Z_{\mathbf{G}}Z^1_{\mathbf{M}}\backslash Z_{\mathbf{M}}$ suitably chosen, we obtain

$$I(\varphi,\psi;\lambda) = m \int_{Z_{\mathbf{M}}M(k)U(\mathbb{A})\backslash \mathbf{G}} m_P(g)^\lambda \psi(g) \sum_{\pi'\in\mathfrak{P},-\chi_{\pi'}=\chi_\pi+\lambda} \overline{\varphi(\pi')}(g)\, dg.$$

The remaining integral is easy to calculate and we finally obtain

$$I(\varphi,\psi;\lambda) = \begin{cases} 0, & \text{if } -(\overline{\pi\otimes\lambda}) \notin \mathfrak{P}, \\ m\langle \varphi(-(\overline{\pi}\otimes\lambda)), \lambda\circ\psi\rangle, & \text{if } -(\overline{\pi\otimes\lambda}) \in \mathfrak{P}, \end{cases}$$

where we refer to the natural product on

$$A_0(M(k)U(\mathbb{A})\backslash\mathbf{G})_{-(\overline{\pi\otimes\lambda})} \times A_0(M(k)U(\mathbb{A})\backslash\mathbf{G})_{\pi\otimes\lambda}.$$

Thus

$$I(f^*\varphi,\psi;\lambda)$$

$$= \begin{cases} 0, & \text{if } -(\overline{\pi\otimes\lambda}) \notin \mathfrak{P}, \\ m\langle f^*(-(\overline{\pi\otimes\lambda}))\varphi(-(\overline{\pi\otimes\lambda})), \lambda\circ\psi\rangle, & \text{if } -(\overline{\pi\otimes\lambda}) \in \mathfrak{P}. \end{cases}$$

By definition (see III.1.2)

$$\begin{aligned}\langle f^*(-(\overline{\pi\otimes\lambda}))\varphi(-(\overline{\pi\otimes\lambda})), \lambda\circ\psi\rangle \\ &= \langle f(\pi\otimes\lambda)^*\varphi(-(\overline{\pi\otimes\lambda})), \lambda\circ\psi\rangle \\ &= \langle \varphi(-(\overline{\pi\otimes\lambda})), f(\pi\otimes\lambda)(\lambda\circ\psi)\rangle.\end{aligned}$$

Using the preceding calculation, we obtain

$$I(f^*\varphi,\psi;\lambda) = \int_{Z_{\mathbf{G}}M(k)U(\mathbb{A})\backslash\mathbf{G}} \overline{\varepsilon F\varphi}(g) f(\pi\otimes\lambda)(\lambda\circ\psi)(g)\, dg.$$

Applying formula (3), we obtain

$$\langle \Delta_\theta(f^*)\theta_\varphi, \phi\rangle =$$

$$\int_{Z_{\mathbf{G}}M(k)U(\mathbb{A})\backslash\mathbf{G}} \sum_{(Q,\pi,\psi)\in D(M,\phi)} \overline{\varepsilon F\varphi}(g)\partial_Q(f(\pi\otimes\lambda)(\lambda\circ\psi)(g))_{|\lambda=0}\, dg,$$

and it suffices to apply the definition of $\Delta_P(f)\phi_P^{\mathrm{cusp}}$ to obtain formula (2). This concludes the proof. □

III.2.3. Remark

The statement III.2.1 can be generalised as follows. If $P = MU$ is a standard parabolic subgroup, we deduce the generalised statement from the statement for M by considering a character η of $Z_{\mathbf{M}}$, functions $\phi \in A(M(k)U(\mathbb{A})\backslash\mathbf{G})_\eta$, $f \in H_\eta^{M,R}$ (the analogue of H_ξ^R when G is replaced by M), θ belonging to the space 'induced' from Θ_{η}^{M} (the analogue of Θ_{ξ} when G is replaced by M) constructed via the spaces $P_{(M',\mathfrak{P}'_M)}$ of II.2.2. We denote this 'induced' space by $\Theta_{\eta}^{M,G}$.

III.2.4. A subalgebra of operators

We will need a subalgebra $H_\xi^{\exp}$ of H_ξ^R which we now introduce. It is the subspace consisting of $f \in H_\xi^R$ such that for every pair (M,π), where M is a standard Levi and π an irreducible cuspidal representation of $\mathbf{M}$ such that $\chi_{\pi|Z_{\mathbf{G}}} = \xi$, there exists an integer n and elements $m_i \in \mathbf{M}$ and $u_i \in \mathrm{End}_{\mathbf{M}}(A_0(M(k)\backslash\mathbf{M})_\pi)$, for $i = 1,\dots,n$, such that

$$f_{M,\pi}(\lambda) = \sum_{i=1}^{n} m_i^\lambda u_i$$

for every $\lambda \in X_M^{\mathbf{G}}$; see III.1.1(i) for the definition of $f_{M,\pi}$.

This subspace $H_\xi^{\exp}$ satisfies a property of stability under restriction. In general, if $f \in H_\xi^R$, if M is a standard Levi and η a character of $Z_{\mathbf{M}}$ whose restriction to $Z_{\mathbf{G}}$ is ξ, we define f_η to be the restriction of f to pairs (M',π') such that $M' \subset M$ and $\chi_{\pi'|Z_{\mathbf{M}}} = \eta$. We have $f_\eta \in H_\eta^{M,R'}$, whenever $R' + \|\mathrm{Re}\,\eta - \mathrm{Re}\,\xi\| < R$. If $f \in H_\xi^{\exp}$, we have $f_\eta \in H_\eta^{M,\exp}$ (the analogue of $H^{\exp}$ when G is replaced by M). More precisely, we show that there exists an integer n and for $i = 1,\dots,n$, a character η_i of $Z_{\mathbf{M}}$ whose restriction to $Z_{\mathbf{G}}$ is ξ, an element $f_i \in H_{\eta_i}^{M,\exp}$ and an element $m_i \in \mathbf{M}$ such that

(1)(i) for every pair (M',π') such that $M' \subset M$ and $\chi_{\pi'|Z_{\mathbf{M}}} = \eta_i$, for every $\nu \in X_M^{\mathbf{M}}$, we have

$$f_i(M',\pi' \otimes \nu) = m_i^\nu \nu \circ f_i(M',\pi') \circ (-\nu);$$

(1)(ii) for every pair (M',π') such that $M' \subset M$ and $\chi_{\pi'|Z_{\mathbf{M}}} = \eta$, we have

$$f_\eta(M',\pi') = \sum_{i=1}^{n} f_{\eta,i}(M',\pi'),$$

where we set

$$f_{\eta,i}(M',\pi') = 0,$$

if $\eta - \eta_i$ is non-trivial on $Z_{\mathbf{M}}^1$;

$$f_{\eta,i}(M',\pi') = m_i^\mu \mu \circ f_i(M',\pi' \otimes (-\mu)) \circ (-\mu),$$

if $\eta - \eta_i$ is trivial on $Z^1_{\mathbf{M}}$, where μ is an element of $X^{\mathbf{G}}_M$ such that $\mu_{|Z_{\mathbf{G}}} = \eta - \eta_i$.

(By (i), the above expression does not depend on the choice of μ.)

Proof By the definition of H_ξ, we can find a finite set J' and for every $j \in J'$, a pair (M_j, π_j) consisting of a standard Levi M_j and an irreducible cuspidal representation π_j of $\mathbf{M}_j$ such that $\chi_{\pi_j|Z_{\mathbf{G}}} = \xi$, such that f is supported on

$$\bigcup_{j\in J'} \{(M_j, \pi_j \otimes \lambda);\ \lambda \in X^{\mathbf{G}}_{M_j}\}.$$

Fix such objects with $|J'|$ minimal and let $J = \{j \in J';\ M_j \subset M\}$. By definition of $H^{\exp}_\xi$, for every $j \in J$, we can choose a finite set L_j and for each $\ell \in L_j$, elements $m_{j,\ell} \in \mathbf{M}^1_j\backslash\mathbf{M}_j$, $u_{j,\ell} \in \mathrm{End}_{\mathbf{M}_j}(A_0(M_j(k)\backslash\mathbf{M}_j)_{\pi_j})$, such that

$$f_{M_j,\pi_j}(\lambda) = \sum_{\ell\in L_j} m^{\lambda}_{j,\ell} u_{j,\ell} \tag{2}$$

for all $\lambda \in X^G_{M_j}$. We suppose, as we may, that the $m_{j,\ell}$ are all distinct for fixed j. Denote by $\mathfrak{M}_j$ the set of images of $m_{j,\ell}$ in $\mathbf{M}^1\backslash\mathbf{M}$. Let us consider the set $\coprod_{j\in J} \mathfrak{M}_j$ which we re-index by the set $\{1,\ldots,n\}$, which defines the integer n. With $i \in \{1,\ldots,n\}$ is thus associated an element $j \in J$ and an element, denoted by $\dot{m}_i$, of $\mathfrak{M}_j \subset \mathbf{M}^1\backslash\mathbf{M}$. We set $\eta_i = \chi_{\pi_j|Z_{\mathbf{M}}}$ and we fix an element m_i of $\mathbf{M}$ of image $\dot{m}_i$ in $\mathbf{M}^1\backslash\mathbf{M}$. It remains to define f_i. Let us begin by defining a function f'_i. For a pair (M', π') consisting of a Levi M' of M and of a cuspidal representation π' of $\mathbf{M}'$ such that $\chi_{\pi'|Z_{\mathbf{M}}} = \eta_i$, we set

$$f'_i(M',\pi') = \begin{cases} \lambda \circ \left(\sum_{\ell\in L_j(\dot{m}_i)} m^{\lambda}_{j,\ell} u_{j,\ell}\right) \circ (-\lambda), & \text{if } M' = M_j \text{ and } \pi' = \pi_j \otimes \lambda, \text{ with } \lambda \in X^{\mathbf{M}}_{M_j}, \\ 0, & \text{otherwise;} \end{cases} \tag{3}$$

where $L_j(\dot{m}_i)$ is the set of $\ell \in L_j$ such that $m_{j,\ell}$ is of image $\dot{m}_i$ in $\mathbf{M}^1\backslash\mathbf{M}$. We check that this definition is correct: in the first case the term thus defined must be independent of the choice of λ, which is determined only modulo $\mathrm{Stab}(\pi_j) := \{\nu \in X^{\mathbf{M}_j}_{M_j};\ \pi_j \otimes \nu \simeq \pi_j\}$. But as f_{M_j,π_j} comes from an element of $H^{\exp}_\xi$, we have the relation

$$(-\nu) \circ f_{M_j,\pi_j}(\lambda+\nu) \circ \nu = f_{M_j,\pi_j}(\lambda)$$

for every $\nu \in \mathrm{Stab}(\pi_j)$. The uniqueness of the decomposition (2) implies the equality

$$(-\nu) \circ m^{\nu}_{j,\ell} u_{j,\ell} \circ \nu = u_{j,\ell}$$

for all $\ell \in L_j$, $\nu \in \mathrm{Stab}(\pi_j)$, which conversely implies the independence of the choice of λ in formula (3). The function f'_i is thus well-defined. As $m^\nu_{j,\ell} = m^\nu_i$ for all $\nu \in X^{\mathbf{M}}_M$, $\ell \in L_j(\dot{m}_i)$, we easily show that f'_i satisfies condition (1)(i). Define a function f_i by

$$f_i(M',\pi') = |W_M(M')|^{-1} \sum_{w \in W_M(M')} \mathrm{ad}(w)^{-1} \circ f'_i(wM'w^{-1}, w\pi') \circ \mathrm{ad}(w)$$

where $W_M(M')$ is the set of elements of the Weyl group of M such that $wM'w^{-1}$ is standard. It is clear that the function f_i thus defined satisfies condition (iii) of III.1.1 with respect to the group M. We easily deduce that $f_i \in H^{M,\exp}_{\eta_i}$. Like f'_i, it satisfies property (1)(i). It remains to show that with these definitions, we have the desired property (1)(ii).

Let us begin by proving the relation

$$(4) \qquad f_\eta(M',\pi') = \sum_{i=1}^{n} f'_{\eta,i}(M',\pi'),$$

for M',π',η as in (1)(ii), where $f'_{\eta,i}$ is defined by replacing f_i by f'_i in the definition of $f_{\eta,i}$. Let M',π',η be as in (1)(ii). We have

$$f_\eta(M',\pi') = f(M',\pi').$$

(a) If there exists no $j \in J$ such that $M' = M_j$ and π' is of the form $\pi_j \otimes \lambda$ with $\lambda \in X^{\mathbf{G}}_{M_j}$, we have $f(M',\pi') = 0$ by definition of the set J. But then for every $i = 1,\dots,n$, either $\eta - \eta_i$ is non-trivial on $\mathbf{M}^1$, or it is trivial but $f'_i(\pi' \otimes \lambda) = 0$ for every $\lambda \in X^{\mathbf{G}}_{M'}$ such that $\lambda_{|Z_{\mathbf{M}}} = \eta - \eta_i$. In any case $f'_{\eta,i}(M',\pi') = 0$, which gives (4).

(b) Suppose on the contrary that there exists $j \in J$ and $\lambda \in X^{\mathbf{G}}_{M_j}$ such that $M' = M_j$ and $\pi' = \pi_j \otimes \lambda$. We fix such a pair (j,λ) (j will actually be unique). We then have

$$(5) \qquad f(M',\pi') = \lambda \circ f_{M_j,\pi_j}(\lambda) \circ (-\lambda) = \lambda \circ \left(\sum_{\ell \in L_j} m^\lambda_{j,\ell} u_{j,\ell} \right) \circ (-\lambda)$$

$$= \sum_{\dot{m} \in \mathfrak{M}_j} \lambda \circ \left(\sum_{\ell \in L_j(\dot{m})} m^\lambda_{j,\ell} u_{j,\ell} \right) \circ (-\lambda)$$

Let $i \in \{1,\dots,n\}$. If the element of J associated with i is different from j, we check as in (a) that $f'_{\eta,i}(M',\pi') = 0$. Suppose that i is associated with the pair $(j,\dot{m}_i)$, with $\dot{m}_i \in \mathfrak{M}_j$. The hypothesis $\pi' = \pi_j \otimes \lambda$ implies that $\eta - \eta_i$ is trivial on $Z^1_{\mathbf{M}}$. Fix an element μ of $X^{\mathbf{G}}_M$ such that $\mu_{|Z_{\mathbf{M}}} = \eta - \eta_i$. Note that $\pi' \otimes (-\mu) = \pi_j \otimes (\lambda - \mu)$. The definitions imply that

$$f'_{\eta,i}(M',\pi') = m^\mu_i \lambda \circ \left(\sum_{\ell \in L_j(\dot{m}_i)} m^{\lambda-\mu}_{j,\ell} u_{j,\ell} \right) \circ (-\lambda).$$

But $m_i^\mu = m_{j,\ell}^\mu$ for every $\ell \in L_j(\dot{m}_i)$, so

$$(6) \qquad f'_{\eta,i}(M',\pi') = \lambda \circ \left(\sum_{\ell \in L_j(\dot{m}_i)} m_{j,\ell}^\lambda u_{j,\ell} \right) \circ (-\lambda).$$

Comparing formulae (5) and (6), we obtain the equality (4) which is now proved in every case.

Let $w \in W_M(M')$. Apply equality (4) to the pair $wM'w^{-1}$, $w\pi'$, then conjugate the equality obtained by $\mathrm{ad}(w)$, and take the weighted sum over $w \in W_M(M')$ of the resulting formulae. On the left we obtain $f_\eta(M',\pi')$ since f_η satisfies condition (iii) of III.1.1 (relative to M). On the right we obtain

$$\sum_{i=1}^n f_{\eta,i}(M',\pi')$$

by definition of the functions f_i. This proves (1).

III.2.5. Proof of Proposition III.2.1

We prove this by induction on the semi-simple rank of G. Suppose the result is true for every proper standard Levi subgroup of G. This implies an 'induced' statement concerning the elements of $A(M(k)U(\mathbb{A})\backslash \mathbf{G})$ if $P = MU$ is a proper standard parabolic, see III.2.3.

Twisting all the objects by $\mathrm{Re}\,\xi$, we reduce to the case where ξ is unitary, which we suppose from now on. Let $\phi \in A(G(k)\backslash \mathbf{G})_\xi$. For every integer N, denote by $H_\xi^R(\phi, N)$ the space of $f \in H_\xi^R$ such that for every standard Levi M and every $\pi \in \Pi_0(\mathbf{M}, \phi)$ (see I.3.5), the function $f_{M,\pi}$ vanishes to the order at least N at $\lambda = 0$; see III.1.1(i) for the definition of $f_{M,\pi}$. By definition of the terms $\Delta_P(f)\phi_P^{\mathrm{cusp}}$, there exists an integer N such that $\Delta_P(f)\phi_P^{\mathrm{cusp}} = 0$ for every P if $f \in H_\xi^R(\phi, N)$. Fix such an integer N. It is a consequence of Lemma III.2.2 that for $f \in H_\xi^R(\phi, N)$, the function $\Delta(f)\phi := 0$ satisfies the equality of the statement of III.2.1. The proposition is thus proved for $f \in H_\xi^R(\phi, N)$.

We check that if $f \in H_\xi^R$, there exists $f' \in H_\xi^{\exp}$ (see III.2.4) such that $f'_{M,\pi}$ has the same expansion to the order N at 0 as $f_{M,\pi}$ for every M and every $\pi \in \Pi_0(\mathbf{M}, \phi)$. In other words, $H_\xi^R = H_\xi^{\exp} + H_\xi^R(\phi, N)$. It thus suffices to prove the proposition under the hypothesis that $f \in H_\xi^{\exp}$.

With this hypothesis, we will show that:

(1) there exists a function ϕ' on $G(k)\backslash \mathbf{G}$, of 'central character' ξ, which is the sum of a square integrable function and of a function of moderate growth, such that

$$\langle \theta, \phi' \rangle = \langle \Delta_\theta(f^*)\theta, \phi \rangle$$

for every $\theta \in \Theta_\xi$.

Fix $T \in \mathrm{Re}\,\mathfrak{a}_{M_0}$, with $\langle T, \check{\alpha}\rangle$ large enough for every $\alpha \in \Delta_0$. For every $g \in \mathbf{G}$ we have the equality

$$\phi(g) = \wedge^T \phi(g) - \sum_{P=MU;P_0 \subset P \subsetneq G} (-1)^{\dim T_M - \dim T_G} \phi[P](g),$$

where we have set

$$\phi[P](g) = \sum_{\gamma \in P(k)\backslash G(k)} \phi_P(\gamma g)\hat{\tau}_P\left(\log_M m_P(\gamma g) - T\right),$$

see I.2.13 and [A2] §1. It suffices to prove the analogues of (1) obtained by replacing ϕ by $\wedge^T \phi$ and by each of the $\phi[P]$.

The case of $\wedge^T \phi$ is easy: as $\wedge^T \phi \in L^2(G(k)\backslash \mathbf{G})_\xi$ and $f \in H_\xi^{\exp} \subset H_{\xi,b}^R$, $\Delta(f) \wedge^T \phi$ is defined (see III.1.4): it suffices to set $\phi' = \Delta(f) \wedge^T \phi$.

Consider the case of $\phi[P]$ for a proper standard parabolic subgroup $P \subsetneq G$. Fix another standard parabolic subgroup $P' = M'U'$ of G (possibly $P' = P$), $\mathfrak{P}'$ an orbit of irreducible cuspidal representations of $\mathbf{M}'$ such that $\chi_{\pi'|Z_\mathbf{G}} = \xi$ for $\pi' \in \mathfrak{P}'$, and let $\varphi \in P_{(M',\mathfrak{P}')}$. We calculate

$$\langle \Delta_\theta(f^*)\theta_\varphi, \phi[P]\rangle = \int_{Z_\mathbf{G} M(k)U(\mathbb{A})\backslash \mathbf{G}} \overline{\theta}_{f^*\varphi,P}(g)\hat{\tau}_P(\log_M m_P(g) - T)\phi_P(g)\, dg.$$

Fix R' and $\mu_0 \in \mathrm{Re}\,X_M^\mathbf{G}$ so as to ensure the validity of Lemmas II.2.2 and II.2.5 (where we set $R = \infty$ and exchange the roles of M and M'). Applying these lemmas, we obtain

(2) $\langle \Delta_\theta(f^*)\theta_\varphi, \phi[P]\rangle =$

$$\int_{Z_\mathbf{G} M(k)U(\mathbb{A})\backslash \mathbf{G}} \sum_{w \in W(M',M)} \int_{D^\bullet_{w,\mu_0}} \overline{\theta^M(f^*\varphi, \pi)(g)}\hat{\tau}_P(\log_M m_P(g) - T)\phi_P(g)\, d\pi\, dg,$$

this expression being absolutely convergent.

Choose a finite set

$$D \subset \mathbb{C}[\mathrm{Re}\,\mathfrak{a}_M^G] \times \mathrm{Hom}(Z_\mathbf{M}, \mathbb{C}^*) \times A(M(k)U(\mathbb{A})\backslash \mathbf{G}),$$

such that

(3)(i) if $(Q, \eta, \psi) \in D$, then $\psi \in A(M(k)U(\mathbb{A})\backslash \mathbf{G})_\eta$,

(ii) for every $g \in \mathbf{G}$ we have the equality (see I.3.2)

$$\phi_P(g) = \sum_{(Q,\eta,\psi)\in D} Q(\log_M^G m_P(g))\psi(g).$$

Temporarily fix $(Q, \eta, \psi) \in D$, $w \in W(M', M)$ and $\pi \in D^\bullet_{w,\mu_0}$, and set

(4) $$J = \int_{Z_\mathbf{G} M(k)U(\mathbb{A})\backslash \mathbf{G}} \overline{\theta^M(f^*\varphi, \pi)(g)}\hat{\tau}_P(\log_M m_P(g) - T) \times Q(\log_M^G m_P(g))\psi(g)\, dg.$$

Denote by χ the restriction of χ_π to $\mathbf{Z_M}$ and define a function b on $\mathbf{G}$ by

$$b(g) = \int_{Z_G(Z_M(k)\cap \mathbf{Z_M})\backslash \mathbf{Z_M}} z^{\eta+\overline{\chi}}\hat{\tau}_P\left(\log_M m_P(zg) - T\right) Q\left(\log_M^G m_P(zg)\right)\, dz.$$

Letting an intermediary integral on $Z_G(Z_M(k)\cap \mathbf{Z_M})\backslash \mathbf{Z_M}$ appear in the definition of J, we obtain the equality

$$(5) \qquad J = \int_{\mathbf{Z_M}M(k)U(\mathbb{A})\backslash \mathbf{G}} \overline{\theta^M(f^*\varphi,\pi)}(g) b(g)\psi(g)\, dg.$$

The function b is left invariant under $\mathbf{M}^1$, right invariant under $\mathbf{K}$, and $\mathbf{Z_M}$ acts on it via the character $-\eta-\overline{\chi}$. If this last is non-trivial on $\mathbf{Z}_\mathbf{M}^1$, we have $b=0$. Otherwise b is a linear combination of functions $g \mapsto m_P(g)^\nu$ where ν runs through the finite subset, called $Y(-\eta-\overline{\chi})$, of elements of $X_M^\mathbf{G}$ whose restriction to $\mathbf{Z_M}$ is $-\eta-\overline{\chi}$. In this case we can introduce uniquely determined coefficients $b_\nu \in \mathbb{C}$ for $\nu \in Y(-\eta-\overline{\chi})$, such that for every $g \in \mathbf{G}$, we have

$$(6) \qquad b(g) = \sum_{\nu \in Y(-\nu-\overline{\chi})} b_\nu m_P(g)^\nu.$$

In any case, the function $b\psi$ belongs to $A(M(k)U(\mathbb{A})\backslash \mathbf{G})_{-\overline{\chi}}$. From $f \in H_\xi^{\exp}$ we deduce an element $f_{-\overline{\chi}} \in H_{-\overline{\chi}}^{M,\exp}$; see III.2.4. Considering definitions II.2.2 and III.1.2, we see that $\theta^M(\varphi,\pi) \in \Theta_\chi^{M,G}$, see III.2.3, and that we have the equality

$$\theta^M(f^*\varphi,\pi) = \Delta_\theta((f_{-\overline{\chi}})^*)\theta^M(\varphi,\pi).$$

Using the induction hypothesis, properly generalised as was explained in III.2.3, we obtain

$$J = \int_{\mathbf{Z_M}M(k)U(\mathbb{A})\backslash \mathbf{G}} \overline{\theta^M(\varphi,\pi)}(g)(\Delta(f_{-\overline{\chi}})(b\psi))(g)\, dg.$$

Let us calculate $\Delta(f_{-\overline{\chi}})(b\psi)$. Introduce the integer n and the elements η_i, f_i, m_i of III.2.4. For $i=1,\ldots,n$, define a function ψ_i as follows: if $\eta-\eta_i$ is non-trivial on $\mathbf{Z}_\mathbf{M}^1$, $\psi_i = 0$; if $\eta-\eta_i$ is trivial on $\mathbf{Z}_\mathbf{M}^1$, we choose $\mu_i \in X_M^\mathbf{G}$ such that $\mu_{i|\mathbf{Z_M}} = \eta-\eta_i$ and we set

$$\psi_i = m_i^{\mu_i}\mu_i \circ \Delta(f_i)((-\mu_i)\circ\psi).$$

This term does not depend on the choice of μ_i: it suffices to check that the cuspidal components $\psi_{i,P'}^{\text{cusp}}$ don't, for every standard parabolic $P' \subset P$. But these components were calculated in Lemma III.2.2 and it suffices to use relation (2)(i) of III.2.4 to prove independence of the choice of μ_i. The function ψ_i is thus well-defined. It belongs to $A(M(k)U(\mathbb{A})\backslash \mathbf{G})_\eta$. Let us show that we have the equality

$$(7)\qquad (\Delta(f_{-\overline{\chi}})(b\psi))(g) = \sum_{i=1}^{n} b(m_i g)\psi_i(g).$$

If $-\eta - \overline{\chi}$ is non-trivial on $Z^1_{\mathbf{M}}$, then $b = 0$ and the equality is trivial. Suppose $-\eta - \overline{\chi}$ is trivial on $Z^1_{\mathbf{M}}$. Using equality (6), we obtain

$$\Delta(f_{-\overline{\chi}})(b\psi) = \sum_{\nu \in Y(-\eta-\overline{\chi})} b_\nu \Delta(f_{-\overline{\chi}})(\nu \circ \psi).$$

Using III.2.4(ii) and comparing cuspidal components of the functions in question as indicated above, we see that

$$\Delta(f_{-\overline{\chi}})(\nu \circ \psi) = \sum_i m_i^{\mu_i'} \mu_i' \circ \Delta(f_i)((\nu - \mu_i') \circ \psi),$$

where the sum is over the $i \in \{1, \dots, n\}$ such that $\eta - \eta_i + \nu_{|Z_{\mathbf{M}}}$ is trivial on $Z^1_{\mathbf{M}}$, and where μ_i' is an element of $X^{\mathbf{G}}_M$ such that $\mu'_{i|Z_{\mathbf{M}}} = \eta - \eta_i + \nu_{|Z_{\mathbf{M}}}$. As $\nu \in X^{\mathbf{G}}_M$, $\nu_{|Z^1_M}$ is trivial and the conditions '$\eta - \eta_i + \nu_{|Z_{\mathbf{M}}}$ is trivial on $Z^1_{\mathbf{M}}$' and '$\eta - \eta_i$ is trivial on $Z^1_{\mathbf{M}}$' are equivalent. When they are satisfied, we can choose $\mu_i' = \nu + \mu_i$, where μ_i is as in the definition of ψ_i. We obtain

$$\Delta(f_{-\overline{\chi}})(\nu \circ \psi) = \sum_i m_i^{\mu_i+\nu}(\mu_i + \nu) \circ \Delta(f_i)((-\mu_i) \circ \psi)$$

summing over the $i \in \{1, \dots, n\}$ such that $\eta - \eta_i$ is trivial on $Z^1_{\mathbf{M}}$, i.e.

$$\Delta(f_{-\overline{\chi}})(\nu \circ \psi) = \sum_{i=1}^{n} m_i^{\nu}(\nu \circ \psi_i).$$

This gives

$$\Delta(f_{-\overline{\chi}})(b\psi) = \sum_{i=1}^{n} \left(\sum_{\nu \in Y(-\eta-\overline{\chi})} b_\nu m_i^{\nu}(\nu \circ \psi_i) \right).$$

At a point g, we have the equalities

$$\sum_{\nu \in Y(-\eta-\overline{\chi})} b_\nu m_i^{\nu} m_P(g)^{\nu} = \sum_{\nu \in Y(-\eta-\overline{\chi})} b_\nu m_P(m_i g)^{\nu} = b(m_i g),$$

which gives the equality (7).

Using this equality, we obtain

$$J = \sum_{i=1}^{n} \int_{Z_{\mathbf{M}} M(k) U(\mathbb{A}) \backslash \mathbf{G}} \overline{\theta^M(\varphi, \pi)(g)} b(m_i g)\psi_i(g)\, dg$$

or even, by the calculation which we used to pass from (4) to (5),

$$J = \sum_{i=1}^{n} \int_{Z_{\mathbf{G}} M(k) U(\mathbb{A}) \backslash \mathbf{G}} \overline{\theta^M(\varphi, \pi)(g)} \hat{\tau}_P(\log_M m_P(m_i g) - T) \times$$

$$Q(\log^G_M m_P(m_i g))\psi_i(g)\, dg.$$

Define a function e on $M(k)U(\mathbb{A})\backslash\mathbf{G}$ by

$$e(g)=\sum_{i=1}^{n}\sum_{(Q,\eta,\psi)\in D}\hat{\tau}_P\left(\log_M m_P(m_ig)-T\right)Q\left(\log_M^G m_P(m_ig)\right)\psi_i(g).$$

The above result and the equalities (2) and (3) (ii) lead to:

(8) $$\langle\Delta_\theta(f^*)\theta_\varphi,\phi[P]\rangle=$$

$$\sum_{w\in W(M',M)}\int_{D^\bullet_{w,\mu_0}}\int_{Z_{\mathbf{G}}M(k)U(\mathbb{A})\backslash\mathbf{G}}\overline{\theta^M(\varphi,\pi)}(g)e(g)\,d\pi\,dg.$$

We check by the definition that $Z_{e\mathbf{G}}$ acts on e via the character ξ and that there exists $T'\in\operatorname{Re}\mathfrak{a}_{M_0}$ and a function e' of moderate growth on $M(k)U(\mathbb{A})\backslash\mathbf{G}$ such that we have the upper bound

(9) $$|e(g)|\leq e'(g)\hat{\tau}_P\left(\log_M m_P(g)-T'\right).$$

By Lemma II.2.5, the expression (8) is absolutely convergent. Thus we can permute the integrals. Thanks to Lemma II.2.2, we obtain

$$\langle\Delta_\theta(f^*)\theta_\varphi,\phi[P]\rangle=\int_{Z_{\mathbf{G}}M(k)U(\mathbb{A})\backslash\mathbf{G}}\overline{\theta}_{\varphi,P}(g)e(g)dg.$$

Define a function $\phi'[P]$ on $G(k)\backslash\mathbf{G}$ by

$$\phi'[P](g)=\sum_{\gamma\in P(k)\backslash G(k)}e(\gamma g).$$

We show using the upper bound (9) that this series converges (it is even a finite sum) and that it defines a function of moderate growth; see I.2.13. With this definition, we obtain

$$\langle\Delta_\theta(f^*)\theta_\varphi,\phi[P]\rangle=\langle\theta_\varphi,\phi'[P]\rangle.$$

The function $\phi'[P]$ does satisfy the analogue of (1) for the function $\phi[P]$, which concludes the proof of (1).

Now choose a function ϕ' satisfying (1) and a smooth, right and left **K**-finite function F on **G** with compact support, such that

$$\delta(F)\phi=\phi.$$

Set $\Delta(f)\phi=\delta(F)\phi'$. Let us show that this function satisfies the conditions of Proposition III.2.1. It is smooth, **K**-finite and of moderate growth; see I.2.5. For $\theta\in\Theta_\xi$, we have the equalities

$$\langle\theta,\Delta(f)\phi\rangle=\langle\theta,\delta(F)\phi'\rangle=\langle\delta(\check{F})\theta,\phi'\rangle,$$

where $\check{F}$ is the function defined by $\check{F}(g)=\overline{F}(g^{-1})$. We have $\delta(\check{F})\theta\in\Theta_\xi$, whence

$$\begin{aligned}\langle\delta(\check{F})\theta,\phi'\rangle&=\langle\Delta_\theta(f^*)\delta(\check{F})\theta,\phi\rangle=\langle\delta(\check{F})\Delta_\theta(f^*)\theta,\phi\rangle\\&=\langle\Delta_\theta(f^*)\theta,\delta(F)\phi\rangle=\langle\Delta_\theta(f^*)\theta,\phi\rangle.\end{aligned}$$

We thus obtain the equality

$$\langle \theta, \Delta(f)\phi \rangle = \langle \Delta_\theta(f^*)\theta, \phi \rangle.$$

This equality implies that $\Delta(f)\phi$ is $\mathfrak{z}$-finite; see I.2.12. Indeed if $z \in \mathfrak{z}$ kills ϕ, we have similarly:

$$\begin{aligned}\langle \theta, \delta(z)\Delta(f)\phi \rangle &= \langle \delta(\check{z})\theta, \Delta(f)\phi \rangle = \langle \Delta_\theta(f^*)\delta(\check{z})\theta, \phi \rangle \\ &= \langle \delta(\check{z})\Delta_\theta(f^*)\theta, \phi \rangle = \langle \Delta_\theta(f^*)\theta, \delta(z)\phi \rangle \\ &= 0.\end{aligned}$$

Thus $\delta(z)\Delta(f)\phi$ is orthogonal to every pseudo-Eisenstein series and thus $\delta(z)\Delta(f)\phi = 0$; see II.1.12, Remark 1. The function $\Delta(f)\phi$ does satisfy the conditions of the proposition. Its uniqueness is immediate by the orthogonality argument just employed.

III.2.6. Decomposition of the space of automorphic forms

Fix a character ξ of Z_G and consider the set of pairs (M, π) where M is a standard Levi subgroup of G and π an irreducible cuspidal representation of $\mathbf{M}$ the restriction of whose central character to Z_G is ξ. We say that two pairs (M, π) and (M', π') are equivalent if there exists $w \in G(k)$ such that $M' = wMw^{-1}$, $\pi' = w\pi$. Denote by $\Xi(\xi)$ the set of equivalence classes under this equivalence relation.

Remark Let $\mathfrak{y} \in \Xi(\xi)$ and set $\mathfrak{X} = \{(M, \mathfrak{P}) \mid \exists (M, \pi) \in \mathfrak{y}$ such that $\mathfrak{P}$ is the orbit of $\pi\}$. Then $\mathfrak{X}$ is an equivalence class for the equivalence relation defined in II.2.1.

For $\mathfrak{y} \in \Xi(\xi)$, we denote by $A(G(k)\backslash\mathbf{G})_\mathfrak{y}$ the space of automorphic forms $\phi \in A(G(k)\backslash\mathbf{G})_\xi$ whose cuspidal support (see I.3.5) is concentrated on $\mathfrak{y}$. That is, for every standard parabolic subgroup $P = MU$ of G, we ask that there exist a set of cuspidal data $D(M, \phi) \subset \mathbb{C}[\operatorname{Re}\mathfrak{a}_M^G] \times \Pi_0(\mathbf{M}) \times A_0(M(k)U(\mathbb{A})\backslash\mathbf{G})$ such that for every $(Q, \pi, \psi) \in D(M, \phi)$, we have $(M, \pi) \in \mathfrak{y}$. Recall from I.3.3 that $D(M, \phi)$ is a set such that

(i) for every $(Q, \pi, \psi) \in D(M, \phi)$, $\psi \in A_0(M(k)U(\mathbb{A})\backslash\mathbf{G})_\pi$;

(ii) or every $g \in \mathbf{G}$,

$$\phi_P^{\text{cusp}}(g) = \sum_{(Q,\pi,\psi)\in D(M,\phi)} Q(\log_M^G m_P(g))\psi(g).$$

It is clear that $A(G(k)\backslash\mathbf{G})_\mathfrak{y}$ is stable under right translation; more precisely, if k is a number field, under the action of $\operatorname{Lie}\mathbf{G}_\infty \times \mathbf{K} \times \mathbf{G}_f$, see I.2.12.

Theorem *We have the following decomposition into a direct sum:*

$$A(G(k)\backslash\mathbf{G})_\xi = \bigoplus_{\mathfrak{y}\in\Xi(\xi)} A(G(k)\backslash\mathbf{G})_\mathfrak{y}.$$

Proof Let $\phi \in A(G(k)\backslash\mathbf{G})_\xi$. For every standard parabolic subgroup $P = MU$ of G, choose a set of cuspidal data $D(M,\phi)$. There exists a finite set of elements of $\Xi(\xi)$, say $\{\mathfrak{y}_1,\ldots,\mathfrak{y}_n\}$ such that

$$\{(M,\pi); M \text{ standard }, \exists(Q,\psi) \text{ such that } (Q,\pi,\psi) \in D(M,\phi)\} \subset \textstyle\bigcup_{i=1}^n \mathfrak{y}_i.$$

Fix an integer N such that $N > \deg(Q)$ for every polynomial Q occurring in a cuspidal datum. For $i = 1,\ldots,n$, we can find an element $f^i \in H^R_{\xi,\mathbb{C}}$, see III.1.1, where R is chosen sufficiently large so that

(i) if $(M,\pi) \in \mathfrak{y}_j$, $j \neq i$, the function $f^i_{M,\pi}$, see III.1.1(i), vanishes to the order N at 0;
(ii) if $(M,\pi) \in \mathfrak{y}_i$, the function $f^i_{M,\pi} - 1$ vanishes to the order N at 0.

It is a consequence of these conditions and of Lemma III.2.2 that

(i) $\Delta(f^i)\phi \in A(G(k)\backslash\mathbf{G})_{\mathfrak{y}_i}$;
(ii) $\phi = \sum_{i=1}^n \Delta(f^i)\phi$.

Thus $\phi \in \sum_{\mathfrak{y}\in\Xi(\xi)} A(G(k)\backslash\mathbf{G})_\mathfrak{y}$, whence the equality

$$A(G(k)\backslash\mathbf{G})_\xi = \sum_{\mathfrak{y}\in\Xi(\xi)} A(G(k)\backslash\mathbf{G})_\mathfrak{y}.$$

That the sum is direct can be proved in a similar way: if $\sum \phi_i = 0$, where $\phi_i \in A(G(k)\backslash\mathbf{G})_{\mathfrak{y}_i}$, we choose functions f^i such that $\Delta(f^i)\phi_j = 0$ if $j \neq i$, $\Delta(f^i)\phi_i = \phi_i$, and we deduce that all the functions ϕ_i are zero.

□

III.3. Cuspidal exponents and square integrable automorphic forms

III.3.1. The result

Let ξ be a unitary character of $Z_\mathbf{G}$. We are interested in

$$A^2(G(k)\backslash\mathbf{G})_\xi = A(G(k)\backslash\mathbf{G})_\xi \cap L^2(G(k)\backslash\mathbf{G})_\xi.$$

Let $\phi \in A(G(k)\backslash\mathbf{G})_\xi$ and $P = MU$ be a standard parabolic subgroup of G. We have defined (see I.3.5) the cuspidal support $\Pi_0(\mathbf{M},\phi)$ of ϕ along P. It is a set of cuspidal representations of $\mathbf{M}$. Recall that if π is an irreducible representation of $\mathbf{M}$, we denote its central character by χ_π.

For every standard Levi subgroup M' of G, denote by M'_{der} the derived group of M', $\mathbf{M}'_{\text{der}}$ the inverse image of $M'_{\text{der}}(\mathbb{A})$ in $\mathbf{G}$ and $T_{\mathbf{M}'}$ the intersection of $Z_{\mathbf{M}'}$ with the inverse image of $T_{M'}(\mathbb{A})$ in $\mathbf{G}$.

In III.3.2 we will define an integer $N(G) \geq 1$.

Proposition *For every element ϕ of $A^2(G(k)\backslash\mathbf{G})_\xi$, for every standard* *parabolic subgroup $P = MU$ of G and for every $\pi \in \Pi_0(\mathbf{M}, \phi)$, the restriction to $T_{\mathbf{M}} \cap \mathbf{G}_{\mathrm{der}}$ of $N(G)\chi_\pi$ is positive real-valued.*

Remarks (a) To explain the notation $N(G)\chi_\pi$, recall that we have written the group law of all groups of characters additively.

(b) Suppose that k is a number field. It is a consequence of the proposition that for ϕ, M and π as in the statement, the restriction to $A_{\mathbf{M}} \cap \mathbf{G}_{\mathrm{der}}$ of χ_π is positive real-valued.

III.3.2. Definition of $N(G)$

The algebraic group $T_G \cap G_{\mathrm{der}}$ is finite. Thus $T_{\mathbf{G}} \cap \mathbf{G}_{\mathrm{der}}$ is a torsion group. Denote by $N'(G)$ the exponent of this group.

Let M and M' be two standard Levi subgroups of G and w an element of the Weyl group W of G. Suppose

(i) M' proper and maximal;
(ii) $M \subset M'$;
(iii) $wMw^{-1} = M$;
(iv) w, acting in $\mathfrak{a}_{M_0}^*$, does not fix the subspace $(\mathfrak{a}_{M'}^G)^*$ pointwise.

For $t \in T_{\mathbf{M}}$ denote by $(w-1)T_{\mathbf{M}}$ the subgroup of $T_{\mathbf{M}}$ consisting of elements $wtw^{-1}t^{-1}$. It is contained in $T_{\mathbf{M}} \cap \mathbf{G}_{\mathrm{der}}$. Let us show that the quotient

$$(1) \qquad (T_{\mathbf{M}} \cap \mathbf{G}_{\mathrm{der}})/((w-1)T_{\mathbf{M}})(T_{\mathbf{M}} \cap \mathbf{M}'_{\mathrm{der}})$$

is torsion. We reduce easily to the case where $\mathbf{G} = G(\mathbb{A})$, in which case the assertion can be deduced from the following algebraic assertion. Denote by $(T_M \cap G_{\mathrm{der}})^\circ$ and $(T_M \cap M'_{\mathrm{der}})^\circ$ the connected components of $T_M \cap G_{\mathrm{der}}$ and $T_M \cap M'_{\mathrm{der}}$; for every split torus T, denote by $X_*(T)$ the group of one-parameter subgroups of T; denote by $(w-1)X_*(T_M)$ the subgroup of $X_*(T_M)$ consisting of elements of the form $wx - x$ for $x \in X_*(T_M)$, where w acts naturally in $X_*(T_M)$. Then the quotient

$$X_*((T_M \cap G_{\mathrm{der}})^\circ)/[(w-1)X_*(T_M) + X_*((T_M \cap M'_{\mathrm{der}})^\circ)]$$

is finite. Tensorising by $\mathbb{C}$, we are reduced to showing that

$$\mathfrak{a}_M^G = (w-1)\mathfrak{a}_M + \mathfrak{a}_M^{M'}$$

or even, by duality, that

$$\mathfrak{a}_G^* = \{\lambda \in \mathfrak{a}_M^*;\ w\lambda = \lambda\} \cap \mathfrak{a}_{M'}^*,$$

which is a consequence of hypothesis (iv).

Denote by $N(M, M', w)$ the exponent of the group (1). We now define $N(G)$ by induction on the semi-simple rank of G. It is the least common multiple of $N'(G)$ and of the integers $N(M, M', w)N(M')$ for all the triples (M, M', w) satisfying hypotheses (i) to (iv).

III.3.3. Reduction of the problem

Let $\phi \in A^2(G(k)\backslash\mathbf{G})_\xi$, R be sufficiently large, and $f \in H^R_{\xi,b}$; see III.1.4, III.1.7. There are two ways to define $\Delta(f)\phi$ according to whether we consider ϕ as an element of $A(G(k)\backslash\mathbf{G})_\xi$ or of $L^2(G(k)\backslash\mathbf{G})_\xi$. These two definitions coincide since the scalar product of $\Delta(f)\phi$ and of a pseudo-Eisenstein series is given by a unique formula. In particular, we still have $\Delta(f)\phi \in A^2(G(k)\backslash\mathbf{G})_\xi$. The same proof as in III.2.6 shows that we have

$$A^2(G(k)\backslash\mathbf{G})_\xi = \bigoplus_{\mathfrak{y}\in\Xi(\xi)} A^2(G(k)\backslash\mathbf{G})_\mathfrak{y},$$

where we have set

$$A^2(G(k)\backslash\mathbf{G})_\mathfrak{y} = A(G(k)\backslash\mathbf{G})_\mathfrak{y} \cap L^2(G(k)\backslash\mathbf{G})_\xi.$$

Fix $\mathfrak{y} \in \Xi(\xi)$. Let $\phi \in A^2(G(k)\backslash\mathbf{G})_\mathfrak{y}$. As the operators $\Delta(f)$ commute with the actions of $\mathbf{K}$ and $\mathfrak{z}$, it is a consequence of the finiteness theorem (see[HC] Theorem 1, [BJ] Theorem 5.6) that there exists a finite-dimensional subspace $V \subset A^2(G(k)\backslash\mathbf{G})_\mathfrak{y}$ containing ϕ and a real number $R > 0$ such that for every $f \in H^R_{\xi,b}$, the operator $\Delta(f)$ is defined on V and preserves V. Let us limit ourselves to the algebra $H^R_{\xi,b,\mathbb{C}} = H^R_{\xi,b} \cap H^R_{\xi,\mathbb{C}}$. The set of operators $\Delta(f)_{|V}$ for $f \in H^R_{\xi,b,\mathbb{C}}$, is a commutative subalgebra of $\mathrm{End}(V)$ which is stable under adjunction. It is thus diagonalisable.

It is a consequence of this discussion that in order to prove Proposition III.3.1, we may fix $\mathfrak{y} \in \Xi(\xi)$ and suppose that ϕ is an element of $A^2(G(k)\backslash\mathbf{G})_\xi$, eigenfunction of all the operators $\Delta(f)$ for $f \in H^R_{\xi,b,\mathbb{C}}$. It is a consequence of III.2.2 that the associated eigenvalue is $f(M, \pi)$ for a pair $(M, \pi) \in \mathfrak{y}$, and this value does not depend on the choice of the pair (M, π) by III.1.1(iii).

III.3.4. Proof of Proposition III.3.1

We prove the proposition by induction on the semi-simple rank of G. We thus suppose it proved for every proper Levi subgroup of G. As usual, we deduce from this an induced statement concerning the elements of $A^2(U(\mathbb{A})M(k)\backslash\mathbf{G})_\eta$ if $P = MU$ is a proper parabolic subgroup of G and η a unitary character of $Z_\mathbf{M}$.

Let $\mathfrak{y} \in \Xi(\xi)$, $\phi \in A^2(G(k)\backslash\mathbf{G})_\mathfrak{y}$ (see III.3.3) be an eigenfunction of the operators $\Delta(f)$ for $f \in H^R_{\xi,b,\mathbb{C}}$. If ϕ is cuspidal, the only Levi subgroup occurring in the proposition is G itself. Now $T_\mathbf{G} \cap \mathbf{G}_{\mathrm{der}}$ is a torsion group, killed by $N(G)$, and the assertion of the proposition is immediate. We thus suppose that ϕ is not cuspidal. Let us consider the set $\mathfrak{t}$ of triples

(M, π, α) consisting of a standard Levi M, an element $\pi \in \Pi_0(\mathbf{M}, \phi)$ and a root $\alpha \in \Delta_M$. It is non-empty. Introduce the basis $\{\varpi_{\check{\beta}}; \beta \in \Delta_0\}$ of $\mathfrak{a}_{M_0}^G$, dual to the basis Δ_0 of $(\mathfrak{a}_{M_0}^G)^*$. Recall (see I.4.11) that for $\alpha \in \Delta_M$, we define $\varpi_{\check{\alpha}} = \varpi_{\check{\beta}}$ where β is the unique element of Δ_0 which is projected onto α; and $\{\varpi_{\check{\alpha}}; \alpha \in \Delta_M\}$ is the basis of $\mathfrak{a}_M^G$ dual to the basis Δ_M of $(\mathfrak{a}_M^G)^*$. By I.4.11, we have

$$\langle \varpi_{\check{\alpha}}, \mathrm{Re}\,\pi \rangle < 0$$

for all $\alpha \in \Delta_M, \pi \in \Pi_0(\mathbf{M}, \phi)$; see I.3.3 for the definition of $\mathrm{Re}\,\pi$. Fix a triple $(M, \pi, \alpha) \in \mathfrak{t}$ such that

$$|\langle \varpi_{\check{\alpha}} \mathrm{Re}\,\pi \rangle| \; \|\varpi_{\check{\alpha}}\|^{-1}$$

is minimal. Let us consider the maximal proper standard parabolic subgroup $P' = M'U'$ of G containing M such that α is the unique simple root of Δ_M whose restriction to $\mathfrak{a}_{M'}$ is non-trivial. Fix a basis $\mathfrak{q}_{M'}^G$ of $\mathbb{C}[\mathrm{Re}\,\mathfrak{a}_{M'}^G]$. Then there exists a unique set

$$D \subset \mathfrak{q}_{M'}^G \times \mathrm{Hom}(Z_{\mathbf{M}'}, \mathbb{C}^*) \times A(U'(\mathbb{A})M'(k)\backslash \mathbf{G})$$

such that

(i) for every $(Q, \eta, \psi) \in D$, we have $\psi \in A(U'(\mathbb{A})M'(k)\backslash \mathbf{G})_\eta$ and $\psi \neq 0$;
(ii) for every $g \in \mathbf{G}$,

$$\phi_{P'}(g) = \sum_{(Q,\eta,\psi)\in D} Q\left(\log_{M'}^G m_{P'}(g)\right) \psi(g).$$

For every standard parabolic subgroup $P'' = M''U'' \subset P'$, we deduce from (ii) the equality

$$\phi_{P''}^{\mathrm{cusp}}(g) = \sum_{(Q,\eta,\psi)\in D} Q(\log_{M'}^G m_{P'}(g))\psi_{P''}^{\mathrm{cusp}}(g)$$

then

(1) $$\Pi_0(\mathbf{M}'', \phi) = \bigcup_{(Q,\eta,\psi)\in D} \Pi_0(\mathbf{M}'', \psi).$$

In particular, for $P'' = P$, there exists $(Q, \eta, \psi) \in D$ such that $\pi \in \Pi_0(\mathbf{M}, \psi)$. Fix one such triple $(Q, \eta, \psi) \in D$. The character η is not unitary in general but we reduce to the unitary case by replacing ψ by $\psi' = (-\mathrm{Re}\,\eta)\psi$. We have $\psi' \in A(U'(\mathbb{A})M'(k)\backslash \mathbf{G})_{\mathrm{Im}\,\eta}$. Let us show that that we also have $\psi' \in L^2(U'(\mathbb{A})M'(k)\backslash \mathbf{G})_{\mathrm{Im}\,\eta}$. Introduce the basis $\{\varpi'_{\check{\beta}}; \beta \in \Delta_0^{M'}\}$ of $\mathfrak{a}_{M_0}^{M'}$, dual to the basis $\Delta_0^{M'}$ of $(\mathfrak{a}_{M_0}^{M'})^*$. It has the same properties as $\{\varpi_{\check{\beta}}; \beta \in \Delta_0\}$: we have simply replaced G by M'. By I.4.11, to show that $\psi' \in L^2(U'(\mathbb{A})M'(k)\backslash \mathbf{G})_{\mathrm{Im}\,\eta}$ it suffices to show that for every standard parabolic subgroup $P'' = M''U''$ of P', for every root $\alpha'' \in \Delta_{M''}^{M'}$ (simple roots of $T_{M''}$ in M'), for every $\pi'' \in \Pi_0(\mathbf{M}'', \psi')$ we have

(2) $$\langle \varpi'_{\check{\alpha}''}, \mathrm{Re}\,\pi'' \rangle < 0.$$

Fix such a P'', α'' and π''. We have $\Delta_{M''}^{M'} \subset \Delta_{M''}$, thus α'' is also an element of $\Delta_{M''}$ and we have an element $\mathfrak{w}_{\check{\alpha}''}$ of our dual basis. It is in general different from $\mathfrak{w}'_{\check{\alpha}''}$, but the difference $\mathfrak{w}_{\check{\alpha}''} - \mathfrak{w}'_{\check{\alpha}''}$ belongs to $\mathfrak{a}_{M'}^G$. Now $\mathrm{Re}\,\pi'' \in (\mathfrak{a}_{M''}^{M'})^*$, since the restriction of π'' to $Z_{\mathbf{M}'}$ is unitary. The relation (2) is thus equivalent to

$$\langle \mathfrak{w}_{\check{\alpha}''}, \mathrm{Re}\,\pi'' \rangle < 0.$$

We have $(\mathrm{Re}\,\eta) \otimes \pi'' \in \Pi_0(\mathbf{M}'', \phi)$ by (1). Thus $(M'', (\mathrm{Re}\,\eta) \otimes \pi'', \alpha'') \in \mathfrak{t}$. By the maximality property of α, we have

(3) $$\langle \mathfrak{w}_{\check{\alpha}''}, \mathrm{Re}\,\eta + \mathrm{Re}\,\pi'' \rangle \|\mathfrak{w}_{\check{\alpha}''}\|^{-1} \le \langle \mathfrak{w}_{\check{\alpha}}, \mathrm{Re}\,\pi \rangle \|\mathfrak{w}_{\check{\alpha}}\|^{-1}.$$

As the restriction of η to $Z_{\mathbf{G}}$ is ξ and ξ is unitary, we have $\mathrm{Re}\,\eta \in (\mathfrak{a}_{M'}^G)^*$. Thus there exists $x \in \mathbb{R}$ such that $\mathrm{Re}\,\eta = x\mathfrak{w}_\alpha$, where $\{\mathfrak{w}_\beta; \beta \in \Delta_0\}$ is the basis of $(\mathfrak{a}_{M_0}^G)^*$ dual to the basis $\{\check{\beta}; \beta \in \Delta_0\}$ of $\mathfrak{a}_{M_0}^G$. Moreover, as $\pi \in \Pi_0(\mathbf{M}, \psi)$, the restriction of χ_π to $Z_{\mathbf{M}'}$ is η, so $\mathrm{Re}\,\pi - \mathrm{Re}\,\eta \in (\mathfrak{a}_{M_0}^{M'})^*$ and

$$\langle \mathfrak{w}_{\check{\alpha}}, \mathrm{Re}\,\pi - \mathrm{Re}\,\eta \rangle = 0.$$

We deduce from this that:

$$\langle \mathfrak{w}_{\check{\alpha}}, \mathrm{Re}\,\pi \rangle = \langle \mathfrak{w}_{\check{\alpha}}, \mathrm{Re}\,\eta \rangle = x\langle \mathfrak{w}_{\check{\alpha}}, \mathfrak{w}_\alpha \rangle.$$

As $\langle \mathfrak{w}_{\check{\alpha}}, \mathrm{Re}\,\pi \rangle < 0$ (since ϕ is square integrable, see I.4.11), we deduce that $x < 0$. Finally

$$\langle \mathfrak{w}_{\check{\alpha}''}, \mathrm{Re}\,\eta \rangle = x\langle \mathfrak{w}_{\check{\alpha}''}, \mathfrak{w}_\alpha \rangle.$$

We then deduce from (3) the inequality

$$\langle \mathfrak{w}_{\check{\alpha}''}, \mathrm{Re}\,\pi'' \rangle \le x\left(\langle \mathfrak{w}_{\check{\alpha}}, \mathfrak{w}_\alpha \rangle \|\mathfrak{w}_{\check{\alpha}''}\| \, \|\mathfrak{w}_{\check{\alpha}}\|^{-1} - \langle \mathfrak{w}_{\check{\alpha}''}, \mathfrak{w}_\alpha \rangle \right).$$

If we identify $\mathfrak{a}_{M_0}$ with $\mathfrak{a}_{M_0}^*$ via the scalar product, $\mathfrak{w}_\alpha$ becomes a positive multiple of $\mathfrak{w}_{\check{\alpha}}$ and the Schwartz inequality shows that the right-hand side of the above inequality is < 0. This proves (2).

We thus have $\psi' \in A^2(U'(\mathbb{A})M'(k)\backslash\mathbf{G})_{\mathrm{Im}\,\eta}$. By the induction hypothesis applied to $\pi \otimes (-\mathrm{Re}\,\eta) \in \Pi_0(\mathbf{M}, \psi')$, we deduce that the restriction of $N(M')\chi_\pi$ to $T_{\mathbf{M}} \cap \mathbf{M}'_{\mathrm{der}}$ is positive real-valued.

Let us show that

(4) there exists $w \in W$ such that $wMw^{-1} = M$ and $-w\overline{\chi}_\pi = \chi_\pi$;

by definition, $w\chi_\pi(z) = \chi_\pi(w^{-1}zw)$ for $z \in Z_{\mathbf{M}}$.

If this condition is not satisfied, *a fortiori* the pair (M, π) and $(M, -\overline{\pi})$ are not conjugate. Then there exists $f \in H_{\xi,b,\mathbb{C}}^R$ such that $f(M, \pi) = 1$, $f(M, -\overline{\pi}) = 0$. The second condition is equivalent to $f^*(M, \pi) = 0$. By the hypothesis on ϕ and what was said in III.3.3, we have $\Delta(f)\phi = \phi$, $\Delta(f^*)\phi = 0$. This contradicts the relation

$$\langle \Delta(f)\phi, \phi \rangle = \langle \phi, \Delta(f^*)\phi \rangle,$$

see III.1.4, and the contradiction proves (4).

Fix w such that (4) is satisfied. Let us show that w, acting in a natural way on $\mathfrak{a}_{M_0}^*$, does not fix the subspace $(\mathfrak{a}_{M'}^G)^*$ pointwise. Indeed, if it did, we would have $w^{-1}\varpi_{\check{\alpha}} = \varpi_{\check{\alpha}}$, and so

$$\begin{aligned}\langle \varpi_{\check{\alpha}}, \operatorname{Re}\pi\rangle &= \langle w^{-1}\varpi_{\check{\alpha}}, \operatorname{Re}\pi\rangle = \langle \varpi_{\check{\alpha}}, w\operatorname{Re}\pi\rangle \\ &= -\langle \varpi_{\check{\alpha}}, \operatorname{Re}\pi\rangle,\end{aligned}$$

by (4). This contradicts the fact that $\langle \varpi_{\check{\alpha}}, \operatorname{Re}\pi\rangle < 0$.

Denote by λ the restriction to $T_{\mathbf{M}}$ of $\operatorname{Im}\chi_\pi$; we define $\operatorname{Im}\chi_\pi$ by $\chi_\pi = \operatorname{Re}\chi_\pi + \operatorname{Im}\chi_\pi$ – see III.3.1 Remark (a) for the interpretation of the sign +, where $\operatorname{Re}\chi_\pi$ is strictly positive real-valued and $\operatorname{Im}\chi_\pi$ is unitary. We then have

(i) $N(M')\lambda_{|T_{\mathbf{M}}\cap\mathbf{M}'_{\mathrm{der}}} = 0$;

(ii) $w\lambda = \lambda$, for a w such that $wMw^{-1} = M$ and w does not fix the subspace $(\mathfrak{a}_{M'}^G)^*$ pointwise.

It is a consequence of (ii) that λ is trivial on the group $(w-1)T_{\mathbf{M}}$; see III.3.2. Thus $N(M')\lambda$ is trivial on $(T_{\mathbf{M}} \cap \mathbf{M}'_{\mathrm{der}})(w-1)T_{\mathbf{M}}$. This last group is contained in $T_{\mathbf{M}} \cap \mathbf{G}_{\mathrm{der}}$: the quotient is a torsion group killed by $N(M, M', w)$ (see III.3.2). Thus $N(M, M', w)N(M')\lambda$ is trivial on $T_{\mathbf{M}} \cap \mathbf{G}_{\mathrm{der}}$. As $N(M, M', w)N(M')$ divides $N(G)$, this proves the statement of Proposition III.3.1 for the representation π. But as $\phi \in A^2(G(k)\backslash\mathbf{G})_{\mathfrak{y}}$, all the elements of its cuspidal support are conjugate under the Weyl group and we deduce the assertion for every element of this support.

□

IV

Continuation of Eisenstein Series

IV.1. The results

IV.1.1. The spaces

Fix a unitary character ξ of $Z_{\mathbf{G}}$. The results of this chapter can be generalised to the non-unitary case by tensorisation with an element of X_G.

We fix a finite set $\mathfrak{F}$ of $\mathbf{K}$-types, which we suppose stable under passage to the contragredient. For every space V on which $\mathbf{K}$ acts, we denote by $V^{\mathfrak{F}}$ the subspace of elements $v \in V$ such that the subspace generated by the elements kv, for $k \in \mathbf{K}$, decomposes into a sum of irreducible subspaces under the action of $\mathbf{K}$, each of whose isomorphism classes belongs to $\mathfrak{F}$.

We consider the following spaces:

$A_\xi = A(G(k)\backslash\mathbf{G})_\xi$, the space of automorphic forms on $G(k)\backslash\mathbf{G}$, having 'central character' ξ;

$L^2_\xi = L^2(G(k)\backslash\mathbf{G})_\xi$, the space of complex-valued functions on $G(k)\backslash\mathbf{G}$, of central character ξ, which are square integrable modulo $Z_{\mathbf{G}}$;

$L^2_{\xi,\mathrm{loc}}$, the space of complex-valued functions on $G(k)\backslash\mathbf{G}$ of central character ξ which are locally square integrable modulo $Z_{\mathbf{G}}$.

We have the inclusions

$$A_\xi \subset L^2_{\xi,\mathrm{loc}} \supset L^2_\xi.$$

We equip L^2_ξ with the topology defined by the L^2 norm and $L^2_{\xi,\mathrm{loc}}$ with the topology defined by the filter of semi-norms

$$f \longmapsto \int_C |f(g)|^2 \, dg,$$

where C runs through the set of compact subsets of $Z_{\mathbf{G}}G(k)\backslash\mathbf{G}$. Then L^2_ξ and $L^2_{\xi,\mathrm{loc}}$ are Fréchet spaces; see I.4.9. The injection $L^2_\xi \to L^2_{\xi,\mathrm{loc}}$ is continuous.

We denote by $\mathscr{H}^{\mathfrak{F}}$ the space of smooth functions $h : \mathbf{G} \to \mathbb{C}$ with compact support, such that the space generated by the translates of h under $\mathbf{K} \times \mathbf{K}$, acting on the right and on the left, decomposes into a sum of irreducible subspaces under the action of $\mathbf{K} \times \mathbf{K}$, each of whose isomorphism classes belongs to $\mathfrak{F} \times \mathfrak{F}$. It is an algebra, which acts on $A^{\mathfrak{F}}_\xi, L^{2,\mathfrak{F}}_\xi$ and so on, via an action which we denote by δ.

IV.1.2. The representations

In this chapter we consider pairs $(M, \mathfrak{P})$ satisfying the following conditions:

> M is a standard Levi subgroup of G; we denote by $\mathbf{M}$ its inverse image in $\mathbf{G}$;
>
> $\mathfrak{P}$ is an orbit under $X^{\mathbf{G}}_M$ of irreducible cuspidal representations of $\mathbf{M}$, the restriction of whose central character to $Z_{\mathbf{G}}$ is ξ.

As usual we denote by $P = MU$ the standard parabolic subgroup of G of Levi M. For $\pi \in \mathfrak{P}$, we defined in I.1.1 a space

$$A(U(\mathbb{A})M(k)\backslash\mathbf{G})_\pi.$$

We denote it here more simply by $A(M, \pi)$.

IV.1.3. Holomorphic functions

Let us fix some definitions which will be useful for the statements of the theorems. Let $(M, \mathfrak{P})$ be as in IV.1.2. For $\pi \in \mathfrak{P}$ and $\lambda \in X^{\mathbf{G}}_M$, there is an isomorphism of vector spaces

$$\underline{\lambda} : A(M,\pi)^{\mathfrak{F}} \to A(M, \pi \otimes \lambda)^{\mathfrak{F}}$$

$$\varphi \longmapsto \lambda\varphi$$

where $\lambda\varphi$ is defined by $(\lambda\varphi)(g) = m_P(g)^\lambda \varphi(g)$.

Let U be an open set of $\mathfrak{P}$, H a Fréchet space and E a function defined on U such that for every $\pi \in U$,

$$E(\pi) \in \mathrm{Hom}_{\mathbb{C}}(A(M,\pi)^{\mathfrak{F}}, H).$$

Since $A(M,\pi)^{\mathfrak{F}}$ is finite-dimensional, the above space is also a Fréchet space. We will say that E is holomorphic if for every $\pi \in U$, the function

$$\lambda \longmapsto E(\pi \otimes \lambda) \circ \underline{\lambda},$$

defined in $X_M^{\mathbf{G}}$ on a neighborhood of 0 with values in $\mathrm{Hom}_{\mathbb{C}}(A(M,\pi)^{\mathfrak{F}}, H)$ is holomorphic. If E is only defined almost everywhere on U, we will say that E is meromorphic if for every $\pi \in U$ and every sufficiently small neighborhood V of π in U, there exist two holomorphic functions $d : V \to \mathbb{C}$, $d \neq 0$, and E_1 defined on V, analogous to E, such that

$$d(\pi')E(\pi') = E_1(\pi')$$

for every $\pi' \in V$ where $E(\pi')$ is defined.

Let ϕ be a function defined on U such that for every $\pi \in U$, $\phi(\pi) \in A(M,\pi)^{\mathfrak{F}}$. We will say that ϕ is holomorphic if for every $\pi \in U$, the function

$$\lambda \longmapsto -\underline{\lambda} \circ \phi(\pi \otimes \lambda)$$

is holomorphic in a neighborhood of 0 in $X_M^{\mathbf{G}}$, see II.1.2. If ϕ and E are holomorphic in U, the function

$$U \longrightarrow H$$

$$\pi \longmapsto E(\pi)(\phi(\pi))$$

is holomorphic. Let $\triangle$ be a holomorphic differential operator in U. We can define the function

$$U \longrightarrow H$$

$$\pi \longmapsto \triangle\big(E(\pi)(\phi(\pi))\big) :$$

we apply the operator $\triangle$ to the preceding function and evaluate the result at π. This function is still holomorphic.

IV.1.4. Holomorphic operators

Denote by $W(M)$ the set of elements $w \in W$ of minimal length in their right class modulo the Weyl group of M, such that wMw^{-1} is a standard Levi of G. Let U be an open set of $\mathfrak{P}$, $w \in W(M)$ and M_w a function defined on U, such that for every $\pi \in U$,

$$M_w(\pi) \in \mathrm{Hom}_{\mathbb{C}}\big(A(M,\pi)^{\mathfrak{F}}, A(wMw^{-1}, w\pi)^{\mathfrak{F}}\big).$$

We say that M_w is holomorphic if for every $\pi \in U$, the function

$$\lambda \longmapsto -\underline{w\lambda} \circ M_w(\pi \otimes \lambda) \circ \underline{\lambda},$$

defined in a neighborhood of 0 in $X_M^{\mathbf{G}}$, with values in the finite-dimensional space

$$\mathrm{Hom}_{\mathbb{C}}(A(M,\pi)^{\mathfrak{F}}, A(wMw^{-1}, w\pi)^{\mathfrak{F}}),$$

is holomorphic. The definition of a meromorphic function is analogous.

IV.1.5. Rationality

In the case where k is a function field, $X_M^{\mathbf{G}}$ is a complex algebraic manifold, by the description given in I.1.4. A polynomial on $X_M^{\mathbf{G}}$ is a linear combination of functions of the form $\lambda \mapsto m^\lambda$, where $m \in \mathbf{M}$. We can algebrise the definitions of IV.1.3 and IV.1.4 so that the expressions 'E is a polynomial', 'E is a rational function' and so on have precise meanings.

IV.1.6. Singularities along hyperplanes

Let $\pi \in \mathfrak{P}$, and $\alpha \in R^+(T_M, G)$ be an indivisible root. We define a function $h : \mathfrak{P} \to \mathbb{C}$ as follows:

(i) if k is a number field, $h(\pi \otimes \lambda) = \langle \lambda, \check{\alpha} \rangle$ for every $\lambda \in X_M^{\mathbf{G}} \simeq \mathfrak{a}_M^G$;
(ii) if k is a function field, denote by $n(\alpha)$ the smallest integer $n > 0$ such that $\alpha^{*n\lambda} = 1$ for every $\lambda \in \mathrm{Fix}_{X_M^{\mathbf{G}}}(\mathfrak{P})$; then $h(\pi \otimes \lambda) = \alpha^{*n(\alpha)\lambda} - 1$ for every $\lambda \in X_M^{\mathbf{G}}$.

Set $H = \{\pi' \in \mathfrak{P};\ h(\pi') = 0\}$. Such a set is called a root hyperplane; this terminology will be modified and made more precise in V.1.2. The function h is determined by H. We denote it more precisely by h_H.

Let D be a set of root hyperplanes and E a function as in IV.1.3, meromorphic on $\mathfrak{P}$. We say that the singularities of E are carried by D if, for every $\pi \in \mathfrak{P}$, we can find a function $n_\pi : D \to \mathbb{N}$, zero almost everywhere, such that the function

$$\pi' \longmapsto \left(\prod_{H \in D} h_H(\pi')^{n_\pi(H)} \right) E(\pi')$$

is holomorphic at π. We say that the singularities are without multiplicity at π if we can choose n_π with values in $\{0, 1\}$. Analogous definitions can be given for the functions M_w of IV.1.4.

Finally, we will say that D is locally finite if for every compact subset $C \subset \mathfrak{P}$, $\{H \in D;\ H \cap C \neq \emptyset\}$ is finite.

IV.1.7. Holomorphic property of Eisenstein series

Denote by $\mathfrak{c}_{\mathfrak{P}} = \{\pi \in \mathfrak{P}; \forall \alpha \in \Delta_M,\ \langle \mathrm{Re}\, \pi, \check{\alpha} \rangle > \langle \rho_P, \check{\alpha} \rangle\}$. For $\pi \in \mathfrak{c}_{\mathfrak{P}}$ and $\varphi \in A(M, \pi)^{\mathfrak{F}}$, we defined the Eisenstein series $E(\varphi, \pi)$ in II.1.5. We have $E(\varphi, \pi) \in A_\xi^{\mathfrak{F}}$, *a fortiori* $E(\varphi, \pi) \in L^{2,\mathfrak{F}}_{\xi,\mathrm{loc}}$. This defines a function on $\mathfrak{c}_{\mathfrak{P}}$:

$$E : \pi \longmapsto E(\pi) \in \mathrm{Hom}_{\mathbb{C}}(A(M, \pi)^{\mathfrak{F}}, L^{2,\mathfrak{F}}_{\xi,\mathrm{loc}}).$$

Lemma *The function E is holomorphic on $\mathfrak{c}_{\mathfrak{P}}$.*

Proof For every compact subset C of $Z_{\mathbf{G}}G(k)\backslash\mathbf{G}$, denote by $L^2_{\xi,C}$ the subspace of elements of L^2_{ξ} with support in C. By a variation of this which we described in I.4.9, it suffices to prove that for every $\pi \in \mathfrak{c}_{\mathfrak{P}}$, every $\varphi \in A(M,\pi)^{\mathfrak{F}}$, every set C as above and every $\psi \in L^2_{\xi,C}$, the map

$$\lambda \longmapsto \int_{Z_{\mathbf{G}}G(k)\backslash\mathbf{G}} \overline{\psi}(g)E(\lambda\varphi, \pi\otimes\lambda)(g)\, dg$$

is holomorphic at 0. We can also consider a compact subset C of $S \cap \mathbf{G}^c$, an element ψ of $L^2(C)$ and the map

$$\lambda \longmapsto \int_C \overline{\psi}(g)E(\lambda\varphi, \pi\otimes\lambda)(g)\, dg.$$

For λ in a neighborhood of 0 and $g \in C$, the series

$$E(\lambda\varphi, \pi\otimes\lambda)(g) = \sum_{\gamma\in P(k)\backslash G(k)} m_P(\gamma g)^{\lambda}\varphi(\gamma g)$$

converges normally. Each term is holomorphic in λ. Consequently the above integral is holomorphic. □

Let $w \in W(M)$. For $\pi \in \mathfrak{c}_{\mathfrak{P}}$, we defined in II.1.6 the intertwining operator

$$M(w,\pi) \in \mathrm{Hom}_{\mathbb{C}}(A(M,\pi)^{\mathfrak{F}}, A(wMw^{-1}, w\pi)^{\mathfrak{F}}).$$

This defines a function $M(w,x)$ on $\mathfrak{c}_{\mathfrak{P}}$ which is holomorphic; see IV.1.4.

For $\pi \in \mathfrak{P}$, the algebra $\mathscr{H}^{\mathfrak{F}}$ acts on $A(M,\pi)^{\mathfrak{G}}$ by a representation which we denote by $i(\pi,.)$.

Remark By definition of $A(M,\pi)$, $i(\pi,.)$ is a multiple of the induced representation of π, or more exactly of the representation of $\mathscr{H}^{\mathfrak{F}}$ which can be deduced from it.

For $\pi \in \mathfrak{c}_{\mathfrak{P}}$ and $h \in \mathscr{H}^{\mathfrak{F}}$, we have the relations

$$(*) \qquad E(\pi)\circ i(\pi,h) = \delta(h)\circ E(\pi)$$

$$(**) \qquad M(w,\pi)\circ i(\pi,h) = i(w\pi,h)\circ M(w,\pi)$$

Remark From the above relations, we can deduce that $i(\pi,.)$ and $i(w\pi,.)$ are equivalent for almost every π; it suffices to show that $M(w,\pi)$ is invertible for almost every π. We prefer not to prove this fact *a priori*; it will be a consequence of Theorem IV.1.10 below.

IV.1.8. Continuation of Eisenstein series

Theorem *Let $(M, \mathfrak{P})$ be as in* IV.1.2.

(a) *The function E can be continued in a unique way to a meromorphic function on $\mathfrak{P}$.*

(b) *Let $w \in W(M)$. The function $M(w,.)$ can be continued in a unique way to a meromorphic function on $\mathfrak{P}$.*

For $\pi \in \mathfrak{P}$ and $\varphi \in A(M,\pi)^{\mathfrak{F}}$, we will often write $E(\varphi,\pi)$ for the term $E(\pi)(\varphi)$.

IV.1.9. Properties of Eisenstein series

Proposition *Let $(M, \mathfrak{P})$ be as in* IV.1.2.

(a) *Let $h \in \mathcal{H}^{\mathfrak{F}}$ and $w \in W(M)$. We have the equalities* IV.1.7(*) *and* (**) *of meromorphic functions on $\mathfrak{P}$.*

(b) *Let U be the open set of $\mathfrak{P}$ where E is holomorphic.*
(i) *For $\pi \in U$ and $\varphi \in A(M,\pi)^{\mathfrak{F}}$, $E(\varphi,\pi) \in A_{\xi}^{\mathfrak{F}}$.*
(ii) *For $\pi \in U$ and $\varphi \in A(M,\pi)^{\mathfrak{F}}$, $E(\varphi,\pi)^{\mathrm{cusp}} = 0$ if $M \neq G$.*

(c) *Let ϕ be a holomorphic function on $\mathfrak{P}$ such that $\phi(\pi) \in A(M,\pi)^{\mathfrak{F}}$ for every π. Then:*
(i) *the function*

$$\mathbf{G} \times U \to \mathbb{C}$$

$$(g,\pi) \longmapsto E(\phi(\pi),\pi)(g)$$

is smooth;
(ii) *the function $\pi \mapsto E(\phi(\pi),\pi)(g)$ is holomorphic on U, meromorphic on $\mathfrak{P}$, where $g \in \mathbf{G}$;*
(iii) *the function $\pi \mapsto E(\phi(\pi),\pi)_{P'}(g)$ is meromorphic on $\mathfrak{P}$ and given by formula* II.1.7*; here $g \in \mathbf{G}$ and P' is a standard parabolic subgroup of G.*

(d) *More generally, let V be an open set of $\mathfrak{P}$, $d : V \to \mathbb{C}$ a holomorphic function such that $\pi \mapsto d(\pi)E(\pi)$ is holomorphic in V; let D be a holomorphic differential operator in V and ϕ a holomorphic function defined in V such that $\phi(\pi) \in A(M,\pi)^{\mathfrak{F}}$ for every $\pi \in V$. Then the function*

$$V \to L^{2,\mathfrak{F}}_{\xi,\mathrm{loc}}$$

$$\pi \longmapsto D(d(\pi)E(\phi(\pi),\pi))$$

satisfies properties analogous to (b) *and* (c).

(e) *The adjunction formula* II.1.8 *can be continued to an equality of meromorphic functions on $\mathfrak{P}$.*

IV.1.10 The functional equation
We write E_M for the function previously denoted by E.

Theorem *Let $(M, \mathfrak{P})$ be as in* IV.1.2 *and $w \in W(M)$.*

(a) *We have the following equality of meromorphic functions on $\mathfrak{P}$*
$$E_M(\pi) = E_{wMw^{-1}}(w\pi) \circ M(w, \pi).$$
(b) *Let $w' \in W(wMw^{-1})$. We have the equality of meromorphic functions on $\mathfrak{P}$ given by*
$$M(w', w\pi) \circ M(w, \pi) = M(w'w, \pi).$$

IV.1.11. Singularities of Eisenstein series

Proposition *Let $(M, \mathfrak{P})$ be as in* IV.1.2.

(a) *There exists a locally finite set of root hyperplanes D such that the singularities of the functions E and $M(w, .)$, for every $w \in W(M)$, are carried by D.*
(b) *Let $\pi \in \mathfrak{P}$ be such that* $\mathrm{Re}\,\pi = 0$. *Then the functions E and $M(w, .)$ are holomorphic at π.*
(c) *Let $\pi \in \mathfrak{P}$ be such that $\langle \mathrm{Re}\,\pi, \check{\alpha} \rangle \geq 0$ for every $\alpha \in \Delta_M$. Then the singularities of the functions E and $M(w, .)$ are without multiplicity at π.*
(d) *There is only a finite number of singular hyperplanes (of E or of $M(w, .)$) which intersect $\{\pi \in \mathfrak{P} \mid \langle \mathrm{Re}\,\pi, \check{\alpha} \rangle \geq 0$ for every $\alpha \in \Delta_M\}$.*

IV.1.12. The function field case

Proposition *Suppose that k is a function field and let $(M, \mathfrak{P})$ be as in* IV.1.2.

(a) *Let $g \in \mathbf{G}$ and denote by ε_g the evaluation at the point g. Then the function $\pi \longmapsto \varepsilon_g \circ E(\pi)$ is rational.*
(b) *Let $w \in W(M)$. Then the function $M(w, .)$ is rational.*

IV.2. Some preparations

IV.2.1. Transformation of the problem
From now on we fix $(M, \mathfrak{P})$ as in IV.1.2 and $\pi \in \mathfrak{P}$, unitary. We modify the definitions of the first section in the following way.

For $\lambda \in X_M^{\mathbf{G}}$, let $i(\lambda, .)$ be the representation of $\mathscr{H}^{\mathfrak{F}}$ into $A(M,\pi)^{\mathfrak{F}}$ defined by

$$i(\lambda, h)\varphi = -\lambda[i(\pi \otimes \lambda, h)(\lambda\varphi)]$$

for all $h \in \mathscr{H}^{\mathfrak{F}}$, $\varphi \in A(M,\pi)^{\mathfrak{F}}$.

Set $\mathfrak{c} = \{\lambda \in X_M^{\mathbf{G}}; \forall \alpha \in \Delta_M, \langle \operatorname{Re}\lambda, \check{\alpha}\rangle > \langle \rho_P, \check{\alpha}\rangle\}$. We define a holomorphic function

$$E : \mathfrak{c} \to \operatorname{Hom}_{\mathbb{C}}(A(M,\pi)^{\mathfrak{F}}, L^{2,\mathfrak{F}}_{\xi,\mathrm{loc}})$$

by $E(\varphi, \lambda) = E(\lambda\varphi, \pi \otimes \lambda)$. Similarly, for $w \in W(M)$ we define

$$M(w, .) : \mathfrak{c} \to \operatorname{Hom}_{\mathbb{C}}(A(M,\pi)^{\mathfrak{F}}, A(wMw^{-1}, w\pi)^{\mathfrak{F}})$$

by $M(w, \lambda)\varphi = (-w\lambda)M(w, \pi \otimes \lambda)(\lambda\varphi)$.

More generally, for $w' \in W(M)$, we define objects $i_{w'}(\lambda, .)$, $E_{w'}$, $M_{w'}(w, .)$ by replacing the triple $(M, \mathfrak{P}, \pi)$ by $(w'Mw'^{-1}, w'\mathfrak{P}, w'\pi)$ in the preceding constructions (the objects in those constructions correspond to $w' = 1$).

The statements IV.1.8 to IV.1.12 have analogues for the functions we just defined. It is clear that these new statements imply those.

Remark The only subtlety concerns the function field case. Suppose for example that we continue the function $E(\lambda)$. For $\pi' \in \mathfrak{P}$, we define $E(\pi')$ as follows: choose λ such that $\pi' = \pi \otimes \lambda$; for $\varphi' \in A(M, \pi')^{\mathfrak{F}}$, and set

$$E(\varphi', \pi') = E((-\lambda)\varphi', \lambda).$$

We must check that this does not depend on the choice of λ, i.e. that for $\mu \in \operatorname{Fix}_{X_{\mathrm{M}}^{\mathbf{G}}}(\mathfrak{P})$, we have $E((-\lambda-\mu)\varphi', \lambda+\mu) = E((-\lambda)\varphi', \lambda)$ for every φ', or even $E((-\mu)\varphi, \lambda+\mu) = E(\varphi, \lambda)$ for every $\varphi \in A(M,\pi)^{\mathfrak{F}}$. Now this is true if $\lambda \in \mathfrak{c}$ since we know that then $E(\varphi', \pi')$ is well-defined. The equality can be meromorphically continued to every λ.

IV.2.2 Choice of functions in the Hecke algebra

For $h \in \mathscr{H}^{\mathfrak{F}}$, we define $h^* \in \mathscr{H}^{\mathfrak{F}}$ by $h^*(g) = \overline{h}(g^{-1})$ for every $g \in \mathbf{G}$.

Lemma *Let $\lambda \in X_M^{\mathbf{G}}$. There exists $h \in \mathscr{H}^{\mathfrak{F}}$ such that*

(i) $h = h^*$;

(ii) $i(\lambda, h) = 1$; *more generally for every $w \in W(M)$, $i_w(w\lambda, h) = 1$;*

(iii) *there exists $\mu \in X_M^{\mathbf{G}}$ such that $\langle \operatorname{Re}\mu, \check{\alpha}\rangle \gg 0$ for every $\alpha \in \Delta_M$, and such that we have $i_w(w\mu, h) = \mathrm{i}$ for every $w \in W(M)$.*

Proof For $\mu \in X_M^{\mathbf{G}}$, denote by $\Pi_1(\mu)$ and $\Pi_2(\mu)$ the following sets of representations of $\mathscr{H}^{\mathfrak{F}}$:

$i(w(\pi \otimes \lambda),.)$, $i(-w(\overline{\pi \otimes \lambda}),.)$, $i(-w(\overline{\pi \otimes \mu}),.)$, where w runs through $W(M)$ and $i(w(\pi \otimes \mu),.)$ respectively.

Denote by $\sum_1(\mu)$ and $\sum_2(\mu)$ the sets of irreducible representations of $\mathscr{H}^{\mathfrak{F}}$ occurring as subquotients of an element of $\Pi_1(\mu)$ and $\Pi_2(\mu)$ respectively. We check that $\sum_1(\mu) \cap \sum_2(\mu) = \emptyset$ for μ in a dense open set of $X_M^{\mathbf{G}}$; for example at a place v of k where π is unramified, we compare this with the Jacquet modules of representations occurring in $\sum_1(\mu)$ and $\sum_2(\mu)$. Fix μ such that this condition is satisfied and moreover $\langle \operatorname{Re}\mu, \check{\alpha} \rangle \gg 0$ for every $\alpha \in \Delta_M$. It is well-known that we can find $h_1 \in \mathscr{H}^{\mathfrak{F}}$ such that

$$\sigma(h_1) = 1 \text{ for every } \sigma \in \Sigma_1(\mu),$$

$$\sigma(h_1) = 0 \text{ for every } \sigma \in \Sigma_2(\mu).$$

For $\tau \in \Pi_2(\mu)$, $\tau(h_1)$ is thus nilpotent. Up to raising h_1 to a power, we can suppose $\tau(h_1) = 0$ for every $\tau \in \Pi_2(\mu)$. For $\tau \in \Pi_1(\mu)$, $\tau(h_1) - 1$ is nilpotent. There exists an integer N such that $(\tau(h_1) - 1)^N = 0$ for every $\tau \in \Pi_1(\mu)$. Write

$$(X-1)^N = \sum_{i=0}^{N} a_i X^i$$

and set

$$h_1' = (-1)^{N+1} \sum_{i=1}^{N} a_i h_1^i.$$

Then $h_i' \in \mathscr{H}^{\mathfrak{F}}$ satisfies:

$$\tau(h_1') = 1 \text{ for every } \tau \in \Pi_1(\mu),$$

$$\tau(h_1') = 0 \text{ for every } \tau \in \Pi_2(\mu).$$

We construct H_2' similarly, satisfying

$$\tau(h_2') = 0 \text{ for every } \tau \in \Pi_1(\mu),$$

$$\tau(h_2') = 1 \text{ for every } \tau \in \Pi_2(\mu).$$

Set $h = (h_1' + ih_2')(h'^*_1 - ih'^*_2)$. Note that, for example, for $h' \in \mathscr{H}^{\mathfrak{F}}$, $i(-(\overline{\pi \otimes \lambda}), h'^*)$ is the adjoint of $i(\pi \otimes \lambda, h')$. Thus (ii) and (iii) are results of the above properties, and (i) is immediate by construction. □

IV.2.3. Compact operators

We recall the definition of a compact operator. Let H be a Hilbert space and T a transformation (an 'operator') defined on every H. We say that T is compact if $\{Th;\ h \in H,\ \|h\| < 1\}$ is relatively compact in H.

Example Let X be a space equipped with a measure and $H = L^2(X)$;

suppose that T is Hilbert-Schmidt, i.e. there exists $t \in L^2(X \times X)$ such that for every $h \in H$, $x \in X$, we have

$$(Th)(x) = \int_X t(x, y)h(y)\, dy.$$

Then T is compact.

Suppose that T is compact and auto-adjoint. Then there exists a countable subset Γ of $\mathbb{R}$, tending to 0 (i.e. $\forall \varepsilon > 0, \{\gamma \in \Gamma;\ |\gamma| > \varepsilon\}$ is finite), and for every $\gamma \in \Gamma$ there exists a closed subspace H_γ of H which is finite-dimensional if $\gamma \neq 0$, such that

(i) $H = \bigoplus_{\gamma\in\Gamma} H_\gamma$ (the Hilbertian direct sum);
(ii) for $x \in H$, write $x = \sum_{\gamma\in\Gamma} x_\gamma$ as in (i); then $Tx = \sum_{\gamma\in\Gamma} \gamma x_\gamma$.

Conversely, if we take Γ and $(H_\gamma)_{\gamma\in\Gamma}$ as above, satisfying (i), the transformation defined by (ii) is compact and auto-adjoint.

IV.2.4. Truncation is an orthogonal projection

Fix $T \in \mathrm{Re}\,\mathfrak{a}_{M_0}$ very positive, i.e. such that $\langle \alpha, T\rangle$ is sufficiently large for every $\alpha \in \Delta_0$. For every locally L^1 function φ on $G(k)\backslash \mathbf{G}$, we defined in I.2.11 a function $\wedge^T \varphi$ on $G(k)\backslash \mathbf{G}$. As the elements of L^2_ξ are locally L^1, we can in particular define $\wedge^T \varphi$ for $\varphi \in L^2_\xi$.

Lemma *If $\varphi \in L^2_\xi$, then $\wedge^T \varphi \in L^2_\xi$. The operator $\wedge^T$ of L^2_ξ is an orthogonal projection.*

Proof Denote by $\mathcal{S}_\xi$ the space of smooth **K**-finite functions on $G(k)\backslash \mathbf{G}$, of central character ξ and compact support mod $Z_{\mathbf{G}}$. From Lemma I.2.16 we see that, if $\varphi \in \mathcal{S}_\xi$, then $\wedge^T \varphi$ is rapidly decreasing if k is a number field, and has compact support mod$Z_{\mathbf{G}}$ if k is a function field. In both cases $\wedge^T \varphi \in L^2_\xi$. From Lemmas I.2.14 and I.2.15 we obtain the equality

$$\|\varphi\|^2 = \| \wedge^T \varphi\|^2 + \|(1 - \wedge^T)\varphi\|^2,$$

where the norm employed is the L^2 norm. *A fortiori* we have $\| \wedge^T \varphi\| \leq \|\varphi\|$. We can then extend $\wedge^T$ by continuity to the closure of $\mathcal{S}_\xi$ in L^2_ξ, which is just L^2_ξ itself, to a bounded operator temporarily denoted by $\tilde{\wedge}^T$. Lemmas I.2.14 and I.2.15 can be extended by continuity to the operator $\tilde{\wedge}^T$, which is thus an orthogonal projection.

It remains to see that $\wedge^T \varphi = \tilde{\wedge}^T \varphi$ for every $\varphi \in L^2_\xi$. Let $\varphi \in L^2_\xi$. Choose a sequence $(\varphi_n)_{n\in\mathbb{N}}$ of elements of $\mathcal{S}_\xi$ such that

$$\lim_{n\to\infty} \|\varphi_n - \varphi\| = 0.$$

Using Fubini's Theorem, we check that for every standard parabolic subgroup Q of G and every compact subset C of $\mathbf{G}$, we have

$$\lim_{n\to\infty} \int_C |\varphi_{n,Q}(g) - \varphi_Q(g)|^2 \, dg = 0.$$

We know that for every set X equipped with a measure and for every sequence $(\psi_n)_{n\in\mathbb{N}}$ of elements of $L^2(X)$ such that $\lim_{n\to\infty} \|\psi_n\| = 0$, we can find a subsequence $\psi_{n_1}, \psi_{n_2}, \ldots$ such that $\lim_{i\to\infty} \psi_{n_i}(x) = 0$ for almost every $x \in X$. We deduce that, up to extracting a subsequence of the sequence $(\varphi_n)_{n\in\mathbb{N}}$, there exists a set of measure zero $\mathfrak{N}$ of $\mathbf{G}$ such that for every standard parabolic subgroup Q and every $g \in \mathbf{G} - \mathfrak{N}$, we have

$$\lim_{n\to\infty} \varphi_{n,Q}(g) = \varphi_Q(g).$$

We can suppose $\mathfrak{N}$ stable on the left under $G(k)$. Then by construction

$$\lim_{n\to\infty} \wedge^T \varphi_n(g) = \wedge^T \varphi(g)$$

for $g \notin \mathfrak{N}$. On the other hand, the continuity of $\tilde{\wedge}^T$ implies that

$$\lim_{n\to\infty} \|\tilde{\wedge}^T \varphi_n - \tilde{\wedge}^T \varphi\| = 0.$$

Up to again extracting a subsequence, there exists a measure zero subset $\tilde{\mathfrak{N}}$ of $\mathbf{G}$ such that for $g \in \mathbf{G} - \tilde{\mathfrak{N}}$,

$$\lim_{n\to\infty} \tilde{\wedge}^T \varphi_n(g) = \tilde{\wedge}^T \varphi(g).$$

As $\tilde{\wedge}^T \varphi_n = \wedge^T \varphi_n$ for every n, the two equalities above imply that

$$\wedge^T \varphi(g) = \tilde{\wedge}^T \varphi(g)$$

for $g \notin \mathfrak{N} \cup \tilde{\mathfrak{N}}$, which concludes the proof. □

IV.2.5. Upper bound of a truncated kernel

Extend ξ to a character of $Z_{\mathbf{G}}G(k)$ by setting $\xi(zg) = \xi(z)$ for all $z \in Z_{\mathbf{G}}$, $g \in G(k)$. For $h \in \mathcal{H}^{\mathfrak{F}}$, define a function $k_h : \mathbf{G} \times \mathbf{G} \to \mathbb{C}$ by

$$k_h(x, y) = \int_{Z_{\mathbf{G}}G(k)} h(x^{-1}\gamma y)\xi(\gamma) \, d\gamma.$$

It is smooth. Denote by $\wedge_1^T k_h(x, y)$ the value at x of the image under the operator $\wedge^T$ of the function $x \longmapsto k_h(x, y)$, for fixed y.

Lemma *Let $h \in \mathcal{H}^{\mathfrak{F}}$.*

(a) *Suppose that k is a number field. For every $r > 0$, there exists $c > 0$ such that for all $x, y \in S \cap \mathbf{G}^1$, we have*

$$| \wedge_1^T k_h(x, y)| \leq c\|x\|^{-r} \|y\|^{-r}.$$

(b) *Suppose that k is a function field. The restriction of $\wedge_1^T k_h$ to $S \cap \mathbf{G}^c \times S \cap \mathbf{G}^c$ is bounded and has compact support.*

Proof Suppose that k is a number field and let $r > 0$. Choose $r' \in \mathbb{R}$ such that the conclusion of Lemma I.2.4(b) is satisfied. Apply Lemma I.2.16 to the pair (r, r'). Then there exists a finite set $\{X_i; i \in I\} \subset \mathcal{U}$ such that for all $x \in S \cap \mathbf{G}^1$ and $y \in \mathbf{G}$:

$$(*) \qquad |\wedge_1^T k_h(x,y)| \leq \|x\|^{-r} \left(\sum_{i \in I} \sup_{x' \in \mathbf{G}^1} |X_i k_h(x',y)| \, \|x'\|^{-r'} \right).$$

We have the equality

$$X_i k_h(x',y) = \int_{Z_G G(k)} X_i h(x'^{-1}\gamma y)\xi(\gamma) d\gamma,$$

where X_i acts on h via left translation. As the support of $X_i h$ is compact, Lemma I.2.4(b) implies that there exists $c > 0$ such that for every $i \in I$, all $x' \in \mathbf{G}^1$ and $y \in S \cap \mathbf{G}^1$, we have

$$(**) \qquad |X_i k_h(x',y)| \leq c\|x'\|^{r'} \|y\|^{-r}.$$

We obtain the desired upper bound from the inequalities (*) and (**).

Suppose now that k is a function field. Note that the function $x \mapsto k_h(x,y)$ is invariant under a compact open subgroup independent of y. Lemma I.2.16 implies that there exists a compact subset $C \subset S \cap \mathbf{G}^c$ such that for every $y \in \mathbf{G}$, the restriction to $S \cap \mathbf{G}^c$ of the function $x \mapsto \wedge_1^T k_h(x,y)$ has support in C. Let $P' = M'U'$ be a standard parabolic subgroup of G. Fix a compact subset $C_{U'} \subset U'(\mathbb{A})$ which projects surjectively onto $U'(k)\backslash U'(\mathbb{A})$. By Lemma I.2.13 the set

$$\{\gamma \in P'(k)\backslash G(k); \exists x \in C, \hat{\tau}_{P'}(\log_{M'} m_{P'}(\gamma x) - T) = 1\}$$

is finite. Let $\Gamma_{P'} \subset G(k)$ be a subfinite set which projects surjectively onto the above set. Set

$$C_{P'} = C_{U'}\Gamma_{P'} C \operatorname{Supp}(h).$$

Then $C_{P'}$ is compact. By the construction of k_h, if $x \in C$, $y \in \mathbf{G}$ and $\gamma \in G(k)$ are such that

$$k_{h,P'}(\gamma x, y)\hat{\tau}_{P'}(\log_{M'} m_{P'}(\gamma x) - T) \neq 0,$$

then $y \in Z_G G(k) C_{P'}$. Set

$$C' = S \cap \mathbf{G}^c \cap \left(\bigcup_{P'} Z_G G(k) C_{P'} \right).$$

Then C' is compact and for $x \in C$, the restriction to $S \cap \mathbf{G}^c$ of the function

$$y \longmapsto \wedge_1^T k_h(x,y)$$

has support in C'. Thus the restriction of $\wedge_1^T k_h$ to $(S \cap \mathbf{G}^c) \times (S \cap \mathbf{G}^c)$ has support in $C \times C'$. As it is invariant under a compact open subgroup, it is also bounded. □

IV.2.6. Compactness of truncated operators

For $h \in \mathscr{H}^{\mathfrak{F}}$, the operator $\delta(h)$ of $L_\xi^{2,\mathfrak{F}}$ is bounded; its adjoint is $\delta(h^*)$.

Lemma *Let $h \in \mathscr{H}^{\mathfrak{F}}$.*

(a) *The operator $\wedge^T \circ \delta(h)$ of $L_\xi^{2,\mathfrak{F}}$ is compact.*

(b) *The operator $\wedge^T \circ \delta(h) \circ \wedge^T$ of $L_\xi^{2,\mathfrak{F}}$ is compact. If $h = h^*$, it is also auto-adjoint.*

Proof The last assertion is immediate. The first assertion of (b) is a consequence of (a) which we now prove. For $\varphi \in L_\xi^{2,\mathfrak{F}}$ and $x \in \mathbf{G}$, we have

$$(\delta(h)\varphi)(x) = \int_{\mathbf{G}} h(y)\varphi(xy)\,dy = \int_{\mathbf{G}} h(x^{-1}y)\varphi(y)\,dy$$

$$= \int_{Z_{\mathbf{G}}G(k)\backslash\mathbf{G}} k_h(x,y)\varphi(y)\,dy.$$

By the construction of $\wedge^T$, we obtain

$$(\wedge^T \circ \delta(h)\varphi)(x) = \int_{Z_{\mathbf{G}}G(k)\backslash\mathbf{G}} \wedge_1^T k_h(x,y)\varphi(y)\,dy.$$

From Lemma IV.2.5 we see that the function $|\wedge_1^T k_h|$ is square integrable on $(Z_{\mathbf{G}}G(k)\backslash\mathbf{G})^2$. The operator $\wedge^T \circ \delta(h)$ is thus (essentially, since we must take the centre into consideration) Hilbert-Schmidt, and so compact.

□

IV.2.7. A geometric lemma

Let us introduce the basis $\{\varpi_\alpha;\ \alpha \in \Delta_0\}$ of $\mathrm{Re}(\mathfrak{a}_{M_0}^G)^*$ dual to the basis $\{\check{\alpha};\ \alpha \in \Delta_0\}$.

Lemma *There exists $c > 0$ such that for all $\alpha \in \Delta_0$, $w \in W$ and $u \in U_0(\mathbb{A})$, we have*

$$m_{P_0}(wu)^{w\varpi_\alpha} \geq c.$$

Proof We can suppose that $\mathbf{G} = G(\mathbb{A})$. Let $\alpha \in \Delta_0$. There exists a character $\chi \in \mathrm{Rat}(M_0)$, an irreducible algebraic representation $\sigma : G \to GL(E)$ and a vector $e \in E$ such that

(i) $|\chi| = r\varpi_\alpha$, with $r \in \mathbb{Q}$, $r > 0$;

(1) (ii) $\sigma(m)e = m^\chi e$ for every $m \in M_0$;

(iii) $\chi_{|T_0}$ is the highest weight of $\sigma_{|T_0}$.

Let $w \in W$. Fix a basis $(e_i)_{i \in I}$ of E consisting of eigenvectors for the action of T_0. We suppose that e is a basis element. As $\sigma(w)e$ is an eigenvector for the action of T_0, we can suppose that $\sigma(w)e$ is proportional to a basis element.

For a vector $e' = \sum_{i \in I} x_i e_i \in E(\mathbb{A})$, we set

• for every place v of k, $\|e'\|_v = \sup\{|x_i|_v;\ i \in I\}$,

• if $\|e'\|_v = 1$ for almost every v, $\|e'\| = \prod_v \|e'\|_v$.

There exists $c_K > 0$ such that for every $k \in K$ and every $e' \in E(\mathbb{A})$ for which $\|e'\|_v = 1$ for almost every v, we have

$$\|\sigma(k)e'\| \geq c_K \|e'\|.$$

Let $u \in U_0(\mathbb{A})$. Set

$$N = \|\sigma(wu)^{-1}\sigma(w)e\|.$$

Write $wu = mu'k$, with $m = m_{P_0}(wu) \in M_0(\mathbb{A})$, $u' \in U_0(\mathbb{A})$, $k \in K$. Then

$$N = \|\sigma(k^{-1})\sigma(u'^{-1})\sigma(m^{-1})\sigma(w)e\| \geq c_K \|\sigma(u'^{-1})\sigma(m^{-1})\sigma(w)e\|.$$

We have

$$\sigma(m^{-1})\sigma(w)e = \sigma(w)\sigma(w^{-1}m^{-1}w)e = (w^{-1}m^{-1}w)^{\chi}\sigma(w)e.$$

Thus

$$N \geq c_K (w^{-1}m^{-1}w)^{|\chi|} \|\sigma(u'^{-1})\sigma(w)e\|. \tag{2}$$

But the action of $\sigma(u'^{-1})$ is unipotent and even upper diagonal if a suitable ordering is put on I. We thus have

$$\sigma(u'^{-1})\sigma(w)e = \sigma(w)e + \sum_{i \in I'} x_i e_i,$$

where $I' \subset I$ is such that if e_i is the vector proportional to $\sigma(w)e$, then $i \notin I'$. Then

$$\|\sigma(u'^{-1})\sigma(w)e\| \geq \|\sigma(w)e\| = 1.$$

By (1)(i) and (2), we obtain

$$N \geq c_K (w^{-1}mw)^{-r\varpi_\alpha} = c_K m_{P_0}(wu)^{-rw\varpi_\alpha}.$$

On the other hand

$$\sigma(wu)^{-1}\sigma(w)e = \sigma(u^{-1})e = e,$$

so $N = 1$. The inequality of the statement is a consequence of this, with $c = c_K^{1/r}$. □

IV.2.8. Remark

We will need the following observation. Let H be a Hilbert space, and suppose we are given a decomposition into a Hilbertian direct sum

$$H = \bigoplus_{\gamma \in \Gamma} H_\gamma.$$

For $x \in H$ and $\gamma \in \Gamma$ denote by x_γ the component of x in H_γ. Let $f : \mathbb{C} \to H$ be a holomorphic function and $D \subset \mathbb{C}$ a compact subset. Then

(*) the series $\sum_{\gamma\in\Gamma} \|f(z)_\gamma\|^2$ is normally convergent for $z \in D$.

Proof Choose a circle C containing D in its interior. By Cauchy's Formula, there exists $c > 0$ such that for all $\gamma \in \Gamma$, $z \in D$, we have

$$\|f(z)_\gamma\|^2 < c \int_C \|f(z')_\gamma\|^2 \, \mathrm{d}z'.$$

Now

$$\sum_{\gamma\in\Gamma} \int_C \|f(z')_\gamma\|^2 \, \mathrm{d}z' = \int_C \|f(z')\|^2 \, \mathrm{d}z' < \infty.$$

□

IV.3. The case of relative rank 1

IV.3.1. The situation of relative rank 1

We suppose here that $\dim X_M^{\mathbf{G}} = 1$. Then $W(M)$ has two elements: 1 for which $M(1, \lambda) = 1$ for every λ; and a non-trivial element which we denote here by w. We denote by ${}^wP$ the standard parabolic of Levi subgroup wMw^{-1} and by ${}^wU$ its radical unipotent (it can happen that ${}^wP = P$). We denote by α the unique element of Δ_M. Then $-w\alpha$ is the unique element of $\Delta_{wMw^{-1}}$.

IV.3.2. An auxiliary series

We define a map

$$E^T : X_M^{\mathbf{G}} \to \mathrm{Hom}_{\mathbb{C}}(A(M,\pi)^{\mathfrak{F}}, L^{2,\mathfrak{F}}_{\xi,\mathrm{loc}})$$

by

$$E^T(\varphi,\lambda)(g) = \sum_{\gamma\in P(k)\backslash G(k)} (\lambda\varphi)(\gamma g)\hat{\tau}_P(\log_M m_P(\gamma g) - T),$$

for all $\varphi \in A(M,\pi)^{\mathfrak{F}}$, $\lambda \in X_M^{\mathbf{G}}$, $g \in \mathbf{G}$. Recall that this sum is actually finite.

Lemma

(i) *The map E^T is holomorphic.*

(ii) *For every compact subset $C \subset X_M^{\mathbf{G}}$, there exists $r > 0$ and for every*

$\varphi \in A(M,\pi)^{\mathfrak{F}}$, there exists $c > 0$ such that for every $\lambda \in C$ and every $g \in \mathbf{G}$, we have the upper bound

$$|E^T(\varphi,\lambda)(g)| \leq c\|g\|^r.$$

(iii) *For every $\lambda \in X_M^{\mathbf{G}}$, $E^T(\lambda)$ is injective.*

(iv) *For every $\lambda \in X_M^{\mathbf{G}}$, $\wedge^T \circ E^T(\lambda) = 0$.*

(v) *There exists $c \in \mathbb{R}$ such that for every $\lambda \in X_M^{\mathbf{G}}$ satisfying*

$$\langle \operatorname{Re}\lambda, \check{\alpha}\rangle < c$$

and every $\varphi \in A(M,\pi)^{\mathfrak{F}}$, we have $E^T(\varphi,\lambda) \in L_\xi^{2,\mathfrak{F}}$.

Remark It can be shown that we can take $c = 0$ in (v). We do not prove this.

Proof For $\varphi \in A(M,\pi)^{\mathfrak{F}}$, the series defining $E^T(\varphi,\lambda)(g)$ is normally convergent when λ and g remain in compact sets (Lemma I.2.13). We prove (i) in a similar way to Lemma IV.1.7. Then (ii) results from Lemma I.2.13.

Let $P' = M'U'$ be a standard parabolic subgroup, $\lambda \in X_M^{\mathbf{G}}$ and $\varphi \in A(M,\pi)^{\mathfrak{F}}$. We calculate the constant term $E^T(\varphi,\lambda)_{P'}$. As in II.1.7, we obtain

$$E^T(\varphi,\lambda)_{P'}(g) = \sum_{v\in W^\bullet_{M,M'}} \sum_{m'\in M'(k)\cap vP(k)v^{-1}} i(v,m'),$$

where

$$i(v,m') = \int_{U'(k)\cap vP(k)v^{-1}\backslash U'(\mathbb{A})} (\lambda\varphi)(v^{-1}u'm'g)\hat{\tau}_P(\log_M m_P(v^{-1}u'm'g) - T)\, du'.$$

We can also integrate over $U'(\mathbb{A}) \cap vM(k)U(\mathbb{A})v^{-1}\backslash U'(\mathbb{A})$, then decompose the integral into

$$\int_{U'(\mathbb{A})\cap vP(\mathbb{A})v^{-1}\backslash U'(\mathbb{A})} \int_{U'(\mathbb{A})\cap vM(k)v^{-1}\backslash U'(\mathbb{A})\cap vM(\mathbb{A})v^{-1}} (\lambda\varphi)(v^{-1}u_1'u_2'm'g) \times$$
$$\hat{\tau}_P(\log_M m_P(v^{-1}u_1'u_2'm'g) - T)\, du_1'\, du_2'.$$

We have

$$\hat{\tau}_P(\log_M m_P(v^{-1}u_1'u_2'm'g) - T) = \hat{\tau}_P(\log_M m_P(v^{-1}u_2'm'g) - T).$$

The usual cuspidality argument shows that $i(v,m') = 0$ if $vMv^{-1} \not\subset M'$. The only parabolic subgroups P' for which such v exist are G, P and ${}^wP$. We obtain

(1) $\quad E^T(\varphi,\lambda)_{P'} = 0$ if $P' \neq G, P, {}^wP$ if $P \neq {}^wP$

(2) $\quad E^T(\varphi,\lambda)_P(g) = (\lambda\varphi)(g)\hat{\tau}_P(\log_M m_P(g) - T),$

(3) $$E^T(\varphi,\lambda)_{{}^wP}(g) = \int\limits_{{}^wU(\mathbb{A})} (\lambda\varphi)(w^{-1}ug)\hat{\tau}_P(\log_M m_P(w^{-1}ug) - T)\,du.$$

If $P = {}^wP$, then $E^T(\varphi,\lambda)_P(g)$ is the sum of the expressions (2) and (3).

Write $g = m_1u_1k_1$, with

$$m_1 = m_{{}^wP}(g) \in w\mathbf{M}w^{-1},\ u_1 \in {}^wU(\mathbb{A}),\ k_1 \in \mathbf{K}.$$

For $u \in {}^wU(\mathbb{A})$, we have

$$w^{-1}ug = w^{-1}um_1u_1k_1 = w^{-1}m_1u_2k_1,$$

where $u_2 = m_1^{-1}um_1u_1$. Set $w^{-1}u_2 = m_3u_3k_3$, with

$$m_3 = m_P(w^{-1}u_2) \in \mathbf{M},\ u_3 \in U(\mathbb{A}),\ k_3 \in \mathbf{K}.$$

Then

$$w^{-1}ug = w^{-1}m_1ww^{-1}u_2k_1 = w^{-1}m_1wm_3u_3k_3k_1.$$

Finally

$$m_P(w^{-1}ug) = w^{-1}m_1wm_3 = w^{-1}m_{{}^wP}(g)wm_P(w^{-1}u_2).$$

This gives

$$\begin{aligned}\langle\varpi_\alpha, \log_M m_P(w^{-1}ug) - T\rangle &= -\langle w^{-1}\varpi_{-w\alpha}, w^{-1}\log_{wMw^{-1}} m_{{}^wP}(g)\rangle \\ &\quad - \langle w^{-1}\varpi_{-w\alpha}, \log_M m_P(w^{-1}u_2)\rangle - \langle\varpi_\alpha, T\rangle \\ &\leq -\langle\varpi_{-w\alpha}, \log_{wMw^{-1}} m_{{}^wP}(g)\rangle - \langle\varpi_\alpha, T\rangle + c\end{aligned}$$

by lemma IV.2.7. Suppose

(4) $$\langle\varpi_{-w\alpha}, \log_{wMw^{-1}} m_{{}^wP}(g)\rangle \leq -\langle\varpi_\alpha, T\rangle + c.$$

Then $\hat{\tau}(\log_M m_P(w^{-1}ug) - T) = 0$ for every $u \in {}^wU(\mathbb{A})$ and expression (3) is zero. Under the hypothesis (4), $E^T(\varphi,\lambda)(g)$ is thus given by formula (2). But then if $\varphi \neq 0$, $E^T(\varphi,\lambda)_P \neq 0$ and *a fortiori* $E^T(\varphi,\lambda) \neq 0$. This proves (iii).

By the above calculations, $\wedge^T E^T(\varphi,\lambda)(g)$ is the sum of three terms:

$$E^T(\varphi,\lambda)(g);$$

$$-\sum_{\gamma\in P(k)\backslash G(k)} (\lambda\varphi)(\gamma g)\hat{\tau}_P(\log_M m_P(\gamma g) - T),$$

which is in fact just $-E^T(\varphi,\lambda)(g)$; and

(5) $$-\sum_{\gamma\in {}^wP(k)\backslash G(k)} \int\limits_{{}^wU(\mathbb{A})} (\lambda\varphi)(w^{-1}u\gamma g)\times$$
$$\hat{\tau}_P(\log_M m_P(w^{-1}u\gamma g) - T)\,du\ \hat{\tau}_{{}^wP}(\log_{wMw^{-1}} m_{{}^wP}(\gamma g) - T).$$

But if $\hat{\tau}_{{}^wP}(\log_{wMw^{-1}} m_{{}^wP}(\gamma g) - T) = 1$, we have

$$\langle\varpi_{-w\alpha}, \log_{wMw^{-1}} m_{{}^wP}(\gamma g)\rangle \geq \langle\varpi_{-w\alpha}, T\rangle \geq -\langle\varpi_\alpha, T\rangle + c,$$

for T is supposed sufficiently small. Thus γg satisfies (4) and the integral in (5) is zero. Thus (5) is zero, as well as $\wedge^T E^T(\varphi,\lambda)$. This proves (iv).

Let $\mu \in \operatorname{Re} X^{\mathbf{G}}_{M_0}$ be such that $\langle \mu, \check\beta \rangle > \langle \rho_0, \check\beta \rangle$ for every $\beta \in \Delta_0$. We will prove (v) by taking for c:

$$c = \inf\{-\langle (w\mu)_M, \check\alpha \rangle;\ w \in W\},$$

where $(w\mu)_M$ is the projection of $w\mu$ onto $\operatorname{Re} X^{\mathbf{G}}_M$. Let $\lambda \in X^{\mathbf{G}}_M$ be such that $\langle \operatorname{Re}\lambda, \check\alpha \rangle < c$ and $\varphi \in A(M,\pi)^{\mathfrak{F}}$. Choose a continuous function $y : \mathbf{G} \to \mathbb{R}^{\times}_{+}$ satisfying

$$y(umg) = m^{\mu+\rho_0} y(g)$$

for all $u \in U_0(\mathbb{A})$, $m \in \mathbf{M}_0$ and $g \in \mathbf{G}$. For $g \in \mathbf{G}$, set

$$E_y(g) = \sum_{\gamma \in P_0(k)\backslash G(k)} y(\gamma g).$$

This series is convergent. From the hypothesis on λ we deduce that for every $g \in \mathbf{G}$ such that $\hat\tau_P(\log_M m_P(g) - T) = 1$, we have

$$|m_P(g)^{\lambda}| \le m_P(g)^{\mu_M},$$

where μ_M is the projection of μ onto $\operatorname{Re} X^{\mathbf{G}}_M$. We then show as in II.1.10 that there exists $C > 0$ such that, for every $g \in \mathbf{G}$, we have

$$|E^T(\varphi,\lambda)(g)| \le C E_y(g).$$

To prove that $E^T(\varphi,\lambda)$ is square integrable, it suffices to prove that the integral

$$\int_{Z_{\mathbf{G}} G(k)\backslash \mathbf{G}} |E^T(\varphi,\lambda)(g)| E_y(g)\, dg$$

is convergent. Indeed, it suffices for the integral

$$\int_{Z_{\mathbf{G}} U(\mathbb{A}) M(k)\backslash \mathbf{G}} |(\lambda\varphi)(g)| \hat\tau_P(\log_M m_P(g) - T) E_{y,P}(g)\, dg$$

to converge. From the calculation of $E_{y,P}$ we deduce as in II.1.10 that it suffices for the integral

$$\int_{A_{\mathbf{G}}\backslash A_{\mathbf{M}}} a^{\lambda + (w\mu)_M} \hat\tau_P(\log_M a - T)\, da$$

to converge for every $w \in W$. This is a consequence of the hypothesis on λ. □

We also define a map

$$E^T_w : X^{\mathbf{G}}_{wMw^{-1}} \longrightarrow \operatorname{Hom}_{\mathbb{C}}(A(wMw^{-1}, w\pi)^{\mathfrak{F}}, L^{2,\mathfrak{F}}_{\xi,\mathrm{loc}})$$

satisfying analogous properties.

IV.3.3. Truncation of auxiliary series

Until IV.3.10, fix a function $h \in \mathscr{H}^{\mathfrak{F}}$ satisfying Lemma IV.2.2 for a point of $X_M^{\mathbf{G}}$. For $\lambda \in X_M^{\mathbf{G}}$ and $\varphi \in A(M,\pi)^{\mathfrak{F}}$, set $F^T(\varphi,\lambda) = \wedge^T \delta(h) E^T(\varphi,\lambda)$. The function $\delta(h)E^T(\varphi,\lambda)$ being smooth, **K**-finite, of moderate growth and even uniformly moderate if k is a number field, it is a consequence of Lemma I.2.16 that $F^T(\varphi,\lambda)$ is rapidly decreasing if k is a number field, and has compact support mod $Z_{\mathbf{G}}G(k)$ if k is a function field. *A fortiori*, $F^T(\varphi,\lambda) \in L_\xi^{2,\mathfrak{F}}$. This defines a function

$$F^T : X_M^{\mathbf{G}} \to \mathrm{Hom}_{\mathbb{C}}(A(M,\pi)^{\mathfrak{F}}, L_\xi^{2,\mathfrak{F}}).$$

Lemma *This function is holomorphic.*

Proof It suffices to show that for all $\varphi \in A(M,\pi)^{\mathfrak{F}}$ and $\psi \in L_\xi^{2,\mathfrak{F}}$, the function

$$\lambda \longmapsto \langle \psi, F^T(\varphi,\lambda)\rangle = \int_{Z_{\mathbf{G}}G(k)\backslash \mathbf{G}} \overline{\psi}(g) F^T(\varphi,\lambda)(g)\, dg$$

is holomorphic. Now

$$\langle \psi, F^T(\varphi,\lambda)\rangle = \int_{Z_{\mathbf{G}}G(k)\backslash \mathbf{G}} \overline{\psi}(g)[\wedge^T \rho(h) E^T(\varphi,\lambda)](g)\, dg$$

$$= \int_{(Z_{\mathbf{G}}G(k)\backslash \mathbf{G})^2} \overline{\psi}(g) \wedge_1^T k_h(g,y) E^T(\varphi,\lambda)(y)\, dy\, dg,$$

see the proof of Lemma IV.2.6. By Lemmas IV.2.5 and IV.3.2(ii), the integral is normally convergent when λ remains in a compact set. As the values of the integrand are holomorphic in λ, the integral is also holomorphic. □

We define similarly

$$F_w^T : X_{wMw^{-1}}^{\mathbf{G}} \to \mathrm{Hom}_{\mathbb{C}}(A(wMw^{-1}, w\pi)^{\mathfrak{F}}, L_\xi^{2,\mathfrak{F}}).$$

IV.3.4. The functional equation for truncation of Eisenstein series

For $\lambda \in \mathfrak{c}$ and $\varphi \in A(M,\pi)^{\mathfrak{F}}$, we define $\wedge^T E(\varphi,\lambda)$. As above, this function belongs to $L_\xi^{2,\mathfrak{F}}$. This defines

$$\wedge^T \circ E(\lambda) \in \mathrm{Hom}_{\mathbb{C}}(A(M,\pi)^{\mathfrak{F}}, L_\xi^{2,\mathfrak{F}}).$$

Recall that for every standard parabolic subgroup P' of G, we have

$$E(\varphi,\lambda)_{P'} = 0 \text{ if } P' \neq G, P, {}^wP;$$

$$\text{if } P \neq {}^wP, \begin{cases} E(\varphi,\lambda)_P = \lambda\varphi, \\ E(\varphi,\lambda)_{{}^wP} = (w\lambda)(M(w,\lambda)\varphi); \end{cases}$$

$$\text{if } {}^wP = P,\ E(\varphi, \lambda)_P = \lambda\varphi + (w\lambda)(M(w, \lambda)\varphi).$$

From the definition of $\wedge^T$, we obtain the following equality in $\mathrm{Hom}_{\mathbb{C}}(A(M, \pi)^{\mathfrak{F}}, L_{\xi}^{2,\mathfrak{F}})$:

$$(*) \qquad \wedge^T \circ E(\lambda) = E(\lambda) - E^T(\lambda) - E_w^T(w\lambda) \circ M(w, \lambda).$$

Lemma *For $\lambda \in \mathfrak{c}$, we have the equality*

$$(\wedge^T \circ \delta(h) \circ \wedge^T)(\wedge^T \circ E(\lambda)) - (\wedge^T \circ E(\lambda)) \circ i(\lambda, h)$$
$$= -F^T(\lambda) - F_w^T(w\lambda) \circ M(w, \lambda).$$

Proof The left-hand side is equal to

$$\wedge^T \circ [\delta(h) \circ \wedge^T \circ E(\lambda) - E(\lambda) \circ i(\lambda, h)],$$

i.e., by (*)

$$\wedge^T \circ [\delta(h) \circ E(\lambda) - \delta(h) \circ E^T(\lambda) - \delta(h) \circ E_w^T(w\lambda) \circ M(w, \lambda) - E(\lambda) \circ i(\lambda, h)].$$

Applying IV.1.7(*), it is simply

$$\wedge^T \circ [-\delta(h) \circ E^T(\lambda) - \delta(h) \circ E_w^T(w\lambda) \circ M(w, \lambda)],$$

which is the right-hand side in the statement. □

IV.3.5. Resolution of a functional equation

Set $V_h = \{\lambda \in X_M^{\mathbf{G}}; \det i(\lambda, h) = 0\}$. The conditions imposed on h imply that for every $z \in \mathbb{R}$, the function $\lambda \mapsto \det(z - i(\lambda, h))$ is not identically zero. In particular, V_h is discrete.

Lemma

(i) *There exists a unique meromorphic function*

$$e^T : X_M^{\mathbf{G}} - V_h \longrightarrow \mathrm{Hom}_{\mathbb{C}}(A(M, \pi)^{\mathfrak{F}}, L_{\xi}^{2,\mathfrak{F}})$$

such that, for every $\lambda \in X_M^{\mathbf{G}} - V_h$, we have the equality

$$(\wedge^T \circ \delta(h) \circ \wedge^T) \circ e^T(\lambda) - e^T(\lambda) \circ i(\lambda, h) = -F^T(\lambda).$$

(ii) *We have the equality $\wedge^T \circ e^T(\lambda) = e^T(\lambda)$ for every $\lambda \in X_M^{\mathbf{G}} - V_h$.*

Proof We use a simple adaptation of the usual resolvent theory. Decompose the space $L_{\xi}^{2,\mathfrak{F}}$ into eigensubspaces for the auto-adjoint compact operator $\wedge^T \circ \delta(h) \circ \wedge^T$:

$$L_{\xi}^{2,\mathfrak{F}} = \bigoplus_{\gamma \in \Gamma} H_{\gamma},$$

see IV.2.3. We have

$$\mathrm{Hom}_{\mathbb{C}}(A(M, \pi)^{\mathfrak{F}}, L_{\xi}^{2,\mathfrak{F}}) = \bigoplus_{\gamma \in \Gamma} \mathrm{Hom}_{\mathbb{C}}(A(M, \pi)^{\mathfrak{F}}, H_{\gamma}).$$

For $x \in \mathrm{Hom}_{\mathbb{C}}(A(M,\pi)^{\mathfrak{F}}, L^{2,\mathfrak{F}}_{\xi})$ and $\gamma \in \Gamma$, we set x_γ to be the component of x in $\mathrm{Hom}_{\mathbb{C}}(A(M,\pi)^{\mathfrak{F}}, H_\gamma)$. Set

$$V'_h = \{\lambda \in X^{\mathbf{G}}_M ; \exists\, \gamma \in \Gamma \cup \{0\},\ \det(\gamma - i(\lambda, h)) = 0\}.$$

From the remark preceding the statement and the properties of Γ we see that $X^{\mathbf{G}}_M - V'_h$ is a dense open set in $X^{\mathbf{G}}_M$. Let U be a relatively compact open set of $X^{\mathbf{G}}_M$ whose closure $\overline{U}$ does not intersect V'_h. For $\lambda \in U$ and $\gamma \in \Gamma$, the equation

$$(\wedge^T \circ \delta(h) \circ \wedge^T) \circ e^T(\lambda)_\gamma - e^T(\lambda)_\gamma \circ i(\lambda, h) = -F^T(\lambda)_\gamma$$

has a unique solution in $\mathrm{Hom}_{\mathbb{C}}(A(M,\pi)^{\mathfrak{F}}, H_\gamma)$, namely

$$e^T(\lambda)_\gamma = -F^T(\lambda)_\gamma (\gamma - i(\lambda, h))^{-1}.$$

The function $(\gamma, \lambda) \longmapsto (\gamma - i(\lambda, h))^{-1}$ is continuous on $(\Gamma \cup \{0\}) \times \overline{U}$, since $\overline{U} \cap V'_h = \emptyset$. As $(\Gamma \cup \{0\}) \times \overline{U}$ is compact, the function is bounded on this set. Thus there exists $c > 0$ such that for every $(\gamma, \lambda) \in \Gamma \times U$, we have

$$(*) \qquad \|e^T(\lambda)_\gamma\| \leq c \|F^T(\lambda)_\gamma\|.$$

We can then define $e^T(\lambda)$ by the convergent series

$$e^T(\lambda) = \sum_{\gamma \in \Gamma} e^T(\lambda)_\gamma.$$

It is the unique solution of the equation of the statement for $\lambda \in U$. By $(*)$ and IV.2.8, the series

$$\sum_{\gamma \in \Gamma} \|e^T(\lambda)_\gamma\|^2$$

is normally convergent for $\lambda \in U$. As each function $\lambda \longmapsto e^T(\lambda)_\gamma$ is clearly holomorphic in U, the sum $e^T(\lambda)$ is also holomorphic.

Now let U be a relatively compact open set of $X^{\mathbf{G}}_M$ whose closure $\overline{U}$ does not intersect V_h. The set

$$\Gamma(U) = \{\gamma \in \Gamma ; \exists\, \lambda \in U \text{ such that } \det(\gamma - i(\lambda, h)) = 0\}$$

is finite. We show as above that the function

$$\lambda \longmapsto \left(\prod_{\gamma \in \Gamma(u)} \det(\gamma - i(\lambda, h)) \right) e^T(\lambda),$$

already defined on a dense open set of U, is holomorphic on U. As

$$\prod_{\gamma \in \Gamma(U)} \det(\gamma - i(\lambda, h))$$

is not identically zero, e^T is meromorphic in U.

Applying the operator $\wedge^T$ to the equation of which $e^T(\lambda)$ is such a solution and using the relation $\wedge^T \circ \wedge^T = \wedge^T$, we see that $\wedge^T \circ e^T(\lambda)$ satisfies the same equation as $e^T(\lambda)$. By uniqueness, this gives (ii). □

Similarly, we introduce a meromorphic function e_w^T. The set V_h must be replaced by

$$V_{w,h} = \{\lambda \in X^{\mathbf{G}}_{wMw^{-1}}; \det i_w(\lambda, h) = 0\}.$$

IV.3.6. A decomposition of Eisenstein series

Define a map

$$\tilde{E} : X^{\mathbf{G}}_M - V_h \longrightarrow \mathrm{Hom}_{\mathbb{C}}(A(M,\pi)^{\mathfrak{F}}, L^{2,\mathfrak{F}}_{\xi,\mathrm{loc}})$$

by $\tilde{E}(\lambda) = E^T(\lambda) + e^T(\lambda)$. It is meromorphic by Lemmas IV.3.2 and IV.3.5. We define $\tilde{E}_w$ similarly.

Lemma *For every $\lambda \in \mathfrak{c} \cap (V_h \cup w^{-1}V_{w,h})$, we have the equality*

$$E(\lambda) = \tilde{E}(\lambda) + \tilde{E}_w(w\lambda) \circ M(w,\lambda).$$

Proof For such a λ, set

$$E_1(\lambda) = e^T(\lambda) + e_w^T(w\lambda) \circ M(w,\lambda).$$

This element of $\mathrm{Hom}_{\mathbb{C}}(A(M,\pi)^{\mathfrak{F}}, L^{2,\mathfrak{F}}_{\xi})$ satisfies the equation

$$(\wedge^T \circ \delta(h) \circ \wedge^T) \circ E_1(\lambda) - E_1(\lambda) \circ i(\lambda, h) = -F^T(\lambda) - F^T_w(w\lambda) \circ M(w,\lambda).$$

This results from the definitions of e^T and e^T_w and the relation IV.1.7(**). We show as in the preceding proof that this equation has only one solution. Now $\wedge^T \circ E(\lambda)$ is a solution by Lemma IV.3.4. This gives the equality $\wedge^T \circ E(\lambda) = E_1(\lambda)$. We apply the relation IV.3.4(*) to conclude. □

IV.3.7. Corollary

For $\lambda \in \mathfrak{c} - \mathfrak{c} \cap (V_h \cup w^{-1}V_{w,h})$, we have the equality

$$\begin{aligned}\delta(h) &\circ \tilde{E}(\lambda) - \tilde{E}(\lambda) \circ i(\lambda, h) \\ &+ [\delta(h) \circ \tilde{E}_w(w\lambda) - \tilde{E}_w(w,\lambda) \circ i_w(w\lambda, h)] \circ M(w,\lambda) \\ &= 0.\end{aligned}$$

Proof Apply IV.1.7(*) and (**) and the preceding lemma. □

IV.3.8. An injectivity lemma

Lemma *There exists $\mu \in X^{\mathbf{G}}_M - (V_h \cup w^{-1}V_{w,h})$ such that the element*

$$\delta(h) \circ \tilde{E}_w(w\mu) - \tilde{E}_w(w\mu) \circ i_w(w\mu, h)$$

of $\mathrm{Hom}_{\mathbb{C}}(A(wMw^{-1}, w\pi)^{\mathfrak{F}}, L^{2,\mathfrak{G}}_{\xi,\mathrm{loc}})$ *is injective.*

Proof Let μ satisfy condition (iii) of Lemma IV.2.2. We have $\mu \in (V_h \cup w^{-1}V_{w,h})$. The element of the statement can be simplified to

$$(\rho(h) - i) \circ \tilde{E}_w(w\mu).$$

By Lemmas IV.3.2(v) and IV.3.5(ii), we have

$$(1 - \wedge^T) \circ \tilde{E}_w(w\mu) = E_w^T(w\mu).$$

Now, $E_w^T(w\mu)$ is injective (see Lemma IV.3.2(iii)), thus $\tilde{E}_w(w\mu)$ is as well. We have $\langle \operatorname{Re} w\mu, -w\check{\alpha}\rangle \ll 0$, thus $E_w^T(w\mu)$ has values in $L_\xi^{2,\mathfrak{F}}$; see Lemma IV.3.2(iv). Then so does

$$\tilde{E}_w(w\mu) = e_w^T(w\mu) + E_w^T(w\mu).$$

Now the restriction to $L_\xi^{2,\sigma}$ of $\rho(h) - i$ is injective since $\rho(h)$ is auto-adjoint. This concludes the proof. □

IV.3.9. Proof of Theorem IV.1.8

Let μ be as in the preceding lemma. We choose, as we may, a continuous map

$$p \in \operatorname{Hom}_{\mathbb{C}}(L_{\xi,\mathrm{loc}}^{2,\mathfrak{F}}, A(wMw^{-1}, w\pi)^{\mathfrak{F}})$$

such that

$$p \circ [\delta(h) \circ \tilde{E}_w(w\mu) - \tilde{E}_w(w\mu) \circ i_w(w\mu, h)]$$

is an isomorphism. Define a meromorphic map

$$R : X_M^{\mathbf{G}} - V_h \longrightarrow \operatorname{Hom}_{\mathbb{C}}(A(M,\pi)^{\mathfrak{F}}, A(wMw^{-1}, w\pi)^{\mathfrak{F}})$$

by

$$R(\lambda) = p \circ [\delta(h) \circ \tilde{E}(\lambda) - \tilde{E}(\lambda) \circ i(\lambda, h)].$$

Define similarly

$$R_w : X_{wMw^{-1}}^{\mathbf{G}} - V_{w,h} \longrightarrow \operatorname{End}_{\mathbb{C}}(A(wMw^{-1}, w\pi)^{\mathfrak{F}}).$$

As $R_w(w\mu)$ is invertible, the function $\det R_w(\lambda)$ is not identically zero and the map

$$X_{wMw^{-1}}^{\mathbf{G}} - V_{w,h} \longrightarrow \operatorname{End}_{\mathbb{C}}(A(wMw^{-1}, w\pi)^{\mathfrak{F}})$$

$$\lambda \longmapsto R_w(\lambda)^{-1}$$

is still meromorphic.

Apply p to the equality of Corollary IV.3.7. We obtain

$$R(\lambda) + R_w(w\lambda) \circ M(w, \lambda) = 0$$

for $\lambda \in \mathfrak{c} - \mathfrak{c} \cap (V_h \cup w^{-1}V_{w,h})$. We then continue the function $M(w, .)$ meromorphically to $X_M^{\mathbf{G}} - (V_h \cup w^{-1}V_{w,h})$ by setting

$$M(w, \lambda) = -R_w(w\lambda)^{-1} \circ R(\lambda).$$

By Lemma IV.3.6, we may continue E to the same domain by setting

$$E(\lambda) = \tilde{E}(\lambda) + \tilde{E}_w(w\lambda) \circ M(w, \lambda).$$

The different continuations obtained by changing the function h can be glued together. By Lemma IV.2.2, for every $\lambda \in X_M^{\mathbf{G}}$, we can find h such that $\lambda \notin V_h \cup w^{-1}V_{w,h}$. We thus obtain continuations to all of $X_M^{\mathbf{G}}$.

IV.3.10. Proof of Proposition IV.1.9

Assertions (a) and (e) are obvious by meromorphic continuation. Let us continue to use the function h fixed in IV.3.3. Denote by U_h the set of $\lambda \in X_M^{\mathbf{G}} - (V_h \cup w^{-1}V_{w,h})$ such that the functions $e^T(.)$, $e_w^T(w,.)$, $M(w,.)$ and $i(.,h)^{-1}$ are holomorphic at λ. For $\lambda \in U_h$ and $\varphi \in A(M,\pi)^{\mathfrak{F}}$, we have by IV.1.7(*):

$$E(\varphi,\lambda) = \delta(h)E(i(\lambda,h)^{-1}\varphi,\lambda),$$

i.e. for almost every g

$$(*) \qquad E(\varphi,\lambda)(g) = \int_{Z_{\mathbf{G}}G(k)\backslash\mathbf{G}} k_h(g,x)E(i(\lambda,h)^{-1}\varphi,\lambda)(x)\,dx.$$

The function k_h is smooth. For every compact set C_1 of $\mathbf{G}$, there exists a compact set C_2 of $Z_{\mathbf{G}}G(k)\backslash\mathbf{G}$ such that, for $g \in C_1$, the support of the function $x \mapsto k_h(g,x)$ is in C_2. Finally, the function

$$U_h \longrightarrow L^{2,\mathfrak{F}}_{\xi,\mathrm{loc}}$$

$$\lambda \longmapsto E(i(\lambda,h)^{-1}\varphi,\lambda)$$

is holomorphic. It is then a consequence of (*) that the function

$$\mathbf{G}\times U_h \longrightarrow \mathbb{C}$$

$$(g,\lambda) \longmapsto E(\varphi,\lambda)(g)$$

is smooth. For fixed g, the function

$$U_h \longrightarrow \mathbb{C}$$

$$\lambda \longmapsto E(\varphi,\lambda)(g)$$

is holomorphic. On the other hand, since $\lambda \in U_h$, $E(i(\lambda,h)^{-1}\varphi,\lambda)$ is by construction the sum of an element of $L_\xi^{2,\mathfrak{F}}$ and of a function having moderate growth. More precisely, using Lemmas I.2.4(2)(ii) and IV.3.2(ii), we show that for every compact $C \subset U_h$, there exists $c > 0$ and $r > 0$ such that for all $\lambda \in C$, $g \in \mathbf{G}$, we have

$$|E(\varphi,\lambda)(g)| \leq c\|g\|^r.$$

Let $\psi \in A_0(G(k)\backslash\mathbf{G})_\xi$. We deduce from this upper bound that the function

$$\lambda \longmapsto \int_{Z_{\mathbf{G}}G(k)\backslash\mathbf{G}} \overline{\psi}(g)E(\varphi,\lambda)(g)\,dg$$

is holomorphic on U_h. It is zero on $\mathfrak{c}$. As U_h is connected, it is thus zero on all of U_h. From this we obtain the equality $E(\varphi,\lambda)^{\text{cusp}} = 0$ for every $\lambda \in U_h$.

Let P' be a proper standard parabolic subgroup of G. For $g \in \mathbf{G}$, $E(\varphi,\lambda)_{P'}(g)$ is the integral of $E(\varphi,\lambda)$ over a compact subset of $Z_{\mathbf{G}}G(k)\backslash\mathbf{G}$. The smoothness and holomorphic properties already proved imply that the map $\lambda \mapsto E(\varphi,\lambda)_{P'}(g)$ is holomorphic on U_h. By holomorphic continuation, this function is given by the formulae recalled in IV.3.4. Indeed, these formulae imply the equalities

$$E(\varphi,\lambda)_P^{\text{cusp}} = E(\varphi,\lambda)_P, E(\varphi,\lambda)_{^wP}^{\text{cusp}} = E(\varphi,\lambda)_{^wP},$$

$$E(\varphi,\lambda)_{P'}^{\text{cusp}} = 0 \text{ if } P' \neq P, {}^wP.$$

All the cuspidal components of $E(\varphi,\lambda)$ along standard parabolics are thus automorphic. By I.3.4, $E(\varphi,\lambda)$ is itself automorphic.

Let λ be any element of $X_M^{\mathbf{G}}$. Choose h such that $\lambda \notin V_h \cup w^{-1}V_{w,h}$. The properties of $E(\lambda')$ established above for $\lambda' \in U_h$ and the calculation of its cuspidal components show that we can find a small neighborhood D of λ in $X_M^{\mathbf{G}}$ such that setting $D' = D \cap U_h$, the hypotheses of Lemma I.4.10 are satisfied. If E is holomorphic at λ, this lemma implies properties (b)(i),(ii),(c)(i),(ii) in a neighborhood of λ. As previously noted, (c)(iii) is a consequence of (c)(i) and (ii). If E is not holomorphic at λ, it suffices to replace E by dE, where d is a holomorphic function which is non-zero on D, such that dE is holomorphic. This proves (b) and (c).

Remark Lemma I.4.10 shows that E is holomorphic at λ if and only if $M(w,.)$ is. More precisely, E has a pole of order $N \geq 0$ at λ if and only if this is true of $M(w,.)$. Let us prove assertion (d) of IV.1.9. We can suppose that V is small enough to be identified with an open set of $\mathbb{C}$ and that $D = \mathrm{d}^k/\mathrm{d}\lambda^k$. Suppose for simplicity that E is holomorphic on V (the general case is similar). Fix $\varphi \in A(M,\pi)^{\mathfrak{F}}$ and consider the function $(\mathrm{d}^k/\mathrm{d}\lambda^k)E(\varphi,\lambda)$. For every function f holomorphic on V, we have the formula

$$\frac{\mathrm{d}^k}{\mathrm{d}\lambda^k}f(\lambda) = \lim_{n\to\infty} n^k \sum_{\ell=0}^{k}(-1)^{k-\ell}\binom{\ell}{k} f\left(\lambda + \frac{\ell}{n}\right).$$

Fix $\lambda \in V$ and consider the sequence of automorphic forms $(\phi_n)_{n\in\mathbb{N}}$ defined by

$$\phi_n = n^k \sum_{\ell=0}^{k}(-1)^{k-\ell}\binom{\ell}{k} E\left(\varphi,\lambda + \frac{\ell}{n}\right).$$

For every standard parabolic subgroup P' of G, we have

$$\phi_{n,P'}^{\mathrm{cusp}} = n^k \sum_{\ell=0}^{k} (-1)^{k-\ell} \binom{\ell}{k} E\left(\varphi, \lambda + \frac{\ell}{n}\right)_{P'}^{\mathrm{cusp}}.$$

The explicit formulae for the functions $E(\varphi, \lambda + \frac{\ell}{n})_{P'}^{\mathrm{cusp}}$ show that $(\phi_{n,P'}^{\mathrm{cusp}})_{n\in\mathbb{N}}$ converges uniformly on every compact set to

$$\frac{\mathrm{d}^k}{\mathrm{d}\lambda^k}(E(\varphi, \lambda)_{P'}^{\mathrm{cusp}}).$$

The same formulae show that the hypotheses of Lemma I.4.4(3) are satisfied. We then deduce from this lemma that $(\mathrm{d}^k/\mathrm{d}\lambda^k)E(\varphi, \lambda)$ is automorphic and

$$(\frac{\mathrm{d}^k}{\mathrm{d}\lambda^k} E(\varphi, \lambda))_{P'}^{\mathrm{cusp}} = \frac{\mathrm{d}^k}{\mathrm{d}\lambda^k}(E(\varphi, \lambda)_{P'}^{\mathrm{cusp}})$$

for every P'. It remains to apply Lemma I.4.10 to obtain properties (b) and (c).

IV.3.11. Proof of Theorem IV.1.10

Let us consider the meromorphic function

$$e : X_M^{\mathbf{G}} \longrightarrow \mathrm{Hom}_{\mathbb{C}}(A(M,\pi)^{\mathfrak{F}}, L_{\xi,\mathrm{loc}}^{2,\mathfrak{F}})$$
$$\lambda \longmapsto E(\lambda) - E_w(w\lambda) \circ M(w, \lambda).$$

Suppose $e \neq 0$. Choose $\lambda \in X_M^{\mathbf{G}}$ such that

(i) e is holomorphic at λ and $e(\lambda) \neq 0$;
(ii) $\langle \mathrm{Re}\,\lambda, \check{\alpha}\rangle < 0$;
(iii) the restriction of $N\chi_{\pi\otimes\lambda}$ to $T_{\mathbf{M}} \cap \mathbf{G}_{\mathrm{der}}$ is not strictly positive real-valued, where N is the integer $N(G)$ of Proposition III.3.1.

It is possible to choose such a λ. Let $\varphi \in A(M,\pi)^{\mathfrak{F}}$ be such that $e(\varphi, \lambda) \neq 0$. Thanks to IV.1.9(b)(ii) and IV.1.9(c)(iii), we check that for every standard parabolic subgroup P' of G we have

$$e(\varphi, \lambda)_{P'}^{\mathrm{cusp}} = 0 \text{ if } P' \neq P,$$

$$e(\varphi, \lambda)_P^{\mathrm{cusp}} = \lambda(\varphi - M_w(w^{-1}, w\lambda)M(w, \lambda)\varphi).$$

By (ii) and I.4.11, $e(\varphi, \lambda)$ is square integrable. We have $\pi \otimes \lambda \in \Pi_0(\mathbf{M}, e(\varphi, \lambda))$, see I.3.5. Then (iii) contradicts Proposition III.3.1. Thus we must have $e = 0$ and the equality

$$E(\lambda) = E_w(w\lambda) \circ M(w, \lambda)$$

for every $\lambda \in X_M^{\mathbf{G}}$. From the equality $e(\varphi, \lambda)_P^{\mathrm{cusp}} = 0$ for all λ, φ, we deduce the equality

$$M_w(w^{-1}, w\lambda)M(w, \lambda) = 1.$$

IV.3.12. Proof of Proposition IV.1.11

Part (a) is trivial in relative rank 1. Let us consider the adjunction relation

$$\langle \varphi', M(w,\lambda)\varphi\rangle = \langle M_w(w^{-1}, -w\bar{\lambda})\varphi', \varphi\rangle$$

for all $\varphi \in A(M,\pi)^{\mathfrak{F}}, \varphi' \in A(wMw^{-1}, w\pi)^{\mathfrak{F}}$. We restrict ourselves to the set

$$\operatorname{Im} X_M^{\mathbf{G}} = \{\lambda \in X_M^{\mathbf{G}};\ \operatorname{Re}\lambda = 0\}.$$

Using the functional equation, we obtain

$$\langle M(w,\lambda)\varphi, M(w,\lambda)\varphi\rangle = \langle \varphi, \varphi\rangle$$

for every $\varphi \in A(M,\pi)^{\mathfrak{F}}$, say for every $\lambda \in \operatorname{Im} X_M^{\mathbf{G}}$ where $M(w,.)$ is holomorphic. Let $\lambda \in \operatorname{Im} X_M^{\mathbf{G}}$. If $M(w,.)$ was not holomorphic in λ, we could find $\varphi \in A(M,\pi)^{\mathfrak{F}}$ such that $\langle\varphi,\varphi\rangle = 1$ and a sequence $(\lambda_n)_{n\in\mathbb{N}}$ of points of $\operatorname{Im} X_M^{\mathbf{G}}$, such that $\lim_{n\to\infty}\lambda_n = \lambda$ and

$$\lim_{n\to\infty}\langle M(w,\lambda_n)\varphi, M(w,\lambda_n)\varphi\rangle = \infty.$$

This would contradict the above relation, which proves (b).

Let $\lambda \in X_M^{\mathbf{G}}$ be such that E and $M(w,.)$ have a pole at λ; we saw in IV.3.10 that E and $M(w,.)$ have the same singularities. Denote by N the order of the pole of E and $M(w,.)$. Write the expansions in a neighborhood of λ:

$$E(\mu) = \sum_{i=1}^{N} \text{`}\mu - \lambda\text{'}^{-i} E_{-i}(\lambda) + E_0(\mu),$$

$$M(w,\mu) = \sum_{i=1}^{N} \text{`}\mu - \lambda\text{'}^{-i} M_{-i}(w,\lambda) + M_0(w,\mu),$$

where we have written '$\mu-\lambda$' for a coordinate function in a neighborhood of λ. Let $\varphi \in A(M,\pi)^{\mathfrak{F}}$, $g \in \mathbf{G}$, $i \geq 1$ and Q be a standard parabolic subgroup of G. We calculate:

$$E_{-i}(\varphi,\lambda)_Q^{\text{cusp}}(g) = 0 \text{ if } Q \neq {}^wP,$$

$$(*)\qquad E_{-i}(\varphi,\lambda)_{{}^wP}^{\text{cusp}}(g) = \sum_{j=i}^{N} \langle -w\alpha, \log_{wMw^{-1}} m_{{}^wP}(g)\rangle^{N-j} \times \frac{1}{(N-j)!} m_{{}^wP}(g)^{w\lambda}(M_{-j}(w,\lambda)\varphi)(g).$$

Suppose $\langle \operatorname{Re}\lambda, \check{\alpha}\rangle > 0$ and $N \geq 2$. Choose φ such that

$$M_{-N}(w,\lambda)\varphi \neq 0.$$

From (∗) and Lemma I.4.11 we deduce that $E_{-j}(\varphi,\lambda)$ is square integrable for every $i \in \{1,\ldots,N\}$. These functions are linearly independent by (∗). Denote by V the space of dimension 2 generated by $E_{-N}(\varphi,\lambda)$ and

$E_{-N+1}(\varphi, \lambda)$. Fix R sufficiently large and set $H^R_{\xi,b,\mathbf{C}} = H^R_{\xi,b} \cap H^R_{\xi,\mathbf{C}}$, see III.1.7. For $f \in H^R_{\xi,b,\mathbf{C}}$, we have the equality of meromorphic functions

$$\Delta(f)E(\mu) = f(\pi \otimes \mu)E(\mu),$$

from which we deduce the equalities

$$(**) \qquad \Delta(f)E_{-N}(\varphi, \lambda) = f(\pi \otimes \lambda)E_{-N}(\varphi, \lambda)$$

and

$$\Delta(f)E_{-N+1}(\varphi, \lambda) = f(\pi \otimes \lambda)E_{-N+1}(\varphi, \lambda) + f'(\pi \otimes \lambda)E_{-N}(\varphi, \lambda)$$

where

$$f'(\pi \otimes \lambda) = \frac{\mathrm{d}}{\mathrm{d}z} f(\pi \otimes \lambda\kappa(z\alpha))|_{z=0}.$$

In particular $\Delta(f)$ leaves V stable. Let us consider the subalgebra of $\mathrm{End}_{\mathbf{C}}(V)$ consisting of restrictions $\Delta(f)|_V$ for $f \in H^R_{\xi,b,\mathbf{C}}$. It is commutative like $H^R_{\xi,\mathbf{C}}$. It is stable under adjunction by III.1.5. It is thus diagonalisable. But as $\langle \mathrm{Re}\,\lambda, \check{\alpha}\rangle > 0$, we have $\pi \otimes \lambda \neq w(\pi \otimes \lambda)$ and by the definitions we can find $f \in H^R_\xi$, such that for example $f(\pi \otimes \lambda) = 0$, $f'(\pi \otimes \lambda) = 1$. By (**), $\Delta(f)|_V$ is nilpotent non-zero, which contradicts the preceding property. This contradiction proves that $N \leq 1$, which is assertion (c) of the proposition.

We continue to suppose that $\langle \mathrm{Re}\,\lambda, \check{\alpha}\rangle > 0$ and also suppose that $N = 1$, $\pi \otimes \lambda \neq -w(\overline{\pi \otimes \lambda})$. We check directly on the definitions that we can find $f \in H^R_{\xi,b,\mathbf{C}}$ such that $f = f^*$ and for example $f(\pi \otimes \lambda) = i$. As $f = f^*$, the operator $\Delta(f)$ of L^2_ξ is auto-adjoint. Its eigenvalues are thus real. The relation

$$\Delta(f)E_{-1}(\varphi, \lambda) = iE_{-1}(\varphi, \lambda)$$

is thus contradictory. From this we deduce that the points λ such that $\langle \mathrm{Re}\,\lambda, \check{\alpha}\rangle \geq 0$ and E or $M(w, .)$ has a pole at λ satisfy $\pi \otimes \lambda = -w(\overline{\pi \otimes \lambda})$; note that if $\langle \mathrm{Re}\,\lambda, \check{\alpha}\rangle = 0$, (b) shows that there is no pole. They also satisfy $\langle \mathrm{Re}\,\lambda, \check{\alpha}\rangle \leq \langle \rho_P, \check{\alpha}\rangle$. But the set of λ satisfying these conditions is compact. Thus there is only a finite number of them, and this concludes the proof of the theorem.

Remark The condition $\pi \otimes \lambda = -w(\overline{\pi \otimes \lambda})$ forces $P = {}^wP$ and $w\pi \in \mathfrak{P}$; if one of these conditions is not satisfied, then there is no pole when $\langle \mathrm{Re}\,\lambda, \check{\alpha}\rangle \geq 0$.

IV.3.13. Proof of Proposition IV.1.12(b)

We suppose that k is a function field. Recall that α^* is a generator of the **Z**-module

$$\mathbf{M} \cap \mathbf{G}^1/\mathbf{M}^1.$$

For $\varphi \in A(M,\pi)^{\mathfrak{F}}$, $g \in \mathbf{G}$ and $\lambda \in C$, we have the formula

$$(M(w,\lambda)\varphi)(g) = m_{^wP}(g)^{-w\lambda} \int_{wU(\mathbb{A})} (\lambda\varphi)(w^{-1}ug)\, du.$$

We check that the right-hand side depends only on the restriction of λ to $\mathbf{G}^1$. Thus $M(w,\lambda)$ is actually a function of $\alpha^{*\lambda} \in \mathbb{C}^\times$ and defines a meromorphic function on $\mathbb{C}^\times$, with values in

$$\mathrm{Hom}_{\mathbb{C}}(A(M,\pi)^{\mathfrak{F}}, A(wMw^{-1}, w\pi)^{\mathfrak{F}}).$$

Thanks to IV.2.7, we see that there exists $\varepsilon > 0, c > 0$ and $N \in \mathbb{N}$ such that for $|\alpha^{*\lambda}| < \varepsilon$, we have

$$\|M(w,\lambda)\| < c|\alpha^{*\lambda}|^{-N}.$$

But then $(\alpha^{*\lambda})^{N+1}M(w,\lambda)$ can be continued to a function continuous at 0, thus it is holomorphic at 0. Also, $M(w,\lambda)$ can be continued to a function of $\alpha^{*\lambda}$ which is meromorphic on $\mathbb{C}$. The same is true of the inverse of this function. Using this result for the function $M_w(w^{-1},.)$, and the functional equation, we see that $M(w,\lambda)$ is also meromorphic when $|\alpha^{*\lambda}|$ tends to ∞, i.e. $M(w,\lambda)$ is meromorphic on $\mathbb{P}^1(\mathbb{C})$. But such a function is rational.

IV.3.14. Proof of Proposition IV.1.12(a)

We use the elegant argument of Harder ([H], Theorem 1.6.6). Let $g \in \mathbf{G}$ and $\varphi \in A(M,\pi)^{\mathfrak{F}}$. We want to show that the function $\lambda \longmapsto E(\varphi,\lambda)(g)$ is rational. By I.2.6, there exists a subset $C \subset G(k)\backslash\mathbf{G}$, of compact image in $Z_{\mathbf{G}}G(k)\backslash\mathbf{G}$, such that if $g \in (G(k)\backslash\mathbf{G}) - C$, there exists a proper standard parabolic subgroup Q of G such that we have $E(\varphi,\lambda)(g) = E(\varphi,\lambda)_Q(g)$. The calculation of $E(\varphi,\lambda)_Q$ and IV.1.12(b) show that this term is rational.

Fix an open subgroup $\mathbf{K}' \subset \mathbf{K}$ such that all elements of $\mathfrak{F}$ are trivial on $\mathbf{K}'$. We can suppose C stable under right multiplication by $\mathbf{K}'$. For every space H on which $\mathbf{K}$ acts, denote by $H^{\mathbf{K}'}$ the subspace of invariants under $\mathbf{K}'$. Denote by $L^2(C)_\xi$ the space of elements of L^2_ξ whose support is in C. The space $L^2(C)_\xi^{\mathbf{K}'}$ is finite-dimensional and the restriction map $P_C : L_\xi^{2,\mathbf{K}'} \to L^2(C)_\xi^{\mathbf{K}'}$ is an orthogonal projection. This implies that for every subspace $H \subset L_\xi^{2,\mathbf{K}'}$, we have the equality $p_C(H) = p_C(\overline{H})$, where $\overline{H}$ is the closure of H. From Theorem II.1.12, we then deduce that for every $f \in L^2(C)_\xi^{\mathbf{K}'}$, there exists a finite set of triples $(M_i, \mathfrak{P}_i, \phi_i)_{1<i<n}$ such that M_i is a standard Levi of G, $\mathfrak{P}_i$ an orbit of irreducible cuspidal representations of $\mathbf{M}_i$ and $\phi_i \in P_{(M_i,\mathfrak{P}_i)}^{\mathbf{K}'}$, such that

$$f = \sum_{i=1}^{n} p_C\theta_{\phi_i}.$$

By II.1.10, the support of

$$f - \sum_{i=1}^{n} \theta_{\phi_i}$$

is compact, contained in the complement of C.

Let $g \in C$. Denote by f_g the function on $G(k)\backslash\mathbf{G}$ with support in $gZ_{\mathbf{G}}\mathbf{K}'$, such that $f_g(gzk') = \xi(z)$ for all $z \in Z_{\mathbf{G}}$, $k \in \mathbf{K}'$, if such a function exists: otherwise set $f_g = 0$. There exists $c > 0$ such that for every $\lambda \in X_M^{\mathbf{G}}$ we have

$$E(\varphi, \lambda)(g) = c\langle f_g, E(\varphi, \lambda)\rangle.$$

Let us apply the above results to f_g. The product

$$\langle f_g - \sum_{i=1}^{n} \theta_{\phi_i}, E(\varphi, \lambda)\rangle$$

is rational since we already saw that $E(\varphi, \lambda)(g')$ is rational for $g' \notin C$. For $i \in \{1, \ldots, n\}$, we calculate

$$(*) \qquad \langle \theta_{\phi_i}, E(\varphi, \lambda)\rangle = \int_{Z_{\mathbf{G}} M_i(k) U_i(\mathbb{A})\backslash \mathbf{G}} \overline{\varepsilon F(\phi_i)(x)} E(\varphi, \lambda)_{P_i}^{\mathrm{cusp}}(x)\, dx$$

with the obvious notation. It is non-zero only if $P_i \in \{P, {}^wP\}$. Suppose for example $P \neq {}^wP$. If $P_i = P$, the above integral is equal to

$$(**) \qquad \int_{Z_{\mathbf{G}} U(\mathbb{A}) M(k)\backslash \mathbf{G}} \overline{\varepsilon F(\phi_i)}(x)(\lambda\varphi)(x)\, dx.$$

As $\varepsilon F(\phi_i)(x)$ has compact support modulo $Z_{\mathbf{G}}U(\mathbb{A})\mathbf{M}^1$, the above function of λ is a finite linear combination of functions of the form m^λ; it is polynomial. If $P_i = {}^wP$, the integral $(*)$ is equal to

$$\int_{Z_{\mathbf{G}}{}^{w}U(\mathbb{A})wM(k)w^{-1}\backslash \mathbf{G}} \overline{\varepsilon F(\phi_i)}(x)(w\lambda(M(w, \lambda)\varphi))(x)\, dx.$$

We can write

$$M(w, \lambda)\varphi = \sum_{j=1}^{N} c_j(\lambda)\varphi_j,$$

where the φ_j form a basis of $A(wMw^{-1}, w\pi)^{\mathfrak{F}}$ and the c_j are rational functions. We are reduced to considering

$$c_j(\lambda) \int_{Z_{\mathbf{G}}{}^{w}U(\mathbb{A})wM(k)w^{-1}\backslash \mathbf{G}} \overline{\varepsilon F(\phi_i)}(x)(w\lambda\varphi_j)(x)\, dx,$$

which can be treated like $(**)$. Finally, $\langle f_g, E(\varphi, \lambda)\rangle$ is rational and $E(\varphi, \lambda)(g)$ as well.

IV.4. The general case

IV.4.1. Continuation of intertwining operators

Let $w \in W(M), \ell = \ell(w)$. Fix a sequence of standard Levis $M_0 = M, M_1, \ldots, M_\ell = wMw^{-1}$, and for every $i \in \{1, \ldots, \ell\}$, fix a root $\alpha_i \in \triangle_{M_{i-1}}$, such that

$$w = s_{\alpha_\ell} \ldots s_{\alpha_1} ,$$

see I.1.8(iii). For $i \in \{1, \ldots, \ell\}$, we set $w(i-1) = s_{\alpha_{i-1}} \ldots s_{\alpha_1}$. For $\lambda \in \mathfrak{c}$, then

$$(*) \qquad M(w,\lambda) = M_{w(\ell-1)}(s_{\alpha_\ell}, w(\ell-1)\lambda) \ldots M(s_{\alpha_1},\lambda)$$

see II.1.6. Each of the operators $M_{w(i-1)}(s_{\alpha_i}, w(i-1)\lambda)$ comes from the case of relative rank 1: we work in the standard Levi $G_i \supset M_{i-1}$ such that $\triangle_{M_{i-1}} = \{\alpha_i\}$. We can thus meromorphically continue each of the operators of the right-hand side of the equality $(*)$ and thus continue $M(w,\lambda)$ by this equality. The continuation, being unique, does not depend on the chosen decomposition of w. The assertions of statements IV.1.9, IV.1.11 and IV.1.12 relative to the operator $M(w,\lambda)$ can be deduced from $(*)$ and from the similar assertions in the relative rank 1 case.

Let $w \in W(M)$ and $s \in W(wMw^{-1})$ be an elementary symmetry. If $\ell(sw) > \ell(w)$, we have $M(sw,\lambda) = M_w(s,w\lambda)M(w,\lambda)$ by construction. Otherwise, we have $\ell(sw) < \ell(w)$, $w = s^{-1}sw$ and

$$M(w,\lambda) = M_{sw}(s^{-1}, sw\lambda)M(sw,\lambda)$$

by construction. Using the functional equation for the operator $M_w(s,.)$, we obtain $M(sw,\lambda) = M_w(s,w\lambda)M(w,\lambda)$. It is then easy, by induction, to obtain the functional equation on $\ell(w')$:

$$M_w(w',w\lambda)M(w,\lambda) = M(w',w,\lambda)$$

for every $w' \in W(wMw^{-1})$.

IV.4.2. The auxiliary series

We reason by induction on the integer $\dim X_M^{\mathbf{G}}$. For every proper standard Levi M' of G and every $w \in W(M,M')$ (see II.1.7), we can define a meromorphic map $\lambda \mapsto E_w^{M'}(w\lambda) \circ M(w,\lambda)$ on $X_M^{\mathbf{G}}$; $E_w^{M'}$ is defined with respect to the Levi wMw^{-1} of M' and to the representation $w\pi$. Let U_0 be the dense open set of $X_M^{\mathbf{G}}$ where all these functions are holomorphic. We define

$$E^T : U_0 \longrightarrow \mathrm{Hom}_{\mathbb{C}}(A(M.\pi)^{\mathfrak{F}}, L^{2,\mathfrak{F}}_{\xi,\mathrm{loc}})$$

by

$$E^T(\varphi,\lambda)(g) = \sum_{P'=M'U';P_0\subset P'\subsetneq G} (-1)^{\mathrm{rg}(G)-\mathrm{rg}(M')} \sum_{\gamma\in P'(k)\backslash G(k)}$$
$$\sum_{w\in W(M,M')} E_w^{M'}(M(w,\lambda)\varphi, w\lambda)(\gamma g)\hat{\tau}_{P'}(\log_{M'} m_{P'}(\gamma g) - T_{M'}).$$

This series is convergent by I.2.13.

Lemma

(i) *The map E^T can be continued to a meromorphic map*

$$X_M^{\mathbf{G}} \longrightarrow \mathrm{Hom}_{\mathbb{C}}(A(M,\pi)^{\mathfrak{F}}, L^{2,\mathfrak{F}}_{\xi,\mathrm{loc}}).$$

(ii) *For every compact subset $C \subset U_0$, there exists $r > 0$ and for every $\varphi \in A(M,\pi)^{\mathfrak{F}}$, there exists $c > 0$, such that for every $\lambda \in C$ and every $g \in \mathbf{G}$, we have the upper bound*

$$|E^T(\varphi,\lambda)(g)| \leq c\|g\|^r.$$

Proof Let $\lambda \in X_M^{\mathbf{G}}$. Choose a holomorphic complex-valued function d defined in a neighborhood of λ and not identically zero, such that all the functions $d(\mu)M(w,\mu)$ and $d(\mu)E_w^{M'}(w\mu)$ are holomorphic at λ. Calculating the cuspidal components of the series $E_w^{M'}(M(w,\mu)\varphi, w\mu)$ and using I.4.4(ii), we show that there exists $r > 0$ and $c > 0$ such that for every μ in a neighborhood of λ, every $g \in \mathbf{G}$ and every pair M', w, we have

$$(*) \qquad |d(\mu)E_w^{M'}(M(w\mu)\varphi, w\mu)(g)| \leq c\|g\|^r.$$

Lemma I.2.13 then shows that the series defining $d(\mu)E^T(\varphi,\mu)(g)$ is normally convergent when μ remains in a neighborhood of λ and g remains in a compact set. Each term of the series is holomorphic at λ by IV.1.9(c) which is known for the functions $E_w^{M'}$ by the induction hypothesis. Then (i) can be proved as in IV.1.7. Part (ii) results from (*) and Lemma I.2.13.

By II.1.7, we have the equality

$$(**) \qquad \wedge^T E(\lambda) = E(\lambda) + E^T(\lambda)$$

for every $\lambda \in \mathfrak{c}$.

IV.4.3. Continuation of Eisenstein series

Fix $h \in \mathscr{H}^{\mathfrak{F}}$ satisfying Lemma IV.2.2 for a point of $X_M^{\mathbf{G}}$. For $\lambda \in U_0$, $\varphi \in A(M,\pi)^{\mathfrak{F}}$, set

$$F^T(\varphi,\lambda) = \wedge^T \delta(h) E^T(\varphi,\lambda).$$

We show as in IV.3.3 that $F^T(\varphi, \lambda) \in L_\xi^{2,\mathfrak{F}}$. This defines a map

$$F^T : U_0 \longrightarrow \mathrm{Hom}_{\mathbb{C}}(A(M,\pi)^{\mathfrak{F}}, L_\xi^{2,\mathfrak{F}})$$

and we show as in IV.3.3 that it can be continued to a meromorphic map

$$F^T : X_M^{\mathbf{G}} \longrightarrow \mathrm{Hom}_{\mathbb{C}}(A(M,\pi)^{\mathfrak{F}}, L_\xi^{2,\mathfrak{F}}).$$

As in IV.3.5, we have

Lemma *There exists a unique meromorphic map*

$$e^T : X_M^{\mathbf{G}} - V_h \longrightarrow \mathrm{Hom}_{\mathbb{C}}(A(M,\pi)^{\mathfrak{F}}, L_\xi^{2,\mathfrak{F}})$$

such that for every $\lambda \in X_M^{\mathbf{G}} - V_h$, we have the equality

$$(\wedge^T \circ \delta(h) \circ \wedge^T) \circ e^T(\lambda) - e^T(\lambda) \circ i(\lambda, h) = F^T(\lambda).$$

It is an easy consequence of IV.4.2(**) that

$$e^T(\lambda) = \wedge^T \circ E(\lambda)$$

for $\lambda \in \mathfrak{c}$. We can then continue E to a meromorphic map

$$E : X_M^{\mathbf{G}} - V_h \longrightarrow \mathrm{Hom}_{\mathbb{C}}(A(M,\pi)^{\mathfrak{F}}, L_{\xi,\mathrm{loc}}^{2,\mathfrak{F}})$$

by setting $E(\lambda) = e^T(\lambda) - E^T(\lambda)$.

As in the case of relative rank 1, we continue E to all of $X_M^{\mathbf{G}}$ by letting h vary. The uniqueness of this continuation, which is used here, results from the fact that $X_M^{\mathbf{G}} - V_h$ is connected: it is a complex manifold from which we remove the zeros of a holomorphic function.

IV.4.4. End of the proof

Denote by U_h the open set where the functions $e^T(.)$, $E_w^{M'}(w.)$ and $M(w,.)$ are all holomorphic. This set is also connected: locally, it is still a complex manifold from which we remove the zeros of a holomorphic function. As in relative rank 1, we prove the properties of $E(\lambda)$ required for Proposition IV.1.9 for $\lambda \in U_h$. We then generalise them to every λ using Lemmas I.4.4(3) and I.4.10.

Remark In relative rank 1, we have shown that the singularities of E are identical to those of $M(w,.)$. Here the situation is more complicated. However Lemma I.4.10 shows the following. Let d be a holomorphic function on $X_M^{\mathbf{G}}$ and suppose that for every $w \in W(M)$, $dM(w,.)$ is holomorphic at λ; then dE is as well. This suffices to make the proofs work. This remark shows that the assertions of Proposition IV.1.11 concerning E can be deduced from those concerning the functions $M(w,.)$.

To prove (a) of Theorem IV.1.10, we calculate the cuspidal components of two terms of the formula and we use (b) of the theorem. Finally, (a) of Proposition IV.1.12 can be deduced from (b) by the same method as in relative rank 1.

IV.4.5. Remark

Let $M \subset L$ be two standard Levis, $\mathfrak{P}^L$ an orbit under $X_M^{\mathbf{G}}$ of irreducible cuspidal representations, $\mathfrak{P} = \{\pi \otimes \lambda;\ \pi \in \mathfrak{P}^L,\ \lambda \in X_M^{\mathbf{G}}\} = \{\pi \otimes \lambda;\ \pi \in \mathfrak{P}^L,\ \lambda \in X_L^{\mathbf{G}}\}$ and $w \in W_L(M)$, the analogue of $W(M)$ when we replace G by L. We have operators

$$M(w,\pi) : A\Big(U(\mathbb{A})M(k)\backslash\mathbf{G}\Big)_\pi \to A\Big({}^wU(\mathbb{A})wM(k)w^{-1}\backslash\mathbf{G}\Big)_{w\pi},$$

for $\pi \in \mathfrak{P}$,

$$M^L(w,\pi) : A((U \cap L)(\mathbb{A})M(k)\backslash\mathbf{L})_\pi \longrightarrow$$
$$A(({}^wU \cap L)(\mathbb{A})wM(k)w^{-1}\backslash\mathbf{L})_{w\pi}, (\text{for } \pi \in \mathfrak{P}^L).$$

It is clear that $M(w,.)$ can be deduced from $M^L(w,.)$ by induction. Let $\pi \in \mathfrak{P}^L, \lambda \in X_L^{\mathbf{G}}$ and f be a holomorphic function defined in a neighborhood of $\pi \otimes \lambda$ in $\mathfrak{P}$. Then the function $\pi' \mapsto f(\pi')M(w,\pi')$ is holomorphic in $\pi' = \pi \otimes \lambda$ if and only if the function

$$\pi'' \mapsto f(\pi'' \otimes \lambda)M^L(w,\pi''),$$

defined in a neighborhood of π in $\mathfrak{P}^L$, is holomorphic at π. In particular, the singular hyperplanes of $M(w,.)$ are the connected components of the manifolds $\{\pi \otimes \lambda;\ \pi \in H,\ \lambda \in X_L^{\mathbf{G}}\}$, where H is a singular hyperplane of $M^L(w,.)$.

V

Construction of the Discrete Spectrum via Residues

V.1. Generalities and the residue theorem

In the whole of this chapter, we fix a unitary character ξ of the centre of $\mathbf{G}$ and an equivalence class $\mathfrak{X}$ of cuspidal data given by pairs $(M, \mathfrak{P})$ where $\mathfrak{P}$ consists of cuspidal representations of $\mathbf{M}$ whose restrictions to $Z_{\mathbf{G}}$ have character ξ. We will constantly use the following notation, for $(\mathbf{M}, \mathfrak{P}) \in \mathfrak{X}$:

$W(M)$ is the set of elements w of W of minimal length in their right class modulo the Weyl group of $\mathbf{M}$ such that wMw^{-1} is still a standard Levi of $\mathbf{G}$.

$$\mathrm{Stab}(M, \mathfrak{P}) := \{w \in W(M) | wMw^{-1} = M, w\mathfrak{P} = \mathfrak{P}\}.$$

We note that $\mathrm{Stab}(M, \mathfrak{P})$ is a group.

V.1.1. Affine subspace

Fix $(M, \mathfrak{P}) \in \mathfrak{X}$. For $\pi \in \mathfrak{P}$, we defined (see I.3.3, notation) $\mathrm{Re}\,\pi$ which is an element of $\mathrm{Re}\,X_M^{\mathbf{G}}$ and $\mathrm{Im}\,\pi$ which is an element of $\mathfrak{P}$. Let $\mathfrak{S}$ be a subset of $\mathfrak{P}$ and set:

$$\mathfrak{S}^0 = \{\lambda \in X_M^{\mathbf{G}} | \mathfrak{S} \otimes \lambda = \mathfrak{S}\}.$$

It is a subgroup of $X_M^{\mathbf{G}}$ which operates naturally on $\mathfrak{S}$ and contains $\mathrm{Fix}_{X_M^{\mathbf{G}}} \mathfrak{P}$.

Definition We say that $\mathfrak{S}$ is an affine subspace of $\mathfrak{P}$ if there exists a vector subspace V of $\mathfrak{a}_M^{*G}$ ($\simeq \mathrm{Re}\,X_M^{\mathbf{G}} \otimes_{\mathbf{R}} \mathbf{C}$) such that $\mathfrak{S} = \pi \otimes \kappa(V)$ (see

I.I.4(3) for the notation), where π is any element of $\mathfrak{S}$. We say that the vector part of $\mathfrak{S}$ is defined over $\mathbb{R}$ if V is defined over $\mathbb{R}$.

Let $\mathfrak{S}$ be an affine subspace of $\mathfrak{P}$ with vector part defined over $\mathbb{R}$ and let V be as in the definition. We note that:

(1) $\mathrm{Re}\,\mathfrak{S} := \{\mathrm{Re}\,\pi | \pi \in \mathfrak{S}\}$ is an affine subspace of $\mathrm{Re}\,X_M^{\mathbf{G}}$ whose vector part, $\mathrm{Re}\,\mathfrak{S}^0 := \{\mathrm{Re}\,\lambda | \lambda \in \mathfrak{S}^0\}$, coincides with $\mathrm{Re}\,V$.

$$V = \mathrm{Re}\,\mathfrak{S}^0 \otimes_{\mathbb{R}} \mathbb{C} \text{ in } \mathfrak{a}_M^{*G};\ \mathrm{Re}\,\mathfrak{S}^0 = \mathrm{Re}\,X_M^{\mathbf{G}} \cap \mathfrak{S}^0$$

and (2) $\mathfrak{S}^0 = \bigcup_{\nu \in \mathrm{Fix}_{X_M^{\mathbf{G}}} \mathfrak{P}} \kappa(V) + \nu.$

If these conditions are satisfied, we denote by $\tilde{\mathfrak{S}}^0$ the image of $\mathrm{Re}\,\mathfrak{S}^0 \otimes_{\mathbb{R}} \mathbb{C}$ in $X_M^{\mathbf{G}}$ and we say that $\tilde{\mathfrak{S}}^0$ is the vector part of $\mathfrak{S}$. In particular $\mathfrak{S}$ is a homogeneous space under $\mathfrak{S}^0$ and under $\tilde{\mathfrak{S}}^0$.

If k is a number field, identifying $\mathfrak{P}$ and $X_M^{\mathbf{G}}$ by the choice of an element of $\mathfrak{P}$ (see II.1.1), we recover the usual notion of affine subspace of the vector space $X_M^{\mathbf{G}}$, whose vector part is defined over $\mathbb{R}$. In particular $\mathfrak{P}$ itself is an affine space in this sense.

If k is a function field, then $\mathfrak{P}$ is a union of affine spaces which are parallel, i.e. which all have the same vector part.

Generally, we say that two affine subspaces of $\mathfrak{P}$ with vector part defined over $\mathbb{R}$ are parallel if they have the same vector part. We say that $\mathfrak{S}$ is an affine hyperplane with vector part defined over $\mathbb{R}$ if the above conditions are satisfied and if $\mathrm{Re}\,\mathfrak{S}^0$ is a hyperplane of $\mathrm{Re}\,X_M^{\mathbf{G}}$. We note that the intersection of two affine subspaces of $\mathfrak{P}$ with vector part defined over $\mathbb{R}$ has the following property: all its irreducible components are also affine subspaces of $\mathfrak{P}$ with vector part defined over $\mathbb{R}$; this intersection is irreducible if k is a number field but not always if k is a function field.

An important class

Let $\alpha \in R(T_M, G)$. We use the notation α^* of I.1.11; in particular if k is a number field then α^* is the coroot associated with α. We set

$$H_{\alpha^*} := \begin{cases} \lambda \in X_M^{\mathbf{G}} | \langle \lambda, \check{\alpha} \rangle = 0 & \text{if } k \text{ is a number field} \\ (\alpha^*)^{\lambda} = 1 & \text{if } k \text{ is a function field.} \end{cases}$$

Let $\pi \in \mathfrak{P}$; then $H_{\pi,\alpha^*} := \{\pi \otimes \lambda | \lambda \in H_{\alpha^*}\}$ is a union of parallel affine hyperplanes of $\mathfrak{P}$ (there is only one of them if k is a number field) of vector part defined over $\mathbb{R}$. This vector part is the image of $\mathrm{Re}\,H_{\alpha^*} \otimes_{\mathbb{R}} \mathbb{C}$ in $X_M^{\mathbf{G}}$.

Let $w \in W(M)$ and let $\phi \in P_{(wM, -w\overline{\mathfrak{P}})}$. We know by IV.1.11 that the set of elements π of $\mathfrak{P}$ on which $M(w^{-1}, -w\overline{\pi})\phi(-w\overline{\pi})$ is singular is a locally finite union of spaces of the form $\pi_0 \otimes H_{\alpha^\vee}$. We denote by $S^h_{(M,\mathfrak{P})}$ the set of affine hyperplanes (in the above sense) of $\mathfrak{P}$ which are singular for $M(w^{-1}, -w\pi)\phi(-w\pi)$ where w varies in $W(M)$ and where ϕ varies in $P_{(wM,-w\mathfrak{P})}$, augmented by the set of conjugates of such hyperplanes under the action of $\mathrm{Stab}(M,\mathfrak{P})$. We denote by $S_{(M,\mathfrak{P})}$ the set of affine spaces of $\mathfrak{P}$ with vector part defined over $\mathbb{R}$ obtained as irreducible components of intersections of elements of $S^h_{(M,\mathfrak{P})}$. Considering an empty intersection, we see that $S_{(M,\mathfrak{P})}$ contains the irreducible components of $\mathfrak{P}$. The family $S_{(M,\mathfrak{P})}$ is no longer locally finite, at least *a priori*; we are thus led to fix a finite set of $\mathbf{K}$-types, denoted in general by $\mathfrak{F}$, and to define $S^{\mathfrak{F}}_{(M,\mathfrak{P})}$ in an analogous way to $S_{(M,\mathfrak{P})}$ but letting the functions ϕ (in the above notation) vary only in the set of elements of $P_{(wM,-w\mathfrak{P})}$ which are acted on via a $\mathbf{K}$-type belonging to $\mathfrak{F}$.

An important property of elements of $S_{(M,\mathfrak{P})}$ is the following:
Let $\mathfrak{S} \in S_{(M,\mathfrak{P})}$; we set

$$R_{\mathfrak{S}} := \{\alpha \in R(T_M, G) | \langle \lambda, \check{\alpha} \rangle = 0, \forall \lambda \in \mathrm{Re}\,\mathfrak{S}^0\};$$

then

$$\mathrm{Re}\,\mathfrak{S}^0 = \{\lambda \in \mathrm{Re} X^{\mathbf{G}}_M | \langle \lambda, \check{\alpha} \rangle = 0, \forall \alpha \in R_{\mathfrak{S}}\}.$$

We set:

$$S_{\mathfrak{X}} := \bigcup_{(M',\mathfrak{P}') \in \mathfrak{X}} S_{(M',\mathfrak{P}')}$$

and

$$S^{\mathfrak{F}}_{\mathfrak{X}} := \bigcup_{(M',\mathfrak{P}') \in \mathfrak{X}} S^{\mathfrak{F}}_{(M',\mathfrak{P}')}.$$

Let $\mathfrak{S} \in S_{\mathfrak{X}}$. There exists $(M', \mathfrak{P}') \in \mathfrak{X}$ such that $\mathfrak{S} \in S_{(M',\mathfrak{P}')}$; we say that $(M', \mathfrak{P}')$ is the cuspidal datum attached to $\mathfrak{S}$. For $\mathfrak{S} \in \mathfrak{S}_{\mathfrak{X}}$ of cuspidal datum $(M', \mathfrak{P}')$, we write

$$\mathrm{Norm}\,\mathfrak{S} := \{w \in \mathrm{Stab}(M', \mathfrak{P}') | w\mathfrak{S} = \mathfrak{S}\}.$$

V.1.2. Meromorphic functions with polynomial singularities on $\mathfrak{S}_{\mathfrak{X}}$

We have equipped $\mathfrak{P}$ with the structure of a complex analytic manifold (see II.1.1); whence the notion of holomorphic and meromorphic functions. We also define the notion of polynomials. We begin by defining polynomials on $X^{\mathbf{G}}_M/\mathrm{Fix}_{X^{\mathbf{G}}_M}\mathfrak{P}$: if k is a number field, we take the usual definition. Suppose that k is a function field; we denote by $\mathrm{Fix}_{X^{\mathbf{G}}_M}\mathfrak{P}^\perp$ the subgroup of $\mathbf{M}$ consisting of elements m such that $m^\lambda = 1$ for every $\lambda \in \mathrm{Fix}_{X^{\mathbf{G}}_M}$. This group contains $\mathbf{M}^1$ and $\mathrm{Fix}_{X^{\mathbf{G}}_M}/\mathbf{M}^1$ is a free $\mathbb{Z}$-module of finite rank which can be identified via $\log_M$ with the orthogonal lattice of $\mathrm{Fix}_{X^{\mathbf{G}}_M}$ in $\mathrm{Re}\,\mathfrak{a}_M$, i.e. the lattice $L_1^\perp$ where L_1 is the lattice of $\mathrm{Rat}(M) \otimes_{\mathbb{Z}} \mathbb{Q}$

such that $(2i\pi/\log q)L_1$ is the inverse image under κ of $\mathrm{Fix}_{X_M^{\mathbf{G}}}$. Then the algebra of polynomials over $X_M^{\mathbf{G}}/\mathrm{Fix}_{X_M^{\mathbf{G}}}\mathfrak{P}$ is the algebra of this group. Let us return to the case of arbitrary k; we identify $\mathfrak{P}$ with the vector space $X_M^{\mathbf{G}}/\mathrm{Fix}_{X_M^{\mathbf{G}}}$ by choosing of an element in $\mathfrak{P}$ and we can then transport the notion of polynomial, since it does not depend on this choice. We do however lose the notion of homogeneous polynomial. Let us give some examples which will be useful to us later on.

First example: let $\pi_0 \in \mathfrak{P}$, let $\alpha \in R(T_M, G)$ and let α^* be as in I.I.11; we define the polynomial P_{π_0,α^*} on $\mathfrak{P}$ by:

$$P_{\pi_0,\alpha^*}(\pi_0 \otimes \lambda) = \begin{cases} \langle \lambda, \alpha^* \rangle & \text{if } k \text{ is a number field,} \\ (\alpha^*)^\lambda - 1 & \text{if } k \text{ is a function field.} \end{cases}$$

Second example: We suppose here that k is a function field and we show how, given parallel but distinct spaces $\mathfrak{S}$ and $\mathfrak{S}' \in S_{(M,\mathfrak{P})}$, we can construct a polynomial on $\mathfrak{P}$ which is zero on $\mathfrak{S}$ and non-zero on $\mathfrak{S}'$. For this, we use the notation L_1 introduced above. We know that $\mathrm{Re}\,\mathfrak{S}^0$ is defined on $\mathbb{Q}$, i.e. generated by its intersection with $\mathrm{Rat}(M) \otimes_{\mathbb{Z}} \mathbb{Q}$, intersection denoted by $\mathfrak{s}$. We identify $X_M^{\mathbf{G}}/\mathrm{Fix}_{X_M^{\mathbf{G}}}\mathfrak{P}$ with $\mathfrak{P}$ by choice of an element of $\mathfrak{P}$ and fix λ_0 in the inverse image of $\mathfrak{S}$ in $\mathfrak{a}_M^*$, λ_0' in the inverse image of $\mathfrak{S}'$ in $\mathfrak{a}_M^*$ and $H \in (\mathfrak{s}+L_1)^\perp$ separating $(2\pi i/\log q)L_1+\lambda_0$ and $(2\pi i/\log q)L_1 + \lambda_0'$. Then the function on $\mathfrak{a}_M^*$ defined by:

$$\lambda \longmapsto q^{\langle \lambda, H \rangle} - q^{\langle \lambda_0, H \rangle}$$

passes to the quotient to give a function on X_M, and then, by restriction, a polynomial on $X_M^{\mathbf{G}}$ invariant under translation by $\mathrm{Fix}_{X_M^{\mathbf{G}}}\mathfrak{P}$; whence a polynomial on $\mathfrak{P}$ which has the desired properties.

Third example: let $\mathfrak{S}$ and $\mathfrak{S}'$ be non-parallel; we will separate them by a polynomial which is zero on $\mathfrak{S}$ and non-constant on $\mathfrak{S}'$; we can proceed as in the second example but we can also, more easily, fix a linear form l over $\mathfrak{a}_M$, defined on $\mathbb{Q}$, zero on $\mathrm{Re}\,\mathfrak{S}^0$ and non-zero on $\mathrm{Re}(\mathfrak{S}')^0$ and fix λ_0 in the inverse image of $\mathfrak{S}$ in $\mathfrak{a}_M^*$, after having identified $X_M^{\mathbf{G}}/\mathrm{Fix}_{X_M^{\mathbf{G}}}\mathfrak{P}$ with $\mathfrak{P}$ as in the second example. Then for N a well-chosen integer the function on $\mathfrak{a}_M^*$ given by:

$$\lambda \mapsto q^{2\pi i N \langle \lambda, H \rangle} - q^{2\pi i N \langle \lambda_0, H \rangle}$$

defines, as above, a polynomial on $\mathfrak{P}$ which works.

Definition of 'polynomial singularities' on $S_{\mathfrak{X}}$ or on $S_{\mathfrak{X}}^{\mathfrak{F}}$

Let $\mathfrak{S} \in S_{(M,\mathfrak{P})}$; by Properties V.1.1(1) and (2) satisfied by $\mathfrak{S}$, we see that

$\mathfrak{S}$ is a complex analytic closed set of $\mathfrak{P}$. We thus also have the notions of holomorphic and meromorphic functions. Let $\pi \in \mathfrak{S} \mapsto A(\pi) \in \mathbb{C}$ be a meromorphic function on $\mathfrak{S}$. We say that A has polynomial singularities on $S_{\mathfrak{X}}$ if for every $\pi \in \mathfrak{S}$, there exists a polynomial P_π on $\mathfrak{P}$ such that $(P_\pi)_{|\mathfrak{S}}A$ is holomorphic at π and P_π is obtained as product of powers of the polynomials $P_{\pi,\alpha^\vee}$ above, where α runs through the elements of $R(T_M, G)$ such that the connected component of $\pi \otimes H_{\alpha^\vee}$ which passes through π (in the notation of V.1.1) is an element of $S_{\mathfrak{X}}$ not containing $\mathfrak{S}$. In particular the set of singularities of the function A is a union of affine hyperplanes of $\mathfrak{S}$ which are elements of $S_{\mathfrak{X}}$. If k is a number field, this notion is quite usual. We obtain an analogous definition by replacing $S_{\mathfrak{X}}$ by $S_{\mathfrak{X}}^{\mathfrak{F}}$ when a finite set of $\mathbf{K}$-types is fixed.

We will use the following property: let A be a meromorphic function on $\mathfrak{S}$ with polynomial singularities $S_{\mathfrak{X}}$ and let $\mathfrak{H}$ be a hyperplane of $\mathfrak{S}$ which is singular for A; then there exists $n \in \mathbb{N}$ and $\alpha \in R(T_M, G)$ such that for every $\pi \in \mathfrak{H}$ the singular hyperplanes passing through π of the function $(P_{\pi,\alpha^\vee}^n)_{|\mathfrak{S}}A$ are the singular hyperplanes of A passing through π with the exception of $\mathfrak{H}$.

V.1.3. Residue data

For every element $\mathfrak{S}$ of $S_{\mathfrak{X}}$, we denote by $d_{\mathfrak{S}}\pi$ the measure on $\mathfrak{S}$ obtained via the scalar product on $\operatorname{Re}\mathfrak{S}^0$ coming from the one fixed on $\operatorname{Re}X_M^{\mathbf{G}}$ in I.1.12. This measure is a Lebesgue measure invariant under $\operatorname{Norm}_W\mathfrak{S}$. It induces a measure on $\mathfrak{S}^0$ and on $\operatorname{Im}\mathfrak{S}$ and $\operatorname{Im}\mathfrak{S}^0$; all these measures are denoted by $d_{\mathfrak{S}}$ and we are only interested in the measures on $\operatorname{Re}\mathfrak{S}^0$, $\operatorname{Im}\mathfrak{S}^0$ and $\operatorname{Im}\mathfrak{S}$. Let $\mathfrak{S}, \mathfrak{S}' \in S_{(M,\mathfrak{P})}$ and $w \in W$. We suppose that $w\mathfrak{S} = \mathfrak{S}'$, which means that w must send the cuspidal datum of $\mathfrak{S}$ onto that of $\mathfrak{S}'$. Then $wd_{\mathfrak{S}} = d_{\mathfrak{S}'}$.

Let $(M, \mathfrak{P}) \in \mathfrak{X}$, $\mathfrak{S}, \mathfrak{S}' \in S_{(M,\mathfrak{P})}$. We suppose that $\mathfrak{S}$ is contained in $\mathfrak{S}'$. A residue datum for $\mathfrak{S}$ into $\mathfrak{S}'$ is a map of the set of meromorphic functions on $\mathfrak{S}'$ with polynomial singularities $S_{\mathfrak{X}}$ into the set of meromorphic functions on $\mathfrak{S}$ with polynomial singularities $S_{\mathfrak{X}}$ which are finite linear combinations with real coefficients of operators of the following type: let $\mathfrak{D} := \{\mathfrak{S} =: \mathfrak{S}_0 \subset \mathfrak{S}_1 \cdots \subset \mathfrak{S}_d := \mathfrak{S}'\}$ be a flag consisting of elements of $S_{(M,\mathfrak{P})}$ such that for $i = 1, \ldots, d$, we have $\dim_{\mathbb{R}}(\operatorname{Re}\mathfrak{S}_i/\operatorname{Re}\mathfrak{S}_{i-1}) = 1$; for $i = 1, \ldots, d$ we fix $\alpha_i \in R(T_M, G)$ such that:

$$\operatorname{Re}\mathfrak{S}_{i-1}^0 = \{\lambda \in \operatorname{Re}\mathfrak{S}_i^0 | \langle \lambda, \check{\alpha}_i \rangle = 0\}.$$

We associate with $\mathfrak{D}$ and the collection of α_i the operator obtained by composition of the operators Res_i for i decreasing from d to 1, from the set of meromorphic functions on $\mathfrak{S}_i$ with polynomial singularities $S_{\mathfrak{X}}$ into the set of meromorphic functions on $\mathfrak{S}_{i-1}$ with polynomial singularities $S_{\mathfrak{X}}$ defined as follows: let $A_i : \pi \in \mathfrak{S}_i \mapsto A_i(\pi) \in \mathbb{C}$ be a meromorphic function on $\mathfrak{S}_i$ with polynomial singularities $S_{\mathfrak{X}}$ and let $\pi_0 \in \mathfrak{S}_{i-1}$. We use the notation $H_{\pi_0,\alpha^\bullet}$ and $P_{\pi_0,\alpha^\bullet}$ of V.1.1 and V.1.2. We check that $\mathfrak{S}_{i-1}$ is the irreducible component of $\mathfrak{S}_i \cap H_{\pi_0,\alpha_i^\bullet}$ passing through π_0 and by V.1.2, we can fix an integer n such that the singular hyperplanes passing through π of the function:

$$\pi \in \mathfrak{S}_i \mapsto P^n_{\pi_0,\alpha_i}(\pi) A_i(\pi)$$

are the singular hyperplanes of A_i passing through π with the exception of $\mathfrak{S}_{i-1}$. We still fix $\epsilon_i \in (\operatorname{Re}\mathfrak{S}^0_{i-1})^\perp \cap \operatorname{Re}\mathfrak{S}^0_i$ and a linear combination of polynomials $P^m_{\pi_0,\alpha_i^\bullet}$, for $m \in [1,n]$, denoted by $Q_{n,\alpha_i^\bullet}$ such that, for $\pi \in \mathfrak{S}_i$, the function of the complex variable z given by $Q_{n,\alpha_i^\bullet}(\pi \otimes z\epsilon_i)$ is of the form $z^n(1+O(z))$. We then define $\operatorname{Res}_i A_i$ by setting:

$$\pi \in \mathfrak{S}_{i-1} \mapsto (\operatorname{Res}_i A_i)(\pi) :=$$
$$((n-1!))^{-1} \left(d^{n-1}(Q_{n,\alpha_i^\bullet} A_i)(\pi \otimes z\epsilon_i)/(dz)^{n-1}\right)_{z=0},$$

this definition depends up to a scalar on the choice of ϵ_i but does not depend on the choice of n nor of $Q_{n,\alpha_i^\bullet}$. Let us check that $\operatorname{Res}_i A_i$ is a meromorphic function on $\mathfrak{S}_{i-1}$ with polynomial singularities $S_{\mathfrak{X}}$. Let $\pi \in \mathfrak{S}_{i-1}$; there exists a polynomial P_π obtained as a power of $\prod P_{\pi,\beta^\bullet}$ where the product is over all roots β of $R(T_M, G)$ such that $\mathfrak{S}_i \cap H_{\pi,\beta^\bullet}$ does not contain $\mathfrak{S}_{i-1}$, such that $(P^n_{\pi,\alpha_i^\bullet} P_\pi)_{|\mathfrak{S}_i} A_i$ is holomorphic at π. We then see that:

$$P^n_{\pi|\mathfrak{S}_{i-1}} \operatorname{Res}_i A_i = P^n_{\pi|\mathfrak{S}_{i-1}} \operatorname{Res}_i(P^{-1}_{\pi|\mathfrak{S}_i}(P_{\pi|\mathfrak{S}_i} A_i))$$

is holomorphic at π by Leibniz' rule.

Convention

It will be very useful for us to employ the convention that for $\mathfrak{S} \in S_{\mathfrak{X}}$, a residue datum for $\mathfrak{S}$ in $\mathfrak{S}$ is given by multiplication by a real number.

Let $\mathfrak{S} \in S_{\mathfrak{X}}$ and let $(M, \mathfrak{P})$ be the cuspidal datum attached to $\mathfrak{S}$. We will speak of residue data for $\mathfrak{S}$ in $\mathfrak{P}$ even though if k is a function field, then in general $\mathfrak{P} \notin S_{\mathfrak{X}}$: in this case the phrase will signify residue data for the component of $\mathfrak{P}$ containing $\mathfrak{S}$.

V.1.4. Remark on the choices made in V.1.3

Let us pick up the notation of V.1.3. To define the operators Res_i we used elements ϵ_i of $\operatorname{Re}\mathfrak{S}^0_i \cap \left(\operatorname{Re}\mathfrak{S}^0_{i-1}\right)^\perp$. Let us no longer make this hypothesis and authorise all elements ϵ_i of $\operatorname{Re}\mathfrak{S}^0_i - \operatorname{Re}\mathfrak{S}^0_{i-1}$. We obtain a bigger set of 'residue datum' operators for $\mathfrak{S}$ in $\mathfrak{S}'$ but we can check that for every operator $\mathrm{Res}_{\mathfrak{S}}^{\mathfrak{S}'}$ in the more general sense above, there exists an operator $\mathrm{Res}'^{\mathfrak{S}'}_{\mathfrak{S}}$ in the sense of V.1.3 such that for every meromorphic function A on $\mathfrak{S}'$ with polynomial singularities $S_{\mathfrak{X}}$, we have:

$\mathrm{Res}_{\mathfrak{S}}^{\mathfrak{S}'}A - \mathrm{Res}'^{\mathfrak{S}'}_{\mathfrak{S}}A$ belongs to the vector subspace of the set of meromorphic functions on $\mathfrak{S}$ obtained as derivatives of a meromorphic function with polynomial singularities $\mathfrak{S}_{\mathfrak{X}}$, in a direction belonging to $\mathrm{Re}\,\mathfrak{S}^0$.

The reader can check that in all future formulae we can replace $\mathrm{Res}_{\mathfrak{S}}^{\mathfrak{S}'}$ by $\mathrm{Res}'^{\mathfrak{S}'}_{\mathfrak{S}}$ without changing the results. However, in practice, it is somewhat painful to calculate $\mathrm{Re}\,\mathfrak{S}_i^0 \cap \mathrm{Re}\,\mathfrak{S}_{i-1}^{0\perp}$ and is thus preferable to be able to use any element ϵ_i of $\mathrm{Re}\,\mathfrak{S}_i^0 - \mathrm{Re}\,\mathfrak{S}_{i-1}^0$.

V.1.5. The residue theorem

In this section, we will recall the residue theorem in the form in which we will frequently make use of it.

(a) **Preparation needed only in the number field case**

In this theorem, there occur integrals over translates of the set $\mathrm{Im}\,\mathfrak{S}$ for $\mathfrak{S} \in S_{\mathfrak{X}}$. In the case of a function field, such a set is compact and there are no serious convergence problems, however if k is a number field, this is unfortunately untrue. To avoid these convergence problems, we truncate as follows: recall the notation $A_M^{\mathbf{G}}$ of I.2.1. Suppose that k is a number field; as $X_M^{\mathbf{G}}$ has no torsion, there exists a unique element π_0 of $\mathfrak{P}$ whose restriction to $A_M^{\mathbf{G}}$ is trivial. This allows us to canonically and even $\mathrm{Stab}(M, \mathfrak{P})$-equivariantly identify $X_M^{\mathbf{G}}$ with $\mathfrak{P}$. The identification is denoted by:

$$\lambda_\pi \in X_M^{\mathbf{G}} \leftrightarrow \pi \simeq \pi_0 \otimes \lambda_\pi \in \mathfrak{P}.$$

This also allows us to transport the scalar product of $X_M^{\mathbf{G}}$ to $\mathfrak{P}$. Let us no longer make any hypothesis on k. With this notation, for every $\mathfrak{S} \in S_{\mathfrak{X}}$ and for every $T \in \mathbb{R}_+^*$, we set:

$$\mathfrak{S}_{\leq T} := \begin{cases} \mathfrak{S} & \text{if } k \text{ is a function field,} \\ \{\pi \in \mathfrak{S} \mid \|\mathrm{Im}\,\lambda_\pi\|^2 \leq T + \|\mathrm{Re}\,\lambda_\pi\|^2\} & \text{if } k \text{ is a number field.} \end{cases}$$

Definition of $=_T$

Suppose that k is a number field. Let $\mathfrak{S}' \in S_{\mathfrak{X}}$ with $\mathfrak{S}' \subset \mathfrak{S}$, (possibly $\mathfrak{S}' = \mathfrak{S}$), U be a measurable set of $\mathfrak{S}'$ and $\mathrm{Res}_{\mathfrak{S}',U}^{\mathfrak{S}}$ a residue datum for $\mathfrak{S}'$ in $\mathfrak{S}$. We consider $\int_U \mathrm{Res}_{\mathfrak{S}',U}^{\mathfrak{S}}$ as a linear form defined on the set of meromorphic functions A on $\mathfrak{S}$ with polynomial singularities $S_{\mathfrak{X}}$ such that $\mathrm{Res}_{\mathfrak{S}',U}^{\mathfrak{S}}A$ is integrable on U; its value at A is then:

$$\int_U (\mathrm{Res}_{\mathfrak{S}',U}^{\mathfrak{S}}A)(\pi)\, d_{\mathfrak{S}'}\pi.$$

We consider the vector space V generated by these linear forms and the vector subspace V' generated by those where we impose the following conditions on U:

U must be in $\mathfrak{S}' - \mathfrak{S}'_{\leq T}$;

U must be relatively compact if $\operatorname{Re}\mathfrak{S}'$ is strictly contained in $\operatorname{Re} X_M^{\mathbf{G}}$.

Let $v, v' \in V$. We will say that $v(A) =_T v'(A)$ for a meromorphic function A on $\mathfrak{S}$ with polynomial singularities $S_{\mathfrak{x}}$ in the domains of definition of v and v' if $(v - v') \in V'$ can be written in the form $\sum \int_U \operatorname{Res}_{\mathfrak{S}',U}^{\mathfrak{S}}$ where the U are sets satisfying the above conditions and such that $\operatorname{Res}_{\mathfrak{S}',U}^{\mathfrak{S}} A$ is holomorphic at every point of U and integrable over U.

(b) **Definition of a general path**
We fix a finite set of **K**-types, denoted by $\mathfrak{F}$, which will only serve for the moment to work with $S_{\mathfrak{x}}^{\mathfrak{F}}$, which is a family of locally finite affine spaces, instead of with $S_{\mathfrak{x}}$, which does not have this property, at least *a priori*. We fix $T \in \mathbb{R}_+^*$ which is only useful if k is a number field. We also fix $\mathfrak{S} \in S_{\mathfrak{x}}^{\mathfrak{F}}$ and $\lambda_1, \lambda_2 \in \operatorname{Re}\mathfrak{S}$. We denote by E_T the set of hyperplanes of $\mathfrak{S}$ belonging to $S_{\mathfrak{x}}^{\mathfrak{F}}$ which intersect $\mathfrak{S}_{\leq T}$ in at least one point of real part belonging to the interior of the segment of $\operatorname{Re}\mathfrak{S}$ joining λ_1 to λ_2. Let Γ be a path in $\operatorname{Re}\mathfrak{S}$ joining λ_1 to λ_2. We say that Γ is $T - \mathfrak{F}$-general if it satisfies the following conditions (some of which are somewhat in contradiction with the notion of generality, for which reason this terminology is not really very suitable):

(1) let $\mathfrak{T}$ and $\mathfrak{T}' \in E_T$ be such that $\{\Gamma \cap \operatorname{Re}\mathfrak{T} \cap \operatorname{Re}\mathfrak{T}'\} \neq \emptyset$: then $\operatorname{Re}\mathfrak{T} = \operatorname{Re}\mathfrak{T}'$; in particular $\mathfrak{T}$ and $\mathfrak{T}'$ are parallel, see V.1.1;

(2) $\forall \mathfrak{T} \in E_T, |\{\Gamma \cap \operatorname{Re}\mathfrak{T}\}| = 1$ (*a priori* this cardinal is ≥ 1) and $\forall \gamma \in \Gamma, \|\gamma\| \leq \sup(\|\lambda_1\|, \|\lambda_2\|)$,

(3) for every hyperplane $\mathfrak{T}'$ of $\mathfrak{S}$ belonging to $S_{\mathfrak{x}}^{\mathfrak{F}}$, intersecting $\mathfrak{S}_{\leq T}$ but not belonging to E_T, we have: $\Gamma \cap \operatorname{Re}\mathfrak{T}' = \emptyset$.

If k is a function field, E_T does not depend on T since $\mathfrak{S}_{\leq T} = \mathfrak{S}$ and the notion of $T - \mathfrak{F}$-general also does not depend on T. For general k, if $T' > T$, then obviously a $T' - \mathfrak{F}$-general path is also $T - \mathfrak{F}$-general and $E_{T'}$ contains E_T, the difference consisting of elements not intersecting $\mathfrak{S}_{\leq T}$.

For $\mathfrak{T} \in E_T$ and for Γ, $T - \mathfrak{F}$-general as above, we set $y_{\mathfrak{T}} = \{\Gamma \cap \operatorname{Re}\mathfrak{T}\}$; it is a point of $\operatorname{Re}\mathfrak{T}$.

(c) **Residue theorem** *Let $\mathfrak{S} \in S_{\mathfrak{x}}^{\mathfrak{F}}$ and $\lambda_1, \lambda_2 \in \operatorname{Re}\mathfrak{S}$; we fix a positive real number T and $T' \in \mathbb{R}$ greater than $3T + 2\sup(\|\lambda_1\|, \|\lambda_2\|)^2$. Let Γ*

be a $T' - \mathfrak{F}$-general path as in (b). For every $\mathfrak{T} \in E_T$, there exists a residue datum for $\mathfrak{T}$ in $\mathfrak{S}$, denoted by $\mathrm{Res}_{\mathfrak{T}}^{\mathfrak{S}}$, depending only on the set $\{y_{\mathfrak{T}}; \mathfrak{T} \in E_T\}$ and not on Γ or T', such that we have:

Let $\pi \in \mathfrak{S} \mapsto A(\pi) \in \mathbb{C}$ be a meromorphic function on $\mathfrak{S}$ with polynomial singularities $S_{\mathfrak{T}}^{\mathfrak{F}}$, holomorphic at every point of $\mathfrak{S}_{\leq T'}$ of real part λ_1 or λ_2. Then $\pi \in \mathfrak{T} \mapsto \mathrm{Res}_{\mathfrak{T}}^{\mathfrak{S}} A(\pi) \in \mathbb{C}$ is holomorphic at every point of $\mathfrak{T}_{\leq T'}$ of real part $y_{\mathfrak{T}}$ and

$$\int_{\substack{\pi \in \mathfrak{S}_{\leq T} \\ \mathrm{Re}\,\pi = \lambda_1}} A(\pi)\, \mathrm{d}_{\mathfrak{S}}\pi - \int_{\substack{\pi \in \mathfrak{S}_{\leq T} \\ \mathrm{Re}\,\pi = \lambda_2}} A(\pi)\, \mathrm{d}_{\mathfrak{S}}\pi - \sum_{\mathfrak{T} \in E_T} \int_{\substack{\pi \in \mathfrak{T}_{\leq T} \\ \mathrm{Re}\,\pi = y_{\mathfrak{T}}}} \mathrm{Res}_{\mathfrak{T}}^{\mathfrak{S}} A(\pi)\, \mathrm{d}_{\mathfrak{T}}\pi$$

$$=_T 0 \text{ if } k \text{ is a number field,}$$
$$= 0 \;\text{ if } k \text{ is a function field.}$$

Note that this version is less precise than the one proved by Langlands. In his version the conclusion is true for every function A as above which is holomorphic at every point of $\mathfrak{S}_{\leq T}$ (and not $\mathfrak{S}_{\leq T'}$) of real part λ_1 or λ_2.

We first reduce the proof to that of the usual residue theorem in dimension 1. For this, we introduce the following terminology: a broken line joining λ_1 to λ_2 is a path formed of segments. To prove the theorem, we can replace Γ by a broken line passing through the points $y_{\mathfrak{T}}$ for $\mathfrak{T} \in E_{\mathfrak{T}}$, and still satisfying (1), (2) and (3) of (b), where we set $T = T'$ ((1) is in fact a consequence of (2) and of property (1) for Γ) and suppose moreover that none of the points $y_{\mathfrak{T}}$ for $\mathfrak{T} \in E_{T'}$ is the extremity of one of the segments of the broken line. This last hypothesis ensures that every function A as in the statement is holomorphic at every point of $\mathfrak{S}_{\leq T'}$ of real part an extremity of one of the segments of the broken line. It will be useful in what follows to note that:

(4) for every $\mathfrak{T} \in E_T$, the direction in which a general path cuts $\mathrm{Re}\,\mathfrak{T}$ is independent of the path, because of hypothesis (2).

It is easy to see that it is sufficient to prove the theorem in the case where Γ is a segment and where $\{y_{\mathfrak{T}}\}_{\mathfrak{T} \in E_{T'}}$ has at most one element, i.e. $E_{T'} = \emptyset$ or all the elements of $E_{T'}$ have the same real part and are thus parallel. Suppose first that $E_{T'} \neq \emptyset$; we still replace Γ by a broken line satisfying the same hypotheses as before but such that the segment of this broken line which contains $y_{\mathfrak{T}}$ is of the form $y_{\mathfrak{T}} + x\epsilon$

with $x \in [-x_0, +x_0]$, $x_0 \in \mathbb{R}_+$ and $\epsilon \in (\mathrm{Re}\,\mathfrak{T}^0)^\perp \cap \mathrm{Re}\,\mathfrak{S}^0$, where $\mathfrak{T}$ is any element of $E_{T'}$. We have reduced the proof to the case of a segment such that either $E_{T'} = \emptyset$ or it is of the particular type described above. We first deal with this last case. Since $y_{\mathfrak{T}}$ and $\mathfrak{T}^0$ are independent of the choice of $\mathfrak{T}$ in $E_{T'}$, we denote them by y_0 and $\mathfrak{Y}$ respectively.

We separate the number field from the function field case; suppose that k is a number field. We identify $X_M^{\mathbf{G}}$ with $\mathfrak{P}$ as was explained in (a) and we have:

$$\mathfrak{S}^0 = \mathfrak{Y} + \mathbb{C}\epsilon$$
$$\mathfrak{S} = \pi_{\mathfrak{S}} + \mathfrak{Y} + \mathbb{C}\epsilon,$$

where $\pi_{\mathfrak{S}}$ is an element of $\mathfrak{S}$ such that $\mathrm{Re}\,\pi_{\mathfrak{S}} = y_0$ and $\mathrm{Im}\,\pi_{\mathfrak{S}} \in (\mathfrak{S}^0)^\perp \cap \mathrm{Im}\,\mathfrak{S}$. Thus we have $\|\mathrm{Im}\,\pi_{\mathfrak{S}}\| \leq \|\mathrm{Im}\,\pi\|$ for every $\pi \in \mathfrak{S}$. Since there exists $\pi \in \mathfrak{S}_{\leq T}$ such that $\mathrm{Re}\,\pi = y_0$, we necessarily have:

$$\|\mathrm{Im}\,\pi_{\mathfrak{S}}\|^2 \leq T + \|y_0\|^2 = T + \|\mathrm{Re}\,\pi_{\mathfrak{S}}\|^2.$$

We set $T_1 := T + \sup(\|\lambda_1\|, \|\lambda_2\|)^2$ and

$$(\mathrm{Im}\,\mathfrak{Y})_{\leq T_1} = \{y \in \mathrm{Im}\,\mathfrak{Y} | \ \|y\|^2 \leq T_1\}.$$

Let $\pi \in \mathfrak{S}_{\leq T}$ be such that $\mathrm{Re}\,\pi \in \Gamma$; we write $\pi = \pi_{\mathfrak{S}} + y + z\varepsilon$ with $y \in \mathfrak{Y}$ and we have:

- by definition of $\Gamma, \mathrm{Re}\,y = 0$ and $\mathrm{Re}\,z \in [-x_0, x_0]$,
- by hypothesis (2) for Γ, $\|y_0 + \mathrm{Re}\ z\ \epsilon\| \leq \sup(\|\lambda_1\|, \|\lambda_2\|)$,
- by definition of $\mathfrak{S}_{\leq T}$ and the orthogonality relations:

$$\|\mathrm{Im}\,\pi_{\mathfrak{S}}\|^2 + \|y\|^2 + \|\mathrm{Im}\ z\ \epsilon\|^2 \leq T + \|y_0 + \mathrm{Re}\,z\ \epsilon\|^2,$$
$$\leq T + \sup(\|\lambda_1\|, \|\lambda_2\|)^2.$$

Thus $y \in (\mathrm{Im}\,\mathfrak{Y})_{\leq T_1}$ and $\|\mathrm{Im}\ z\ \epsilon\|^2 \leq T_1$. We thus obtain:

$$\{\pi \in \mathfrak{S}_{\leq T} | \mathrm{Re}\,\pi \in \Gamma\} \subset D :=$$
$$\{\pi_{\mathfrak{S}} + y + z\epsilon | y \in (\mathrm{Im}\,\mathfrak{Y})_{\leq T_1}, \|\mathrm{Im}\ z\ \epsilon\|^2 \leq T_1\} \subset \mathfrak{S}_{\leq T'},$$

by hypothesis on T'. The hypotheses on Γ ensure that every function A as in the statement is holomorphic at such a point, except possibly if π belongs to one of the elements of $E_{T'}$ and is of real part y_0. In this case the function becomes holomorphic after multiplication by a polynomial invariant under $\mathfrak{Y}$, denoted by Q. For every function A as in the statement, we define a function on a neighborhood of $\{z \in \mathbb{C} | \mathrm{Re}\,z \in [x_0, x_0]$ and $\|\mathrm{Im}\ z\ \epsilon\|^2 \leq T_1\}$ by:

$$\tilde{A}(z) = \int_{(\mathrm{Im}\,\mathfrak{Y})_{\leq T_1}} A(\pi_{\mathfrak{S}} + y + z\varepsilon)\ \mathrm{d}_{\mathfrak{Y}}y,$$

where $\mathrm{d}_{\mathfrak{Y}}y$ is $\mathrm{d}_{\mathfrak{T}}y$ for $\mathfrak{T}$ any element of $E_{T'}$, see V.1.3. By replacing A by

QA in this definition, we see that $\tilde{A}$ is meromorphic in a neighborhood of the set described above. The set of poles is contained in the set $\{z_{\mathfrak{T}})_{\mathfrak{T}\in E_{T'}}$ where for $\mathfrak{T} \in E_{T'}, z_{\mathfrak{T}}$ is the unique element satisfying $\pi_{\mathfrak{S}} + z_{\mathfrak{T}}\epsilon + \mathfrak{Y} = \mathfrak{T}$. For every $\gamma \in \Gamma$, we have:

$$\int_{\substack{\pi\in\mathfrak{S}_{\leq T}\\ \mathrm{Re}\,\pi=\gamma}} A(\pi) d_{\mathfrak{S}}\pi =_T \int_{\substack{\pi\in D\\ \mathrm{Re}\,\pi=\gamma}} A(\pi)\ \mathrm{d}_{\mathfrak{S}}\pi = \int_{\substack{z\in\mathbb{C}\\ \|\mathrm{Im}\, z\epsilon\|^2\leq T_1\\ \mathrm{Re}\, z=x_\gamma}} \tilde{A}(z)\ \mathrm{d}z,$$

where $\mathrm{d}z$ is a Lebesgue measure which can be calculated as a function of $\mathrm{d}_{\mathfrak{S}}$ and of $\mathrm{d}y$ and where the x_γ satisfy $\gamma = y_0 + x_\gamma\epsilon$. This gives:

$$(5)\quad \begin{aligned}&\int_{\substack{\pi\in\mathfrak{S}_{\leq T}\\ \mathrm{Re}\,\pi=\lambda_1}} A(\pi)\ \mathrm{d}_{\mathfrak{S}}\pi - \int_{\substack{\pi\in\mathfrak{S}_{\leq T}\\ \mathrm{Re}\,\pi=\lambda_2}} A(\pi)\ \mathrm{d}_{\mathfrak{S}}\pi\\ &=_T \int_{\substack{z\in\mathbb{C}\\ \|\mathrm{Im}\ z\|^2\leq T_1\\ \mathrm{Re}\, z=x_0}} \tilde{A}(z)\ \mathrm{d}z - \int_{\substack{z\in\mathbb{C}\\ \|\mathrm{Im}\ z\|^2\leq T_1\\ \mathrm{Re}\, z=-x_0}} \tilde{A}(z)\ \mathrm{d}z.\end{aligned}$$

But we can also replace T_1 by a number very slightly larger, denoted by T_1', which is such that the set

$$D_\pm := \{z|\mathrm{Im}\, z = \pm i(T_1')^{1/2}/\|\epsilon\| \text{ and } \mathrm{Re}\, z \in [-x_0, x_0]\}$$

contains none of the points $z_{\mathfrak{T}}$ where $\mathfrak{T}$ runs through $E_{T'}$; this ensures that for every function A as above $\tilde{A}$ is holomorphic on the set $D_\pm$. We fix such a T_1' and we note that the two integrals:

$$\int_{\substack{z\in\mathbb{C}\\ \mathrm{Im}\, z=\pm iT_1'/\|\epsilon\|^2\\ \mathrm{Re}\, z\in[-x_0,x_0]}} \tilde{A}(z)\ \mathrm{d}z$$

are $=_T 0$. By the usual residue theorem, the right-hand side of (5) is thus equal to the sum of the residues of the function $\tilde{A}$ at the points z which are among its poles and which satisfy $\|\mathrm{Im}\ z\|^2 < T_1'/\|\epsilon\|^2$. By what we just saw, such a point is of the form $z_{\mathfrak{T}}$ with $\mathfrak{T} \in E_{T'}$ and thus the left hand side of (5):

$$=_T \sum_{\mathfrak{T}\in E_{T'}} \mathrm{Res}_{z_{\mathfrak{T}}} \tilde{A}.$$

Now for $\mathfrak{T} \in E_{T'}$, there exists a residue datum for $\mathfrak{T}$ in $\mathfrak{S}$, independent of A, such that for every function A as above, we have:

$$\mathrm{Res}_{z_{\mathfrak{T}}} \tilde{A} = \int_{(\mathrm{Im}\,\mathfrak{Y})_{T_1}} \mathrm{Res}_{\mathfrak{T}}^{\mathfrak{S}} A(\pi_{\mathfrak{S}} + y + z_{\mathfrak{T}}\epsilon)\ \mathrm{d}_{\mathfrak{Y}} y.$$

For every $y \in \mathfrak{Y}$, we have $\pi_y := \pi_{\mathfrak{S}} + y + z_{\mathfrak{T}}\epsilon \in \mathfrak{T}$. As above, we have:

$$\{\pi \in \mathfrak{T}_{\leq T}|\mathrm{Re}\,\pi = y_0\} \subset \{\pi_y|y \in (\mathrm{Im}\,\mathfrak{Y})_{\leq T}\},$$

whence

$$\operatorname{Res}_{z_{\mathfrak{T}}} \tilde{A} =_T \int_{\substack{\pi \in \mathfrak{T}_{\leq T} \\ \operatorname{Re}\pi = y_0}} \operatorname{Res}_{\mathfrak{T}}^{\mathfrak{S}} A(\pi) \, d_{\mathfrak{T}}\pi,$$

this integral being zero if $\mathfrak{T} \in E_{T'} - E_T$. This gives the desired result. The independence of residue data with respect to paths Γ is a consequence of (4).

Now suppose that k is a function field. We fix $\pi_{\mathfrak{S}} \in \mathfrak{S}$ and we have a surjective map:

$$\mathfrak{Y} \times \mathbb{C}\epsilon \to \mathfrak{S}, \tag{6}$$
$$(y, z\epsilon) \mapsto \pi_{\mathfrak{S}} \otimes y \otimes z\epsilon.$$

We suppose that $\operatorname{Re}\pi_{\mathfrak{S}} = y_0$. As in the number field case, we define a meromorphic function on a neighborhood of $\{z \in \mathbb{C} | \operatorname{Re} z \in [x_0, x_0]\}$, by setting:

$$\tilde{A}(z) = \int_{y \in \operatorname{Im}\mathfrak{Y}} A(\pi_{\mathfrak{S}} \otimes y \otimes z\epsilon) \, d_{\mathfrak{Y}} y.$$

The map (6) can be easily studied: let $(y_1, z_1) \in \mathfrak{Y} \times \mathbb{C}$ and $(y_2, z_2) \in \mathfrak{Y} \times \mathbb{C}$ be such that:

$$\pi_{\mathfrak{S}} \otimes y_1 \otimes z_1\epsilon = \pi_{\mathfrak{S}} \otimes y_2 \otimes z_2\epsilon.$$

Then we have:

$$\operatorname{Re} y_1 = \operatorname{Re} y_2, \operatorname{Re} z_1 = \operatorname{Re} z_2, \; (z_1^{-1} z_2)\epsilon \in y_2^{-1} y_1 \operatorname{Fix}_{X_M^G} \mathfrak{P} \subset \mathfrak{Y}.$$

Thus $L := \{\omega \in \mathbb{R} | i\omega\epsilon \in \mathfrak{Y}\}$ is a lattice of $\mathbb{R}$ under which $\tilde{A}$ is invariant. For $\gamma \in \Gamma$, we define x_γ by $\gamma = y_0 + x_\gamma \epsilon$ and we have:

$$\int_{\substack{\pi \in \mathfrak{S} \\ \operatorname{Re}\pi = \gamma}} A(\pi) \, d_{\mathfrak{S}}\pi = \int_{\substack{z \in \mathbb{C}/iL \\ \operatorname{Re} z = x_\gamma}} \tilde{A}(z) \, dz.$$

We use the periodic residue theorem in dimension 1 and we conclude as in the number field case.

The case where $E_{T'}$ is empty can be dealt with in a similar but simpler way.

V.2. Decomposition of the scalar product of two pseudo-Eisenstein series

V.2.1. Notation

We fix $(M, \mathfrak{P}) \in \mathfrak{X}$ and a finite set $\mathfrak{F}$ of **K**-types; we write an $\mathfrak{F}$ in the exponent to denote the vector space generated by the functions on which

$\mathbf{K}$ acts via one of the $\mathbf{K}$-types of $\mathfrak{F}$, for example $P_{\mathfrak{X}}^{\mathfrak{F}}, P_{(M,\mathfrak{P})}^{R,\mathfrak{F}}$ etc. Let $\mathfrak{S} \in S_{\mathfrak{X}}^{\mathfrak{F}}$; we define the origin of $\operatorname{Re}\mathfrak{S}$, denoted by $o(\mathfrak{S})$, by setting:

$$o(\mathfrak{S}) = \operatorname{Re}\mathfrak{S} \cap (\operatorname{Re}\mathfrak{S}^0)^{\perp},$$

where the orthogonal is taken in $\operatorname{Re} X_M^{\mathbf{G}}$. Let $T \in \mathbb{R}_+^*$: it is only needed when k is a number field. Let $z(\mathfrak{S}) \in \operatorname{Re}\mathfrak{S}$; we say that $z(\mathfrak{S})$ is $T - \mathfrak{F}$-general but near $o(\mathfrak{S})$ if for every element $\mathfrak{S}'$ of $S_{\mathfrak{X}}^{\mathfrak{F}}$ strictly contained in $\mathfrak{S}$, we have:

(1) $$z(\mathfrak{S}) \notin \{\operatorname{Re}\pi \mid \pi \in \mathfrak{S}'_{\leq T}\},$$

(2) the set $\{\operatorname{Re}\pi \mid \pi \in \mathfrak{S}'_{\leq T}\}$ does not intersect the ball of $\operatorname{Re}\mathfrak{S}$ of centre $o(\mathfrak{S})$ and of radius $\|o(\mathfrak{S}) - z(\mathfrak{S})\|$ unless this set contains $o(\mathfrak{S})$.

It is useful to make the following remarks:

(3) having fixed a relatively compact neighborhood of $o(\mathfrak{S})$, there exists only a finite number of elements $\mathfrak{S}'$ of $S_{\mathfrak{X}}^{\mathfrak{F}}$ such that $\mathfrak{S}'_{\leq T}$ contains an element whose real part is in this neighborhood. This shows that the set of $T - \mathfrak{F}$-general points near $o(\mathfrak{S})$ contains a non-empty open and even dense set in a relatively compact neighborhood of $o(\mathfrak{S})$.

(4) Let $\pi \in \mathfrak{S} \mapsto A(\pi) \in \mathbb{C}$ be a meromorphic function on $\mathfrak{S}$ with polynomial singularities $S_{\mathfrak{X}}^{\mathfrak{F}}$ and let $z(\mathfrak{S})$ be a $T - \mathfrak{F}$-general point near $o(\mathfrak{S})$. Then A is holomorphic at every point of the set $\{\pi \in \mathfrak{S}_{\leq T} | \operatorname{Re}\pi = z(\mathfrak{S})\}$.

We have already defined $W(M)$; see the introduction in V.1. For $w \in W(M)$, we have $(wMw^{-1}, w\mathfrak{P}) \in \mathfrak{X}$ and for $(M', \mathfrak{P}') \in \mathfrak{X}$, we set:

$$W((M,\mathfrak{P}),(M',\mathfrak{P}')) := \{w \in W(M) | (wMw^{-1}, w\mathfrak{P}) = (M', \mathfrak{P}')\}.$$

From this we obtain:

$$W(M) = \cup_{(M',\mathfrak{P}')\in\mathfrak{X}} W((M,\mathfrak{P}),(M',\mathfrak{P}')).$$

Let $\phi' \in P_{\mathfrak{X}}^{\mathfrak{F}}$, i.e. $\phi' = (\phi'_{(M',\mathfrak{P}')})_{(M',\mathfrak{P}')\in\mathfrak{X}}$ where $\phi'_{(M',\mathfrak{P}')} \in P_{(M',\mathfrak{P}')}^{\mathfrak{F}}$ and let $\phi \in P_{(M,\mathfrak{P})}^{\mathfrak{F}}$. We define a meromorphic function on $\mathfrak{P}$ with polynomial singularities on $S_{\mathfrak{X}}^{\mathfrak{F}}$, by setting for $\pi \in \mathfrak{P}$:

$$A(\phi',\phi)(\pi) = \langle \sum_{w\in W(M)} M(w^{-1}, -w\overline{\pi})\phi'(-w\overline{\pi}), \phi(\pi)\rangle,$$

where $\sum_{w\in W(M)} M(w^{-1}, -w\overline{\pi})\phi'(-w\overline{\pi})$ signifies:

$$\sum_{(M',\mathfrak{P}')\in\mathfrak{X}} \left(\sum_{w\in W((M,\mathfrak{P}),(M',\mathfrak{P}'))} M(w^{-1}, -w\overline{\pi})\phi'_{(M',\mathfrak{P}')}(-w\overline{\pi}) \right).$$

Let $R \in \mathbb{R}$. This definition can be generalised to $\phi' \in P_{\mathfrak{X}}^{R,\mathfrak{F}}$ and to $\phi \in P_{M,\mathfrak{P}}^{R,\mathfrak{F}}$; in this case $A(\phi',\phi)$ is defined meromorphically on $\mathfrak{P}_R$. By II.2.1, there exists $\lambda_0 \in \operatorname{Re} X_M^{\mathbf{G}}$ (very positive) such that:

$$\langle \theta_{\phi'}, \theta_\phi \rangle = \int_{\substack{\pi\in\mathfrak{P} \\ \operatorname{Re}\pi=\lambda_0}} A(\phi',\phi)(\pi)\, d\pi.$$

In what follows we fix $R \in \mathbb{R}$ such that there exists $\lambda_0 \in \operatorname{Re} X_M^{\mathbf{G}}$ of norm less than R satisfying the above equality.

V.2.2. Statement of the theorem

In addition to $(M,\mathfrak{P})$ and R, we fix a positive real number T and T' greater than $3T + 2R^2$, which intervene only when k is a number field. Let $\mathfrak{S} \in S_{(M,\mathfrak{P})}^{\mathfrak{F}}$; we set:

$$\operatorname{Norm}\mathfrak{S} := \{w \in \operatorname{Stab}(M,\mathfrak{P}) | w\mathfrak{S} = \mathfrak{S}\}.$$

For every $\mathfrak{S} \in S_{(M,\mathfrak{P})}^{\mathfrak{F}}$, we fix $z(\mathfrak{S}) \in \operatorname{Re}\mathfrak{S}$ such that for every $w \in \operatorname{Norm}\mathfrak{S}$, $wz(\mathfrak{S})$ is $T' - \mathfrak{F}$-general but near $o(\mathfrak{S})$. To simplify, we also require that $z(\mathfrak{S}) = z(\mathfrak{S}')$ if $\operatorname{Re}\mathfrak{S} = \operatorname{Re}(\mathfrak{S}')$.

Theorem *For every $\mathfrak{S} \in S_{(M,\mathfrak{P})}^{\mathfrak{F}}$, there exists a residue datum for $\mathfrak{S}$ in $\mathfrak{P}$ which is zero for almost all $\mathfrak{S}$ and in particular whenever $\|o(\mathfrak{S})\| > R$, denoted by* $\operatorname{Res}_{\mathfrak{S}}$*, such that:*

$$\forall \phi' \in P_{\mathfrak{X}}^{\mathfrak{F}},\ \forall \phi \in P_{(M,\mathfrak{P})}^{R,\mathfrak{F}}$$

$$\langle \theta_{\phi'}, \theta_\phi \rangle =_T \sum_{\mathfrak{S}\in S_{(M,\mathfrak{P})}^{\mathfrak{F}}} |\operatorname{Norm}\mathfrak{S}|^{-1} \int_{\substack{\pi\in\mathfrak{S}_{\leq T} \\ \operatorname{Re}\pi=z(\mathfrak{S})}} \sum_{w\in\operatorname{Norm}\mathfrak{S}} \operatorname{Res}_{\mathfrak{S}} A(\phi',\phi)(w\pi)\, d_{\mathfrak{S}}\pi.$$

Remarks, notation The residue data in the above statement are not unique. They depend on the choice of points $z(\mathfrak{S})$. Suppose that these points satisfy the following condition:

(*) $\quad \forall w \in \operatorname{Stab}(M,\mathfrak{P}), \forall \mathfrak{S} \in S_{(M,\mathfrak{P})}^{\mathfrak{F}}, wz(\mathfrak{S}) \in (\operatorname{Norm} w\mathfrak{S})z(w\mathfrak{S})$.

Then we will see later in V.3.12 that the values of the operators $\operatorname{Res}_{\mathfrak{S}}$ on the set of functions of the form $\pi \mapsto A(\phi',\phi)(\pi)$ for $\phi' \in P_{\mathfrak{X}}^{\mathfrak{F}}, \phi \in P_{(M,\mathfrak{P})}^{\mathfrak{F}}$ is uniquely determined. This explains why under the hypothesis (*) we will adopt the notation $\operatorname{Res}_{\mathfrak{S}}^G$. The formula (1) can be rewritten under this hypothesis as follows:

$$\langle\theta_{\phi'},\theta_\phi\rangle =_T \sum_{\mathfrak{C}\in S^{\mathfrak{F}}_{(M,\mathfrak{P})}/\mathrm{Stab}(M,\mathfrak{P})} |\mathrm{Norm}\,\mathfrak{S}_{\mathfrak{C}}|^{-1}$$

$$\int_{\substack{\pi\in\mathfrak{S}_{\mathfrak{C},\leq T}\\ \mathrm{Re}\,\pi=z(\mathfrak{S}_{\mathfrak{C}})}} \sum_{w\in\mathrm{Stab}(M,\mathfrak{P})} \left(\mathrm{Res}^G_{w\mathfrak{S}_{\mathfrak{C}}} A(\phi',\phi)(\pi')\right)_{\pi'=w\pi}\, \mathrm{d}_{\mathfrak{S}_{\mathfrak{C}}}\pi,$$

where $\mathfrak{S}_{\mathfrak{C}}$ is a fixed element of $\mathfrak{C}$. The formula is obviously independent of this choice. We must point out here a certain ambiguity in some of the notation used later on; the notation $(\mathrm{Res}^G_{w\mathfrak{S}_{\mathfrak{C}}} A(\phi',\phi))(w\pi)$ signifies

$$(\mathrm{Res}^G_{w\mathfrak{S}_{\mathfrak{C}}} A(\phi',\phi)(\pi'))_{\pi'=w\pi}.$$

V.2.3. Beginning of the proof, first step

Lemma *For every $\mathfrak{S}\in S^{\mathfrak{F}}_{(M,\mathfrak{P})}$, there exists a residue datum, zero for almost all $\mathfrak{S}$ and in particular whenever $\|o(\mathfrak{S})\|>R$, denoted by $\mathrm{Res}'_{\mathfrak{S}}$ for $\mathfrak{S}$ in $\mathfrak{P}$, such that for every $\phi'\in P^{\mathfrak{F}}_{\mathfrak{X}}$ and for every $\phi\in P^{R,\mathfrak{F}}_{(M,\mathfrak{P})}$, we have:*

$$\langle\theta_{\phi'},\theta_\phi\rangle =_T \sum_{\mathfrak{S}\in S^{\mathfrak{F}}_{(M,\mathfrak{P})}} \int_{\substack{\pi\in\mathfrak{S}_{\leq T}\\ \mathrm{Re}\,\pi=z(\mathfrak{S})}} (\mathrm{Res}'_{\mathfrak{S}} A(\phi',\phi))(\pi)\, \mathrm{d}_{\mathfrak{S}}\pi.$$

For $d\in\{0,\ldots,\dim_{\mathbb{R}}\mathrm{Re}\,X^{\mathbf{G}}_M\}$, we set

$$S(d) := \{\mathfrak{S}\in S^{\mathfrak{F}}_{(M,\mathfrak{P})} \mid \dim_{\mathbb{R}}\mathfrak{S}^0 = d\}.$$

We construct the residue data $\mathrm{Res}'_{\mathfrak{S}}$ for the elements of $S(d)$, by descending induction by simultaneously introducing the following auxiliary objects for every $\mathfrak{T}\in S(d-1)$:

a set of points of $\mathrm{Re}\,\mathfrak{T}$, denoted by $P(\mathfrak{T})$, possibly empty, without multiplicity;

for every $y\in P(\mathfrak{T})$, a residue datum for $\mathfrak{T}$ in $X^{\mathbf{G}}_M$, denoted by $\mathrm{Res}^y_{\mathfrak{T}}$, satisfying (2) and (3) below:

(2) $$\forall\mathfrak{T}\in S(d-1), \forall y\in P(\mathfrak{T}), \forall\mathfrak{S}'\subsetneq\mathfrak{T}, \mathfrak{S}'\in S^{\mathfrak{F}}_{(M,\mathfrak{P})},\quad \{\pi\in\mathfrak{T}_{\leq T'} \mid \mathrm{Re}\,\pi = y\}\cap\mathfrak{S}' = \emptyset.$$

In particular, (2) implies that every meromorphic function on $\mathfrak{S}'$ with polynomial singularities $S^{\mathfrak{F}}_{\mathfrak{X}}$ is holomorphic at every point of $\mathfrak{S}'_{\leq T'}$ whose real part $y \in P(\mathfrak{S}')$.

$$\langle \theta_{\phi'}, \theta_\phi \rangle =_T \sum_{\substack{\mathfrak{S}\in S(d)\\ d'\geq d}} \int_{\substack{\pi\in\mathfrak{S}_{\leq T}\\ \mathrm{Re}\,\pi=z(\mathfrak{S})}} (\mathrm{Res}'_{\mathfrak{S}} A(\phi',\phi))(\pi)\, \mathrm{d}_{\mathfrak{S}}\pi \ +$$

(3)

$$\sum_{\mathfrak{T}\in S(d-1)} \left(\sum_{y\in P(\mathfrak{T})} \int_{\substack{\pi\in\mathfrak{T}_{\leq T}\\ \mathrm{Re}\,\pi=y}} (\mathrm{Res}^y_{\mathfrak{T}} A(\phi',\phi))(\pi)\, \mathrm{d}_{\mathfrak{T}}\pi \right).$$

Let us prove this, starting with the case where $d = \dim_{\mathbb{R}} \mathrm{Re}\, X^{\mathbf{G}}_M$. In this case $S(d)$ contains the irreducible components of $\mathfrak{P}$; we take for residue datum the identity (i.e. multiplication by the number 1). To obtain (3), we have to calculate, considering V.2.1(1):

(4) $$\int_{\substack{\pi\in\mathfrak{P}\\ \mathrm{Re}\,\pi=\lambda_0}} A(\phi',\phi)(\pi)\, \mathrm{d}\pi - \int_{\substack{\pi\in\mathfrak{P}_{\leq T}\\ \mathrm{Re}\,\pi=z(\mathfrak{P})}} A(\phi',\phi)(\pi)\, \mathrm{d}\pi.$$

Here $z(\mathfrak{P})$ is the point $z(\mathfrak{S})$ for $\mathfrak{S}$ any one of the irreducible components of $\mathfrak{P}$.

Up to $=_T$, we can replace the domain of integration of the first integral by $\{\pi \in \mathfrak{P}_{\leq T} | \mathrm{Re}\,\pi = \lambda_0\}$. This being done, we can apply the residue theorem as in V.1.5(c); there we defined a set $E_{T'}$, which here consists of hyperplanes of $\mathfrak{P}$, for $\mathfrak{T} \in E_{T'}$ a point $y_{\mathfrak{T}}$ which satisfies (2) above. For $\mathfrak{T} \in S(d-1) = S(\dim_{\mathbb{R}} \mathrm{Re}\, X^{\mathbf{G}}_M - 1)$, we thus set:

$$P(\mathfrak{T}) = \begin{cases} \emptyset & \text{if } \mathfrak{T} \notin E_T \\ \{y_{\mathfrak{T}}\} & \text{if } \mathfrak{T} \in E_T. \end{cases}$$

The residue theorem also gives a residue datum, here denoted by $\mathrm{Res}^y_{\mathfrak{T}}$, such that (3) is true. Hence the assertion in this case. Fix d and suppose we have done all the constructions for $d' > d$; we will do them for d. Let $\mathfrak{S} \in S(d)$; by the induction hypothesis, $P(\mathfrak{S})$ is defined as well as $\mathrm{Res}^y_{\mathfrak{S}}$ for $y \in P(\mathfrak{S})$. We then set:

$$\mathrm{Res}'_{\mathfrak{S}} = \sum_{y\in P(\mathfrak{S})} \mathrm{Res}^y_{\mathfrak{S}}.$$

Considering (3) for $d+1$, we have to calculate:

$$\sum_{\mathfrak{S}\in S(d)} \sum_{y\in P(\mathfrak{S})} \left(\int_{\substack{\pi\in\mathfrak{S}_{\leq T}\\ \mathrm{Re}\,\pi=y}} (\mathrm{Res}^y_{\mathfrak{S}} A(\phi',\phi))(\pi)\, \mathrm{d}_{\mathfrak{S}}\pi - \int_{\substack{\pi\in\mathfrak{S}_{\leq T}\\ \mathrm{Re}\,\pi=z(\mathfrak{S})}} (\mathrm{Res}^y_{\mathfrak{S}} A(\phi',\phi))(\pi)\, \mathrm{d}_{\mathfrak{S}}\pi \right).$$

Each difference can be calculated with the help of the residue theorem in the form given in V.1.5(c) and the constructions are done as above (to keep the hypothesis that $P(\mathfrak{T})$ is without multiplicity, we may have to regroup terms intervening with different $\mathfrak{S} \in S(d)$ or $y \in P(\mathfrak{S})$). It is amusing to note that the case $d = 1$ is the end of the construction, i.e. (3) for $d = 1$ is (1) if for $\mathfrak{S} \in S(0)$, we set $\mathrm{Res}'_{\mathfrak{S}} = \mathrm{Res}^{y}_{\mathfrak{S}}$ if $P(\mathfrak{S}) \neq \emptyset$ and 0 otherwise. It is an immediate consequence of this that for $\mathfrak{S} \in S(0)$, $\mathfrak{S}$ is reduced to a single element and necessarily $o(\mathfrak{S}) = z(\mathfrak{S})$; it is the only element of $P(\mathfrak{S})$ if $P(\mathfrak{S}) \neq \emptyset$.

V.2.4. Second step

The statement below, somewhat more general than necessary, will be useful in what follows: we fix T, T' as above.

Lemma *Let $\mathfrak{S} \in S^{\mathfrak{F}}_{(M,\mathfrak{P})}$ and let $z(\mathfrak{S})$ and $z'(\mathfrak{S})$ be $T' - \mathfrak{F}$-general points near $o(\mathfrak{S})$. We denote by $E_{\mathfrak{S},T}$ the set of elements of $S^{\mathfrak{F}}_{(M,\mathfrak{P})}$ which are hyperplanes of $\mathfrak{S}$ containing at least one point in $\mathfrak{S}_{\leq T}$ of real part $o(\mathfrak{S})$. This set is finite. Then for every $\mathfrak{T} \in E_{\mathfrak{S},T}$, we have $o(\mathfrak{T}) = o(\mathfrak{S})$ and there exists $y_{\mathfrak{T}} \in \mathrm{Re}\,\mathfrak{T}$ and a residue datum, $\mathrm{Res}^{\mathfrak{S}}_{\mathfrak{T}}$ for $\mathfrak{T}$ in $\mathfrak{S}$, dependent on $z(\mathfrak{S})$ and $z'(\mathfrak{S})$, which for every function A, meromorphic on $\mathfrak{S}$ with polynomial singularities $S^{\mathfrak{F}}_{\mathfrak{T}}$, satisfies:*

(1) *$y_{\mathfrak{T}}$ is a $T' - \mathfrak{F}$-general point but near $o(\mathfrak{S})$*

(2) $$\int_{\substack{\pi\in\mathfrak{S}_{\leq T}\\ \mathrm{Re}\,\pi=z(\mathfrak{S})}} A(\pi)\,d_{\mathfrak{S}}\pi - \int_{\substack{\pi\in\mathfrak{S}_{\leq T}\\ \mathrm{Re}\,\pi=z'(\mathfrak{S})}} A(\pi)\,d_{\mathfrak{S}}\pi =_T \sum_{\mathfrak{T}\in E_{\mathfrak{S},T}} \int_{\substack{\pi\in\mathfrak{T}_{\leq T}\\ \mathrm{Re}\,\pi=y_{\mathfrak{T}}}} (\mathrm{Res}^{\mathfrak{S}}_{\mathfrak{T}}A)(\pi)\,d_{\mathfrak{T}}\pi.$$

We will obtain this lemma as a direct corollary of the residue theorem. We first check that the set E_T defined in V.1.5(b), where we set $\lambda_1 = z(\mathfrak{S}), \lambda_2 = z'(\mathfrak{S})$, is contained in $E_{\mathfrak{S},T}$.

Let $\mathfrak{T} \in E_T$. By definition, there exists $\pi \in \mathfrak{T}_{\leq T}$ of real part belonging to the segment joining $z(\mathfrak{S})$ to $z'(\mathfrak{S})$. Now this segment is contained in the ball of centre $o(\mathfrak{S})$ and radius r where $r = \sup(\|o(\mathfrak{S})-z(\mathfrak{S})\|, \|o(\mathfrak{S})-z'(\mathfrak{S})\|)$. Using the hypothesis of generality either for $z(\mathfrak{S})$ or for $z'(\mathfrak{S})$, we conclude that:

$$\{\pi \in \mathfrak{T}_{\leq T} | \mathrm{Re}\,\pi = o(\mathfrak{S})\} \neq \emptyset,$$

i.e. $\mathfrak{T} \in E_{\mathfrak{S},T}$.

Let us show that for $\mathfrak{T} \in E_{\mathfrak{S},T}$, we have $o(\mathfrak{T}) = o(\mathfrak{S})$: we have $\operatorname{Re}\mathfrak{T}^0 \subset \operatorname{Re}\mathfrak{S}^0$ and $o(\mathfrak{S}) \in \operatorname{Re}\mathfrak{T}$. The assertion is then a consequence of:

$$\begin{aligned} o(\mathfrak{S}) &\in (\operatorname{Re}\mathfrak{T}) \cap ((\operatorname{Re}\mathfrak{S}) \cap (\operatorname{Re}\mathfrak{S}^0)^{\perp}) \\ &= (\operatorname{Re}\mathfrak{T}) \cap (\operatorname{Re}\mathfrak{S}^0)^{\perp} \\ &\subset (\operatorname{Re}\mathfrak{T}) \cap (\operatorname{Re}\mathfrak{T}^0)^{\perp} = \{o(\mathfrak{T})\}. \end{aligned}$$

Let T' be as in the statement. We fix a $T' - \mathfrak{F}$-general path Γ joining $z(\mathfrak{S})$ to $z'(\mathfrak{S})$. As we may, we require it to be contained in the ball of $\operatorname{Re}\mathfrak{S}$ of centre $o(\mathfrak{S})$ and of radius r. We define $y_{\mathfrak{T}}$ for $\mathfrak{T} \in E_{\mathfrak{T}}$ as in V.1.5(b), i.e. $\{y_{\mathfrak{T}}\} = \{\operatorname{Re}\mathfrak{T} \cap \Gamma\}$. Let us show that $y_{\mathfrak{T}}$ then satisfies (1) of the statement.

Suppose that $\mathfrak{S}' \in S^{\mathfrak{F}}_{(M,\mathfrak{P})}$ does not contain $\mathfrak{T}$ and is strictly contained in $\mathfrak{S}$. We suppose that $y_{\mathfrak{T}}$ is the real part of an element of $\mathfrak{S}'_{<T'}$. By the definition of $\mathfrak{S} \in S^{\mathfrak{F}}_{(M,\mathfrak{P})}$, we can suppose that $\mathfrak{S}'$ is a hyperplane of $\mathfrak{S}$; the condition (3) of V.1.5(b) satisfied by Γ ensures that $\mathfrak{S}' \in E_{T'}$ and the condition (1) of V.1.5(b) excludes this hypothesis. This gives a contradiction and the first property necessary to obtain (1). We prove the second property:

> the ball of centre $o(\mathfrak{S}) = o(\mathfrak{T})$ and of radius $\|o(\mathfrak{S}) - y_{\mathfrak{T}}\|$ is contained in the ball of centre $o(\mathfrak{S})$ and of radius r, by choice of Γ. The desired property is thus a result of the analogous property for $z(\mathfrak{S})$ or $z'(\mathfrak{S})$.

For $\mathfrak{T} \in E_{\mathfrak{S},T} - E_T$, we take for $y_{\mathfrak{T}}$ any point satisfying (1) and we set $\operatorname{Res}^{\mathfrak{S}}_{\mathfrak{T}} = 0$. The lemma is now an immediate corollary of V.1.5(c).

V.2.5. Corollary

Let T, T' be as before and let $\mathfrak{S} \in S^{\mathfrak{F}}_{(M,\mathfrak{P})}$. We set:

$$E_{\mathfrak{S},T} = \{\mathfrak{T} \in S^{\mathfrak{F}}_{(M,\mathfrak{P})} | \mathfrak{T} \subsetneq \mathfrak{S} \text{ and } \exists \pi \in \mathfrak{T}_{\leq T} \text{ with } \operatorname{Re}\pi = o(\mathfrak{S})\};$$

that is, contrary to V.2.4, we do not ask that $E_{\mathfrak{S},T}$ consist of hyperplanes of $\mathfrak{S}$. For $\mathfrak{T} \in E_{\mathfrak{S},T}$, we have $o(\mathfrak{T}) = o(\mathfrak{S})$. Let $z(\mathfrak{S})$ and $z'(\mathfrak{S})$ in $\operatorname{Re}\mathfrak{S}$ *be $T' - \mathfrak{F}$-general points but near $o(\mathfrak{S})$, and for every $\mathfrak{T} \in E_{\mathfrak{S},T}$, let $z'(\mathfrak{T})$ be a $T' - \mathfrak{F}$-general point near $o(\mathfrak{T})$. Then for every $\mathfrak{T} \in E_{\mathfrak{S},T}$ there exists a residue datum* $\operatorname{Res}^{\mathfrak{S}}_{\mathfrak{T}}$ *for $\mathfrak{T}$ in $\mathfrak{S}$ such that for every meromorphic function A on $\mathfrak{S}$ with polynomial singularities on $S^{\mathfrak{F}}_{\mathfrak{T}}$, we have:*

$$\int_{\substack{\pi\in\mathfrak{S}_{\le T}\\ \mathrm{Re}\,\pi=z(\mathfrak{S})}} A(\pi)\,\mathrm{d}_{\mathfrak{S}}\pi - \int_{\substack{\pi\in\mathfrak{S}_{\le T}\\ \mathrm{Re}\,\pi=z'(\mathfrak{S})}} A(\pi)\,\mathrm{d}_{\mathfrak{S}}\pi$$

$$=_T \sum_{\mathfrak{T}\in E_{\mathfrak{S},T}} \int_{\substack{\pi\in\mathfrak{T}_{\le T}\\ \mathrm{Re}\,\pi=z'(\mathfrak{T})}} (\mathrm{Res}_{\mathfrak{T}}^{\mathfrak{S}} A)(\pi)\,\mathrm{d}_{\mathfrak{T}}(\pi).$$

We iterate the application of the preceding lemma.

V.2.6. Corollary

For every $\mathfrak{S}\in S_{(M,\mathfrak{P})}^{\mathfrak{F}}$ let $z(\mathfrak{S})$ be as in V.2.2. Fix $\mathfrak{S}\in S_{(M,\mathfrak{P})}^{\mathfrak{F}}$ and define $E_{\mathfrak{S},T}$ as in V.2.5. Then for every $\mathfrak{T}\in E_{\mathfrak{S},T}$, there exists a residue datum for $\mathfrak{T}$ in $\mathfrak{S}$, denoted by $\mathrm{Res}_{\mathfrak{T}}^{\mathfrak{S}}$, such that for every meromorphic function A on $\mathfrak{S}$ with polynomial singularities on $S_{\mathfrak{X}}^{\mathfrak{F}}$, we have:

$$\int_{\substack{\pi\in\mathfrak{S}_{\le T}\\ \mathrm{Re}\,\pi=z(\mathfrak{S})}} A(\pi)\,\mathrm{d}_{\mathfrak{S}}\pi - |\mathrm{Norm}\,\mathfrak{S}|^{-1} \int_{\substack{\pi\in\mathfrak{S}_{\le T}\\ \mathrm{Re}\,\pi=z(\mathfrak{S})}} \sum_{w\in\mathrm{Norm}\,\mathfrak{S}} A(w\pi)\,\mathrm{d}_{\mathfrak{S}}\pi$$

$$=_T \sum_{\mathfrak{T}\in E_{\mathfrak{S},T}} \int_{\substack{\pi\in\mathfrak{T}_{\le T}\\ \mathrm{Re}\,\pi=z(\mathfrak{T})}} (\mathrm{Res}_{\mathfrak{T}}^{\mathfrak{S}} A)(\pi)\,\mathrm{d}_{\mathfrak{T}}\pi.$$

Since we supposed that $wz(\mathfrak{S})$ is $T'-\mathfrak{F}$-general for every $w\in\mathrm{Norm}\,\mathfrak{S}$, the function A is holomorphic at every point $w\pi$ for $\pi\in\mathfrak{S}_{\le T}$ and Re $\pi = z(\mathfrak{S})$. Thus in the second integral we can exchange the sum and the integral and make variable changes. In fact, we have to calculate:

$$|\mathrm{Norm}\,\mathfrak{S}|^{-1} \sum_{w\in\mathrm{Norm}\,\mathfrak{S}} \left(\int_{\substack{\pi\in\mathfrak{S}_{\le T}\\ \mathrm{Re}\,\pi=z(\mathfrak{S})}} A(\pi)\,\mathrm{d}_{\mathfrak{S}}\pi - \int_{\substack{\pi\in\mathfrak{S}_{\le T}\\ \mathrm{Re}\,\pi=wz(\mathfrak{S})}} A(\pi)\,\mathrm{d}_{\mathfrak{S}}\pi \right).$$

This can be calculated thanks to V.2.5.

V.2.7. End of the proof of V.2.2

By iterating V.2.6, we see that with the hypotheses and notation of V.2.6, evidently modifying $\mathrm{Res}_{\mathfrak{T}}^{\mathfrak{S}}$ for $\mathfrak{T}\in E_{\mathfrak{S},T}$, we have:

$$\int_{\substack{\pi\in\mathfrak{S}_{\le T}\\ \mathrm{Re}\,\pi=z(\mathfrak{S})}} A(\pi)\,\mathrm{d}_{\mathfrak{S}}\pi - |\mathrm{Norm}\,\mathfrak{S}|^{-1} \int_{\substack{\pi\in\mathfrak{S}_{\le T}\\ \mathrm{Re}\,\pi=z(\mathfrak{S})}} \sum_{w\in\mathrm{Norm}\,\mathfrak{S}} A(w\pi)\,\mathrm{d}_{\mathfrak{S}}\pi =_T$$

$$\sum_{\mathfrak{T}\in E_{\mathfrak{S},T}} |\mathrm{Norm}\,\mathfrak{T}|^{-1} \int_{\substack{\pi\in\mathfrak{T}_{\le T}\\ \mathrm{Re}\,\pi=z(\mathfrak{T})}} \sum_{w\in\mathrm{Norm}\,\mathfrak{T}} (\mathrm{Res}_{\mathfrak{T}}^{\mathfrak{S}} A)(w\pi)\,\mathrm{d}_{\mathfrak{T}}\pi. \tag{1}$$

By V.2.1(1) with the notation already introduced, we have:

$$\langle\theta_{\phi'}\theta_\phi\rangle =_T \sum_{\mathfrak{S}\in S^{\mathfrak{F}}_{(M,\mathfrak{P})}} \int_{\substack{\pi\in\mathfrak{S}_{\le T}\\ \mathrm{Re}\,\pi=z(\mathfrak{S})}} (\mathrm{Res}'_{\mathfrak{S}}A(\phi',\phi))(\pi)\,\mathrm{d}_{\mathfrak{S}}\pi$$

$$=_T \sum_{\mathfrak{S}\in S^{\mathfrak{F}}_{(M,\mathfrak{P})}} \Bigg(|\mathrm{Norm}\,\mathfrak{S}|^{-1} \int_{\substack{\pi\in\mathfrak{S}_{\le T}\\ \mathrm{Re}\,\pi=z(\mathfrak{S})}} \sum_{w\in\mathrm{Norm}\,\mathfrak{S}} (\mathrm{Res}'_{\mathfrak{S}}A(\phi',\phi))(w\pi)\,\mathrm{d}_{\mathfrak{S}}\pi$$

$$+\sum_{\mathfrak{T}\in E_{\mathfrak{T},T}} |\mathrm{Norm}\,\mathfrak{T}|^{-1} \int_{\substack{\pi\in\mathfrak{T}_{\le T}\\ \mathrm{Re}\,\pi=z(\mathfrak{T})}} \sum_{w\in\mathrm{Norm}\,\mathfrak{T}} (\mathrm{Res}^{\mathfrak{S}}_{\mathfrak{T}}\mathrm{Res}'_{\mathfrak{S}}A(\phi',\phi))(w\pi)\,\mathrm{d}_{\mathfrak{T}}\pi\Bigg).$$

For every $\mathfrak{S}\in S^{\mathfrak{F}}_{(M,\mathfrak{P})}$, we set:

$$\mathrm{Res}_{\mathfrak{S}} = \mathrm{Res}'_{\mathfrak{S}} + \sum_{\mathfrak{S}'|\mathfrak{S}\in E_{\mathfrak{S}',T}} \mathrm{Res}^{\mathfrak{S}'}_{\mathfrak{S}}\mathrm{Res}'_{\mathfrak{S}'},$$

and we obtain exactly V.2.2.

V.2.8. Remark on the number field case

In this section, suppose that k is a number field. V.2.2 is not satisfactory: we would like a true equality. We identify $X^{\mathbf{G}}_M$ with $\mathfrak{P}$ as in V.1.5(a). This allows us to define a function on $\mathfrak{P}$, denoted by f_0, by $f_0(\pi) = (\lambda_\pi,\lambda_\pi)$. Moreover, with the notation of III.1.4, we set $\Delta = \Delta(f_0)$. Langlands' idea is to use the resolvent of this operator. Recalling the notation p_T of Remark III.1.6, we set $q_T := 1 - p_{-T}$. This is a projection into $L^2(G(k)\backslash\mathbf{G})_\xi$. We have the following theorem, whose proof will be completed only at the end of this chapter, in V.3.11:

Theorem *With the hypotheses and notation of V.2.2 and V.2.1,*

$$\langle\theta_{\phi'},q_T\theta_\phi\rangle =$$

$$\sum_{\mathfrak{S}\in S^{\mathfrak{F}}_{(M,\mathfrak{P})}} |\mathrm{Norm}\,\mathfrak{S}|^{-1} \int_{\substack{\pi\in\mathfrak{S}_{\le T}\\ \mathrm{Re}\,\pi=z(\mathfrak{S})}} \left(\sum_{w\in\mathrm{Norm}\,\mathfrak{S}} (\mathrm{Res}^{\mathbf{G}}_{\mathfrak{S}}A(\phi',\phi))(w\pi)\right)\mathrm{d}_{\mathfrak{S}}\pi.$$

In particular the integrated functions are holomorphic at every point of the domain of integration.

We cannot prove this yet, but we explain here exactly what is needed to obtain the formula. This will justify certain proofs given later.

V.2.9. Lemma

Let $T, T' > 3T + 2R^2$ and let $\phi'\in P^{\mathfrak{F}}_{\mathfrak{X}}$ and $\phi\in P^{R,\mathfrak{F}}_{(M,\mathfrak{P})}$; we suppose that for every $\mathfrak{S}\in S^{\mathfrak{F}}_{(M,\mathfrak{P})}$:

$\sum_{w\in\mathrm{Stab}(M,\mathfrak{P})}(\mathrm{Res}^G_{w.\mathfrak{S}}A(\phi',\phi))(w\pi)$ *is holomorphic at every point of* $\{\pi \in \mathfrak{S}_{\leq T'} | \mathrm{Re}\,\pi = o(\mathfrak{S})\}$,
for every $f \in H^R$; H^R *is the notation* $H^R_{\mathfrak{X},\mathbb{C}}$ *of III.1.1:*

$$\mathrm{Res}^G_{\mathfrak{S}}A(\phi', f\phi) = f_{|\mathfrak{S}}\mathrm{Res}^G_{\mathfrak{S}}A(\phi',\phi),$$

$-T$ *is a point of continuity for the operator* Δ.

Then

$$\langle\theta_{\phi'}, q_T\theta_\phi\rangle = \sum_{\mathfrak{S}\in S^{\mathfrak{F}}_{(M,\mathfrak{P})}} |\mathrm{Norm}\,\mathfrak{S}|^{-1} \int_{\substack{\pi\in\mathfrak{S}_{\leq T}\\ \mathrm{Re}\,\pi=o(\mathfrak{S})}} \sum_{w\in\mathrm{Norm}\,\mathfrak{S}} (\mathrm{Res}^G_{\mathfrak{S}}A(\phi',\phi))(w\pi)\, d_{\mathfrak{S}}\pi.$$

We introduce an auxiliary complex variable z. For $\phi \in P^{\mathfrak{F}}_{(M,\mathfrak{P})}$ and for $z \in \mathbb{C}$, we set

$$\phi_z(\pi) = (z - (\lambda_\pi, \lambda_\pi))^{-1}\phi(\pi);$$

this is an element of $P^{R,\mathfrak{F}}_{(M,\mathfrak{P})}$ for $R \in \mathbb{R}^*_+$ whenever $\mathrm{Re}\,z > R$. We have $(z - \Delta) = \Delta(z - f_0)$ and, see III.1.2:

$$(1) \qquad (z-\Delta)\theta_{\phi_z} = \theta_\phi.$$

For every z of real part greater than R, by the first hypothesis of the lemma and by the residue theorem, we have for every $\mathfrak{S} \in S^{\mathfrak{F}}_{(M,\mathfrak{P})}$:

$$\int_{\substack{\pi\in\mathfrak{S}_{\leq T}\\ \mathrm{Re}\,\pi=z(\mathfrak{S})}} \sum_{w\in\mathrm{Norm}\,\mathfrak{S}} (\mathrm{Res}^G_{\mathfrak{S}}A(\phi',\phi_z))(w\pi)\, d_{\mathfrak{S}}\pi =_T$$

$$\int_{\substack{\pi\in\mathfrak{S}_{\leq T}\\ \mathrm{Re}\,\pi=o(\mathfrak{S})}} \sum_{w\in\mathrm{Norm}\,\mathfrak{S}} (\mathrm{Res}^G_{\mathfrak{S}}A(\phi',\phi_z))(w\pi)\, d_{\mathfrak{S}}\pi,$$

whence:

$$\langle\theta_{\phi'}, \theta_{\phi_z}\rangle =_T \sum_{\mathfrak{S}\in S^{\mathfrak{F}}_{(M,\mathfrak{P})}} |\mathrm{Norm}\,\mathfrak{S}|^{-1} \int_{\substack{\pi\in\mathfrak{S}_{\leq T}\\ \mathrm{Re}\,\pi=o(\mathfrak{S})}} \sum_{w\in\mathrm{Norm}\,\mathfrak{S}} (\mathrm{Res}^G_{\mathfrak{S}}A(\phi',\phi_z))(w\pi)\, d_{\mathfrak{S}}\pi.$$

Let us show that for every $\phi' \in P^{\mathfrak{F}}_{\mathfrak{X}}$, the function of z, which is *a priori* defined only if $\mathrm{Re}\,z > R$, given by:

$$(2) \qquad \langle\theta_{\phi'}, \theta_{\phi_z}\rangle - \sum_{\mathfrak{S}\in S^{\mathfrak{F}}_{(M,\mathfrak{P})}} |\mathrm{Norm}\,\mathfrak{S}|^{-1} \int_{\substack{\pi\in\mathfrak{S}_{\leq T}\\ \mathrm{Re}\,\pi=o(\mathfrak{S})}} \sum_{w\in\mathrm{Norm}\,\mathfrak{S}} (\mathrm{Res}^G_{\mathfrak{S}}A(\phi',\phi_z))(w\pi)\, d_{\mathfrak{S}}\pi$$

is holomorphic on the open set $\{z \in \mathbb{C} | \mathrm{Re}\,z > -T\}$.

Indeed, by V.2.2 and the definition of $=_T$ (see V.1.5(a)), for z with

$\operatorname{Re} z > R$, there exists a finite family of triples $(\mathfrak{S}, U, \operatorname{Res}_{\mathfrak{S},U})$ where $\mathfrak{S} \in S^{\mathfrak{F}}_{(M,\mathfrak{P})}$, U is a measurable subset of $\mathfrak{S} - \mathfrak{S}_{\leq T}$ and $\operatorname{Res}_{\mathfrak{S},U}$ is a residue datum for $\mathfrak{S}$ in $\mathfrak{P}$ such that (2) is the sum of the terms:

$$I_U(\phi', \phi_z) := \int_{\pi \in U} (\operatorname{Res}_{\mathfrak{S},U} A(\phi', \phi_z))(\pi) \, \mathrm{d}_{\mathfrak{S}}\pi,$$

and $\operatorname{Res}_{\mathfrak{S},U} A(\phi', \phi_z)$ is holomorphic at every point of U and integrable over U. It is easy to see that we can choose the same family for every z.

Moreover U is relatively compact if $\operatorname{Re} \mathfrak{S}$ is strictly contained in $\operatorname{Re} X^G_M$.

Suppose first that $\operatorname{Re} \mathfrak{S}$ is strictly contained in $\operatorname{Re} X^G_M$: by the definition of residue data, we have an expansion of

$$\operatorname{Res}_{\mathfrak{S},U} A(\phi', \phi_z)$$

as a finite linear combination of terms of the form

$$(z - (\lambda_\pi, \lambda_\pi))^{-m} A_m(\pi),$$

where $A_m(\pi)$ is a function independent of z. We check by interpolation, using elements z with $\operatorname{Re} z > R$), that each function $A_m(\pi)$ is holomorphic at every point of U. Now if $\operatorname{Re} z > -T$, the functions $(z - (\lambda_\pi, \lambda_\pi))$ do not vanish at any point of the closure of U and the assertion is clear.

Suppose now that $\mathfrak{S}$ is an irreducible component of $\mathfrak{P}$. Here $\operatorname{Res}_{\mathfrak{S},U}$ is multiplication by a scalar, and $z \mapsto A(\phi', \phi_z)(\pi)$ is holomorphic at z for $\pi \in U$, at every point where $z \mapsto \phi_z(\pi)$ is holomorphic. This is the case if $z - (\lambda_\pi, \lambda_\pi)$ does not vanish, i.e. whenever $\operatorname{Re} z \geq -T$ for every $\pi \in U$. The integrals $I_U(\phi', \phi_z)$ thus depend holomorphically on z whenever $\operatorname{Re} z > -T$, which gives the desired result.

We apply III.1.6(viii) to calculate $\langle \theta_{\phi'}, q_T \theta_\phi \rangle$ which, by (1), is equal to:

$$\lim_{\epsilon \to 0} (-1/\pi \mathrm{i}) \int_{\Gamma(-T,R',\epsilon)} \langle \theta_{\phi'}, \theta_{\phi_z} \rangle,$$

where $\Gamma(-T, R', \epsilon)$ is described in III.1.6. Since $-T$ is a point of continuity for $\Delta := \Delta(f_0)$, we can replace q_T by $q_{T-\eta}$ and $\Gamma(-T, R', \epsilon)$ by $\Gamma(-T + \eta, R', \epsilon)$, and let η tend to zero. Since (2) is holomorphic and with the usual residue theorem, we obtain:

$$\langle \theta_{\phi'}, q_{T-\eta} \theta_{\phi_z} \rangle = \lim_{\epsilon \to 0} (-1/\pi \mathrm{i}) \int_{\Gamma(-T+\eta,R',\epsilon)}$$

$$\sum_{\mathfrak{S} \in S^{\mathfrak{F}}_{(M,\mathfrak{P})}} |\operatorname{Norm} \mathfrak{S}|^{-1} \int_{\substack{\pi \in \mathfrak{S}_{\leq T} \\ \operatorname{Re} \pi = o(\mathfrak{S})}} \sum_{w \in \operatorname{Norm} \mathfrak{S}} (\operatorname{Res}^G_{\mathfrak{S}} A(\phi', \phi_z))(w\pi) \, \mathrm{d}_{\mathfrak{S}}\pi.$$

Using the second hypothesis of the lemma, we can replace $\operatorname{Res}^G_{\mathfrak{S}} A(\phi', \phi_z)$

by $1/(f_0(\lambda_\pi) - z)\ \mathrm{Res}_{\mathfrak{S}}^{G} A(\phi', \phi)$ in the above expression. The function of z equal to $1/(f_0(\lambda_\pi) - z)$ is holomorphic on the domains of integration for every ϵ since on these sets $f_0(\lambda_\pi)$ has the form of a constant, denoted by C, plus a real-valued function of π. We can thus exchange the two integrals and the calculation is quite easy: we find 0 if C is not real and the assertion of the lemma otherwise. Since finding 0 is absurd, we prove simultaneously that $o(\mathfrak{S}) \otimes \pi_0$ is an element of $\mathfrak{S}$, which we will also show by a very similar method but one which can also be applied to the function field case, and the lemma.

V.2.10. Lemma

Let $\mathfrak{S} \in S_{(M,\mathfrak{P})}^{\mathfrak{F}}, \phi \in P_{(M,\mathfrak{P})}^{R,\mathfrak{F}}, \phi' \in P_{\mathfrak{X}}^{\mathfrak{F}}$. Then for every function $f \in H^R$ we have:

$$\mathrm{Res}_{\mathfrak{S}}^{G} A(\phi', f\phi) = f_{|\mathfrak{S}} \mathrm{Res}_{\mathfrak{S}}^{G} A(\phi', \phi),$$

if and only if for every function $f \in H^R$ which vanishes on $\mathfrak{S}$, we have:

$$\mathrm{Res}_{\mathfrak{S}}^{G} A(\phi', f\phi) = 0, \text{ on } \{\pi \in \mathfrak{S} |\ \|\pi\| < R\}.$$

The implication is clear. Let us prove the other direction.

Let m be an integer; for $m' < m$, we denote by $D_{m'}^{+}$ the set of elements of the symmetric algebra of $\mathrm{Re}(\mathfrak{S}^0)^{\perp} \otimes_{\mathbf{R}} \mathbf{C}$ which are homogeneous of degree m', and we set $D_{\leq m}^{+} := \bigoplus_{0<m'\leq m} D_{m'}^{+}$; we have excluded $m' = 0$. We identify $D_{\leq m}^{+}$ with a vector space of operators on the set of polynomial functions $\mathfrak{P}$, by identifying $\mathrm{Re}(\mathfrak{S}^0)^{\perp} \otimes_{\mathbf{R}} \mathbf{C}$ and the transverse derivations with $\mathfrak{S}$.

We return to the definition of residue datum operators, see V.1.3. We see that there exist polynomials $P_1, \ldots, P_Q$ on $\mathfrak{P}$, which can be chosen independently of ϕ' and ϕ, such that:

> $\forall \phi' \in P_{\mathfrak{X}}^{\mathfrak{F}}, \phi \in P_{(M,\mathfrak{P})}^{R,\mathfrak{F}}$ $\mathrm{Res}_{\mathfrak{S}}^{G} A(\phi', \phi)$ is the restriction to $\mathfrak{S}$ of a linear combination of successive derivatives in directions of $\mathrm{Re} X_M^G$ belonging to $(\mathrm{Re}\,\mathfrak{S}^0)^{\perp}$ (independent of ϕ' and ϕ), of functions $P_i A(\phi', \phi)$.

Together with Leibniz's rule, this shows that if m is large enough, there exist operators $\delta_1, \ldots, \delta_Q \in D_{\leq m}^{+}$ such that for every $f \in H^R$, for every $\phi' \in P_{\mathfrak{X}}^{\mathfrak{F}}$ and for every $\phi \in P_{(M,\mathfrak{P})}^{\mathfrak{F}}$ we have:

$$(1) \qquad \mathrm{Res}_{\mathfrak{S}}^{G} A(\phi', f\phi) = f_{|\mathfrak{S}} \mathrm{Res}_{\mathfrak{S}}^{G} A(\phi', \phi) + \sum_{1\leq i\leq Q} (\delta_i(f))_{|\mathfrak{S}} r_i(\phi', \phi),$$

where the $r_i(\phi', \phi)$ are meromorphic functions on $\mathfrak{S}$ with polynomial singularities $S_{\mathfrak{X}}^{\mathfrak{F}}$. We can evidently suppose that the elements $\delta_1, \ldots, \delta_Q$ form a basis of $D_{\leq m}^{+}$.

We suppose that there exists $\phi' \in P^{\mathfrak{F}}_{\mathfrak{X}}, \phi \in P^{\mathfrak{F}}_{(M,\mathfrak{P})}$ and $f \in H^R$ such that:

(2) $$\mathrm{Res}^G_{\mathfrak{S}} A(\phi', f\phi) f_{|\mathfrak{S}} \mathrm{Res}^G_{\mathfrak{S}} A(\phi', \phi) \not\equiv 0,$$

and we will show that we can require that f vanish on $\mathfrak{S}$.

We fix ϕ', ϕ and f satisfying (2). With the notation of (1), this means that there exists i between 1 and Q inclusive, such that $r_i(\phi', \phi) \not\equiv 0$. We will prove the following assertion:

V.2.11. Sublemma

There exists a polynomial P on $\mathfrak{P}$, invariant under $\mathrm{Stab}(M, \mathfrak{P})$ *such that:*

(i) $P_{|\mathfrak{S}} \not\equiv 0, \quad \delta P_{|\mathfrak{S}} \equiv 0, \forall \delta \in D^+_{\leq m}$.

(ii) *The image of the restriction map of H^R into the set of holomorphic functions on $\{\pi \in \mathfrak{S} | \mathrm{Re}\,\pi \leq R\}$ invariant under* $\mathrm{Norm}\,\mathfrak{S}$ *contains the set of functions of the form $P_{|\mathfrak{S}} g_1$, where g_1 runs through the set of polynomial functions $\mathfrak{S}$ invariant under* $\mathrm{Norm}\,\mathfrak{S}$.

(iii) *For every $f \in H^R$, there exists $h \in H^R$ in the kernel of the restriction map to $\mathfrak{S}$ such that:*

$$P_{|\mathfrak{S}} \delta f = \delta h, \quad \forall \delta \in D^+_{\leq m}.$$

We fix $\pi_0 \in \mathfrak{P}$: if k is a number field, π_0 is fixed by the property of being trivial on A^G_M (it is the element which gives the identification of $\mathfrak{P}$ with X^G_M which we use); if k is a function field, we take for π_0 any element of $\mathrm{Im}\,\mathfrak{S} := \{\mathrm{Im}\,\pi; \pi \in \mathfrak{S}\}$, which gives a surjective map of $X^{\mathbf{G}}_M$ onto $\mathfrak{P}$ which factors into a bijection of $X^G_M / \mathrm{Fix}_{X^G_M} \mathfrak{P}$.

In the number field case, we denote by $\tilde{\mathfrak{S}}$ the inverse image of $\mathfrak{S}$ in $X^{\mathbf{G}}_M$ and $H := \mathrm{Stab}(M, \mathfrak{P})$. In the function field case we denote by $\tilde{\mathfrak{S}}$ the connected component of the inverse image of $\mathfrak{S}$ in X^G_M which contains $o(\mathfrak{S})$; we denote by H the subgroup of affine automorphisms of $X^{\mathbf{G}}_M$ normalising $\mathrm{Fix}_{X^{\mathbf{G}}_M} \mathfrak{P}$ and whose images after passage to the quotient by $\mathrm{Fix}_{X^{\mathbf{G}}_M} \mathfrak{P}$ can be identified with the action of an element of $\mathrm{Stab}(M, \mathfrak{P})$, i.e. we have an exact sequence:

(1) $$1 \to \mathrm{Fix}_{X^{\mathbf{G}}_M} \mathfrak{P} \to H \to \mathrm{Stab}(M, \mathfrak{P}) \to 1,$$

where $\mathrm{Fix}_{X^{\mathbf{G}}_M} \mathfrak{P}$ can be identified with a subgroup of H consisting of translations. We set:

(2) $$K := \mathrm{Stab}_H \tilde{\mathfrak{S}}.$$

We note that $H.\tilde{\mathfrak{S}}$ is the inverse image of $\mathrm{Stab}(M, \mathfrak{P})\mathfrak{S}$ in $X^{\mathbf{G}}_M$ and that $\mathrm{Re}\,\tilde{\mathfrak{S}} = \mathrm{Re}\,\mathfrak{S}$.

We begin by constructing a polynomial P' on $X^{\mathbf{G}}_M$, invariant under K, such that:

(3) $$P'_{|\tilde{\mathfrak{S}}} \not\equiv 0 \text{ but } (\delta P')_{|\tilde{\mathfrak{S}}} \equiv 0, \forall \delta \in D^+_{\leq m}.$$

(4) $\forall h \in H - K, \quad P'_{|h\tilde{\mathfrak{S}}} \equiv 0$ to a high order.

Fix $h \in H - K$ and let us construct P_h satisfying (3) and vanishing to a high order on $h\tilde{\mathfrak{S}}$.

First case: $h\tilde{\mathfrak{S}}$ and $\tilde{\mathfrak{S}}$ are parallel. Let $\lambda_0 \in \tilde{\mathfrak{S}}$ and let ℓ be a polynomial on $X_M^{\mathbf{G}}$ which is invariant under translation by $\operatorname{Re}\mathfrak{S}^0$ and takes different values on $h\tilde{\mathfrak{S}}$ and $\tilde{\mathfrak{S}}$, i.e. on λ_0 and $h\lambda_0$; for the function field case, the construction is analogous to that of V.1.2, second example. We fix a large integer N; the polynomials $(X-\ell(\lambda_0))^N$ and $(X-\ell(h\lambda_0))^N$ in the variable X are relatively prime; by Bezout's theorem, there exist polynomials $p(X)$ and $q(X)$, such that:

$$p(X)(X-\ell(\lambda_0))^N + q(X)(X-\ell(h\lambda_0))^N = 1.$$

We set:

$$P_h = (\ell - \ell(h\lambda_0))^N (q \circ \ell) = 1 - (p \circ \ell)(\ell - \ell(\lambda_0))^N,$$

and P_h is suitable if N is large enough.

Second case: $h\tilde{\mathfrak{S}}$ and $\tilde{\mathfrak{S}}$ are not parallel: we fix a linear form ℓ, $\mathbb{Q}$-rational on $\operatorname{Re} X_M^{\mathbf{G}}$ (see I.1.4 and V.1.2), zero on the vector part of $\operatorname{Re} h\tilde{\mathfrak{S}}$ and non-constant on that of $\operatorname{Re}\tilde{\mathfrak{S}}$; recall that these vector parts are defined on $\mathbb{Q}$. We write $\ell = \ell_1 - \ell_2$ where ℓ_1 and ℓ_2 are $\mathbb{Q}$-rational linear forms, ℓ_1 is invariant under translation by $\operatorname{Re}\mathfrak{S}^0$ and ℓ_2 is killed by every $\delta \in D_1^+$. If k is a number field, we extend ℓ_1 and ℓ_2 to polynomials on $X_M^{\mathbf{G}}$ i.e. on $\mathfrak{P}$, which we denote by p_1 and p_2 respectively, while if k is a function field, we construct polynomials p_1 and p_2 on $X_M^{\mathbf{G}}$ from ℓ_1 and ℓ_2 by a procedure analogous to V.1.2, third example. We still have that p_1 is constant on $\tilde{\mathfrak{S}}$, that $p_1 - p_2$ is constant on $h\tilde{\mathfrak{S}}$, and that p_2 is killed by every $\delta \in D_1^+$. Up to adding a constant to p_1 and another to p_2, we can suppose that p_1 is zero on $\tilde{\mathfrak{S}}$ and that $p_1 - p_2$ is zero on $h\tilde{\mathfrak{S}}$. We then set:

$$P_h := (p_1^N - p_2^N)^N,$$

where N is a large integer, and this polynomial works.

Then we set:

$$P'' := \prod_{h \in H-K} P_h.$$

This polynomial satisfies (3) and (4) except for the invariance under K. We set:

$$P' := \prod_{k \in K} k\, P''.$$

Taking into account the fact that the elements of K act naturally on $D_{\leq m}^+$, it is clear that P' satisfies (3) and (4).

Let g be a function on $X_M^{\mathbf{G}}$, invariant under K. We set:

$$\tilde{g} := \sum_{h^\bullet \in H/K} h^\bullet(P'g)$$

and this function satisfies:

$$\tilde{g} \text{ is invariant under } H, \tag{5}$$

(6) $$\tilde{g}_{|\tilde{\mathfrak{S}}} = P'_{|\tilde{\mathfrak{S}}} g_{|\tilde{\mathfrak{S}}}$$

(7) $$(\delta\tilde{g})_{|\tilde{\mathfrak{S}}} = P'_{\tilde{\mathfrak{S}}}(\delta g)_{|\tilde{\mathfrak{S}}}, \forall \delta \in D^+_{\leq m}.$$

In particular, $\tilde{g}$ is invariant under $\mathrm{Fix}_{X_M^{\mathbf{G}}} \mathfrak{P}$ and naturally defines a function on $\mathfrak{P}$. We denote by P the function on $\mathfrak{P}$ which can be naturally deduced from $\tilde{1}$, where 1 is the constant function equal to 1. It is clear that P satisfies (i) of the sublemma. Let g_1 be a polynomial function on $\mathfrak{S}$, invariant under $\mathrm{Norm}\,\mathfrak{S}$: then there exists a polynomial g on $X_M^{\mathbf{G}}$, invariant under K, such that we have:

$$g(\lambda) = g_1(\pi_0 \otimes \lambda), \forall \lambda \in \tilde{\mathfrak{S}}.$$

We construct $\tilde{g}$ as above, considered as a polynomial on $\mathfrak{P}$. Thanks to (6), part (ii) of the sublemma is satisfied. Now let g_1 be the restriction to $\mathfrak{S}$ of a holomorphic function on $\{\pi \in \mathfrak{P} |\ \|\mathrm{Re}\,\pi\| \leq R\}$ still denoted by g_1. We will show below that there exists a holomorphic function g' on $\{\lambda \in X_M^{\mathbf{G}} \mid \|\mathrm{Re}\,\lambda\| < R\}$ such that we have:

$$g'(\lambda) = 0, \quad \forall \lambda \in \tilde{\mathfrak{S}},$$
$$(\delta g')(\lambda) = (\delta g_1)(\pi_0 \otimes \lambda), \forall \lambda \in \tilde{\mathfrak{S}}, \quad \forall \delta \in D^+_{\leq m}.$$

Moreover if g_1 is invariant under $\mathrm{Norm}\,\mathfrak{S}$, then replacing g' by

$$|K|^{-1} \sum_{k \in K} k.g',$$

we can require g' to be invariant under K. Part (iii) of the sublemma is then a consequence of the construction of $\tilde{g}'$ which corresponds to these conditions. It remains to show the existence of g'.

We denote by $\alpha_1, \ldots, \alpha_R$ a maximal set of linearly independent roots such that the maps $\lambda \mapsto \langle \lambda, \check{\alpha}_i \rangle$ vanish on the vector part of $\mathrm{Re}\,\tilde{\mathfrak{S}}$. Recall the notation α_i^* of I.1.11. We fix $\lambda_0 \in \tilde{\mathfrak{S}}$ and for $i = 1, \ldots, R$, we set:

$$\ell_i(\lambda) := \begin{cases} ((\alpha_i^*)^\lambda - (\alpha_i^*)^{\lambda_0})(\alpha_i^*)^{-\lambda_0}, & \text{if } k \text{ is a function field,} \\ \langle \lambda, \alpha_i^* \rangle - \langle \lambda_0, \alpha_i^* \rangle, & \text{if } k \text{ is a number field.} \end{cases}$$

Evidently $\ell_i(\lambda) = 0, \forall \lambda \in \tilde{\mathfrak{S}}$. We denote by $\omega_1, \ldots, \omega_R$ the basis of $(\mathrm{Re}\,\mathfrak{S}^0)^\perp$ dual to $(\alpha_1^*, \ldots, \alpha_R^*)$ i.e. for $1 \leq i, j \leq R$ we have:

$$\langle \omega_i, \alpha_j^* \rangle = \delta_{i,j}.$$

We also denote by J the set of R-tuples of integers $(j_1, \ldots, j_R)$ such that $0 < j_1 + \cdots + j_R \leq m$. For $(j) \in J$, we write $\omega^{(j)} := \omega_1^{j_1} \cdots \omega_R^{j_R}$: this is an element of $D^+_{\leq m}$. We also write $\ell^{(j)}(\lambda) = (\ell_1(\lambda))^{j_1} \cdots (\ell_R(\lambda))^{j_R}$: it is a polynomial on $X^{\mathbf{G}}_M$, $(j)! = j_1! \cdots j_R!$. For every i and j between 1 and R inclusive, we have

$$(\omega_i \ell_j)_{|\mathfrak{S}} = \begin{cases} 1, & \text{if } i = j, \\ 0, & \text{if } i \neq j. \end{cases}$$

Thus for every $(i), (j)$ elements of J:

$$(\omega^{(i)}\ell^{(j)}/(j)!)_{|\mathfrak{S}} + \begin{cases} 1, & \text{if } (i) = (j), \\ 0, & \text{if } (i) \neq (j). \end{cases}$$

Let us return to g_1 and denote by g_1' the function on $X^{\mathbf{G}}_M$ satisfying $g_1'(\lambda) = g_1(\pi_0 \otimes \lambda)$. Let $\{a_{(j)}\} \in \mathbb{C}^{|J|}$ and set

$$g'(\{a_{(j)}\}, \lambda) = \sum_{(j)\in J} a_{(j)}(j)!^{-1}\ell^{(j)}(\lambda)(\omega^{(j)}g_1')(\lambda).$$

Let $(i) \in J$; we denote by $<$ the partial order on the set of R-tuples of integers which can be deduced from the natural order on $\mathbb{N}$ and we calculate:

$$\omega^{(i)}g'(\{a_{(j)}\}, \lambda) = \omega^{(i)}g_1'(\lambda) \sum_{\substack{(j)\in J \\ (j)\leq(i)}} \binom{(i)}{(j)} a_{(j)},$$

where $\binom{(i)}{(j)}$ are numbers coming from Leibnitz's rule. It is then clear that we can correctly fix the numbers $a_{(j)}$ in such a way that

$$\sum_{\substack{(j)\in J \\ (j)\leq(i)}} \binom{(i)}{(j)} a_{(j)} = 1$$

for every (i). This concludes the proof of the sublemma.

The proof of the lemma is now an immediate consequence of assertion (iii) of the sublemma: indeed, we return to f and fix an element f' of H^R such that $f'_{|\mathfrak{S}} = 0$ and $(\delta f')_{|\mathfrak{S}} = (\delta f)_{|\mathfrak{S}} P_{|\mathfrak{S}}, \forall \delta \in D^+_{\leq m}$, where P is a polynomial on $\mathfrak{P}$ having the properties of the sublemma. By (1) and (2), we have:

$$\operatorname{Res}^{G}_{\mathfrak{S}} A(\phi', f'\phi) = \sum_{1\leq i\leq Q} ((\delta_i(f))P)_{|\mathfrak{S}} r_i(\phi', \phi) \not\equiv 0,$$

while $f'_{|\mathfrak{S}} \equiv 0$. This is the desired assertion and concludes the proof of the lemma.

V.3. Decomposition along the spectrum of the operators $\Delta(f)$

We again fix an equivalence class $\mathfrak{X}$ of cuspidal data. In the greater part of this chapter, we also fix a finite set of **K**-types, denoted by $\mathfrak{F}$. We denote by $L^2(G(k)\backslash\mathbf{G})_{\mathfrak{X},\mathfrak{F}}$ the vector subspace of $L^2(G(k)\backslash\mathbf{G})_{\mathfrak{X}}$ generated by the functions on which $\mathbf{K}$ acts via one of the $\mathbf{K}$-types contained in $\mathfrak{F}$ and we fix $T \in \mathbb{R}_+^*$; we define q_T as in V.2.8 and we begin by decomposing $q_T L^2(G(k)\backslash \mathbf{G})_{\mathfrak{X},\mathfrak{F}}$. To do this we prove a certain number of results on certain residues of Eisenstein series, see for example the results concerning $A^{\mathfrak{F}}_{\mathfrak{C},\pi}$ (notation of V.3.2) which will be completed in chapter VI (see VI.1.2 and VI.2.4), and which seem somewhat more important than the actual spectral decomposition given in V.3.14.

V.3.1. Hypotheses and notation for this part

We fix $\mathfrak{F}$ as we just explained, R as in V.2.1 and we let T' denote a (large) positive real number which is only needed if k is a number field, and T'' a real number greater than $3T + 2R^2$; the notation is not exactly the same as in V.2.2. We have fixed an equivalence class $\mathfrak{X}$ of cuspidal data. Let $\phi', \phi \in P^{\mathfrak{F}}_{\mathfrak{X}}$; we define a meromorphic function on $\cup_{(M,\mathfrak{P})\in\mathfrak{X}}\mathfrak{P}$, denoted by $A(\phi', \phi)$, by setting:

(1) $$A(\phi', \phi)(\pi) := \sum_{w\in W(M_\pi)} \langle M(w^{-1}, -w\overline{\pi})\phi'(-w\overline{\pi}), \phi(\pi)\rangle,$$

where M_π is the standard Levi of which π is a cuspidal representation and where $W(M_\pi)$ is as defined in V.2.1. This expression means that if $\phi = (\phi_{(M',\mathfrak{P}')})_{(M',\mathfrak{P}')\in\mathfrak{X}}$ with $\phi_{(M',\mathfrak{P}')} \in P^{\mathfrak{F}}_{(M',\mathfrak{P}')}$, we have:

$$A(\phi', \phi) := \sum_{(M',\mathfrak{P}')\in\mathfrak{X}} A(\phi', \phi_{(M',\mathfrak{P}')})$$

where $A(\phi', \phi_{(M',\mathfrak{P}')})$ was defined in V.2.1. As in V.2.1, this function $A(\phi', \phi)$ has polynomial singularities on $S^{\mathfrak{F}}_{\mathfrak{X}}$.

For every $(M, \mathfrak{P}) \in \mathfrak{X}$ and for every $\mathfrak{S} \in S^{\mathfrak{F}}_{(M,\mathfrak{P})}$, we fix $z(\mathfrak{S})$, a $T'' - \mathfrak{F}$-general point near $o(\mathfrak{S})$, and a residue datum for $\mathfrak{S}$ in $\mathfrak{P}$, denoted by $\mathrm{Res}^G_{\mathfrak{S}}$, satisfying:

(*) $\forall (M, \mathfrak{P}) \in \mathfrak{X}, \forall \mathfrak{S} \in S^{\mathfrak{F}}_{\mathfrak{X}}, \forall w \in W(M),\ w(z(\mathfrak{S})) \in (\mathrm{Norm}\, w\mathfrak{S}).z(w\mathfrak{S}),$

$$\forall \phi', \phi \in P^{\mathfrak{F}}_{\mathfrak{X}},\ \langle \theta_{\phi'}, \theta_\phi \rangle =_{T'}$$

(**) $$\sum_{\mathfrak{S}\in S^{\mathfrak{F}}_{(M,\mathfrak{P})}} |\mathrm{Norm}\,\mathfrak{S}|^{-1} \int_{\substack{\pi\in\mathfrak{S}_{\le T}\\ \mathrm{Re}\,\pi = z(\mathfrak{S})}} \sum_{w\in\mathrm{Norm}\,\mathfrak{S}} (\mathrm{Res}^G_{\mathfrak{S}} A(\phi', \phi))(w\pi)\ d_{\mathfrak{S}}\pi.$$

This is the obvious generalisation of V.2.2, up to a change of notation. We suppose, as we may, that $\mathrm{Res}^G_{\mathfrak{S}} = 0$ if $\{\pi \in \mathfrak{S}_{\le T} | \mathrm{Re}\,\pi = z(\mathfrak{S})\} = \emptyset$ or if $\|\mathrm{Re}\, o(\mathfrak{S})\| > R$. This formula is also true when we replace $P^{\mathfrak{F}}_{\mathfrak{X}}$ by $P^{R,\mathfrak{F}}_{\mathfrak{X}}$. Thanks to (*), we can rewrite this formula regrouping the elements of $S^{\mathfrak{F}}_{\mathfrak{X}}$ which are conjuguate by an element of W, as follows.

Let $\mathfrak{S}, \mathfrak{S}' \in S^{\mathfrak{F}}_{\mathfrak{X}}$; we denote by $(M, \mathfrak{P})$ and $(M', \mathfrak{P}')$ the cuspidal data attached to $\mathfrak{S}$ and $\mathfrak{S}'$. We say that $\mathfrak{S}$ and $\mathfrak{S}'$ are equivalent if there exists

$w \in W((M,\mathfrak{P}),(M',\mathfrak{P}'))$ such that $w\mathfrak{S} = \mathfrak{S}'$. We denote by $[S_{\mathfrak{X}}^{\mathfrak{F}}]$ the set of these equivalence classes. For every $\mathfrak{C} \in [S_{\mathfrak{X}}^{\mathfrak{F}}]$, we fix $\mathfrak{S}_{\mathfrak{C}} \in \mathfrak{C}$ and, to simplify the expression, we suppose that all these elements $\mathfrak{S}_{\mathfrak{C}}$ have the same cuspidal data, $(M,\mathfrak{P})$ (which is clearly possible). Then (**) can be rewritten using (*) as:

$$\forall \phi', \phi \in P_{\mathfrak{X}}^{R,\mathfrak{F}},\ \langle \theta_{\phi'}, \theta_{\phi} \rangle =_T$$

$$(2) \quad \sum_{\mathfrak{C}\in[S_{\mathfrak{X}}^{\mathfrak{F}}]} |\mathrm{Norm}\,\mathfrak{S}_{\mathfrak{C}}|^{-1} \int_{\substack{\pi\in\mathfrak{S}_{\mathfrak{C},\leq T}\\ \mathrm{Re}\,\pi = z(\mathfrak{S}_{\mathfrak{C}})}} \sum_{w\in W(M)} ((\mathrm{Res}\,{}^{G}_{w\mathfrak{S}_{\mathfrak{C}}} A)(\phi',\phi))(w\pi)\ d_{\mathfrak{S}}\pi.$$

Because of the importance which the functions under the integral sign will have for us, we give them a name: for every $\mathfrak{C} \in [S_{\mathfrak{X}}^{\mathfrak{F}}]$, for every $\phi', \phi \in P_{\mathfrak{X}}^{R,\mathfrak{F}}$, we define a meromorphic function on $\mathfrak{P}$ by:

$$r_{\mathfrak{C}}(\phi',\phi)(\pi) := |\mathrm{Norm}\,\mathfrak{S}_{\mathfrak{C}}|^{-1} \sum_{w\in W(M)} ((\mathrm{Res}\,{}^{G}_{w\mathfrak{S}_{\mathfrak{C}}} A)(\phi',\phi))(w\pi)$$

The function $r_{\mathfrak{C}}(\phi',\phi)$ has polynomial singularities on $S_{\mathfrak{X}}^{\mathfrak{F}}$. When this function is holomorphic at every point of $\{\pi \in \mathfrak{S}_{\mathfrak{C},\leq T} | \mathrm{Re}\,\pi = o(\mathfrak{S}_{\mathfrak{C}})\}$, we set:

$$m_{\mathfrak{C},T}(\phi',\phi) = \int_{\substack{\pi\in\mathfrak{S}_{\mathfrak{C},\leq T}\\ \mathrm{Re}\,\pi = o(\mathfrak{S}_{\mathfrak{C}})}} r_{\mathfrak{C}}(\phi',\phi)(\pi)\ d_{\mathfrak{S}_{\mathfrak{C}}}\pi.$$

V.3.2. Statement

To simplify, we ask that T' satisfy:

$$T' \text{ is very large with respect to } R.$$

The goal of this section is to associate with every equivalence class $\mathfrak{C}$ of $[S_{\mathfrak{X}}^{\mathfrak{F}}]$ and with every positive real number T satisfying:

$$(*) \qquad T < T'/(4)^{\dim_{\mathbb{R}}(\mathrm{Re}\,X_M^G/\mathrm{Re}\,\mathfrak{S}_{\mathfrak{C}}^0)+1} \text{ and } T > R,$$

– T is a point of continuity for the operator Δ (see V.2.8),

a projection, denoted by $\mathrm{proj}_{\mathfrak{C},T}^{\mathfrak{F}}$ onto a closed subspace of $L^2(G(k)\backslash \mathbf{G})_{\mathfrak{X},\mathfrak{F}}$, which satisfies:

(0) if $\forall \mathfrak{S} \in \mathfrak{C}, \{\pi \in \mathfrak{S}_{\leq T} | \mathrm{Re}\,\pi = o(\mathfrak{S})\} = \emptyset$, then $\mathrm{proj}_{\mathfrak{C},T}^{\mathfrak{F}} = 0$;

(1) $\mathrm{proj}_{\mathfrak{C},T}^{\mathfrak{F}}$ is orthogonal to $\mathrm{proj}_{\mathfrak{C}',T}^{\mathfrak{F}}, \forall \mathfrak{C} \neq \mathfrak{C}' \in [S_{\mathfrak{X}}^{\mathfrak{F}}]$, for every T such that $\mathrm{proj}_{\mathfrak{C},T}^{\mathfrak{F}}$ and $\mathrm{proj}_{\mathfrak{C}',T}^{\mathfrak{F}}$ are defined;

(2) $\forall \phi \in P_{\mathfrak{X}}^{\mathfrak{F}}, T < T'/(4^{\dim \mathrm{Re}\,X_M^G+1}),\ q_T\theta_\phi = \sum_{\mathfrak{C}\in[S_{\mathfrak{X}}^{\mathfrak{F}}]} \mathrm{proj}_{\mathfrak{C},T}^{\mathfrak{F}}\theta_\phi$,

(3) $\forall \mathfrak{C} \in [S_{\mathfrak{X}}^{\mathfrak{F}}], \forall \phi', \phi \in P_{\mathfrak{X}}^{\mathfrak{F}}$:
(i) the function $r_{\mathfrak{C}}(\phi', \phi)$ is holomorphic at every point of the set $\{\pi \in \mathfrak{S}_{\mathfrak{C},\leq T} | \operatorname{Re}\pi = o(\mathfrak{S}_{\mathfrak{C}})\}$,
(ii) $\langle \theta_{\phi'}, \operatorname{proj}_{\mathfrak{C},T}^{\mathfrak{F}} \theta_{\phi} \rangle = m_{\mathfrak{C},T}(\phi', \phi)$.

In V.3.3 we will describe these projections in an *ad hoc* fashion, but we will show that they have the following simple definition (which characterises them):

Let $\mathfrak{C} \in [S_{\mathfrak{X}}^{\mathfrak{F}}]$ and T be as above. For $\phi \in P_{\mathfrak{X}}^{R,\mathfrak{F}}$, we define almost everywhere:

$$\pi \in \mathfrak{S}_{\mathfrak{C}} \mapsto e_{\mathfrak{C}}(\phi, \pi) := |\operatorname{Norm} \mathfrak{S}_{\mathfrak{C}}|^{-1} \sum_{w \in W(M)} \operatorname{Res}_{w\mathfrak{S}_{\mathfrak{C}}}^{\mathbf{G}} E(\phi, w\pi)$$

where $E(\phi, \pi)$ is the Eisenstein series, which is a function meromorphically dependent on π, with polynomial singularities $S_{\mathfrak{X}}^{\mathfrak{F}}$ and with values in the set of automorphic forms on $G(k)\backslash\mathbf{G}$; evidently the residue is calculated after evaluation at the elements of $\mathbf{G}$ and we obtain

$$(\operatorname{Res}_{w\mathfrak{S}_{\mathfrak{C}}}^{G} E(\phi', \pi'))_{\pi' = w\pi}.$$

With this notation, we will show that we have:

(4) (i) $e_{\mathfrak{C}}(\phi, \pi)$ is holomorphic at every point of $\{\pi \in \mathfrak{S}_{\mathfrak{C},\leq T} | \operatorname{Re}\pi = o(\mathfrak{S}_{\mathfrak{C}})\}$;
(ii) $\operatorname{proj}_{\mathfrak{C},T}^{\mathfrak{F}} \theta_{\phi} = \int_{\substack{\pi \in \mathfrak{S}_{\mathfrak{C},\leq T} \\ \operatorname{Re}\pi = o(\mathfrak{S}_{\mathfrak{C}})}} e_{\mathfrak{C}}(\phi, \pi) \, d_{\mathfrak{S}_{\mathfrak{C}}} \pi.$

We set $L_{\mathfrak{C},T}^{2\mathfrak{F}} := \operatorname{proj}_{\mathfrak{C},T}^{\mathfrak{F}} L^2(G(k)\backslash\mathbf{G})_{\mathfrak{X},\mathfrak{F}}$. In the course of the proof we will describe a first spectral decomposition of $L_{\mathfrak{C},T}^{2\mathfrak{F}}$. For every element π of

$$\{\pi \in \mathfrak{S}_{\mathfrak{C},\leq T} | \operatorname{Re}\pi = o(\mathfrak{S}_{\mathfrak{C}})\},$$

we denote by $A_{\mathfrak{C},\pi}^{\mathfrak{F}}$ the space of automorphic forms on $G(k)\backslash\mathbf{G}$ generated by the elements $e_{\mathfrak{C}}(\phi, \pi)$ defined above; $A_{\mathfrak{C},\pi}^{\mathfrak{F}}$ is naturally equal to $A_{\mathfrak{C},w\pi}^{\mathfrak{F}}$ for every $w \in \operatorname{Norm} \mathfrak{S}$. We will check that we have:

(5) (i) For every $\pi \in \mathfrak{S}_{\leq T}$ of real part $o(\mathfrak{S})$, $A_{\mathfrak{C},\pi}^{\mathfrak{F}}$ is equipped with a positive definite scalar product satisfying:

$$\forall \phi', \phi \in P_{\mathfrak{X}}^{R,\mathfrak{F}}, \langle e_{\mathfrak{C}}(\phi', \pi), e_{\mathfrak{C}}(\phi, \pi) \rangle = r_{\mathfrak{C}}(\phi', \phi)(\pi).$$

This product is given by integration on $G(k)\backslash\mathbf{G}$ only if $A_{\mathfrak{C},\pi}^{\mathfrak{F}}$ consists of square integrable automorphic forms, which is equivalent to saying that $\mathfrak{C}$ consists of points; see V.1.6.
(ii) The map $\pi \in \{\mathfrak{S}_{\mathfrak{C},\leq T} | \operatorname{Re}\pi = o(\mathfrak{S}_{\mathfrak{C}})\} \mapsto A_{\mathfrak{C},\pi}^{\mathfrak{F}}$ is a Hilbertian stack and $L_{\mathfrak{C},T}^{2\mathfrak{F}}$ can be identified with the Hilbert space consisting of measurable functions F on $\{\mathfrak{S}_{\mathfrak{C},\leq T} | \operatorname{Re}\pi = o(\mathfrak{S}_{\mathfrak{C}})\}$ such that:

$F(\pi) \in A^{\mathfrak{F}}_{\mathfrak{C},\pi}$ almost everywhere, $F(w\pi) = F(\pi)$ $\forall w \in \mathrm{Norm}\,\mathfrak{S}_{\mathfrak{C}}$ and

$$\int_{\substack{\pi\in\mathfrak{S}_{\mathfrak{C},\leq T}\\ \mathrm{Re}\,\pi=0(\mathfrak{S}_{\mathfrak{C}})}} \|F(\pi)\|^2 \, d_{\mathfrak{S}_{\mathfrak{C}}}\pi < +\infty;$$

where $\|\ \|^2$ is the square of the norm in $A^{\mathfrak{F}}_{\mathfrak{C},\pi}$ defined in (i).

V.3.3. Definition of the family of projections

We first want to replace $S^{\mathfrak{F}}_{\mathfrak{C}}$ by a finite set. Recall from V.2.2 that if $\mathrm{Res}^{G}_{\mathfrak{S}}$ is non-zero we have the two following properties: $\|o(\mathfrak{S})\| < R$ and $\{\pi \in \mathfrak{S}_{\leq T'} | \mathrm{Re}\,\pi = o(\mathfrak{S})\} \neq \emptyset$, which shows that the space $\mathfrak{S}$ intersects the compact set

$$\{\pi; \|\mathrm{Re}\,\pi\| < R \text{ and if } k \text{ is a number field } \|\mathrm{Im}\,\lambda_\pi\|^2 < T' + R^2\}.$$

We denote by $[S'_{\mathfrak{X}}]$ the set of equivalence classes whose elements satisfy these two properties; there are only a finite number since $S^{\mathfrak{F}}_{\mathfrak{X}}$ is a locally finite family. Note the following:

(1) let $\mathfrak{S}, \mathfrak{S}' \in S^{\mathfrak{F}}_{\mathfrak{X}}$; we suppose that the equivalence class of $\mathfrak{S}$ is in $[S'_{\mathfrak{X}}]$, that $\mathfrak{S}' \subset \mathfrak{S}$ and that $\mathfrak{S}' \cap \{\pi \in \mathfrak{S}_{\leq T'} | \mathrm{Re}\,\pi = o(\mathfrak{S})\} \neq \emptyset$. Then $o(\mathfrak{S}) = o(\mathfrak{S}')$ and the equivalence class of $\mathfrak{S}'$ is in $[S'_{\mathfrak{X}}]$.

We equip the set $[S'_{\mathfrak{X}}]$ with a total order on which we impose only the following condition:

$$\forall \mathfrak{C}, \mathfrak{C}' \in [S'_{\mathfrak{X}}], \dim_{\mathbf{R}} \mathrm{Re}\,\mathfrak{S}^0_{\mathfrak{C}} < \dim_{\mathbf{R}} \mathrm{Re}\,\mathfrak{S}^0_{\mathfrak{C}'} \Rightarrow \mathfrak{C} < \mathfrak{C}'.$$

If k is a number field, the maximal element of $[S'_{\mathfrak{X}}]$ is $\mathfrak{X}$ itself. If k is a function field, the maximal element for the fixed total order is the equivalence class of an irreducible component of $\mathfrak{P}$; in general there are several possibilities. Let $\mathfrak{C} \in [S'_{\mathfrak{X}}]$; we set:

$$P^{\mathfrak{F}}_{\mathfrak{C},T'} := \{\phi \in P^{\mathfrak{F}}_{\mathfrak{X}} | \phi \text{ is zero to a high order on every } \mathfrak{S}' \in \mathfrak{C}' \text{ with } \mathfrak{C}' < \mathfrak{C}\}.$$

$$P^{R,\mathfrak{F}}_{\mathfrak{C},T'} := \{\phi \in P^{R,\mathfrak{F}}_{\mathfrak{X}} | \phi \text{ is zero to a high order on every } \mathfrak{S}' \in \mathfrak{C}' \text{ with } \mathfrak{C}' < \mathfrak{C}\}.$$

(2) In particular, let $\mathfrak{S} \in \mathfrak{C}$ and let $\mathfrak{S}' \in S^{\mathfrak{F}}_{\mathfrak{X}}$; if $\mathfrak{S}$ and $\mathfrak{S}'$ satisfy the hypotheses of (1) with $\mathfrak{S} \neq \mathfrak{S}'$, every $\phi \in P^{R,\mathfrak{F}}_{\mathfrak{C},T'}$ vanishes to a high order on $\mathfrak{S}'$.

We denote by $\tilde{L}^{2\mathfrak{F}}_{\geq\mathfrak{C},T'}$ the closed subspace of $L^2(G(k)\backslash\mathbf{G})_{\mathfrak{X},\mathfrak{F}}$ generated by the set of elements θ_ϕ with $\phi \in P^{R,\mathfrak{F}}_{\mathfrak{C},T'}$. Let $T \leq T'$; we still set: $L^{2\mathfrak{F}}_{\geq\mathfrak{C},T} = q_T \tilde{L}^{2\mathfrak{F}}_{\geq\mathfrak{C},T'}$; if k is a function field there is no T' nor T and

the two spaces, with and without tilda, coincide: they are denoted by $L^{2\mathfrak{F}}_{\geq\mathfrak{C}}$ in what follows;
$L^{2\mathfrak{F}}_{>\mathfrak{C},T} = \bigcup_{\mathfrak{C}'>\mathfrak{C}} L^{2\mathfrak{F}}_{\geq\mathfrak{C}',T}$: this is a subspace of $L^{2\mathfrak{F}}_{\geq\mathfrak{C},T}$ since for $\mathfrak{C}' > \mathfrak{C}$, $P^{R,\mathfrak{F}}_{\mathfrak{C}',T'}$ contains $P^{R,\mathfrak{F}}_{\mathfrak{C},T'}$;
$L^{2\mathfrak{F}}_{\mathfrak{C},T} = (L^{2\mathfrak{F}}_{>\mathfrak{C},T})^{\perp} \cap L^{2\mathfrak{F}}_{\geq\mathfrak{C},T}$;
$\mathrm{proj}^{\mathfrak{F}}_{\mathfrak{C},T}$ is the orthogonal projection of $L^2(G(k)\backslash\mathbf{G})_{\mathfrak{X},\mathfrak{F}}$ onto $L^{2\mathfrak{F}}_{\mathfrak{C},T}$.

We will need to know that

(3) $$L^{2\mathfrak{F}}_{\geq\mathfrak{C},T} \subset \tilde{L}^{2\mathfrak{F}}_{\geq\mathfrak{C},T}.$$

Indeed, by [S] Theorem 7.15, the equivalence of (1) and (2), to prove that $\tilde{L}^{2\mathfrak{F}}_{\geq\mathfrak{C},T}$ is stable under q_T, it suffices to prove that $\tilde{L}^{2\mathfrak{F}}_{\geq\mathfrak{C},T}$ 'reduces' the operator Δ, in the sense of Stone, Definition 4.5. By [S] Theorem 4.27, it still suffices to prove that there exists z not belonging to the spectrum of Δ such that $\tilde{L}^{2\mathfrak{F}}_{\geq\mathfrak{C},T}$ is stable under the resolvent $(\Delta - z)^{-1}$. Now choose a real number z greater than R^2. The resolvent $(\Delta - z)^{-1}$ exists and by definition of Δ we have:

$$(\Delta - z)^{-1}\theta_{\phi} = \theta_{\phi'},$$

where $\phi'(\pi) = (z - (\lambda_{\pi}, \lambda_{\pi}))^{-1}\phi(\pi)$ for every $\phi \in P^{R,\mathfrak{F}}_{\mathfrak{X}}$. The map $\phi \mapsto \phi'$ preserves $P^{R,\mathfrak{F}}_{\mathfrak{C},T}$, thus $(\Delta - z)^{-1}$ preserves the space generated by the θ_{ϕ} for $\phi \in P^{R,\mathfrak{F}}_{\mathfrak{C},T}$. As $(\Delta - z)^{-1}$ is continuous, it also preserves the closure of this space, i.e. $\tilde{L}^{2\mathfrak{F}}_{\geq\mathfrak{C},T}$. This gives the assertion.

By construction, for $\mathfrak{C}', \mathfrak{C} \in [S'_{\mathfrak{X}}]$, the projections $\mathrm{proj}^{\mathfrak{F}}_{\mathfrak{C}',T}$ and $\mathrm{proj}^{\mathfrak{F}}_{\mathfrak{C},T}$ are orthogonal; we set $\mathrm{proj}^{\mathfrak{F}}_{\mathfrak{C},T} = 0$ if $\mathfrak{C} \in [S^{\mathfrak{F}}_{\mathfrak{X}}] - [S'_{\mathfrak{X}}]$ and we immediately obtain V.3.2(0) and (1).

V.3.4. Plan and beginning of the proof

The proof of the assertions of V.3.2 works by descending induction on the elements of $[S'_{\mathfrak{X}}]$ equipped with the total order fixed in V.3.3. Fix $\mathfrak{C} \in [S'_{\mathfrak{X}}]$; we suppose that V.3.2(3) and (4) are proved for every class $\mathfrak{C}' > \mathfrak{C}$. We then show that V.3.2(3) and (4) are true for every element ϕ of $P^{R,\mathfrak{F}}_{\mathfrak{C},T'}$ (this is the object of the following sections until V.3.7), then we prove V.3.2(5) (see V.3.9 and 11) for $\mathfrak{C}$ by a passage to the limit, using V.3.2(5)(ii), V.3.2(3) and (4) for every element of $P^{R,\mathfrak{F}}_{\mathfrak{X}}$. In passing, while proving V.3.2(3) for $\phi \in P^{R,\mathfrak{F}}_{\mathfrak{C},T'}$, we prove that if $\phi \in P^{R,\mathfrak{F}}_{\mathfrak{C},T'}$, then the hypotheses of Lemma V.2.9 are satisfied (the proof of this is below for the first hypothesis and in V.3.5 for the second hypothesis); when we have reached the minimal element of $[S'_{\mathfrak{X}}]$, V.3.2(2) becomes a consequence of V.2.9. This will conclude the proof.

The disagreeable aspect of this proof, in the number field case, comes from the fact that we prove results on real numbers T which, once T' is fixed, depend on the equivalence class denoted by $\mathfrak{C}$ above. As we will see, this comes from the version of the residue theorem which we proved above: Langlands, who proves a more precise version (he can deal with $T = T'$), does not have any such problem.

From now on, we fix $\mathfrak{C}$ as above and T satisfying V.3.2(*) and (**). We will show here what the induction hypothesis implies and prove V.3.2(3) for $\phi \in P_{\mathfrak{C},T'}^{R,\mathfrak{F}}$. We also fix T_1 just slightly less than $T'/4^{\dim_{\mathbf{R}} \operatorname{Re} X_M^G/\operatorname{Re} \mathfrak{S}_{\mathfrak{C}}^0}$, such that T_1 is sufficiently large compared to T, so as to be able to apply V.2.2, where we set $T' = T_1$.

Let $\mathfrak{C}' > \mathfrak{C}$ and let $\phi' \in P_{\mathfrak{X}}^{\mathfrak{F}}, \phi \in P_{\mathfrak{C},T'}^{R,\mathfrak{F}}$. We first want to check that $r_{\mathfrak{C}'}(\phi',\phi)(\pi)$ is holomorphic at every point of $\mathfrak{S}'_{\mathfrak{C},\leq T_1}$ whose real part belongs to the segment $[o(\mathfrak{S}_{\mathfrak{C}'}), z(\mathfrak{S}_{\mathfrak{C}'})]$. Thanks to the definition of generality of the points $z(\mathfrak{S}_{\mathfrak{C}'})$ we know that if the function $r_{\mathfrak{C}'}(\phi',\phi)(\pi)$ has singularities at such a point, that point belongs to a hyperplane of $\mathfrak{S}_{\mathfrak{C}'}$ singular for the function and intersecting the set $\{\pi \in \mathfrak{S}_{\mathfrak{C}',\leq T_1} | \operatorname{Re} \pi = o(\mathfrak{S}_{\mathfrak{C}'})\}$. Now, if k is a function field V.3.2(3)(i) excludes the existence in general of such singular hyperplanes; on the other hand if k is a number field the existence of such a hyperplane is not excluded by V.3.2(3)(i) unless $\dim_{\mathbf{R}} \operatorname{Re} \mathfrak{S}_{\mathfrak{C}'}^0 > \dim_{\mathbf{R}} \operatorname{Re} \mathfrak{S}_{\mathfrak{C}}^0$, because of the inequality satisfied by T_1. But as $\phi \in P_{\mathfrak{C},T'}^{R,\mathfrak{F}}$, the vanishing properties of ϕ on the possibly singular planes ensure the $r_{\mathfrak{C}}(\phi',\phi)$ is holomorphic at every point of $\mathfrak{S}_{\mathfrak{C}',T'}$ if $\mathfrak{C}' \geq \mathfrak{C}$ and $\dim_{\mathbf{R}} \operatorname{Re} \mathfrak{S}_{\mathfrak{C}'}^0 = \dim_{\mathbf{R}} \operatorname{Re} \mathfrak{S}_{\mathfrak{C}}^0$; this reasoning can also be applied to the function field case when $\mathfrak{C}' = \mathfrak{C}$, which we also need. Using the residue theorem (see V.1.5, the hypotheses of which are satisfied, but here T_1 replaces the T' of V.1.5, R replaces $\sup(\|\lambda_1\|, \|\lambda_2\|)$ and we use the fact that T' is very large compared to R) we see that we have:

$$\forall \phi' \in P_{\mathfrak{X}}^{R,\mathfrak{F}}, \phi \in P_{\mathfrak{C},T'}^{R,\mathfrak{F}}\ m_{\mathfrak{C}',T}(\phi',\phi) =_T \int\limits_{\substack{\pi \in \mathfrak{S}_{\mathfrak{C}',\leq T} \\ \operatorname{Re}\pi = z(\mathfrak{S}_{\mathfrak{C}'})}} r_{\mathfrak{C}'}(\phi',\phi)(\pi)\, d_{\mathfrak{S}_{\mathfrak{C}'}}\pi.$$

Thus for $\phi \in P_{\mathfrak{C},T'}^{R,\mathfrak{F}}$, it is a consequence of V.3.1(2) that we have:

(1) $$\forall \phi' \in P_{\mathfrak{X}}^{R,\mathfrak{F}},\ \langle \theta_{\phi'}, \theta_\phi \rangle =_T \sum_{\mathfrak{C}' \geq \mathfrak{C}} m_{\mathfrak{C}',T}(\phi',\phi).$$

Suppose that k is a function field. Taking into account the induction hypothesis for V.3.2(3)(ii), we have (T disappears)

$(2)_f$ $$\langle \theta_{\phi'}, \sum_{\mathfrak{C}' > \mathfrak{C}} \operatorname{proj}_{\mathfrak{C}'}^{\mathfrak{F}} \theta_\phi \rangle = \sum_{\mathfrak{C}' > \mathfrak{C}} m_{\mathfrak{C}'}(\phi',\phi).$$

Taking the definition of $\text{proj}_{\mathfrak{C}}^{\mathfrak{F}}$ into account, we deduce that

$(3)_f$ $$\forall \phi \in P_{\mathfrak{C}}^{R,\mathfrak{F}}, \forall \phi' \in P_{\mathfrak{X}}^{R,\mathfrak{F}}, \quad \langle \theta_{\phi'}, \text{proj}_{\mathfrak{C}}^{\mathfrak{F}} \theta_\phi \rangle = m_{\mathfrak{C}}(\phi', \phi),$$

i.e. V.3.2(3).

Suppose now that k is a number field. The induction hypothesis on V.3.2(3) gives, for every $\phi', \phi \in P_{\mathfrak{X}}^{\mathfrak{F}}$:

$(2)_n$ $$\langle \theta_{\phi'}, \sum_{\mathfrak{C}' > \mathfrak{C}} \text{proj}_{\mathfrak{C}',T}^{\mathfrak{F}} \theta_\phi \rangle = \sum_{\mathfrak{C}' > \mathfrak{C}} m_{\mathfrak{C}',T}(\phi', \phi).$$

From this, with (1), we see that for every $\phi' \in P_{\mathfrak{X}}^{R,\mathfrak{F}}$ and for every $\phi \in P_{\mathfrak{C},T'}^{R,\mathfrak{F}}$:

$$\langle \theta_{\phi'}, (1 - \sum_{\mathfrak{C}' > \mathfrak{C}} \text{proj}_{\mathfrak{C}',T}^{\mathfrak{F}}) \theta_\phi \rangle =_T m_{\mathfrak{C},T}(\phi', \phi).$$

We denote by $\widetilde{\text{proj}}_{\mathfrak{C},T}^{\mathfrak{F}}$ the orthogonal projection onto $\tilde{L}_{\geq \mathfrak{C},T'}^{2\mathfrak{F}} \cap (L_{\geq \mathfrak{C},T}^{2\mathfrak{F}})^\perp$; we have (see V.3.3(3)):

$$\text{proj}_{\mathfrak{C},T}^{\mathfrak{F}} = q_T \circ \widetilde{\text{proj}}_{\mathfrak{C},T}^{\mathfrak{F}}$$

and for every $\phi \in P_{\mathfrak{C},T'}^{R,\mathfrak{F}}$;

$(*)$ $$\widetilde{\text{proj}}_{\mathfrak{C},T}^{\mathfrak{F}}(\theta_\phi) = (1 - \sum_{\mathfrak{C}' > \mathfrak{C}} \text{proj}_{\mathfrak{C}',T}^{\mathfrak{F}})(\theta_\phi).$$

Hence, for every $\phi \in P_{\mathfrak{C},T'}^{R,\mathfrak{F}}$ and for every $\phi' \in P_{\mathfrak{X}}^{R,\mathfrak{F}}$:

$(**)$ $$\langle \theta_{\phi'}, \widetilde{\text{proj}}_{\mathfrak{C},T}^{\mathfrak{F}} \theta_\phi \rangle =_T m_{\mathfrak{C},T}(\phi', \phi).$$

By V.3.2(3), if ϕ', ϕ satisfy the hypotheses of V.2.9, we have:

$$\langle \theta_{\phi'}, q_T \theta_\phi \rangle = \sum_{\mathfrak{C}' \geq \mathfrak{C}} m_{\mathfrak{C}',T}(\phi', \phi).$$

We conclude from $(*)$ that:

$$\text{proj}_{\mathfrak{C},T}^{\mathfrak{F}} \theta_\phi = (q_T - \sum_{\mathfrak{C}' > \mathfrak{C}} \text{proj}_{\mathfrak{C}',T}^{\mathfrak{F}}) \theta_\phi,$$

whence:

$(3)_n$ $$\langle \theta_{\phi'}, \text{proj}_{\mathfrak{C},T}^{\mathfrak{F}} \theta_\phi \rangle = m_{\mathfrak{C},T}(\phi', \phi).$$

Thus V.3.2(3) will be a consequence of V.3.5 below, since we just saw that the first hypothesis of V.2.9 is satisfied.

V.3.5. Lemma

Let $\mathfrak{C} \in [S_{\mathfrak{X}}^{\mathfrak{F}}]$; we suppose that the assertions of V.3.2(3) and (4) are true for every $\mathfrak{C}' > \mathfrak{C}$. Let $\phi', \phi \in P_{\mathfrak{X}}^{R,\mathfrak{F}}$ and let $\mathfrak{S} \in \mathfrak{C}$. Then for every function $f \in H^R$ (in the notation of V.2.9), we have the equality of meromorphic functions on $\{\pi \in \mathfrak{S} \mid \|\text{Re}\,\pi\| < R\}$:

$$\text{Res}_{\mathfrak{S}}^{G} A(\phi', f\phi) = f_{|\mathfrak{S}} \text{Res}_{\mathfrak{S}}^{G}(\phi', \phi).$$

We begin with a general technical remark:

V.3.6(a). Remark
Let $(M, \mathfrak{P}) \in \mathfrak{X}$ and let $\mathfrak{S} \in S^{\mathfrak{F}}_{(M,\mathfrak{P})}$. Also, let $\pi \mapsto B(\phi)(\pi)$ denote a family of meromorphic functions on $\{\pi \in \mathfrak{S} | \ \|\mathrm{Re}\,\pi\| < R\}$, dependent on $\phi \in P^{R,\mathfrak{F}}_{\mathfrak{X}}$. We suppose that for π in a dense open set of $\{\pi \in \mathfrak{S} | \ \|\mathrm{Re}\,\pi\| < R\}$ (open independently of ϕ), all the functions $B(\phi)$ are holomorphic at π and the function $\phi \mapsto B(\phi)(\pi)$ can be calculated thanks to the derivatives of ϕ at π of order $\leq a$, where $a \in \mathbb{N}$, see below. Then:

(i) $B(\phi) \equiv 0$ *for every* ϕ *if and only if this is true for every element* ϕ *of* $P^{R,\mathfrak{F}}_{\mathfrak{C},T'}$ *which vanishes to a high order on all the elements of* $\mathfrak{C} - \{\mathfrak{S}\}$, *as well as on* $\cup_{(M',\mathfrak{P}') \in \mathfrak{X} - \{(M,\mathfrak{P})\}} \mathfrak{P}'$.

(ii) *We fix* ϕ *and we suppose that* $B(\phi) \not\equiv 0$. *Then, up to multiplying* ϕ *by a polynomial on* $\mathfrak{P}$, *we can require that:*
(a) $B(\phi)$ *be holomorphic at every point of* $\{\pi \in \mathfrak{S}_{\leq T} | \ \|\mathrm{Re}\,\pi\| < R\}$;
(b) $B(g\phi) = g_{|\mathfrak{S}} B(\phi)$ *for every function* g *on* $\{\pi \in \mathfrak{P} | \ \|\mathrm{Re}\,\pi\| < R\}$ *with polynomial growth;*
(c) $\displaystyle\int_{\substack{\pi \in \mathfrak{S}_{\leq T} \\ \mathrm{Re}\,\pi = o(\mathfrak{S})}} B(\phi)(\pi)\, \mathrm{d}_{\mathfrak{S}}\pi \neq 0.$

For the notion of the derivatives of ϕ we must adopt the second point of view of II.1.2, i.e. fix $\rho \in \mathfrak{P}$ and consider functions on X^G_M with values in the fixed space $A(U(\mathbb{A})M(k)\backslash \mathbf{G})_\rho$ which satisfy V.1.2(3) and (4). Then we can differentiate ϕ without difficulty and we say that the function of $B(\phi)(\pi)$ can be calculated via the derivatives of ϕ of order less than or equal to a, if $B(\phi)(\pi)$ is the value at π of a linear combination of derivatives of ϕ of order $\leq a$ whose coefficients are functions of π, independent of ϕ. Our typical examples are the functions $\mathrm{Res}\,A(\phi', \phi)$ for fixed ϕ' and residue datum Res.

(i) We fix some general π_0 in $\mathfrak{S}$; let $\phi' \in P^{R,\mathfrak{F}}_{\mathfrak{X}}$. Then there exists $\phi \in P^{R,\mathfrak{F}}_{\mathfrak{C},T'}$ having the vanishing properties of the statement of the Remark such that $\phi - \phi'$ is zero in π_0 as well as all its derivatives of order less than or equal to a. We thus have $B(\phi)(\pi_0) = B(\phi')(\pi_0)$ and (i) is clear.

We recall the notation $D^+_{\leq m}$ of V.2.10. As there, we see that there exists an integer m and for $\delta_1, \ldots, \delta_Q$, a basis of $D^+_{\leq m}$ of meromorphic functions r_i on $\{\pi \in \mathfrak{S} | \ \|\mathrm{Re}\,\pi\| < R\}$ such that for every function with polynomial growth g on $\{\pi \in \mathfrak{P} | \ \|\mathrm{Re}\,\pi\| < R\}$, we have:

$$B(g\phi) = g_{|\mathfrak{S}} B(\phi) + \sum_{1 \leq i \leq Q} (\delta_i g)_{|\mathfrak{S}} r_i.$$

We denote by a the smallest integer such that if $r_i \not\equiv 0$ then $\delta_i \in D^+_{\leq a}$ (possibly $a = 0$, if $r_i \equiv 0$ for every i). We fix a polynomial h on $\mathfrak{P}$ such that h vanishes to a very high order on the hyperplanes of $\mathfrak{S}_{\leq T}$ which are singular for $B(\phi)$, and if $a > 0$ and $h_{|\mathfrak{S}} = 0$ for every $\delta \in D^+_{<a}$ then $\delta h_{|\mathfrak{S}} = 0$ and there exists only one index $i \in \{1,\ldots,Q\}$ such that $r_i \neq 0$ and $(\delta_i h)_{|\mathfrak{S}} \not\equiv 0$; we denote this index by i_0. We have:

$$B(hg\phi) = (hg)_{|\mathfrak{S}}B(\phi) + \sum_{1\leq i\leq Q} (h\delta_i g)_{|\mathfrak{S}} r_i + (g\delta_{i_0}h)_{|\mathfrak{S}} r_{i_0} = g_{|\mathfrak{S}}(\delta_{i_0}h)_{|\mathfrak{S}} r_{i_0}$$

and this is still equal to

$$g_{|\mathfrak{S}}B(h\phi),$$

since

$$B(h\phi) = (\delta_{i_0}h)_{|\mathfrak{S}} r_{i_0}.$$

Thus (a) and (b) are satisfied by replacing ϕ by $h\phi$, which we do in what follows. On the other hand, by the choice of i_0, we still have $B(h\phi) \not\equiv 0$. Moreover (a) and (b) remain true if we multiply ϕ by a holomorphic function h' with polynomial growth on $\{\pi \in \mathfrak{P} | \mathrm{Re}\,\pi < R\}$. To obtain (c), we will show that the contrary is absurd. Suppose that for every h' as above, we have:

$$\int_{\substack{\pi\in\mathfrak{S}_{\leq T}\\ \mathrm{Re}\,\pi=o(\mathfrak{S})}} B(h'\phi)(\pi)\, \mathrm{d}_{\mathfrak{S}}\pi = 0.$$

We denote the function $B(\phi)$ by r. It is holomorphic at every point of $\mathfrak{S}_{\leq T}$ of real part $o(\mathfrak{S})$ and not identically zero. Then $B(h'\varphi) = h'_{|\mathfrak{S}}r$. The hypothesis can thus be rewritten, for every h' as above:

$$\int_{\substack{\pi\in\mathfrak{S}_{\leq T}\\ \mathrm{Re}\,\pi=o(\mathfrak{S})}} h'_{|\mathfrak{S}}(\pi)r(\pi)\, \mathrm{d}_{\mathfrak{S}}\pi = 0.$$

The domain of integration is compact and every continuous function on it is a uniform limit of restriction of functions h' as above. We thus have:

$$\int_{\substack{\pi\in\mathfrak{S}_{\leq T}\\ \mathrm{Re}\,\pi=o(\mathfrak{S})}} g(\pi)r(\pi)\, \mathrm{d}_{\mathfrak{S}}\pi = 0,$$

for every function g defined and continuous on the domain of integration. Let π_0 be a point of real part $o(\mathfrak{S})$ where $r(\pi_0) \neq 0$. If the support of g is concentrated in a neighborhood of π_0, we obtain the contradiction:

$$\int_{\substack{\pi\in\mathfrak{S}_{\leq T}\\ \mathrm{Re}\,\pi=o(\mathfrak{S})}} g(\pi)r(\pi)\, \mathrm{d}_{\mathfrak{S}}\pi \neq 0.$$

Let us return to the proof of the lemma. We suppose, as we may, that $(M, \mathfrak{P})$ is the cuspidal datum of $\mathfrak{S}$ and that ϕ has the vanishing properties of Remark (i). Taking V.2.11 into account, it suffices to prove that we have:

(1) $\quad \mathrm{Res}_{\mathfrak{S}}^{G} A(\phi', f\phi) = 0, \forall f \in H^R$ such that $f_{|\mathfrak{S}} \equiv 0$.

To exploit V.3.4(**), we must write $m_{\mathfrak{C},T}(\phi', f\phi)$ precisely, taking into account the vanishing hypotheses we made on ϕ. We have:

$$r_{\mathfrak{C}}(\phi', f\phi)(\pi) = |\mathrm{Norm}\,\mathfrak{S}|^{-1} \sum_{w \in \mathrm{Stab}(M,\mathfrak{P})} \mathrm{Res}_{w\mathfrak{S}}^{G}$$

$$< \sum_{\tau \in W(M)} \langle M(\tau^{-1}, -\tau w\overline{\pi})\phi'(-\tau w\overline{\pi}), (f\phi)(w\pi)\rangle$$

$$= |\mathrm{Norm}\,\mathfrak{S}|^{-1} \sum_{w \in \mathrm{Norm}\,\mathfrak{S}} (\mathrm{Res}_{\mathfrak{S}}^{G} A(\phi', f\phi))(w\pi).$$

whence

$$m_{\mathfrak{C},T}(\phi', \phi) = |\mathrm{Norm}\,\mathfrak{S}|^{-1} \int_{\substack{\pi \in \mathfrak{S}_{\leq T} \\ \mathrm{Re}\ \pi = o(\mathfrak{S})}} \sum_{w \in \mathrm{Norm}\,\mathfrak{S}} (\mathrm{Res}_{\mathfrak{S}}^{G} A(\phi', f\phi))(w\pi)\ \mathrm{d}_{\mathfrak{S}}\pi.$$

Each term of the sum is holomorphic at every point of the domain of integration, thanks to the vanishing properties of ϕ on the elements $\mathfrak{S}'$ of $S_{\mathfrak{X}}^{\mathfrak{F}}$ contained in $\mathfrak{S}$ and intersecting the domain of integration. We can thus take the sum outside the integral and change variables to make the elements w disappear from $\mathrm{Norm}\,\mathfrak{S}$. We obtain:

$$m_{\mathfrak{C},T}(\phi', f\phi) = \int_{\substack{\pi \in \mathfrak{S}_{\leq T} \\ \mathrm{Re}\pi = 0(\mathfrak{S})}} (\mathrm{Res}_{\mathfrak{S}}^{G} A(\phi', f\phi))(\pi)\ \mathrm{d}_{\mathfrak{S}}\pi.$$

If f vanishes to a high order on $\mathfrak{S}$, (1) is clear. We recall the notation $D_{\leq m}^{+}$ of V.2.11. We denote by a the largest integer, if it exists, such that there exists $f \in H^R$ zero on $\mathfrak{S}$, $\phi' \in P_{\mathfrak{X}}^{R,\mathfrak{F}}$ and $\phi \in P_{\mathfrak{C},T}^{R,\mathfrak{F}}$ with the vanishing properties required above, such that:

- $\forall \delta \in D_{<a}^{+},\ \delta f_{|\mathfrak{S}} = 0,$
- $\mathrm{Res}_{\mathfrak{S}}^{G} A(\phi', f\phi) \not\equiv 0.$

If a does not exist the lemma is true; suppose thus that a exists and fix ϕ', ϕ, f satisfying the above conditions. We have:

$$\theta_{f\phi} = D(f)\theta_{\phi},$$

with the notation of III.1.4, and by construction, for every $f' \in H$, $D(f')$ commutes with $\mathrm{proj}_{\mathfrak{C},T}^{\mathfrak{F}}$. We will first suppose that k is a function field, which simplifies everything while indicating the plan of the proof.

Taking into account V.3.4(3)$_f$, we have:

$$(2) \qquad \langle \theta_{\phi'}, \operatorname{proj}_{\mathfrak{C}}^{\mathfrak{F}} \theta_{f\phi} \rangle = \int_{\substack{\pi \in \mathfrak{S} \\ \operatorname{Re} pi = o(\mathfrak{S})}} (\operatorname{Res}_{\mathfrak{S}}^{G} A(\phi', f\phi))(\pi) \, d_{\mathfrak{S}}\pi.$$

It is a consequence of V.3.6(ii) that up to modifying ϕ, without losing the hypotheses on the triple (f, ϕ', ϕ), the right-hand side above is non zero, i.e.:

$$\langle \theta_{\phi'}, \operatorname{proj}_{\mathfrak{C}}^{\mathfrak{F}} \theta_{f\phi} \rangle \neq 0.$$

In particular $\operatorname{proj}_{\mathfrak{C}}^{\mathfrak{F}} \theta_{f\phi} \neq 0$. The positivity of the scalar product on

$$L^2(G(k)\backslash \mathbf{G})_{\mathfrak{X},\mathfrak{F}}$$

then ensures that we have:

$$\langle \operatorname{proj}_{\mathfrak{C}}^{\mathfrak{F}} \theta_{f\phi}, \operatorname{proj}_{\mathfrak{C}}^{\mathfrak{F}} \theta_{f\phi} \rangle \neq 0.$$

Now, we saw that $\operatorname{proj}_{\mathfrak{C}}^{\mathfrak{F}} \theta_{f\phi} = D(f) \operatorname{proj}_{\mathfrak{C}}^{\mathfrak{F}} \theta_{\phi}$; whence:

$$0 \neq \langle \operatorname{proj}_{\mathfrak{C}}^{\mathfrak{F}} \theta_{\phi}, D(f^* f) \operatorname{proj}_{\mathfrak{C}}^{\mathfrak{F}} \theta_{\phi} \rangle = \langle \theta_{\phi}, \operatorname{proj}_{\mathfrak{C}}^{\mathfrak{F}} \theta_{f^* f\phi} \rangle.$$

In particular $\operatorname{Res}_{\mathfrak{S}}^{G} A(\phi, f^* f\phi) \neq 0$. If we show that $f^*_{|\mathfrak{S}} = 0$, we will have

$$\delta(f^*)_{|\mathfrak{S}} = 0, \ \forall \delta \in D^+_{\leq a},$$

the desired contradiction. Note that $f^*_{|\mathfrak{S}} = 0$ if $-\overline{\mathfrak{S}} \in \operatorname{Stab}(M, \mathfrak{P}).\mathfrak{S}$ i.e. $-\overline{\mathfrak{S}} \in \mathfrak{C}$. Let us show this.

V.3.6(b). Remark

(For any k). *Let $\mathfrak{S} \in \mathfrak{C}$ be such that $\operatorname{Res}_{\mathfrak{S}}^{G}$ does not kill all the functions $A(\phi', \phi)$. Then $-\overline{\mathfrak{S}} \in \mathfrak{C}$; by conjugation the same property is then true for every $\mathfrak{S}' \in \mathfrak{C}$.*

We only give the proof here under the hypothesis that k is a function field. By hypothesis, there exists $\phi' \in P_{\mathfrak{X}}^{R,\mathfrak{F}}$ and $\phi \in P_{(M,\mathfrak{P})}^{R,\mathfrak{F}}$ such that $\operatorname{Res}_{\mathfrak{S}}^{G} A(\phi', \phi) \neq 0$. By Remark V.3.6(a), even while keeping the non-vanishing property, we can require that φ satisfy the vanishing conditions already required above and moreover vanish to a high order on every element $-\overline{\mathfrak{S}}'$ with $\mathfrak{S}' \in \mathfrak{C} \cap S_{(M,\mathfrak{P})}^{\mathfrak{F}}$ and $-\overline{\mathfrak{S}}' \notin \mathfrak{C} \cap S_{(M,\mathfrak{P})}^{\mathfrak{F}}$. The calculation in (2) above gives:

$$\langle \theta_{\phi'}, \operatorname{proj}_{\mathfrak{C}}^{\mathfrak{F}} \theta_{\phi} \rangle = \int_{\substack{\pi \in \mathfrak{S} \\ \operatorname{Re} \pi = o(\mathfrak{S})}} (\operatorname{Res}_{\mathfrak{S}}^{G}(\phi', \phi))(\pi) \, d_{\mathfrak{S}}\pi.$$

By V.3.6(a)(ii), up to modifying ϕ without losing the vanishing hypotheses, we can moreover require that the right-hand side be non-zero. Thus, $\operatorname{proj}_{\mathfrak{C}}^{\mathfrak{F}} \theta_{\phi} \neq 0$. We calculate its L^2 norm which is also equal to

$\langle\theta_\phi, \mathrm{proj}_{\mathfrak{C}}^{\mathfrak{F}}\theta_\phi\rangle$. Thus:

$$0 \neq \langle\theta_\phi, \mathrm{proj}_{\mathfrak{C}}^{\mathfrak{F}}\theta_\phi\rangle = \int_{\substack{\pi\in\mathfrak{S}\\ \mathrm{Re}\,\pi=o(\mathfrak{S})}} \mathrm{Res}_{\mathfrak{S}}^{G}A(\phi,\phi)(\pi)\; \mathrm{d}_{\mathfrak{S}}\pi$$

and

$$\mathrm{Res}_{\mathfrak{S}}^{G}A(\phi,\phi) \not\equiv 0.$$

Now we have:

$$\mathrm{Res}_{\mathfrak{S}}^{G}A(\phi,\phi) = \sum_{w\in W(M)} \langle M(w^{-1}, -w\overline{\pi})\phi(-w\overline{\pi}), \phi(\pi)\rangle.$$

The vanishing properties of ϕ ensure that we can limit the sum to those elements w of $\mathrm{Stab}(M,\mathfrak{P})$ with $-w\overline{\mathfrak{S}} \in \mathfrak{C}$. This set is thus non-empty, which concludes the proof of the Remark and of V.3.5 in the function field case.

Now suppose that k is a number field. Thanks to V.3.6(ii), up to modifying ϕ without losing the vanishing hypotheses, we can suppose that we have:

(3) $$\mathrm{Res}_{\mathfrak{S}}^{G}A(\phi', f\phi) \not\equiv 0,$$

for every holomorphic function g with polynomial growth on $\{\pi \in \mathfrak{P} \mid \|\mathrm{Re}\,\pi\| < R\}$:

(4) $$\mathrm{Res}_{\mathfrak{S}}^{G}A(\phi', gf\phi) = g_{|\mathfrak{S}}\, \mathrm{Res}_{\mathfrak{S}}^{G}A(\phi', f\phi), \quad \text{(equality of meromorphic functions)}$$

(5) $$\int_{\substack{\pi\in\mathfrak{S}_{\leq T}\\ \mathrm{Re}\,\pi=o(\mathfrak{S})}} \mathrm{Res}_{\mathfrak{S}}^{G}A(\phi', f\phi)(\pi)\; \mathrm{d}_{\mathfrak{S}}\pi \neq 0.$$

A fortiori, (4) implies that the hypotheses of V.2.9 are satisfied for ϕ' and $f\phi$. We thus have V.3.5$(3)_n$, i.e.

$$\langle\theta_{\phi'}, \mathrm{proj}_{\mathfrak{C},T}^{\mathfrak{F}}\theta_{f\phi}\rangle = m_{\mathfrak{C},T}(\phi', f\phi).$$

By the above calculation:

$$\langle\theta_{\phi'}, \mathrm{proj}_{\mathfrak{C},T}^{\mathfrak{F}}\theta_{f\phi}\rangle = \int_{\substack{\pi\in\mathfrak{S}_{\leq T}\\ \mathrm{Re}\,\pi=o(\mathfrak{S})}} \mathrm{Res}_{\mathfrak{S}}^{G}A(\phi', f\phi)(\pi)\; \mathrm{d}_{\mathfrak{S}}\pi.$$

By (5) this is non-zero; whence:

$$\mathrm{proj}_{\mathfrak{C},T}^{\mathfrak{F}}\theta_{f\phi} \neq 0.$$

We conclude as in the case of function fields; in particular, we must prove Remark V.3.6(b). However, taking the foregoing into account, the necessary modifications to the proof in the function field case are obvious (one must add property (4) with $f \equiv 1$ to the conditions imposed on ϕ).

V.3.7. Proof of the assertions of V.3.2(4), for every $\phi \in P^{R,\mathfrak{F}}_{\mathfrak{C},T'}$
We already saw V.3.2(1) and (3), also showing (see V.3.5) that

$$q_T\theta_\phi = \sum_{\mathfrak{C}' \geq \mathfrak{C}} \mathrm{proj}^{\mathfrak{F}}_{\mathfrak{C}',T}\theta_\phi, \ \forall \phi \in P^{R,\mathfrak{F}}_{\mathfrak{C},T'}.$$

Let us prove V.3.2(4); V.3.2(4)(i) is clear by the fact that ϕ vanishes to a high order on the elements of $S^{\mathfrak{F}}_{\mathfrak{X}}$ contained in $\mathfrak{S}$ and intersecting $\{\pi \in \mathfrak{S}_{\leq T} | \mathrm{Re}\,\pi = o(\mathfrak{S})\}$ (this set contains all the singular hyperplanes of residues along $\mathfrak{S}$ of Eisenstein series). Let $\mathfrak{X}'$ be an equivalence class (possibly different from $\mathfrak{X}$) of cuspidal data relative to the central character ξ. For every $\phi' \in P^{\mathfrak{F}}_{\mathfrak{X}'}$ and every $\phi \in P^{R,\mathfrak{F}}_{\mathfrak{C},T'}$, we will directly calculate:

$$\langle \theta_{\phi'}, \int_{\substack{\pi \in \mathfrak{S}_{\leq T} \\ \mathrm{Re}\,\pi = o(\mathfrak{S})}} e_{\mathfrak{C}}(\phi, \pi)\, \mathrm{d}_{\mathfrak{S}}\pi \rangle,$$

having noted that by I.4.4(b), the function

$$\int_{\substack{\pi \in \mathfrak{S}_{\leq T} \\ \mathrm{Re}\,\pi = o(\mathfrak{S})}} e_{\mathfrak{C}}(\phi, \pi)\, \mathrm{d}_{\mathfrak{S}}\pi$$

on $G(k)\backslash \mathbf{G}$ is slowly increasing.

This is equal to

$$\int_{\substack{\pi \in \mathfrak{S}_{\leq T} \\ \mathrm{Re}\,\pi = o(\mathfrak{S})}} \langle \theta_{\phi'}, e_{\mathfrak{C}}(\phi, \pi) \rangle\, \mathrm{d}_{\mathfrak{S}}\pi.$$

We will calculate $\langle \theta_{\phi'}, e_{\mathfrak{C}}(\phi, \pi) \rangle$, fixing π in $\mathfrak{S}_{\leq T}$ with real part $o(\mathfrak{S})$. We also fix $(M', \mathfrak{P}') \in \mathfrak{X}'$ and it suffices to do the calculation supposing that $\phi' \in P^{\mathfrak{F}}_{(M',\mathfrak{P}')}$. Exchanging residues and integration directly gives:

$$\langle \theta_{\phi'}, e_{\mathfrak{C}}(\phi, \pi) \rangle = \sum_{w \in W(M)} \mathrm{Res}^{\mathbf{G}}_{w\mathfrak{S}_{\mathfrak{C}}} \langle \theta_{\phi'}, E(\phi, w\pi) \rangle.$$

An easy calculation using the computation of constant terms of Eisenstein series done in II.1.7 gives:

$$\forall w \in W(M),\ \langle \theta_{\phi'}, E(\phi, w\pi) \rangle = A(\phi', \phi)(w\pi).$$

Hence

(1) $$\langle \theta_{\phi'}, e_{\mathfrak{C}}(\phi, \pi) \rangle = \begin{cases} 0 & \text{if } \mathfrak{X}' \neq \mathfrak{X} \\ r_{\mathfrak{C}}(\phi', \phi)(\pi), & \text{if } \mathfrak{X}' = \mathfrak{X}. \end{cases}$$

We deduce from this that:

$$\langle \theta_{\phi'}, \int_{\substack{\pi \in \mathfrak{S}_{\leq T} \\ \mathrm{Re}\,\pi = o(\mathfrak{S})}} e_{\mathfrak{C}}(\phi, \pi)\, \mathrm{d}_{\mathfrak{S}}\pi \rangle = \begin{cases} 0 & \text{if } \mathfrak{X}' \neq \mathfrak{X} \\ m_{\mathfrak{C},T}(\phi', \phi)(\pi), & \text{if } \mathfrak{X}' = \mathfrak{X}. \end{cases}$$

Taking into account V.3.2(3) (which was already proved) and the fact that $\theta_{\phi'}$ is orthogonal to $L^2(G(k)\backslash\mathbf{G})_{\mathfrak{X},\mathfrak{F}}$ if $\mathfrak{X}' \neq \mathfrak{X}$, we have:

$$\forall \phi' \in P_{\mathfrak{X}'}^{\mathfrak{F}}, \forall \phi \in P_{\mathfrak{C},T'}^{R,\mathfrak{F}},$$

$$\langle \theta_{\phi'}, \int_{\substack{\pi \in \mathfrak{S}_{\leq T} \\ \mathrm{Re}\,\pi = o(\mathfrak{S})}} e_{\mathfrak{C}}(\phi, \pi)\, \mathrm{d}_{\mathfrak{S}}\pi \rangle = \langle \theta_{\phi'}, \mathrm{proj}_{\mathfrak{C},T}^{\mathfrak{F}} \theta_\phi \rangle.$$

Thus the function:

$$\mathrm{proj}_{\mathfrak{C},T}^{\mathfrak{F}} \theta_\phi - \int_{\substack{\pi \in \mathfrak{S}_{\leq T} \\ \mathrm{Re}\,\pi = o(\mathfrak{S})}} e_{\mathfrak{C}}(\phi, \pi)\, \mathrm{d}_{\mathfrak{S}}\pi,$$

is orthogonal to every pseudo-Eisenstein series; it is thus zero by the Remark in II.1.12.

V.3.8. Another lemma

(This generalises the Remark of V.3.6(b).) *We fix* $\mathfrak{C} \in [S_{\mathfrak{X}}^{\mathfrak{F}}]$; *we suppose that the assertions of V.3.2(3) and (4) are true for every* $\mathfrak{C}' > \mathfrak{C}$ *and that there exists* $\phi' \in P_{\mathfrak{X}}^{R,\mathfrak{F}}$ *and* $\phi \in P_{(M,\mathfrak{P})}^{R,\mathfrak{F}}$ *such that* $\mathrm{Res}_{\mathfrak{S}}^{G} A(\phi', \phi) \not\equiv 0$. *Let* $\mathfrak{S} \in \mathfrak{C}$ *have cuspidal datum* $(M, \mathfrak{S})$. *There exists* $w_{\mathfrak{S}} \in \mathrm{Stab}(M, \mathfrak{P})$ *such that:*

$$\forall \pi \in \mathfrak{S} | \mathrm{Re}\,\pi = o(\mathfrak{S}), \quad w_{\mathfrak{S}}\pi = -\bar{\pi}.$$

In other words $w_{\mathfrak{S}} o(\mathfrak{S}) = -o(\mathfrak{S})$ *and the restriction of* $w_{\mathfrak{S}}$ *to* $\mathrm{Im}\,\mathfrak{S}$ *and to* $\mathrm{Re}\,\mathfrak{S}^0$ *is the identity.*

Fix ϕ', ϕ as in the statement. Thanks to V.3.5(a), we suppose moreover that $\phi \in P_{\mathfrak{C},T'}^{R,\mathfrak{F}}$ has the same vanishing properties as in Remark V.3.6(a) and satisfies:

$$\int_{\substack{\pi \in \mathfrak{S}_{\leq T} \\ \mathrm{Re}\,\pi = o(\mathfrak{S})}} \mathrm{Res}_{\mathfrak{S}}^{G} A(\phi', \phi)(\pi)\, \mathrm{d}_{\mathfrak{S}}\pi \neq 0.$$

Now this integral equals $m_{\mathfrak{C},T}(\phi', \phi)$ (see the calculations of V.3.6(2)) whence, with V.3.2(3), which we have already proved:

$$\langle \theta_{\phi'}, \mathrm{proj}_{\mathfrak{C},T}^{\mathfrak{F}} \theta_\phi \rangle \neq 0.$$

In particular:

$$0 \neq \|\mathrm{proj}_{\mathfrak{C},T}^{\mathfrak{F}} \theta_\phi\|^2 = \langle \theta_\phi, \mathrm{proj}_{\mathfrak{C},T}^{\mathfrak{F}} \theta_\phi \rangle$$
$$= \int_{\substack{\pi \in \mathfrak{S}_{\leq T} \\ \mathrm{Re}\,\pi = o(\mathfrak{S})}} \mathrm{Res}_{\mathfrak{S}}^{G} A(\phi, \phi)(\pi)\, \mathrm{d}_{\mathfrak{S}}\pi,$$

with

$$\mathrm{Res}_{\mathfrak{S}}^{G} A(\phi, \phi)(\pi) = \sum_{\substack{\tau \in W(M) \\ \tau\mathfrak{S} = -\mathfrak{S}}} \mathrm{Res}_{\mathfrak{S}}^{G} \langle M(\tau^{-1}, -\tau\bar{\pi})\phi(-\tau\bar{\pi}), \phi(\pi) \rangle.$$

In V.3.5, we proved the equality:

$$\forall h \in H^R,\ \mathrm{Res}^G_{\mathfrak{S}} A(\phi', h\phi) = h_{|\mathfrak{S}} \mathrm{Res}^G_{\mathfrak{S}} A(\phi', \phi).$$

We show in a similar way that

$$\forall h \in H^R,\ \mathrm{Res}^G_{\mathfrak{S}} A(h\phi', \phi) = h^*_{|\mathfrak{S}} \mathrm{Res}^G_{\mathfrak{S}} A(\phi', \phi).$$

We deduce from this that for all $h, h' \in H^R$ we have as meromorphic functions on $\mathfrak{S}$:

$$\mathrm{Res}^G_{\mathfrak{S}} A(h'\phi, h\phi) = (h'^* h)_{|\mathfrak{S}} \mathrm{Res}^G_{\mathfrak{S}} A(\phi', \phi).$$

Thus the Hermitian form on the set of polynomials on $\mathfrak{P}$ invariant under $\mathrm{Stab}(M, \mathfrak{P})$ defined by

$$b(h', h) = \int_{\substack{\pi \in \mathfrak{S}_{\leq T} \\ \mathrm{Re}\,\pi = o(\mathfrak{S})}} (h'^* h)(\pi) \mathrm{Res}^G_{\mathfrak{S}}(\phi, \phi)(\pi)\ \mathrm{d}_{\mathfrak{S}}\pi,$$

is non-zero positive semi-definite since $b(h, h) = \|\mathrm{proj}^{\mathfrak{F}}_{\mathfrak{C},T} \theta_{hf\phi}\|^2$. By an approximation argument already used, this Hermitian form can be extended to a positive semi-definite Hermitian form on the set of continuous functions on $\mathfrak{P}$ invariant under $\mathrm{Stab}(M, \mathfrak{P})$. By the expression of $b(h', h)$, for every dense open set U of $\{\pi \in \mathfrak{S}_{\leq T}; \mathrm{Re}\,\pi = o(\mathfrak{S})\}$, we can find h with support in U and h' such that $b(h', h) \neq 0$. It is this which allows us to choose a general point π_0 of $\mathfrak{S}_{\leq T}$ with real part $o(\mathfrak{S})$, and a continuous function h, invariant under $\mathrm{Stab}(M, \mathfrak{P})$, concentrated on a small neighborhood of every point of the orbit $\mathrm{Stab}(M, \mathfrak{P})\pi_0$ such that $b(h, h) \neq 0$. The non-vanishing shows that there exists $\sigma \in \mathrm{Stab}(M, \mathfrak{S})$ such that

$$-\sigma\overline{\pi}_0 \in \mathrm{Stab}(M, \mathfrak{P})\sigma\pi_0.$$

Hence:

$$-\overline{\pi}_0 \in \mathrm{Stab}(M, \mathfrak{P})\pi_0.$$

The generality of π_0 proves the lemma.

V.3.9. Proof of V.3.2(5)(i) for general π

More precisely, we fix $\mathfrak{C}$ as in V.3.4. We denote $\mathfrak{S}_{\mathfrak{C}}$ by $\mathfrak{S}$.

(a) We have defined $r_{\mathfrak{C}}(\phi', \phi)$ for $\phi', \phi \in P^{R,\mathfrak{F}}_{\mathfrak{X}}$ as a meromorphic function on $\mathfrak{S}$; this came down to using $\mathfrak{S}$ as a base point for $\mathfrak{C}$. But, by changing the base point, we can obviously define $r_{\mathfrak{C}}(\phi', \phi)$ in an analogous way on every element $\mathfrak{S}'$ of $\mathfrak{C}$. Let us restrict ourselves to the elements $\mathfrak{C}$ having $(M, \mathfrak{P})$ as cuspidal datum. We easily check that thus $r_{\mathfrak{C}}(\phi', \phi)$ becomes a $\mathrm{Stab}(M, \mathfrak{P})$-invariant meromorphic function on the subvariety of $\mathfrak{P}$ equal to

$$\bigcup_{w^\bullet \in \mathrm{Stab}(M,\mathfrak{P})/\mathrm{Norm}\,\mathfrak{S}} w^\bullet\mathfrak{S}.$$

Taking V.3.5(b) into account, we know that $-\overline{\mathfrak{S}}$ is conjugate to $\mathfrak{S}$ by an element of $\mathrm{Stab}(M, \mathfrak{S})$, thus $r_{\mathfrak{C}}(\phi', \phi)$ is defined, in particular, on $\mathfrak{S} \cup -\overline{\mathfrak{S}}$ (as a meromorphic function).

Let us show that we have:

(1) $$\forall \phi', \phi \in P_{\mathfrak{X}}^{R,\mathfrak{F}}, \quad r_{\mathfrak{C}}(\phi', \phi)(\pi) = \overline{r_{\mathfrak{C}}(\phi, \phi')(-\overline{\pi})},$$

equality of meromorphic functions on $\mathfrak{S}$.

Because of Remark V.3.6(a), it suffices to prove (1) supposing that ϕ' and $\phi \in P_{\mathfrak{C},T'}^{R,\mathfrak{F}}$ and it suffices to prove it at every point of $\mathfrak{S}$ of real part $o(\mathfrak{S})$. For a such point, $-\overline{\pi}$ is conjugate to π by an element of $\mathrm{Stab}(M, \mathfrak{P})$ and by invariance of $r_{\mathfrak{C}}(\phi, \phi')$ under this group, it still suffices to show that we have:

$$r_{\mathfrak{C}}(\phi', \phi)(\pi) = \overline{r_{\mathfrak{C}}(\phi, \phi')(\pi)},$$

at every general point of $\mathfrak{S}$ of real part $o(\mathfrak{S})$. Under the hypotheses we made on ϕ and ϕ', thanks to V.3.2(3) which was proved in V.3.4, we have:

$$\begin{aligned}\langle \mathrm{proj}_{\mathfrak{C},T}^{\mathfrak{F}} \theta_{\phi'}, \mathrm{proj}_{\mathfrak{C},T}^{\mathfrak{F}} \theta_{\phi} \rangle &= \langle \theta_{\phi'}, \mathrm{proj}_{\mathfrak{C},T}^{\mathfrak{F}} \theta_{\phi} \rangle = m_{\mathfrak{C},T}(\phi', \phi) \\ &= \int_{\substack{\pi \in \mathfrak{S}_{\leq T} \\ \mathrm{Re}\,\pi = o(\mathfrak{S})}} r_{\mathfrak{C}}(\phi', \phi)(\pi) \, \mathrm{d}_{\mathfrak{S}}\pi;\end{aligned}$$

$$\begin{aligned}\langle \mathrm{proj}_{\mathfrak{C},T}^{\mathfrak{F}} \theta_{\phi}, \mathrm{proj}_{\mathfrak{C},T}^{\mathfrak{F}} \theta_{\phi'} \rangle &= \langle \theta_{\phi}, \mathrm{proj}_{\mathfrak{C},T}^{\mathfrak{F}} \theta_{\phi'} \rangle = m_{\mathfrak{C},T}(\phi, \phi') \\ &= \int_{\substack{\pi \in \mathfrak{S}_{\leq T} \\ \mathrm{Re}\,\pi = o(\mathfrak{S})}} r_{\mathfrak{C}}(\phi, \phi')(\pi) \, \mathrm{d}_{\mathfrak{S}}\pi.\end{aligned}$$

Hence

$$m_{\mathfrak{C},T}(\phi', \phi) = \overline{m_{\mathfrak{C},T}(\phi, \phi')}$$

and

$$\int_{\substack{\pi \in \mathfrak{S}_{\leq T} \\ \mathrm{Re}\,\pi = o(\mathfrak{S})}} (r_{\mathfrak{C}}(\phi', \phi)(\pi) - \overline{r_{\mathfrak{C}}(\phi, \phi')(\pi)}) \, \mathrm{d}_{\mathfrak{S}}\pi = 0.$$

Let $h \in H$. By V.3.5, we have:

$$r_{\mathfrak{C}}(\phi', h\phi) = h_{|\mathfrak{S}} r_{\mathfrak{C}}(\phi', \phi),$$

$$r_{\mathfrak{C}}(h\phi, \phi') = r_{\mathfrak{C}}(\phi, h^{*}\phi') = h^{*}_{|\mathfrak{S}} r_{\mathfrak{C}}(\phi, \phi'),$$

using the fact that by definition, $h^{*}(\pi) = \overline{h(-\overline{\pi})}$. From V.3.8, we see that h^{*} coincides with $\overline{h}$ on $\{\pi \in \mathfrak{S} | \mathrm{Re}\,\pi = o(\mathfrak{S})\}$. Thus, on this set:

(2) $$r_{\mathfrak{C}}(\phi', h\phi) - \overline{r_{\mathfrak{C}}(h\phi, \phi')} = h(r_{\mathfrak{C}}(\phi', \phi) - \overline{r_{\mathfrak{C}}(\phi, \phi')}).$$

We fix P_0 satisfying the conditions of Lemma V.2.11 and we know by this lemma that for every polynomial g on $\mathfrak{P}$ invariant under Norm $\mathfrak{S}$, there exists h as above such that

$$(gP_0)_{|\mathfrak{S}} = h_{|\mathfrak{S}}.$$

Thus for every polynomial g on $\mathfrak{P}$ invariant under Norm $\mathfrak{S}$, we have:

$$\forall \phi', \phi \in P^{R,\mathfrak{F}}_{\mathfrak{C},T'}, \int_{\substack{\pi\in\mathfrak{S}_{\leq T}\\ \mathrm{Re}\,\pi=o(\mathfrak{S})}} g(\pi)P_0(\pi)(r_{\mathfrak{C}}(\phi',\phi)(\pi) - \overline{r_{\mathfrak{C}}(\phi,\phi')(\pi)})\, d_{\mathfrak{S}}\pi = 0.$$

Since $P_0(r_{\mathfrak{C}}(\phi',\phi) - \overline{r_{\mathfrak{C}}(\phi',\phi)})$ is a continuous function on the integration set invariant under Norm $\mathfrak{S}$, the above equality is still true for every polynomial g on $\mathfrak{P}$; use the projection operator on the invariants. By approximation, this remains true for every continuous function g on the integration set. This gives the vanishing of the function $(P_0)_{|\mathfrak{S}}(r_{\mathfrak{C}}(\phi',\phi) - \overline{r_{\mathfrak{C}}(\phi,\phi')})$ on the integration set. As P_0 is not zero on this set, we must have:

(3) $$r_{\mathfrak{C}}(\phi',\phi)(\pi) = \overline{r_{\mathfrak{C}}(\phi,\phi')(\pi)},$$

for every π belonging to $\mathfrak{S}_{\leq T}$ of real part $o(\mathfrak{S})$ and this gives the desired result.

(b) Proof of the semi-positivity of $(\phi',\phi) \mapsto r_{\mathfrak{C}}(\phi',\phi)$

Let $\phi \in P^{R,\mathfrak{F}}_{\mathfrak{X}}$; we will show that at every point π_0 of $\mathfrak{S}_{\leq T}$ of real part $o(\mathfrak{S})$, we have $r_{\mathfrak{C}}(\phi,\phi)(\pi_o) \geq 0$ if this is defined.

As in V.3.6(a), we reduce to the case where $\phi \in P^{\mathfrak{F}}_{\mathfrak{C},T'}$. In this case $r_{\mathfrak{C}}(\phi,\phi)$ is holomorphic, in particular at every point of $\mathfrak{S}_{\leq T}$ of real part $o(\mathfrak{S})$. For every $\phi \in P^{R,\mathfrak{F}}_{\mathfrak{C},T'}$, and every $h \in H^R$, we thus have as above:

$$\begin{aligned} 0 \leq \|\mathrm{proj}^{\mathfrak{F}}_{\mathfrak{C},T}\theta_{h\phi}\|^2 &= \langle\theta_{h\phi}, \mathrm{proj}^{\mathfrak{F}}_{\mathfrak{C},T}\theta_{h\phi}\rangle \\ &= \int_{\substack{\pi\in\mathfrak{S}_{\leq T}\\ \mathrm{Re}\,\pi=o(\mathfrak{S})}} r_{\mathfrak{C}}(h\phi,h\phi)(\pi)\, d_{\mathfrak{S}}\pi \\ &= \int_{\substack{\pi\in\mathfrak{S}_{\leq T}\\ \mathrm{Re}\,\pi=o(\mathfrak{S})}} (h^* h)(\pi) r_{\mathfrak{C}}(\phi,\phi)(\pi)\, d_{\mathfrak{C}}\pi. \end{aligned}$$

As in the proof of V.3.2.(5)(i) above, we can replace h^* by $\overline{h}$, whence:

$$\leq \int_{\substack{\pi\in\mathfrak{S}_{\leq T}\\ \mathrm{Re}\,\pi=o(\mathfrak{S})}} (\overline{h}h)(\pi) r_{\mathfrak{C}}(\phi,\phi)(\pi)\, d_{\mathfrak{S}}\pi.$$

We recall the notation P_0 introduced above and, by the approximation argument already used, we find that we have, for every continuous function g on

$$\{\pi \in \mathfrak{S}_{\leq T} | \mathrm{Re}\,\pi = o(\mathfrak{S})\}$$

which is invariant under Norm $\mathfrak{S}$:

$$0 \leq \int_{\substack{\pi \in \mathfrak{S}_{\leq T} \\ \mathrm{Re}\,\pi = o(\mathfrak{S})}} (\bar{g}g)(\pi)(\overline{P}_0 P_0)(\pi) r_{\mathfrak{C}}(\phi, \phi)(\pi)\, \mathrm{d}_{\mathfrak{S}}\pi.$$

Thus

$$(\overline{P}_0 P_0)(\pi) r_{\mathfrak{C}}(\phi, \phi)(\pi) \geq 0, \quad \forall \pi \in \mathfrak{S}_{\leq T} \text{ of real part } o(\mathfrak{S}),$$

which implies the desired result.

(c) Proof of V.3.2(5)(i)

As we still have not proved V.3.2(4)(i), the space of automorphic forms $A^{\mathfrak{F}}_{\mathfrak{C},\pi}$ is still not defined for every $\pi \in \mathfrak{S}_{\leq T}$ with real part $o(\mathfrak{S})$; thus we will only prove that $A^{\mathfrak{F}}_{\mathfrak{C},\pi}$ is equipped with a positive Hermitian product at the points π_0, where $e_{\mathfrak{C}}(\phi, \pi)$ is holomorphic for every $\phi \in P^{R,\mathfrak{F}}_{\mathfrak{X}}$. It is easy to see that the set of these points contains the complement of a finite union of hyperplanes of $\{\pi \in \mathfrak{S}_{\leq T} | \mathrm{Re}\ \pi = o(\mathfrak{S})\}$. In particular it contains:

$$\mathring{\mathfrak{S}}_{\leq T} := \{\pi \in \mathfrak{S}_{\leq T} | \mathrm{Re}\ \pi = o(\mathfrak{S})\} - \bigcup_{\mathfrak{S}' \subsetneq \mathfrak{S},\ \mathfrak{S}' \in S^{\mathfrak{F}}_{\mathfrak{X}}} \mathfrak{S}'.$$

Let $\mathfrak{X}'$ be an equivalence class of cuspidal data and let $\phi' \in P^{\mathfrak{F}}_{\mathfrak{X}'}, \phi \in P^{R,\mathfrak{F}}_{\mathfrak{X}}$; we show that we have the following equality of meromorphic functions on $\mathfrak{S}$:

$$\langle \theta_{\phi'}, e_{\mathfrak{C}}(h\phi, \pi) \rangle = \begin{cases} 0 & \text{if } \mathfrak{X}' \neq \mathfrak{X}, \\ r_{\mathfrak{C}}(\phi', \phi)(\pi), & \text{if } \mathfrak{X}' = \mathfrak{X}. \end{cases}$$

As in V.3.6(a), it suffices to prove this equality for $\phi \in P^{R,\mathfrak{F}}_{\mathfrak{C},T'}$; in this case the calculation was done in V.3.7.

Thus (see II.1.12, Remark) let $\phi \in P^{R,\mathfrak{F}}_{\mathfrak{X}}$ and let π_0 be a point where $e_{\mathfrak{C}}(\phi, \pi)$ is holomorphic: we have the equivalence

$$e_{\mathfrak{C}}(\phi, \pi_0) = 0 \Leftrightarrow r_{\mathfrak{C}}(\phi', \phi)(\pi_0) = 0, \forall \phi' \in P^{R,\mathfrak{F}}_{\mathfrak{X}}.$$

Suppose moreover that $\pi_0 \in \mathfrak{S}_{\leq T}$ is of real part $o(\mathfrak{S})$; taking (3) above into account, we also have:

$$e_{\mathfrak{C}}(\phi, \pi_0) = 0 \Leftrightarrow r_{\mathfrak{C}}(\phi, \phi')(\pi_0) = 0, \forall \phi' \in P^{\mathfrak{F}}_{\mathfrak{X}}.$$

We still require that π_0 not be a singular point for any of the functions $e_{\mathfrak{C}}(\phi, \pi)$ when ϕ runs through $P^{R,\mathfrak{F}}_{\mathfrak{X}}$; then $A^{\mathfrak{F}}_{\mathfrak{C},\pi_0}$ is defined and, by what precedes, this space is equipped with a sesquilinear form whose kernel is trivial, by setting

$$\forall \phi', \phi \in P^{R,\mathfrak{F}}_{\mathfrak{X}}, \quad \langle e_{\mathfrak{C}}(\phi', \pi_0), e_{\mathfrak{C}}(\phi, \pi_0) \rangle := r_{\mathfrak{C}}(\phi', \phi)(\pi_0).$$

It is positive definite and Hermitian by (a)(3) and (b). This concludes the proof.

V.3.10. Remark

Let $h \in H^R$ and let $\phi \in P_{\mathfrak{X}}^{\mathfrak{F}}$. We have:

$$e_{\mathfrak{C}}(h\phi, \pi) = D(h)e_{\mathfrak{C}}(\phi, \pi) = h(\pi)e_{\mathfrak{C}}(\phi, \pi),$$

equality of meromorphic functions on $\mathfrak{S}$.

To prove this, we can as usual suppose that $\phi \in P_{\mathfrak{C},T'}^{R,\mathfrak{F}}$. For every equivalence class of cuspidal data $\mathfrak{X}'$ and for every $\phi' \in P_{\mathfrak{X}'}^{\mathfrak{F}}$, we have

$$\langle \theta_{\phi'}, e_{\mathfrak{C}}(h\phi, \pi)\rangle = \begin{cases} 0 & \text{if } \mathfrak{X}' \neq \mathfrak{X}, \\ r_{\mathfrak{C}}(\phi', h\phi)(\pi), & \text{if } \mathfrak{X}' = \mathfrak{X}. \end{cases}$$

Thanks to V.3.5, we know that:

$$r_{\mathfrak{C}}(\phi', h\phi)(\pi) = h(\pi)r_{\mathfrak{C}}(\phi', \phi)(\pi).$$

Thus, for every ϕ' as above:

$$\langle \theta_{\phi'}, e_{\mathfrak{C}}(h\phi, \pi) - h(\pi)e_{\mathfrak{C}}(\phi, \pi)\rangle = 0$$

and (see II.1.12):

$$e_{\mathfrak{C}}(h\phi, \pi) = h(\pi)e_{\mathfrak{C}}(\phi, \pi).$$

We also have:

$$\langle \theta_{h^*\phi'}, e_{\mathfrak{C}}(h\phi, \pi)\rangle = \begin{cases} 0 & \text{if } \mathfrak{X}' \neq \mathfrak{X}, \\ r_{\mathfrak{C}}(h^*\phi', \phi)(\pi), & \text{if } \mathfrak{X}' = \mathfrak{X}. \end{cases}$$

Suppose that $\mathfrak{X}' = \mathfrak{X}$. Then

$$\begin{aligned} \langle D(h^*)\theta_{\phi'}, e_{\mathfrak{C}}(\phi, \pi)\rangle &= r_{\mathfrak{C}}(h^*\phi', \phi)(\pi) = \overline{r_{\mathfrak{C}}(\phi, h^*\phi')(-\overline{\pi})} \\ &= h(\pi)\overline{r_{\mathfrak{C}}(\phi, \phi')(-\overline{\pi})} = h(\pi)r_{\mathfrak{C}}(\phi', \phi)(\pi) = \langle \theta_{\phi'}, e_{\mathfrak{C}}(h\phi, \pi)\rangle. \end{aligned}$$

Thus:

$$\langle \theta_{\phi'}, D(h)e_{\mathfrak{C}}(\phi, \pi)\rangle = \langle D(h^*)\theta_{\phi'}, e_{\mathfrak{C}}(\phi, \pi)\rangle = \langle \theta_{\phi'}, e_{\mathfrak{C}}(h\phi, \pi)\rangle$$

and, as above, the equality:

$$D(h)e_{\mathfrak{C}}(\phi, \pi) = e_{\mathfrak{C}}(h\phi, \pi).$$

V.3.11. Spectral decomposition of $L_{\mathfrak{C},T}^{2\mathfrak{F}}$

We adopt the notation $\mathring{\mathfrak{S}}_{\leq T}$ of V.3.9; we will prove V.3.2(5)(ii) by replacing $\{\pi \in \mathfrak{S}_{\leq T} | \operatorname{Re}\pi = o(\mathfrak{S})\}$ by $\{\pi \in \mathring{\mathfrak{S}}_{\leq T} | \operatorname{Re}\pi = o(\mathfrak{S})\}$, as we may. For every π in this set, $A_{\mathfrak{C},\pi}^{\mathfrak{F}}$ is well defined and equipped with a Hilbert space structure thanks to V.3.9.

We first prove that the map

$$\pi \in \{\pi \in \mathring{\mathfrak{S}}_{\leq T} | \operatorname{Re}\, \pi = o(\mathfrak{S})\} \mapsto A_{\mathfrak{C},\pi}^{\mathfrak{F}}$$

is a Hilbertian stack.

We fix a unitary cuspidal representation $\rho \in \mathfrak{P}$; using the second point of view of II.1.2, we realise $P_{\mathfrak{X}}^{R,\mathfrak{F}}$ as a space of functions with values in

$$\bigoplus_{w \in W(M)} \left(A(U^w(\mathbb{A})wMw^{-1}(k)\backslash\mathbf{G})\right)_{w\rho},$$

where U^w is the radical unipotent of the standard parabolic of Levi wMw^{-1}. We then see that there exists an integer N independent of $\pi \in \mathring{\mathfrak{S}}_{\leq T}$ and an equivariant surjective map $\mathbf{K}$:

$$\mathfrak{M} := \bigoplus_{w \in W(M)} \left(A(U^w(\mathbb{A})wMw^{-1}(k)\backslash\mathbf{G})_{w\rho}\right)^N \xrightarrow{p} A_{\mathfrak{C},\pi}^{\mathfrak{F}}.$$

In other words $\mathfrak{M}$ can be identified with

$$P_{\mathfrak{X}}^{R,\mathfrak{F}}/\{\phi \in P_{\mathfrak{X}}^{R,\mathfrak{F}} \mid \phi \text{ vanishes at } \pi \text{ to a certain order}\}.$$

On the other hand the map $\phi', \phi \mapsto r_{\mathfrak{C}}(\phi', \phi)(\pi)$ can be naturally realised as a positive semi-definite sesquilinear form on the space $\mathfrak{M}$ whose kernel is exactly the kernel of p. We fix a $\mathbf{K}$-type and we restrict p to the space of this $\mathbf{K}$-type; all the spaces become finite-dimensional, and $r_{\mathfrak{C}}(\ ,\ )(\pi)$ induces a sesquilinear form on the subspace of $\mathfrak{M}$ corresponding to the fixed $\mathbf{K}$-type, whose kernel can be realised as solutions of a linear system with analytic coefficients, and thus varies analytically on an open set U of $\mathring{\mathfrak{S}}_{\leq T}$. Thus, on this open set, the map $\pi \in U \mapsto A_{\mathfrak{C},\pi}^{\mathfrak{F}}$ is locally trivial and the local trivialisation respects the scalar product. This proves the assertion.

We denote by $\mathrm{Hilb}_{\mathfrak{C},T}^{\mathfrak{F}}$ the set of measurable functions F on $\mathring{\mathfrak{S}}_{\leq T}$ such that

(i) $F(\pi) \in A_{\mathfrak{C},\pi}^{\mathfrak{F}}$, for almost every π,
(ii) $F(\pi) = F(w\pi), \forall w \in \mathrm{Norm}\,\mathfrak{S}$, almost everywhere,

(iii) $$\int_{\substack{\pi \in \mathring{\mathfrak{S}}_{\leq T} \\ \mathrm{Re}\,\pi = o(\mathfrak{S})}} \langle F(\pi), F(\pi)\rangle_{A_{\mathfrak{C},\pi}^{\mathfrak{F}}}\ \mathrm{d}_{\mathfrak{S}}\pi < +\infty.$$

This set is naturally a Hilbert space, whose norm is given by (iii). For $\phi \in P_{\mathfrak{X}}^{R,\mathfrak{F}}$, define a function on $\mathring{\mathfrak{S}}_{\leq T}$, denoted by F_ϕ, by:

$$F_\phi(\pi) = e_{\mathfrak{C}}(\phi, \pi).$$

It satisfies (i) and (ii) and we have:

$$\langle F_\phi(\pi), F_\phi(\pi))\rangle_{A_{\mathfrak{C},\pi}^{\mathfrak{F}}} = r_{\mathfrak{C}}(\phi, \phi)(\pi).$$

Suppose moreover that $\phi \in P_{\mathfrak{C},T'}^{R,\mathfrak{F}}$. Then $r_{\mathfrak{C}}(\phi, \phi)(\pi)$ is defined, continuous for every $\pi \in \mathfrak{S}_{\leq T}$ of real part $o(\mathfrak{S})$; it is thus integrable on $\{\pi \in$

$\overset{\circ}{\mathfrak{S}}_{\leq T} | \mathrm{Re}\,\pi = o(\mathfrak{S})\}$. Thus F_ϕ also satisfies (iii) in this case, i.e.

$$F_\phi \in \mathrm{Hilb}^{\mathfrak{F}}_{\mathfrak{C},T}, \quad \forall \phi \in P^{R,\mathfrak{F}}_{\mathfrak{C},T'}.$$

We set:

$$\mathrm{Hilb}^{0\mathfrak{F}}_{\mathfrak{C},T} := \{F_\phi, \phi \in P^{R,\mathfrak{F}}_{\mathfrak{C},T'}\}.$$

Proposition *There exists a unique isometry, denoted by j, of* $\mathrm{Hilb}^{\mathfrak{F}}_{\mathfrak{C},T}$ *onto* $L^{2\mathfrak{F}}_{\mathfrak{C},T}$, *such that we have:*

(*) $$\forall \phi \in P^{R,\mathfrak{F}}_{\mathfrak{C},T'}, \; jF_\phi = \mathrm{proj}^{\mathfrak{F}}_{\mathfrak{C},T}\theta_\phi.$$

In particular $\mathrm{Hilb}^{0\mathfrak{F}}_{\mathfrak{C},T}$ *is dense in* $\mathrm{Hilb}^{\mathfrak{F}}_{\mathfrak{C},T}$.

We first show:

(1) $$\overline{\mathrm{Hilb}^{0\mathfrak{F}}_{\mathfrak{C},T}} = \mathrm{Hilb}^{\mathfrak{F}}_{\mathfrak{C},T}.$$

We denote by $\langle , \rangle$ the scalar product in $\mathrm{Hilb}^{\mathfrak{F}}_{\mathfrak{C},T}$. Let $F \in \mathrm{Hilb}^{\mathfrak{F}}_{\mathfrak{C},T}$. We must prove that if F is orthogonal to $\mathrm{Hilb}^{0\mathfrak{F}}_{\mathfrak{C},T}$ then F is zero. Suppose thus that F is orthogonal to $\mathrm{Hilb}^{0\mathfrak{F}}_{\mathfrak{C},T}$. This can be interpreted as follows:

$$\forall \phi \in P^{R,\mathfrak{F}}_{\mathfrak{C},T'}, \quad 0 = \int_{\substack{\pi \in \overset{\circ}{\mathfrak{S}}_{\leq T} \\ \mathrm{Re}\,\pi = o(\mathfrak{S})}} \langle F(\pi), e_{\mathfrak{C}}(\phi, \pi)\rangle \, d_{\mathfrak{S}}\pi.$$

Thus in particular for every $g \in H^R$:

$$0 = \int_{\substack{\pi \in \overset{\circ}{\mathfrak{S}}_{\leq T} \\ \mathrm{Re}\,\pi = o(\mathfrak{S})}} g(\pi)\langle F(\pi), e_{\mathfrak{C}}(\phi, \pi)\rangle \, d_{\mathfrak{S}}\pi.$$

We fix P_0 satisfying the conditions of the sublemma of V.2.11. Taking V.3.9 into account, the approximation argument used in V.3.9(a) and the lines following (2), we see that for any continuous function g on $\{\pi \in \mathfrak{S}_{\leq T} | \mathrm{Re}\,\pi = o(\mathfrak{S})\}$, we have:

$$0 = \int_{\substack{\pi \in \overset{\circ}{\mathfrak{S}}_{\leq T} \\ \mathrm{Re}\,\pi = o(\mathfrak{S})}} P_0(\pi) g(\pi)\langle F(\pi), e_{\mathfrak{C}}(\phi, \pi)\rangle \, d_{\mathfrak{S}}\pi,$$

whence:

$$\langle F(\pi), e_{\mathfrak{C}}(\phi, \pi)\rangle = 0,$$

almost everywhere on the integration set, i.e. $F = 0$ in $\mathrm{Hilb}^{\mathfrak{F}}_{\mathfrak{C},T}$ and (1).

We obviously have:

(2) $$\forall \phi', \phi \in P^{R,\mathfrak{F}}_{\mathfrak{C},T'}, F_{\phi'} = F_\phi \Leftrightarrow e_{\mathfrak{C}}(\phi, \pi) = e_{\mathfrak{C}}(\phi', \pi).$$

Thus, (2) shows that there exists a unique linear map of $\mathrm{Hilb}_{\mathfrak{C},T}^{0\mathfrak{F}}$ into $L_{\mathfrak{C},T}^{2\mathfrak{F}}$, denoted by j^0, characterized by:

(3) $$j^0(F_\phi) = \mathrm{proj}_{\mathfrak{C},T}^{\mathfrak{F}}\theta_\phi, \quad \forall\phi \in P_{\mathfrak{C},T'}^{R,\mathfrak{F}}.$$

We show that that j^0 is an isometry.

Let $\phi', \phi \in P_{\mathfrak{C},T'}^{R,\mathfrak{F}}$. We have

$$\langle F_{\phi'}, F_\phi\rangle = \int_{\substack{\pi\in\mathring{\mathfrak{S}}_{\leq T}\\ \mathrm{Re}\,\pi=o(\mathfrak{S})}} \langle e_{\mathfrak{C}}(\phi',\pi), e_{\mathfrak{C}}(\phi,\pi)\rangle \, \mathrm{d}_{\mathfrak{S}}\pi$$

$$= \int_{\substack{\pi\in\mathring{\mathfrak{S}}_{\leq T}\\ \mathrm{Re}\,\pi=o(\mathfrak{S})}} r_{\mathfrak{C}}(\phi',\phi)(\pi)\, \mathrm{d}_{\mathfrak{S}}\pi$$

$$= m_{\mathfrak{C},T}(\phi',\phi) = \langle \theta_{\phi'}, \mathrm{proj}_{\mathfrak{C},T}^{\mathfrak{F}}\theta_\phi\rangle,$$

by V.3.2(3), which we have already proven. This is still equal to

$$\langle \mathrm{proj}_{\mathfrak{C},T}^{\mathfrak{F}}\theta_{\phi'}, \mathrm{proj}_{\mathfrak{C},T}^{\mathfrak{F}}\theta_\phi\rangle = \langle j^0F_{\phi'}, j^0F_\phi\rangle.$$

This gives the result.

Thus j^0 can be uniquely continued to an isometry of $\mathrm{Hilb}_{\mathfrak{C},T}^{\mathfrak{F}}$, denoted by j. The image of j contains the closure of the image of j^0 in $L_{\mathfrak{C},T}^{2\mathfrak{F}}$, which is all of $L_{\mathfrak{C},T}^{2\mathfrak{F}}$, by definition. This gives the proposition.

V.3.12. Proof of V.3.2

We fix $\mathfrak{C}$ as in V.3.4, supposing that V.3.2 is true for every $\mathfrak{C}' > \mathfrak{C}$. Let us summarize what we are going to prove in a lemma; we pick up the notation $\mathring{\mathfrak{S}}_{\leq T}$ of V.3.9.

Lemma *Let $\phi \in P_{\mathfrak{X}}^{R,\mathfrak{F}}$; for every $\pi \in \mathring{\mathfrak{S}}_{\leq T}$ of real part $o(\mathfrak{S})$, we have defined $F_\phi(\pi) = e_{\mathfrak{C}}(\phi,\pi) \in A_{\mathfrak{C},\pi}$.*

(i) *The map $\pi \in \mathring{\mathfrak{S}}_{\leq T} \mapsto e_{\mathfrak{C}}(\phi,\pi)$ can be homomorphically continued to every element π of $\mathfrak{S}_{\leq T}$ of real part in a small neighborhood of $o(\mathfrak{S})$ if and only if:*

$$\int_{\substack{\pi\in\mathring{\mathfrak{S}}_{\leq T}\\ \mathrm{Re}\,\pi=o(\mathfrak{S})}} e_{\mathfrak{C}}(\phi,\pi)\, \mathrm{d}_{\mathfrak{S}}\pi$$

exists as a function on $G(k)\backslash\mathbf{G}$. Then it is equal to:

$$\int_{\substack{\pi\in\mathring{\mathfrak{S}}_{\leq T}\\ \mathrm{Re}\,\pi=o(\mathfrak{S})}} e_{\mathfrak{C}}(\phi,\pi)\, \mathrm{d}_{\mathfrak{S}}\pi.$$

(ii) *Let* $\phi' \in P_{\mathfrak{X}}^{R,\mathfrak{F}}$; *the map* $\pi \in \mathring{\mathfrak{S}}_{\leq T} \mapsto r_{\mathfrak{C}}(\phi',\phi)(\pi)$, *can be holomorphically continued to every element* π *of* $\mathfrak{S}_{\leq T}$ *of real part in a small neighborhood of* $o(\mathfrak{S})$ *if and only if*

$$\int_{\substack{\pi\in\mathring{\mathfrak{S}}_{\leq T}\\ \mathrm{Re}\,\pi=o(\mathfrak{S})}} r_{\mathfrak{C}}(\phi',\phi)(\pi)\ \mathrm{d}_{\mathfrak{S}}\pi$$

exists. This integral then equals:

$$\int_{\substack{\pi\in\mathfrak{S}_{\leq T}\\ \mathrm{Re}\,\pi=o(\mathfrak{S})}} r_{\mathfrak{C}}(\phi',\phi)(\pi)\ \mathrm{d}_{\mathfrak{S}}\pi.$$

(iii) *We have* $F_\phi \in \mathrm{Hilb}_{\mathfrak{C},T}^{\mathfrak{F}}$, $\mathrm{proj}_{\mathfrak{C},T}^{\mathfrak{F}}\theta_\phi = jF_\phi$ *and the conditions of (i) and (ii) are satisfied, which proves V.3.2(3)(i) and V.3.2(4)(i). We also have V.3.2(3)(ii), V.3.2(4)(ii) and V.3.2(5)(i).*

(iv) *V.2.8 and V.3.2(2) are true.*

This will complete the proof of V.3.2.

(i) and (ii) have the same proof; we know, *a priori*, that if one of the functions we consider has singularities on $\{\pi \in \mathfrak{S}_{\leq T} | \mathrm{Re}\,\pi = o(\mathfrak{S})\}$, there exists a hyperplane of $\mathfrak{S}$ intersecting this set such that the function is singular at every general point of the intersection. If this is the case the function is not integrable on $\mathring{\mathfrak{S}}_{\leq T}$. On the other hand if the function has no singularities, it is evidently integrable on the compact set $\{\pi \in \mathfrak{S}_{\leq T} | \mathrm{Re}\,\pi = o(\mathfrak{S})\}$ and on every dense open subset of this set. This gives (i) and (ii).

(iii) Let $\phi \in P_{\mathfrak{X}}^{R,\mathfrak{F}}$ be as in the statement. Let $\phi' \in P_{\mathfrak{C},T'}^{R,\mathfrak{F}}$. We have:

$$\begin{aligned} m_{\mathfrak{C},T}(\phi,\phi') &= \langle \theta_\phi, \mathrm{proj}_{\mathfrak{C},T}^{\mathfrak{F}}\theta_{\phi'}\rangle \\ &= \langle \mathrm{proj}_{\mathfrak{C},T}^{\mathfrak{F}}\theta_\phi, \mathrm{proj}_{\mathfrak{C},T}^{\mathfrak{F}}\theta_{\phi'}\rangle \\ &\leq \|\mathrm{proj}_{\mathfrak{C},T}^{\mathfrak{F}}\theta_\phi\|\ \|\mathrm{proj}_{\mathfrak{C},T}^{\mathfrak{F}}\theta_{\phi'}\|. \end{aligned}$$

We recall the notation $\mathrm{Hilb}_{\mathfrak{C},T}^{\mathfrak{F}}$ and $\mathrm{Hilb}_{\mathfrak{C},T}^{0\mathfrak{F}}$ of V.3.11. Considering V.3.10(*), we know that for $\phi' \in P_{\mathfrak{C},T}^{\mathfrak{F}}$, $\mathrm{proj}_{\mathfrak{C},T}^{\mathfrak{F}}\theta_{\phi'} = j(F_{\phi'})$. What precedes proves that the linear map:

$$F_{\phi'} \in \mathrm{Hilb}_{\mathfrak{C},T}^{0\mathfrak{F}} \mapsto m_{\mathfrak{C},T}(\phi,\phi')$$

is continuous. Thus it can be continued to a continuous map on $\mathrm{Hilb}_{\mathfrak{C},T}^{\mathfrak{F}}$ and there exists $F^\phi \in \mathrm{Hilb}_{\mathfrak{C},T}^{\mathfrak{F}}$ such that this map coincides with:

$$F \in \mathrm{Hilb}_{\mathfrak{C},T}^{\mathfrak{F}} \mapsto \langle F^\phi, F\rangle.$$

In particular, we have:

$$\forall \phi' \in P^{R,\mathfrak{F}}_{\mathfrak{C},T'}, \quad m_{\mathfrak{C},T}(\phi,\phi') = \langle F^{\phi}, F_{\phi'}\rangle.$$

Hence:

$$\int\limits_{\substack{\pi\in\overset{\circ}{\mathfrak{S}}_{\leq T}\\ \mathrm{Re}\,\pi=o(\mathfrak{S})}} (r_{\mathfrak{C}}(\phi,\phi')(\pi) - \langle F^{\phi}(\pi), F_{\phi'}(\pi)\rangle)\, \mathrm{d}_{\mathfrak{S}}\pi = 0.$$

Now for $\pi \in \overset{\circ}{\mathfrak{S}}_{\leq T}$ of real part $o(\mathfrak{S})$, we have:

$$r_{\mathfrak{C}}(\phi,\phi')(\pi) = \langle e_{\mathfrak{C}}(\phi,\pi), e_{\mathfrak{C}}(\phi',\pi)\rangle_{A^{\mathfrak{F}}_{\mathfrak{C},\pi}},$$

$$\langle F^{\phi}(\pi), F_{\phi'}(\pi)\rangle = \langle F^{\phi}(\pi), e_{\mathfrak{C}}(\phi',\pi)\rangle_{A^{\mathfrak{F}}_{\mathfrak{C},\pi}},$$

i.e.

$$\int\limits_{\substack{\pi\in\overset{\circ}{\mathfrak{S}}_{\leq T}\\ \mathrm{Re}\,\pi=o(\mathfrak{S})}} \langle e_{\mathfrak{C}}(\phi,\pi) - F^{\phi}(\pi), e_{\mathfrak{C}}(\phi',\pi)\rangle_{A_{\mathfrak{C},\pi}}\, \mathrm{d}_{\mathfrak{S}}\pi = 0, \quad \forall \phi' \in P^{R,\mathfrak{F}}_{\mathfrak{C},T'}.$$

As in the proof of V.3.10(1), this implies that almost everywhere, we have:

$$\langle e_{\mathfrak{C}}(\phi,\pi) - F^{\phi}(\pi), e_{\mathfrak{C}}(\phi',\pi)\rangle_{A^{\mathfrak{F}}_{\mathfrak{C},\pi}} = 0, \quad \forall \phi' \in P^{R,\mathfrak{F}}_{\mathfrak{C},T'}.$$

Now $\{e_{\mathfrak{C}}(\phi',\pi), \phi' \in P^{R,\mathfrak{F}}_{\mathfrak{C},T'}\} = A_{\mathfrak{C},\pi}$ when $\pi \in \overset{\circ}{\mathfrak{S}}_{\leq T}$; we deduce from this that:

$$e_{\mathfrak{C}}(\phi,\pi) = F^{\phi}(\pi),$$

almost everywhere, i.e.

(1) $$F_{\phi} = F^{\phi} \in \mathrm{Hilb}^{\mathfrak{F}}_{\mathfrak{C},T}.$$

Moreover, for every $\phi \in P^{R,\mathfrak{F}}_{\mathfrak{X}}$ and every $\phi' \in P^{R,\mathfrak{F}}_{\mathfrak{C},T'}$, using the fact that j is an isometry for the third equality, we have:

$$\begin{aligned}\langle \theta_{\phi}, \mathrm{proj}^{\mathfrak{F}}_{\mathfrak{C},T}\theta_{\phi'}\rangle &= m_{\mathfrak{C},T}(\phi,\phi') = \langle F_{\phi}, F_{\phi'}\rangle \\ &= \langle jF_{\phi}, jF_{\phi'}\rangle = \langle jF_{\phi}, \mathrm{proj}^{\mathfrak{F}}_{\mathfrak{C},T}\theta_{\phi'}\rangle.\end{aligned}$$

Thus we also have:

$$\langle \mathrm{proj}^{\mathfrak{F}}_{\mathfrak{C},T}\theta_{\phi} - jF_{\phi}, \mathrm{proj}^{\mathfrak{F}}_{\mathfrak{C},T}\theta_{\phi'}\rangle = 0, \quad \forall \phi' \in P^{R,\mathfrak{F}}_{\mathfrak{C},T'}.$$

By definition of $L^{2\mathfrak{F}}_{\mathfrak{C},T}$, the set of elements $\mathrm{proj}^{\mathfrak{F}}_{\mathfrak{C},T}\theta_{\phi'}$ when ϕ' runs through $P^{\mathfrak{F}}_{\mathfrak{C},T}$ is dense in $L^{2\mathfrak{F}}_{\mathfrak{C},T}$, which implies that the element $\mathrm{proj}^{\mathfrak{F}}_{\mathfrak{C},T}\theta_{\phi} - jF_{\phi}$ of $L^{2\mathfrak{F}}_{\mathfrak{C},T}$ is zero, i.e.:

(2) $$\mathrm{proj}^{\mathfrak{F}}_{\mathfrak{C},T}\theta_{\phi} = jF_{\phi}.$$

Now let ϕ' be any element of $P^{R,\mathfrak{F}}_{\mathfrak{X}}$; again using the fact that j is an isometry, we see as a consequence of (2) that:

$$\langle \mathrm{proj}^{\mathfrak{F}}_{\mathfrak{C},T}\theta_{\phi'}, \mathrm{proj}^{\mathfrak{F}}_{\mathfrak{C},T}\theta_{\phi}\rangle = \langle jF_{\phi'}, jF_{\phi}\rangle$$
$$= \int\limits^{\circ}_{\substack{\pi\in\mathfrak{S}_{\leq T}\\ \mathrm{Re}\,\pi=o(\mathfrak{S})}} r_{\mathfrak{C}}(\phi',\phi)(\pi)\,\mathrm{d}_{\mathfrak{S}}\pi.$$

Thus the equivalent conditions of (ii) are satisfied and we have:

$$\forall \phi',\phi\in P^{R,\mathfrak{F}}_{\mathfrak{X}},\quad \langle\theta_{\phi'}, \mathrm{proj}^{\mathfrak{F}}_{\mathfrak{C},T}\theta_{\phi}\rangle = m_{\mathfrak{C},T}(\phi',\phi),$$

i.e. V.3.2(3).

Let us prove that the equivalent conditions of (i) are satisfied. Suppose the contrary, i.e. that there exists a hyperplane $\mathfrak{S}'$ of $\mathfrak{S}$ with $\mathfrak{S}'\in S^{\mathfrak{F}}_{\mathfrak{X}}$, intersecting $\{\pi\in\mathfrak{S}_{\leq T}|\mathrm{Re}\,\pi = o(\mathfrak{S})\}$ and singular for $\pi\mapsto e_{\mathfrak{C}}(\phi,\pi)$. Thus there exists a residue datum for $\mathfrak{S}'$ in $\mathfrak{S}$, denoted by $\mathrm{Res}^{\mathfrak{S}}_{\mathfrak{S}'}$ such that:

$$\mathrm{Res}^{\mathfrak{S}}_{\mathfrak{S}'}e_{\mathfrak{C}}(\phi,\pi)\not\equiv 0.$$

Let π_0 be a point of $\mathfrak{S}'$ where $\mathrm{Res}^{\mathfrak{S}}_{\mathfrak{S}'}e_{\mathfrak{C}}(\phi,\pi)$ is defined and non-zero. The value at π_0 is then a slowly increasing function on $G(k)\backslash\mathbf{G}$. By a direct calculation already performed, see V.3.7, it satisfies:

$$\langle\theta_{\phi'}, \mathrm{Res}^{\mathfrak{S}}_{\mathfrak{S}'}e_{\mathfrak{C}}(\phi,\pi_0)\rangle = \mathrm{Res}^{\mathfrak{S}}_{\mathfrak{S}'}r_{\mathfrak{C}}(\phi',\phi)(\pi_0).$$

This is zero, since we just saw that $r_{\mathfrak{C}}(\phi',\phi)$ is holomorphic at every point of $\mathfrak{S}_{\leq T}$ of real part $o(\mathfrak{S})$. As in V.3.7, we see that this implies the vanishing of $\mathrm{Res}^{\mathfrak{S}}_{\mathfrak{S}'}e_{\mathfrak{C}}(\phi,\pi_0)$, which gives the desired contradiction. Thus we obtain V.3.2(4)(i) and V.3.2(5)(i), by V.3.9(c).

As in V.3.7, we next prove that we have:

$$\mathrm{proj}^{\mathfrak{F}}_{\mathfrak{C},T}\theta_{\phi} = \int\limits_{\substack{\pi\in\mathfrak{S}_{\leq T}\\ \mathrm{Re}\,\pi=o(\mathfrak{S})}} e_{\mathfrak{C}}(\phi,\pi)\,\mathrm{d}_{\mathfrak{S}}\pi.$$

This gives V.3.2(4)(ii).

(iv) By (iii) and what was said at the beginning of V.3.4, the statements V.3.2, except V.3.2(2), are now proved for every $\mathfrak{S}$. Then V.3.5 shows that the hypotheses of Lemma V.2.9 are satisfied for every $\phi\in P^{R,\mathfrak{F}}_{\mathfrak{X}}$ and for every $T < T'/4^{\dim_{\mathbf{R}}\mathrm{Re}\,X^{\mathbf{G}}_{M}+1}$. For T as above and for every $\phi',\phi\in P^{R,\mathfrak{F}}_{\mathfrak{X}}$, we thus have:

$$\langle\theta_{\phi'}, q_T\theta_{\phi}\rangle = \sum_{\mathfrak{C}\in[S^{\mathfrak{F}}_{\mathfrak{X}}]} m_{\mathfrak{C},T}(\phi',\phi).$$

But we also just proved that the right-hand side is equal to:

$$\sum_{\mathfrak{C}\in[S^{\mathfrak{F}}_{\mathfrak{X}}]}\langle\theta_{\phi'}, \mathrm{proj}^{\mathfrak{F}}_{\mathfrak{C},T}\theta_{\phi}\rangle.$$

whence:

$$q_T = \sum_{\mathfrak{C}\in[S_{\mathfrak{X}}^{\mathfrak{F}}]} \mathrm{proj}_{\mathfrak{C},T}^{\mathfrak{F}}.$$

V.3.13. Decomposition of $q_T L^2(G(k)\backslash \mathbf{G})_{\mathfrak{X},\mathfrak{F}}$.

Theorem

(i) *Fix* $T \in \mathbb{R}_+^*$ *and recall the general notation of V.3.1. For every* $\mathfrak{S} \in S_{\mathfrak{X}}^{\mathfrak{F}}$, *there exists a residue datum, denoted by* $\mathrm{Res}_{\mathfrak{S}}^G$, *whose value on the functions of the form* $A(\phi', \phi)$, *where* $\phi', \phi \in P_{\mathfrak{X}}^{R,\mathfrak{F}}$ *is uniquely determined whenever* $\mathfrak{S}_{\leq T}$ *contains an element of real part* $o(\mathfrak{S})$ *and which is zero for almost all* $\mathfrak{S} \in S_{\mathfrak{X}}^{\mathfrak{F}}$, *in particular for* $\mathfrak{S} \in S_{\mathfrak{X}}^{\mathfrak{F}}$ *if* $\|o(\mathfrak{S})\| > R$, *such that:*

$$\pi \in \mathfrak{S}_{\mathfrak{C}} \longmapsto \sum_{w\in W(M)} (\mathrm{Res}_{w\mathfrak{S}_{\mathfrak{C}}}^G A(\phi', \phi))(w\pi)$$

is holomorphic at every point of $\mathfrak{S}_{\mathfrak{C},\leq T}$ *of real part* $o(\mathfrak{S})$ *and*

$$\langle \theta_{\phi'}, q_T\theta_\phi\rangle = \sum_{\mathfrak{C}\in[S_{\mathfrak{X}}^{\mathfrak{F}}]} |\mathrm{Norm}\,\mathfrak{S}_{\mathfrak{C}}|^{-1} \times$$

$$\int_{\substack{\pi\in\mathfrak{S}_{\mathfrak{C},\leq T}\\ \mathrm{Re}\,\pi=o(\mathfrak{S}_{\mathfrak{C}})}} \sum_{w\in W(M)} (\mathrm{Res}_{w\mathfrak{S}_{\mathfrak{C}}}^G A(\phi', \phi))(w\pi)\, \mathrm{d}_{\mathfrak{S}_{\mathfrak{C}}}\pi.$$

In particular the family of projections $\mathrm{proj}_{\mathfrak{C},T}^{\mathfrak{F}}$ *of V.3.3 is uniquely determined by the property V.3.2(4)(ii). We denote by* $\mathrm{Sing}_T^{G,\mathfrak{F}}$ *the set of elements* $\mathfrak{S}$ *of* $S_{\mathfrak{X}}^{\mathfrak{F}}$ *such that* $\{\pi \in \mathfrak{S}_{\leq T} | \mathrm{Re}\,\pi = o(\mathfrak{S})\} \neq \emptyset$ *and there exists* $\phi', \phi \in P_{\mathfrak{X}}^{R,\mathfrak{F}}$ *satisfying* $\mathrm{Res}_{\mathfrak{S}}^G A(\phi', \phi) \neq 0$.

(ii) *Let* $T_1 > T$ *be elements of* $\mathbb{R}_+^*$. *Then* $\mathrm{Sing}_T^{G,\mathfrak{F}}$ *is contained in* $\mathrm{Sing}_{T_1}^{G,\mathfrak{F}}$ *and consists of elements* $\mathfrak{S}$ *of* $\mathrm{Sing}_{T_1}^{G,\mathfrak{F}}$ *for which* $\{\pi \in \mathfrak{S}_{\leq T} | \mathrm{Re}\,\pi = o(\mathfrak{S})\}$ *is non-empty. We set* $\mathrm{Sing}^{G,\mathfrak{F}} := \bigcup_{T\in\mathbb{R}_+^*} \mathrm{Sing}_T^{G,\mathfrak{F}}$. *Let* $\mathfrak{S} \in \mathrm{Sing}^{G,\mathfrak{F}}$ *and let* $\phi', \phi \in P_{\mathfrak{X}}^{R,\mathfrak{F}}$; *then* $\mathrm{Res}_{\mathfrak{S}}^G A(\phi', \phi)$ *is independent of the choice of* $T \in \mathbb{R}_+^*$ *such that* $\mathfrak{S} \in \mathrm{Sing}_T^{G,\mathfrak{F}}$. *In particular* $\mathfrak{S} \in \mathrm{Sing}^{G,\mathfrak{F}}$ *if and only if* $\mathfrak{S} \in \mathrm{Sing}_{T_{\mathfrak{S}}}^{G,\mathfrak{F}}$ *where* $T_{\mathfrak{S}}$ *is the smallest element of* $\mathbb{R}_+^*$ *such that* $\{\pi \in \mathfrak{S}_{\leq T_{\mathfrak{S}}} | \mathrm{Re}\ \pi = o(\mathfrak{S})\}$ *is non-empty.*

(iii) The discrete spectrum of $L^2(G(k)\backslash\mathbf{G})_{\mathfrak{X},\mathfrak{F}}$ *is the closure of the vector space generated by the functions on* $G(k)\backslash\mathbf{G}$ *of the form:*

$$\sum_{w\in W(M)} \mathrm{Res}_{w\mathfrak{S}}^G E(\phi, w\pi)$$

or equivalently,

$$\mathrm{Res}_{\mathfrak{S}}^G E(\phi, \pi)$$

where $\mathfrak{S} \in \mathrm{Sing}\, G, \mathfrak{S}$, *of cuspidal data* $(M, \mathfrak{B})$ *varying in* $\mathfrak{X}$, *is reduced to* *a point and π is this point. Each of the above functions is an eigenvector for the action of the centre of the enveloping algebra at the archimedean places, and for the action of the Bernstein centre at the finite places.*

Remark

(i) For $\mathfrak{S} \in S_{\mathfrak{X}}^{\mathfrak{F}}$, let $\mathrm{Res}_{\mathfrak{S}}$ denote a residue datum and suppose that for every $\phi \in P_{\mathfrak{X}}^{R,\mathfrak{F}}$, we have the following equality of functions depending meromorphically on $\pi \in \mathfrak{S}$:

$$\mathrm{Res}_{\mathfrak{S}} E(\phi, \pi) = \mathrm{Res}_{\mathfrak{S}}^{G} E(\phi, \pi).$$

Then:

$$(\mathrm{Res}_{\mathfrak{S}} A(\phi', \phi))(\pi) = (\mathrm{Res}_{\mathfrak{S}}^{G} A(\phi', \phi))(\pi), \forall \phi', \phi \in P_{\mathfrak{X}}^{R,\mathfrak{F}}.$$

(ii) The value of the operators $\mathrm{Res}_{\mathfrak{S}}^{G}$ on the functions $A(\phi', \phi)$ where $\phi', \phi \in P_{\mathfrak{X}}^{R,\mathfrak{F}}$ for every $\mathfrak{S}$ is already uniquely determined by the property satisfied by these operators:

$$\forall T \in \mathbb{R}_{+}^{*}, \quad \forall \phi', \phi \in P_{\mathfrak{X}}^{R,\mathfrak{F}}, \ \langle \theta_{\phi'}, \theta_{\phi} \rangle =_T$$

$$\sum_{\mathfrak{S} \in S_{\mathfrak{X}}^{\mathfrak{F}}} |\mathrm{Norm}\ \mathfrak{S}|^{-1} \int_{\substack{\pi \in \mathfrak{S}_{\leq T} \\ \mathrm{Re}\,\pi = z(\mathfrak{S})}} \sum_{w \in \mathrm{Norm}\ \mathfrak{S}} (\mathrm{Res}_{\mathfrak{S}}^{G} A(\phi', \phi))(w\pi)\ d_{\mathfrak{S}}\pi,$$

where $\{z(\mathfrak{S})\}_{\mathfrak{S} \in \mathfrak{C}}$ is a family of points satisfying V.3.1(*).

Part (i) of this Remark is an immediate corollary of part (i) of the Theorem, taking into account the fact that:

$$\mathrm{Res}_{\mathfrak{S}} E(\phi, \pi) = \mathrm{Res}_{\mathfrak{S}}^{G} E(\phi, \pi) \Leftrightarrow$$

$$\forall \mathfrak{X}', \forall \phi' \in P_{\mathfrak{X}'}^{\mathfrak{F}} \ \langle \theta_{\phi'}, \mathrm{Res}_{\mathfrak{S}} E(\phi, \pi) \rangle = \langle \theta_{\phi'}, \mathrm{Res}_{\mathfrak{S}}^{G} E(\phi, \pi) \rangle,$$

where $\mathfrak{X}'$ runs through the set of equivalence classes of cuspidal data, and of the direct calculation:

$$\langle \theta_{\phi'}, \mathrm{Res}_{\mathfrak{S}} E(\phi, \pi) \rangle = \begin{cases} 0 & \text{if } \mathfrak{X}' \neq \mathfrak{X}, \\ (\mathrm{Res}_{\mathfrak{S}} A(\phi', \phi))(\pi), & \text{if } \mathfrak{X}' = \mathfrak{X}, \end{cases}$$

see V.3.7, lines preceding (1). Part (ii) of the Remark is an immediate consequence of V.3.2 and part (i) of the Theorem.

Let us prove the theorem.

(i) The existence results from V.2.8 and V.3.2; we must thus prove uniqueness. Suppose that for $i = 1, 2$ and for every $\mathfrak{S} \in S_{\mathfrak{X}}^{\mathfrak{F}}$ we have a residue datum, depending on i, denoted by $\mathrm{Res}_{\mathfrak{S}_i}^{G}$, whose set satisfies the hypotheses of (i). We fix a subfinite set of $[S_{\mathfrak{X}}^{\mathfrak{F}}]$ such that every element of $\mathrm{Sing}_{1,T}^{G,\mathfrak{F}} \cup \mathrm{Sing}_{2,T}^{G,\mathfrak{F}}$ belongs to one of the elements of this set; we totally order this set (see V.3.3) by supposing that for $\mathfrak{C}, \mathfrak{C}'$ in it, we have $\mathfrak{C} < \mathfrak{C}'$

if $\dim_{\mathbb{R}} \mathrm{Re}\, \mathfrak{S}^0_{\mathfrak{C}} < \dim_{\mathbb{R}} \mathrm{Re}\, \mathfrak{S}^0_{\mathfrak{C}'}$. Suppose there exists $\mathfrak{C} \in [S^{\mathfrak{F}}_{\mathfrak{X}}]$, $\mathfrak{S} \in \mathfrak{C}$ such that $\{\pi \in \mathfrak{S}_{\leq T} | \mathrm{Re}\,\pi = o(\mathfrak{S})\} \neq \emptyset$ and $\phi', \phi \in P^{R,\mathfrak{F}}_{\mathfrak{X}}$ such that:

(1) $$\mathrm{Res}^G_{\mathfrak{S},1} A(\phi', \phi) \not\equiv \mathrm{Res}^G_{\mathfrak{S},2} A(\phi', \phi).$$

Then $\mathfrak{C}$ is necessarily in the above set and we fix it to be maximal for the chosen order. For every $\mathfrak{C}' \geq \mathfrak{C}$ we fix an element $\mathfrak{S}_{\mathfrak{C}'}$ by supposing that $\mathfrak{S} = \mathfrak{S}_{\mathfrak{C}}$. We recall the notation $e_{\mathfrak{C}}(\phi, \pi)$ of V.3.2(4), but here we write $e^i_{\mathfrak{C}}(\phi, \pi)$ for $i = 1, 2$ to differentiate the two residue data; we also fix T' very large with respect to T and we recall the notation $P^{R,\mathfrak{F}}_{\mathfrak{C},T'}$ of V.3.3. The hypothesis and V.3.6(ii) ensure that there exists $\phi' \in P^{R,\mathfrak{F}}_{\mathfrak{X}}$ and $\phi \in P^{\mathfrak{F}}_{\mathfrak{C},T'}$ vanishing to a high order on every element of $\mathfrak{C}$ different from $\mathfrak{S} = \mathfrak{S}_{\mathfrak{C}}$ such that (1) is still satisfied. Fix ϕ' and ϕ with these vanishing properties and for $i = 1, 2$ and for every $\mathfrak{C}' \geq \mathfrak{C}$, write:

$$E^i_{\mathfrak{C}'}(\phi) = \int_{\substack{\pi \in \mathfrak{S}_{\mathfrak{C}'\leq T} \\ \mathrm{Re}\,\pi = o(\mathfrak{S}'_{\mathfrak{C}})}} e^i_{\mathfrak{C}'}(\phi, \pi)\, \mathrm{d}_{\mathfrak{S}_{\mathfrak{C}'}} \pi.$$

For $i = 1, 2$, direct calculation of the scalar product of $\sum_{\mathfrak{C}' \geq \mathfrak{C}} E^i_{\mathfrak{C}'}(\phi)$ against all the pseudo-Eisenstein series proves with V.3.2(2) that we have:

$$q_T \theta_\phi = \sum_{\mathfrak{C}' \geq \mathfrak{C}} E^i_{\mathfrak{C}'}(\phi).$$

Using the maximality of $\mathfrak{C}$, we obtain:

(2) $$E^1_{\mathfrak{C}}(\phi) = E^2_{\mathfrak{C}}(\phi).$$

We calculate the scalar product of this slowly increasing function against $\theta_{\phi'}$ using the vanishing properties of ϕ and we find:

$$\begin{aligned} \langle \theta_{\phi'}, E^i_{\mathfrak{C}}(\phi) \rangle &= |\mathrm{Norm}\, \mathfrak{S}|^{-1} \int_{\substack{\pi \in \mathfrak{S}_{\leq T} \\ \mathrm{Re}\,\pi = o(\mathfrak{S})}} \sum_{w \in \mathrm{Norm}\, \mathfrak{S}} (\mathrm{Res}^G_{\mathfrak{S},i} A(\phi', \phi))(w\pi)\, \mathrm{d}_{\mathfrak{S}} \pi \\ &= \int_{\substack{\pi \in \mathfrak{S}_{\leq T} \\ \mathrm{Re}\,\pi = o(\mathfrak{S})}} (\mathrm{Res}^G_{\mathfrak{S},i} A(\phi', \phi))(\pi)\, \mathrm{d}_{\mathfrak{S}} \pi; \end{aligned}$$

for the second equality, we use the fact that each term of the sum is holomorphic on the domain of integration, since $\phi \in P^{R,\mathfrak{F}}_{\mathfrak{C},T'}$.

Thus (2) can be translated by:

$$\int_{\substack{\pi \in \mathfrak{S}_{\leq T} \\ \mathrm{Re}\,\pi = o(\mathfrak{S})}} \left((\mathrm{Res}^G_{\mathfrak{S},1} - \mathrm{Res}^G_{\mathfrak{S},2}) A(\phi', \phi) \right)(\pi)\, \mathrm{d}_{\mathfrak{S}} \pi = 0,$$

while for ϕ', ϕ well chosen, (1) gives:

$$(\mathrm{Res}^G_{\mathfrak{S},1} - \mathrm{Res}^G_{\mathfrak{S},2})(A(\phi', \phi)) \not\equiv 0.$$

This is a contradiction by Lemma V.3.6. Now (i) is clear.

(ii) Fix T and T_1 as in the statement; we write $\mathrm{Res}^G_{\mathfrak{S},T}$ and $\mathrm{Res}^G_{\mathfrak{S},T_1}$ to differentiate these residue data. We also fix T' very large with respect to T and T_1, and residue data denoted by $\mathrm{Res}^G_{\mathfrak{S},T'}$ such that for every $\phi', \phi \in P_{\mathfrak{X}}^{R,\mathfrak{F}}$, we have:

$$\langle \theta_{\phi'}, \theta_{\phi} \rangle =_{T'}$$

$$\sum_{\mathfrak{C}\in[S_{\mathfrak{X}}^{\mathfrak{F}}]} |\mathrm{Norm}\, \mathfrak{S}_{\mathfrak{C}}|^{-1} \int_{\substack{\pi\in\mathfrak{S}_{\mathfrak{C},\leq T'} \\ \mathrm{Re}\,\pi=z(\mathfrak{S}_{\mathfrak{C}})}} \sum_{w\in W(M)} (\mathrm{Res}^G_{w\mathfrak{S}_{\mathfrak{C}},T'} A(\phi',\phi))(w\pi)\, d_{\mathfrak{S}_{\mathfrak{C}}}\pi.$$

It is a consequence of V.3.2 that:

$$\langle \theta_{\phi'}, q_T\theta_{\phi} \rangle =$$

$$\sum_{\mathfrak{C}\in[S_{\mathfrak{X}}^{\mathfrak{F}}]} |\mathrm{Norm}\, \mathfrak{S}_{\mathfrak{C}}|^{-1} \int_{\substack{\pi\in\mathfrak{S}_{\mathfrak{C},\leq T} \\ \mathrm{Re}\,\pi=o(\mathfrak{S}_{\mathfrak{C}})}} \sum_{w\in W(M)} (\mathrm{Res}^G_{w\mathfrak{S}_{\mathfrak{C}},T'} A(\phi',\phi))(w\pi)\, d_{\mathfrak{S}_{\mathfrak{C}}}\pi$$

$$\langle \theta_{\phi'}, q_{T_1}\theta_{\phi} \rangle =$$

$$\sum_{\mathfrak{C}\in[S_{\mathfrak{X}}^{\mathfrak{F}}]} |\mathrm{Norm}\, \mathfrak{S}_{\mathfrak{C}}|^{-1} \int_{\substack{\pi\in\mathfrak{S}_{\mathfrak{C},\leq T_1} \\ \mathrm{Re}\,\pi=o(\mathfrak{S}_{\mathfrak{C}})}} \sum_{w\in W(M)} (\mathrm{Res}^G_{w\mathfrak{S}_{\mathfrak{C}},T'} A(\phi',\phi))(w\pi)\, d_{\mathfrak{S}_{\mathfrak{C}}}\pi.$$

We suppose that $\mathfrak{S}$ is such that $\{\pi \in \mathfrak{S}_{\leq T}; \mathrm{Re}\,\pi = o(\mathfrak{S})\} \neq \emptyset$. The uniqueness proved in (i) indicates that we have the following equalities of meromorphic functions:

$$\forall \phi', \phi \in P_{\mathfrak{X}}^{R,\mathfrak{F}}, \forall \mathfrak{S} \in S_{\mathfrak{X}}^{\mathfrak{F}},\ \mathrm{Res}^G_{\mathfrak{S},T} A(\phi',\phi) = \mathrm{Res}^G_{\mathfrak{S},T'} A(\phi',\phi)$$
$$= \mathrm{Res}^G_{\mathfrak{S},T_1} A(\phi',\phi).$$

This gives (ii).

(iii) The assertion concerning the action of the Bernstein centre or the enveloping algebra is clear and implies the first part of (iii).

Corollary *Suppose that k is a number field.*

(i) *The assertions of V.2.2 are true for every T; in particular $e_{\mathfrak{C}}(\phi,\pi)$ is holomorphic at every point of $o(\mathfrak{S}) + \mathrm{Im}\,\mathfrak{S}$.*

(ii) *Let $\mathfrak{C} \in [S_{\mathfrak{X}}^{\mathfrak{F}}]$ and let $T < T_1$ be positive real numbers. Then $L^{2\mathfrak{F}}_{\mathfrak{C},T}$ is contained in $L^{2\mathfrak{F}}_{\mathfrak{C},T_1}$ and we denote by $L^2(G(k)\backslash \mathbf{G})_{\mathfrak{C},\mathfrak{F}}$ the closure of the space $\bigcup_{T\in\mathbf{R}^*_+} L^{2\mathfrak{F}}_{\mathfrak{C},T}$. It is a Hilbert space which admits a spectral decomposition exactly as in V.3.2(5) with $\mathfrak{S}_{\mathfrak{C},T}$ simply replaced by $\mathfrak{S}_{\mathfrak{C}}$.*

(iii) $$L^2(G(k)\backslash \mathbf{G})_{\mathfrak{X},\mathfrak{F}} = \hat{\bigoplus}_{\mathfrak{C}\in[S_{\mathfrak{X}}^{\mathfrak{F}}]} L^2(G(k)\backslash \mathbf{G})_{\mathfrak{C},\mathfrak{F}},$$

the sum being orthogonal.

The (i) of the corollary is an immediate consequence of assertion (ii) of the theorem.

With the notation of (ii), the inclusion of $L^{2\mathfrak{F}}_{\mathfrak{C},T_1}$ is a consequence of V.3.11 and the end of (ii) is an immediate consequence of V.3.2(5) by passage to the limit.

By assertion (iii), we note that:

$$L^2(G(k)\backslash\mathbf{G})_{\mathfrak{X},\mathfrak{F}} = \lim_{T\to\infty} q_T\ L^2(G(k)\backslash\mathbf{G})_{\mathfrak{X},\mathfrak{F}}$$

and (iii) is thus a consequence of (ii) and V.3.2(2).

Remark Let k be a function field and let $\mu \in X_G^{\mathbf{G}}$. Then:

$$\mathfrak{S} \in \operatorname{Sing}^{G,\mathfrak{F}} \Leftrightarrow \mathfrak{S} \otimes \mu \in \operatorname{Sing}^{G,\mathfrak{F}},$$

$\forall \phi', \phi \in P_{\mathfrak{X}}^{R,\mathfrak{F}}$,

$$(\operatorname{Res}_{\mathfrak{S}}^{G} A(\phi',\phi))(\pi) = \left(\operatorname{Res}_{\mathfrak{S}\otimes\mu}^{G} A(\phi',\phi)(\pi\otimes\mu^{-1})\right)_{\pi'=\pi\otimes\mu}$$

$$\operatorname{Res}_{\mathfrak{S}}^{G} E(\phi,\pi) = \left(\operatorname{Res}_{\mathfrak{S}\otimes\mu}^{G} E(\phi,\pi'\otimes\mu^{-1})\right)_{\pi'=\pi\otimes\mu}.$$

Let μ be as in the statement and let $\phi \in P_{\mathfrak{X}}^{R,\mathfrak{F}}$. The function $\pi \mapsto \mu\phi(\pi\otimes\mu^{-1})$ is still an element of $P_{\mathfrak{X}}^{R,\mathfrak{F}}$, denoted by $\tilde{T}_\mu\phi$ and we have the equality of Eisenstein series:

$$E(\tilde{T}_\mu\phi,\pi) = \mu E(\phi,\pi\otimes\mu^{-1}),$$

where in the second term, μ is considered as a function on $G(k)\backslash\mathbf{G}$. From II.1.11 we deduce the equality of pseudo-Eisenstein series:

$$\theta_{\tilde{T}_\mu\phi} = \mu\theta_\phi.$$

For every $\phi', \phi \in P_{\mathfrak{X}}^{R,\mathfrak{F}}$, we have:

$$\langle\theta_{\phi'},\theta_\phi\rangle = \langle\theta_{\tilde{T}_\mu\phi'},\theta_{\tilde{T}_\mu\phi'}\rangle$$

whence,

$$\langle\theta_{\phi'},\theta_\phi\rangle$$

$$= \sum_{\mathfrak{C}\in[S_{\mathfrak{X}}^{\mathfrak{F}}]} |\operatorname{Norm}\mathfrak{S}_{\mathfrak{C}}|^{-1} \int_{\substack{\pi\in\mathfrak{S}_{\mathfrak{C}}\\ \operatorname{Re}\pi=o(\mathfrak{S}_{\mathfrak{C}})}} \sum_{w\in W(M)} \left(\operatorname{Res}_{w\mathfrak{S}_{\mathfrak{C}}}^{G} A(\phi',\phi)\right)(w\pi)\ d_{\mathfrak{S}_{\mathfrak{C}}}\pi$$

$$= \sum_{\mathfrak{C}\in[S_{\mathfrak{X}}^{\mathfrak{F}}]} |\operatorname{Norm}\mathfrak{S}_{\mathfrak{C}}|^{-1} \int_{\substack{\pi\in\mathfrak{S}_{\mathfrak{C}}\\ \operatorname{Re}\pi=o(\mathfrak{S}_{\mathfrak{C}})}} \sum_{w\in W(M)} (\operatorname{Res}_{w\mathfrak{S}_{\mathfrak{C}}}^{G} A(\tilde{T}_\mu\phi',\tilde{T}_\mu\phi))(w\pi)\ d_{\mathfrak{S}_{\mathfrak{C}}}\pi.$$

We check directly from the definitions that for every $\mathfrak{S} \in S_{\mathfrak{X}}^{\mathfrak{F}}$, for every residue datum for $\mathfrak{S}$, denoted by $\operatorname{Res}_{\mathfrak{S}}$, and for every $\phi', \phi \in P_{\mathfrak{X}}^{R,\mathfrak{F}}$, we have:

$$(\operatorname{Res}_{\mathfrak{S}} A(\tilde{T}_\mu\phi',\tilde{T}_\mu\phi))(\pi) = \operatorname{Res}_{\mathfrak{S}} A(\phi',\phi)(\pi\otimes\mu^{-1}).$$

By a change of variables, we obtain for every $\mathfrak{C} \in [S_{\mathfrak{X}}^{\mathfrak{F}}]$:

$$\int_{\substack{\pi\in\mathfrak{S}_{\mathfrak{C}}\\ \mathrm{Re}\,\pi=o(\mathfrak{S}_{\mathfrak{C}})}} \sum_{w\in W(M)} (\mathrm{Res}^{G}_{w\mathfrak{S}_{\mathfrak{C}}} A(\tilde{T}_{\mu}\phi', \tilde{T}_{\mu}\phi))(w\pi)\, \mathrm{d}_{\mathfrak{S}_{\mathfrak{C}}}\pi =$$

$$\int_{\substack{\pi\in\mathfrak{S}_{\mathfrak{C}}\otimes\mu^{-1}\\ \mathrm{Re}\,\pi=o(\mathfrak{S}_{\mathfrak{C}}\otimes\mu^{-1})}} \sum_{w\in W(M)} \left(\mathrm{Res}^{G}_{w\mathfrak{S}_{\mathfrak{C}}} A(\phi', \phi)(\pi'\otimes\mu^{-1})\right)_{\pi'=w\pi\otimes\mu} \mathrm{d}_{\mathfrak{S}_{\mathfrak{C}}\otimes\mu^{-1}}\pi,$$

noting that $w\mu = \mu$. For every $\mathfrak{S} \in S_{\mathfrak{X}}^{\mathfrak{F}}$, the uniqueness of V.3.12(i) gives the following equality of meromorphic functions on $\mathfrak{S}\otimes\mu^{-1}$:

$$\forall\phi', \phi \in P_{\mathfrak{X}}^{R,\mathfrak{F}},\quad (\mathrm{Res}^{G}_{\mathfrak{S}} A(\phi', \phi)(\pi'\otimes\mu^{-1}))_{\pi'=\pi\otimes\mu}$$
$$= (\mathrm{Res}^{G}_{\mathfrak{S}\otimes\mu^{-1}} A(\phi', \phi))(\pi)$$

which gives the second formula of the remark and thus the first assertion, by the definition of $\mathrm{Sing}^{G,\mathfrak{F}}$. The third formula can be deduced from the second in the usual way by calculating the scalar product against every pseudo-Eisenstein series.

V.3.14. Decomposition of $L^2(G(k)\backslash \mathbf{G})_{\mathfrak{X}}$

Let $\mathfrak{F}$ and $\mathfrak{F}'$ be finite sets of $\mathbf{K}$-types; we suppose that $\mathfrak{F} \subset \mathfrak{F}'$. We have a certain number of obvious inclusions:

$$S_{\mathfrak{X}}^{\mathfrak{F}} \subset S_{\mathfrak{X}}^{\mathfrak{F}'},\quad [S_{\mathfrak{X}}^{\mathfrak{F}}] \subset [S_{\mathfrak{X}}^{\mathfrak{F}'}],$$
$$P_{\mathfrak{X}}^{\mathfrak{F}} \subset P_{\mathfrak{X}}^{\mathfrak{F}'},\quad P_{\mathfrak{X}}^{R,\mathfrak{F}} \subset P_{\mathfrak{X}}^{R,\mathfrak{F}'}.$$

Let $\phi \in P_{\mathfrak{X}}^{R,\mathfrak{F}}$; we have defined $e_{\mathfrak{C}}(\phi)$ without indicating $\mathfrak{F}$ in the notation; this is justified by V.3.12(i), i.e. $e_{\mathfrak{C}}(\phi)$ does not depend on whether ϕ is considered as an element of $P_{\mathfrak{X}}^{R,\mathfrak{F}}$ or $P_{\mathfrak{X}}^{R,\mathfrak{F}'}$. Also, we have $\mathrm{proj}^{\mathfrak{F}}\theta_{\phi} = \mathrm{proj}^{\mathfrak{F}'}\theta_{\phi}$ which allows us to suppress the exponent in the notation. Thus, we obtain the inclusion:

$$A_{\mathfrak{C},\pi}^{\mathfrak{F}} \subset A_{\mathfrak{C},\pi}^{\mathfrak{F}'},$$

for every $\pi \in \mathfrak{S}_{\mathfrak{C}}$ of real part $o(\mathfrak{S})$ and

$$L_{\mathfrak{C}}^{2\mathfrak{F}} \subset L_{\mathfrak{C}}^{2\mathfrak{F}'}.$$

It is then natural to set:

$$\mathrm{Sing}\,G := \bigcup_{\mathfrak{F}} \mathrm{Sing}\,G\mathfrak{F}.$$

We define $\mathrm{Res}^{G}_{\mathfrak{S}}$ on the set of functions $\pi \in \mathfrak{S} \mapsto A(\phi', \phi)(\pi)$ and $\pi \in \mathfrak{S} \mapsto E(\phi, \pi)$ for $\phi', \phi \in P_{\mathfrak{X}}^{R}$ in such a way that V.3.12(i) is satisfied for every choice of $\mathfrak{F}$. For $\mathfrak{C} \in [S_{\mathfrak{X}}]$, $\mathfrak{S} \in \mathfrak{C}$ and for every $\pi \in \mathfrak{S}$ of real part $o(\mathfrak{S})$, set:

$$A_{\mathfrak{C},\pi} := \bigcup_{\mathfrak{F}\ \mathfrak{C}\in[S_{\mathfrak{X}}^{\mathfrak{F}}]} A_{\mathfrak{C},\pi}^{\mathfrak{F}},$$

$$L^2(G(k)\backslash\mathbf{G}) := \bigcup_{\mathfrak{F},\ \mathfrak{C}\in[S_{\mathfrak{X}}^{\mathfrak{F}}]} L_{\mathfrak{C}}^{2\mathfrak{F}},$$

The union can be completed to obtain a Hilbert space. We easily check that $r_{\mathfrak{C}}(\ ,\ )(\pi)$ equips $A_{\mathfrak{C},\pi}$ with the structure of a pre-Hilbert space; let $\overline{A}_{\mathfrak{C},\pi}$ denote the completion. For every $w \in \operatorname{Norm}\mathfrak{S}$, the space $\overline{A}_{\mathfrak{C},\pi}$ can be identified with $\overline{A}_{\mathfrak{C},w\pi}$. We immediately obtain:

Corollary (i) *Let $\mathfrak{C} \in [S_{\mathfrak{X}}]$; fix $\mathfrak{S} \in \mathfrak{C}$. Then the map $\pi \in o(\mathfrak{S})+\operatorname{Im}\mathfrak{S} \mapsto \overline{A}_{\mathfrak{C},\pi}$ is a Hilbertian stack and $L^2(G(k)\backslash\mathbf{G})_{\mathfrak{C}}$ is isomorphic to the following Hilbert space, denoted by* $\operatorname{Hilb}_{\mathfrak{C}}$*:*

$$\left\{\begin{array}{l}\text{The measurable functions } F \text{ on } o(\mathfrak{S})+\operatorname{Im}\mathfrak{S} \text{ such that:}\\ \text{for almost every } \pi \in o(\mathfrak{S})+\operatorname{Im}\mathfrak{S},\\ F(\pi)\in\overline{A}_{\mathfrak{C},\pi} \text{ and } F(w\pi)=F(\pi)\ \forall w\in\operatorname{Norm}\mathfrak{S},\\ \displaystyle\int_{\pi\in o(\mathfrak{S})+\operatorname{Im}\mathfrak{S}} \|F(\pi)\|^2 < +\infty, \text{ where the norm is that of } \overline{A}_{\mathfrak{C},\pi}.\end{array}\right\}$$

This isomorphism, denoted by $j_{\mathfrak{C}}$, is characterized by the equalities, for every $\phi \in P_{\mathfrak{X}}$:

$$\operatorname{proj}_{L^2(G(k)\backslash\mathbf{G})_{\mathfrak{C}}}\theta_\phi = j_{\mathfrak{C}}F_\phi,$$

where F_ϕ is defined as in V.3.11. Moreover $L^2(G(k)\backslash\mathbf{G})_{\mathfrak{C}}$ and $\overline{A}_{\mathfrak{C},\pi}$ are naturally equipped with a **G***-action; so there is a* **G***-action on* $\operatorname{Hilb}_{\mathfrak{C}}$ *and the above isomorphism is* **G***-equivariant.*

(ii) *$L^2(G(k)\backslash\mathbf{G})_{\mathfrak{X}}$ is the completion of the orthogonal sum:*

$$\bigoplus_{\mathfrak{C}\in[S_{\mathfrak{X}}]} L^2(G(k)\backslash\mathbf{G})_{\mathfrak{C}}.$$

V.3.15. Adjunction formula for residues of intertwining operators

We still fix $\mathfrak{X}$ and $\mathfrak{S} \in S_{\mathfrak{X}}^{\mathfrak{F}}$; we denote by $(M,\mathfrak{P})$ the cuspidal datum attached to $\mathfrak{S}$. We fix a large $T \in \mathbb{R}_+^*$ such that $\{\pi \in \mathfrak{S}_{\le T} | \operatorname{Re}\ \pi = o(\mathfrak{S})\} \ne \emptyset$. We set:

$$\operatorname{Fix}\mathfrak{S} := \{w \in W(M) | w((M,\mathfrak{P})) = (M,\mathfrak{P}) \text{ and } \forall\pi\in\mathfrak{S}, w\pi=\pi\}.$$

We also fix for every $\mathfrak{S}' \in S_{\mathfrak{X}}^{\mathfrak{F}}$ a residue datum such that that V.3.12(i) is satisfied; then we have symbolically (but ambiguously, see below):

$$\forall w \in W(M),\ (\mathrm{Res}^G_{\mathfrak{S}} \sum_{\sigma\in \mathrm{Fix}\,\mathfrak{S}} M(w\sigma,\pi))^*$$

$$= \left(\mathrm{Res}^G_{-w\overline{\mathfrak{S}}}\Big(\sum_{\sigma\in \mathrm{Fix}\,\mathfrak{S}} M((w\sigma)^{-1},\pi')\right)_{\pi'=-w\overline{\pi}},$$

which has the following meaning.

Proposition *For every $\phi', \phi \in P^{R,\mathfrak{F}}_{\mathfrak{X}}$ and for every $w \in W(M)$ we have the following equality of meromorphic functions on $\mathfrak{S}$:*

$$\mathrm{Res}^G_{\mathfrak{S}}\langle \sum_{\sigma\in \mathrm{Fix}\,\mathfrak{S}} \phi'(-w\sigma\overline{\pi}), M(w\sigma,\pi)\phi(\pi)\rangle =$$

$$\left(\mathrm{Res}^G_{-w\overline{\mathfrak{S}}}\langle \sum_{\sigma\in \mathrm{Fix}\,\mathfrak{S}} M((w\sigma)^{-1},\pi')\phi'(\pi'),\phi(-(w\sigma)^{-1}\overline{\pi}'))\rangle\right)_{\pi'=-w\overline{\pi}}.$$

In particular the first term is zero if $-w\overline{\mathfrak{S}} \notin \mathrm{Sing}^{G,\mathfrak{F}}$.

We denote by $\mathfrak{C}$ the element of $[S^{\mathfrak{F}}_{\mathfrak{X}}]$ containing $\mathfrak{S}$. It suffices to prove the equality at a general point of $\mathfrak{S}_{\leq T}$ (where $T \in \mathbb{R}^*_+$ is fixed and large) of real part $o(\mathfrak{S})$. We still fix $T' \in \mathbb{R}^*_+$ large with respect to T and we recall the notation $P^{R,\mathfrak{F}}_{\mathfrak{C},T'}$ of V.3.3. It suffices to prove the proposition under the supplementary hypotheses (see V.3.6):

$\phi \in P^{R,\mathfrak{F}}_{\mathfrak{C},T'}$ and ϕ vanishes to a high order on every element of $\mathfrak{C}$ different from $\mathfrak{S}$,

$\phi' \in P^{R,\mathfrak{F}}_{\mathfrak{C},T'}$ (here we use the fact that $-w\overline{\mathfrak{S}} \in \mathfrak{C}$, see V.3.5(b)) and ϕ' vanishes to a high order on all elements of $\mathfrak{C}$ different from $-w\overline{\mathfrak{S}}$.

Under these vanishing hypotheses, we have:

$$r_{\mathfrak{C}}(\phi',\phi)(\pi) = |\mathrm{Norm}\,\mathfrak{S}|^{-1} \sum_{\tau\in \mathrm{Norm}\,\mathfrak{S}} \left(\mathrm{Res}^G_{\mathfrak{S}} A(\phi',\phi)(\pi')\right)_{\pi'=\tau\pi}$$

and for every $\pi \in \mathfrak{S}$, using the product formula for the second equality:

$$(\mathrm{Res}^G_{\mathfrak{S}} A(\phi',\phi))(\pi)$$

$$= \sum_{\substack{\sigma\in \mathrm{Stab}(M,\mathfrak{P})\\ \sigma\mathfrak{S}=w\mathfrak{S}}} \mathrm{Res}^G_{\mathfrak{S}}\langle M(\sigma^{-1},-\sigma\overline{\pi})\phi'(-\sigma\overline{\pi}),\phi(\pi)\rangle$$

$$= \sum_{\sigma\in \mathrm{Norm}\,\mathfrak{S}} \mathrm{Res}^G_{\mathfrak{S}}\langle M(\sigma^{-1},-\sigma\overline{\pi})M(w^{-1},-w\sigma\overline{\pi})\phi'(-w\sigma\overline{\pi}),\phi(\pi)\rangle.$$

Thus

$$|\mathrm{Norm}\,\mathfrak{S}|\, r_{\mathfrak{C}}(\phi',\phi)(\pi) = \sum_{\tau,\sigma\in \mathrm{Norm}\,\mathfrak{S}} (\mathrm{Res}^G_{\mathfrak{S}} \tag{1}$$

$$\langle M(\sigma^{-1},-\sigma\overline{\pi'})M(w^{-1},-w\sigma\overline{\pi'})\phi'(-w\sigma\overline{\pi'}),\phi(\pi')\rangle)_{\pi'=\tau\pi}.$$

We saw (see V.3.9(a)) that for every $\pi \in \mathfrak{S}$, we have:

$$r_{\mathfrak{C}}(\phi,\phi')(\pi) = \overline{r_{\mathfrak{C}}(\phi,\phi')(-\overline{\pi})}. \tag{2}$$

To express $r_{\mathfrak{C}}(\phi,\phi')(-\overline{\pi})$, we take $-w\overline{\mathfrak{S}}$ as a base point for $\mathfrak{C}$ and by a calculation similar to the preceding one, we find:

$$(3)\qquad \begin{aligned} &|\mathrm{Norm}\,\mathfrak{S}|\; r_{\mathfrak{C}}(\phi,\phi')(-\overline{\pi}) \sum_{\sigma,\tau\in\mathrm{Norm}\,\mathfrak{S}} (\mathrm{Res}^{G}_{-w\overline{\mathfrak{S}}} \\ &\langle M(w,-w^{-1}\overline{\pi'})M(\sigma^{-1},\sigma w^{-1}(-\overline{\pi'}))\phi(\sigma w^{-1}(-\overline{\pi'})),\phi'(\pi')\rangle)_{\pi'=-w\tau\overline{\pi}}. \end{aligned}$$

We want to prove the proposition at a π_0-general point. We can thus still suppose that ϕ vanishes to a high order at every point of $(\mathrm{Norm}\,\mathfrak{S})\pi_0$ different from π_0 and that this is also true of ϕ' at every point of $-w(\mathrm{Norm}\,\mathfrak{S})\overline{\pi}_0$ different from $-w\overline{\pi}_0$. With this hypothesis in (1) and (3), at the point π_0, the sums only run over the stabiliser of $\mathfrak{S}$ and we can replace $\tau\pi_0$ by π_0 in (1) and (3) for every $\tau \in \mathrm{Fix}\,\mathfrak{S}$. Putting together (1), (2) and (3), we find:

$$\begin{aligned} &\sum_{\sigma\in\mathrm{Fix}\,\mathfrak{S}} \mathrm{Res}^{G}_{\mathfrak{S}}\langle M(\sigma^{-1},-\sigma\overline{\pi}_0)M(w^{-1},-w\sigma\overline{\pi}_0)\phi'(-w\sigma\overline{\pi}_0),\phi(\pi_0)\rangle = \\ &\sum_{\sigma\in\mathrm{Fix}\,\mathfrak{S}} \overline{\mathrm{Res}^{G}_{-w\overline{\mathfrak{S}}}\langle M(w,-w^{-1}\overline{\pi'})M(\sigma,-(w\sigma)^{-1}\overline{\pi'})\phi(-(w\sigma)^{-1}\overline{\pi'}),} \\ &\qquad\qquad \overline{\phi'(\pi')\rangle_{\pi'=-w\overline{\pi}_0}}. \end{aligned}$$

We obtain the desired formula using the adjunction properties of intertwining operators, see II.1.8, using the fact that, by hypothesis, the residue data have real coefficients; see V.1.3.

V.3.16. Cuspidal exponents of residues of Eisenstein series

Corollary *Let* $\phi \in P^{R,\mathfrak{F}}_{\mathfrak{X}}$, $\mathfrak{S} \in S^{\mathfrak{F}}_{\mathfrak{X}}$ *and* $\pi_0 \in \mathfrak{S}$. *We adopt the notation* $\mathrm{Res}^{G}_{\mathfrak{S}}$ *of V.3.12 and denote by* $(M,\mathfrak{P})$ *the cuspidal datum attached to* $\mathfrak{S}$. *We suppose that* $\mathrm{Res}^{G}_{\mathfrak{S}}\, E(\phi,\pi)$ *is not identically* 0 *and is holomorphic at* π_0. *Then the set of cuspidal components of* $\mathrm{Res}^{G}_{\mathfrak{S}}\, E(\phi,\pi)$ *is in the set:*

$$\{w\pi_0 | w \in W(M), -w\overline{\mathfrak{S}} \in \mathrm{Sing}^{G,\mathfrak{F}}\}.$$

By II.1.12, we must calculate $\langle \theta_{\phi'}, \mathrm{Res}^{G}_{\mathfrak{S}} E(\phi,\pi_0)\rangle$ for every pseudo-Eisenstein series. This calculation is done as in V.3.7: it is equal to

0 if $\phi' \notin P_{\mathfrak{X}}^{\mathfrak{F}}$; otherwise it is equal to:

$$\begin{aligned}
&(\mathrm{Res}_{\mathfrak{S}}^{G} A(\phi',\phi))(\pi_0) \\
&\quad = \sum_{w\in W(M)} \mathrm{Res}_{\mathfrak{S}}^{G} \langle M(w^{-1}, -w\overline{\pi}_0)\phi'(-w\overline{\pi}_0), \phi(\pi_0)\rangle \\
&\quad = \sum_{\tilde{w}\in W(M)/\mathrm{Fix}\,\mathfrak{S}} \Bigg(\sum_{\sigma\in \mathrm{Fix}\,\mathfrak{S}} \mathrm{Res}_{\mathfrak{S}}^{G} \langle M((\tilde{w}\sigma)^{-1}, -\tilde{w}\sigma\overline{\pi}_0) \\
&\qquad\qquad \phi'(-\tilde{w}\sigma\overline{\pi}_0), \phi(\pi_0)\rangle \Bigg).
\end{aligned}$$

By V.3.15, for w fixed, the term in parentheses is identically zero on $\mathfrak{S}$, thus at π_0, if $-w\overline{\mathfrak{S}} \notin \mathrm{Sing}^{G,\mathfrak{F}}$. Thus, V.3.16 results from the Remark in II.1.12.

V.3.17. Generalisation: decomposition of $L^2(U_L(\mathbb{A})L(k)\backslash \mathbf{G})_{\xi_L}$

Let L be a standard Levi subgroup of G. We denote by U_L the radical unipotent of the standard parabolic subgroup Q of Levi L. We also fix a unitary character ξ_L of the centre of $\mathbf{L}$. We denote by

$$L^2(U_L(\mathbb{A})L(k)\backslash \mathbf{G})_{\xi_L}$$

the set of functions on $U_L(\mathbb{A})L(k)\backslash \mathbf{G}$ on which $Z_{\mathbf{L}}$ acts on the left via the character $\xi_L \cdot {}^{\rho_L}_{L}$ and which are square integrable modulo $Z_{\mathbf{L}}$, i.e.
$f \in L^2(U_L(\mathbb{A})L(k)\backslash \mathbf{G})_{\xi_L} \Leftrightarrow$

$$\left\{ f(zg) = \xi_L(z) z^{\rho_L} f(g), \forall z \in Z_{\mathbf{L}}, \forall g \in \mathbf{G}, \int_{U_L(\mathbb{A})L(k)Z_{\mathbf{L}}\backslash \mathbf{G}} f(g)\overline{f(g)}\, dg < \infty. \right\}$$

We can generalise the results of this chapter to obtain the spectral decomposition of this space of functions. We begin by generalising the notion of equivalence class of cuspidal data, of Paley-Wiener functions and of their associated pseudo-Eisenstein series. A cuspidal datum for L relative to ξ_L is a pair $(M, \mathfrak{P}_L)$ where M is a standard Levi of L and $\mathfrak{P}_L$ an orbit under X_M^L of cuspidal representations of M, the restriction of whose central character to Z_L is ξ_L. We define the set of Paley-Wiener functions on $X_M^{\mathbf{L}}$, denoted by $P(X_M^{\mathbf{L}})$, to be the set of Fourier transforms of smooth functions with compact support on $\mathbf{M}/\mathbf{M}^1 Z_{\mathbf{L}}$. We denote by U the radical unipotent of the standard parabolic Levi subgroup M. The Paley-Wiener functions $P_{(M,\mathfrak{P}_L)}$ are functions f on $\mathfrak{P}_L$ with values in the functions on $\mathbf{G}$ satisfying:

$$\forall \pi \in \mathfrak{P}_L,\ \ f(\pi) \in A(U(\mathbb{A})M(k)\backslash \mathbf{G})_{\pi},$$

and π_0 being fixed in $\mathfrak{P}_L$, the function f' on $\mathbf{G} \times X_M^{\mathbf{L}}$ defined by:

$$f'(g, \lambda) = m_P(g)^{-\lambda} f(\pi_0 \otimes \lambda)(g)$$

is an element of $A(U(\mathbb{A})M(k)\backslash\mathbf{G})_{\pi_0} \otimes P(X_M^{\mathbf{L}})$. We denote by $\mathfrak{P}$ the orbit of $\mathfrak{P}_L$ under $X_M^{\mathbf{G}}$; it is easy to check that $P_{M,\mathfrak{P}_L}$ is the set of restrictions to $X_M^{\mathbf{L}}$ of the elements of $P_{M,\mathfrak{P}}$ defined in II.1.2. For a large positive real number R, we define $P_{M,\mathfrak{P}_L}^R$ similarly. For $\phi \in P_{M,\mathfrak{P}_L}^R$, we define the pseudo-Eisenstein series:

$$\theta_\phi(g) = \sum_{\gamma \in P(k)\cap L(k)\backslash L(k)} \epsilon F(\phi)(\gamma g), \forall g \in \mathbf{G},$$

where $\epsilon F(\phi)$ is defined as in II.1.3; the proof of II.1.10 can be applied word for word to show that the above series converges absolutely and uniformly on every compact set of $\mathbf{G}$ and defines an element of $L^2(U_L(\mathbb{A})L(k)\backslash\mathbf{G})_{\xi_L}$. If we suppose that $\phi \in P_{M,\mathfrak{P}_L}$, we check similarly that θ_ϕ is rapidly decreasing in the Siegel domain S^Q of I.2.1; Q is the standard parabolic subgroup of G of Levi L.

We denote by $\mathfrak{X}_L$ the equivalence class of $(M, \mathfrak{P}_L)$, i.e. the set of pairs $(wMw^{-1}, w\mathfrak{P}_L)$ where $w \in W_L(M)$. We define $L^2(U(\mathbb{A})L(k)\backslash\mathbf{G})_{\mathfrak{X}_L}$ to be the closure in $L^2(U_L(\mathbb{A})L(k)\backslash\mathbf{G})_{\xi_L}$ of the vector space consisting of elements θ_ϕ where $(M', \mathfrak{P}'_L)$ runs through $\mathfrak{X}_L$. We obtain as in II.2.4 the first decomposition of $L^2(U_L(\mathbb{A})L(k)\backslash\mathbf{G})_{\xi_L}$:

Theorem (i) *Let $(M, \mathfrak{P}_L)$ and $(M', \mathfrak{P}'_L)$ be cuspidal data, $\phi \in P_{M,\mathfrak{P}_L}^R$ and $\phi' \in P_{M',\mathfrak{P}'_L}$. We also fix a very positive element λ_0 of $X_M^{\mathbf{L}}$ and we suppose that R is very large with respect to λ_0. Then:*

$$\langle\theta_{\phi'}, \theta_\phi\rangle = \begin{cases} 0 & \text{if } (M, \mathfrak{P}_L) \not\sim (M', \mathfrak{P}'_L) \\ \int_{\pi\in\mathfrak{P}_L, \mathrm{Re}\,\pi=\lambda_0} A^L(\phi', \phi)(\pi)\, d\pi, & \text{otherwise} \end{cases}$$

where, if $W_L((M, \mathfrak{P}_L), (M', \mathfrak{P}'_L))$ denotes the set of elements of $W_L(M)$ taking $(M, \mathfrak{P}_L)$ into $(M', \mathfrak{P}'_L)$, then:

$$A^L(\phi', \phi) := \sum_{w\in W_L\left((M,\mathfrak{P}_L),(M',\mathfrak{P}'_L)\right)} \langle M(w^{-1}, -w\overline{\pi})\phi'(-w\overline{\pi}), \phi(\pi)\rangle.$$

(ii)

$$L^2(U_L(\mathbb{A})L(k)\backslash\mathbf{G})_{\xi_L} = \hat{\oplus} L^2(U_L(\mathbb{A})L(k)\backslash\mathbf{G})_{\mathfrak{X}_L},$$

where $\mathfrak{X}_L$ runs through the set of equivalence classes of cuspidal data.

Fix $\mathfrak{X}_L$. We can decompose $L^2(U_L(\mathbb{A})L(k)\backslash\mathbf{G})_{\mathfrak{X}_L}$ in a finer way. We fix $(M, \mathfrak{P}_L) \in \mathfrak{X}_L$ and show that there exists a set of affine subspaces of $\cup_{w\in W_L(M)} w\mathfrak{P}_L$, with vector part defined over $\mathbb{R}$, equipped with residue data, denoted by $\mathrm{Res}_{\mathfrak{S}_L}^L$ satisfying the theorem below. We say that two affine spaces $\mathfrak{S}_L$ and $\mathfrak{S}'_L$ are conjugate if they are conjugate by an element of W_L. Let

$$\phi = \bigoplus(\phi_{M',\mathfrak{P}'_L}) \in \bigoplus_{(M',\mathfrak{P}'_L)\in\mathfrak{X}_L} P_{M',\mathfrak{P}'_L}^R;$$

we define

$$\theta_\phi := \sum_{(M',\mathfrak{P}'_L)\in\mathfrak{X}_L} \theta_{\phi_{M',\mathfrak{P}'_L}},$$

the sum being in $L^2(U_L(\mathbb{A})L(k)\backslash\mathbf{G})_{\mathfrak{X}_L}$.

Theorem *Let $\phi, \phi' \in \oplus_{(M',\mathfrak{P}'_L)\in\mathfrak{X}_L} P^R_{M',\mathfrak{P}'_L}$. If k is a number field, define λ_π to be the restriction of the central character of π to $A^{\mathbf{L}}_M$ and we have:*

$$\langle\theta_{\phi'}, \theta_\phi\rangle = \lim_{T\to\infty} \sum_{\mathfrak{C}_L} \int_{\substack{\pi\in\mathfrak{S}_{\mathfrak{C}_L} \\ \mathrm{Re}\,\pi=o(\mathfrak{S}_{\mathfrak{C}_L}) \\ (\lambda_\pi,\lambda_\pi)<T}} m_{\mathfrak{C}_L}(\phi',\phi)(\pi)\, d\pi,$$

where

$$m_{\mathfrak{C}_L}(\phi',\phi)(\pi) = |\mathrm{Norm}_{W_L(M)}\mathfrak{S}_{\mathfrak{C}_L}|^{-1} \sum_{w\in W_L(M)} (\mathrm{Res}^L_{w\mathfrak{S}_{\mathfrak{C}_L}} A^L(\phi',\phi))(w\pi)$$

where $\mathfrak{C}_L$ runs through the conjugacy classes of affine subspaces mentioned above and $\mathfrak{S}_{\mathfrak{C}_L}$ is an fixed element of $\mathfrak{C}_L$ contained in $\mathfrak{P}_L$. The limit is taken in the L^2 sense and for ϕ', ϕ and T fixed, almost all terms in the sum are zero. If k is a function field, we have more simply:

$$\langle\theta_{\phi'}, \theta_\phi\rangle = \sum_{\mathfrak{C}_L} \int_{\substack{\pi\in\mathfrak{S}_{\mathfrak{C}_L} \\ \mathrm{Re}\,\pi=o(\mathfrak{S}_{\mathfrak{C}_L})}} m_{\mathfrak{C}_L}(\phi',\phi)(\pi)\, d\pi,$$

with the preceding notation.

We denote by Sing^L the set of affine spaces for which there exist ϕ and ϕ' as above such that $\mathrm{Res}^L_{\mathfrak{S}_L} A^L(\phi',\phi) \not\equiv 0$. We deduce in particular from this theorem that the automorphic forms contained in

$$L^2(U_L(\mathbb{A})L(k)\backslash\mathbf{G})_{\xi_L}$$

are the residues of the Eisenstein series for the group L, which are linear combinations of elements:

$$\mathrm{Res}^L_{\mathfrak{S}_L} E^L(\phi,\pi),$$

where E^L is the Eisenstein series for L, $\mathfrak{S}_L$ is an element of $\mathrm{Sing}\,L$ which is reduced to a point, π is this point and $\phi \in P_{M,\mathfrak{P}_L}$ if $(M,\mathfrak{P}_L)$ is the cuspidal datum of $\mathfrak{S}_L$. The scalar product of two such elements $\mathrm{Res}^L_{\mathfrak{S}'_L}(E^L(\phi',\pi'))$ and $\mathrm{Res}^L_{\mathfrak{S}_L}(E^L(\phi,\pi))$ is zero if $\mathfrak{S}_L$ and $\mathfrak{S}'_L$ are not in the same conjugacy class and is otherwise equal to:

$$\sum \mathrm{Res}_{\mathfrak{S}_L}(M(w^{-1},-w\overline{\pi})\phi'(-w\overline{\pi}),\phi(\pi)),$$

where the sum runs over the elements w such that $-w\mathfrak{S}_L = \overline{\mathfrak{S}'_L}$. We note here the fact that every element $\mathfrak{S}_L$ of Sing^L is conjugate to $-\overline{\mathfrak{S}_L}$; see V.3.6(b).

VI

Spectral Decomposition via the Discrete Levi Spectrum

In Chapter II, we gave an orthogonal decomposition

$$L^2(G(k)\backslash \mathbf{G})_\xi = \hat{\oplus}_{\mathfrak{X}} L^2(G(k)\backslash \mathbf{G})_{\mathfrak{X}},$$

where $\mathfrak{X}$ runs through the set of equivalence classes of cuspidal data $(M, \mathfrak{P})$ where M is a standard Levi of G and $\mathfrak{P}$ is an orbit under $X_M^{\mathbf{G}}$ in the set of automorphic cuspidal representations of the inverse image $\mathbf{M}$ of $M(\mathbb{A})$ in $\mathbf{G}$, whose central character restricted to the centre of $\mathbf{G}$ is ξ. In IV, for fixed $\mathfrak{X}$, we decomposed $L^2(G(k)\backslash \mathbf{G})_{\mathfrak{X}}$ in a finer way using residues of Eisenstein series. Denoting by $[S_{\mathfrak{X}}]$ the conjugacy classes of singular spaces for the Eisenstein series (see V.1.2 and V.3.1 for a precise definition), we obtained

$$L^2(G(k)\backslash \mathbf{G})_{\mathfrak{X}} = \hat{\oplus}_{\mathfrak{C}\in[S_{\mathfrak{X}}]} L^2(G(k)\backslash \mathbf{G})_{\mathfrak{C}}.$$

We gave a relatively technical spectral decomposition of each of these spaces $L^2(G(k)\backslash \mathbf{G})_{\mathfrak{C}}$ (which is basically a tool for proofs). In this chapter we will associate a discrete parameter with each class $\mathfrak{C}$, i.e. an association class of pairs (L, δ) where L is a standard Levi of G and δ is a representation of the discrete spectrum of the inverse image $\mathbf{L}$ of $L(\mathbb{A})$ in $\mathbf{G}$. (In general δ is of infinite length but δ can be calculated explicitly via residues of Eisenstein series for $\mathbf{L}$; in particular the centre of the enveloping algebra acts on δ via a character at the infinite places and the Bernstein centre does so at the finite places; δ is admissible and probably all the subrepresentations of δ are in the same Arthur packet, if k is a number field.) Then we show that $L^2(G(k)\backslash \mathbf{G})_{\mathfrak{C}}$ is a Hilbertian integral on the imaginary axis of $X_L^{\mathbf{G}}$ of Eisenstein series constructed from automorphic forms of $\mathbf{L}$ belonging to the space of δ, where (L,δ) is as above.

In the whole of this chapter, we fix $\mathfrak{X}$ and we adopt the notation

$$L^2(G(k)\backslash\mathbf{G})_{\mathfrak{X}}, P_{\mathfrak{X}}, S_{\mathfrak{X}}, [S_{\mathfrak{X}}]$$

of V. For every $\mathfrak{S} \in S_{\mathfrak{X}}$, we fix a residue datum $\mathrm{Res}_{\mathfrak{S}}^{G}$ for $\mathfrak{S}$ satisfying V.3.13(a). Moreover for every $(M,\mathfrak{P}) \in \mathfrak{X}$ and for every standard Levi L containing M we fix a unitary character of the centre of $\mathbf{L}$, denoted by ξ_L, extending ξ, trivial on $A_L^{\mathbf{G}}$ (see I.2.1) and such that:

$$\mathfrak{P}_L := \{\pi \in \mathfrak{P} \text{ having } \xi_L \text{ for restriction to } Z_L\} \neq \emptyset.$$

We remark that $\mathfrak{P}_L$ is a class of cuspidal representations of $\mathbf{M}$, relative to $\mathbf{L}$, i.e. an orbit under $X_M^{\mathbf{L}}$ contained in $\mathfrak{P}$. We always have existence of ξ_L and thus of $\mathfrak{P}_L$. But if k is a function field, we do not in general have uniqueness. However, if k is a number field, we do have uniqueness and we thus have the following compatibility relation: for every $w \in W(M)$ such that wLw^{-1} is still a standard Levi:

$$w.\mathfrak{P}_L = (w.\mathfrak{P})_{wLw^{-1}}.$$

VI.1. Discrete parameter

VI.1.1. The standard Levi associated with an element of $S_{\mathfrak{X}}$

Let $\mathfrak{S}$ be an element of $S_{\mathfrak{X}}$. We denote by $(M,\mathfrak{P})$ the cuspidal datum attached to $\mathfrak{S}$. Recall (see V.1.1(3)) that there exists a set of elements of $R(T_M,G)$, momentarily denoted by $R_{\mathfrak{S}}$, such that $\mathrm{Re}\,\mathfrak{S}^0$ is defined by the equations $\langle\lambda,\check{\alpha}\rangle = 0$ for $\alpha \in R_{\mathfrak{S}}$. We denote by $M_{\mathfrak{S}}$ the smallest standard Levi of $\mathbf{G}$ containing $\mathbf{M}$ such that $R_{\mathfrak{S}} \subset R(T_M, M_{\mathfrak{S}})$. We have a more intrinsic description of $M_{\mathfrak{S}}$:

> $M_{\mathfrak{S}}$ is the smallest standard Levi subgroup of G containing M such that $\mathfrak{S}$ is stable under translation by the elements of $\mathrm{Re}\,X_{M_{\mathfrak{S}}}^{\mathbf{G}}$.

For every standard Levi L containing $M_{\mathfrak{S}}$, we have:

$$\mathrm{Re}\,\mathfrak{S}^0 = (\mathrm{Re}\,\mathfrak{S}^0) \cap \mathrm{Re}\,X_M^{\mathbf{L}} + \mathrm{Re}\,X_L^{\mathbf{G}}.$$

VI.1.2. Singularities of residues of Eisenstein series

Proposition *Let $\mathfrak{S},(M,\mathfrak{P}),M_{\mathfrak{S}}$ be as in VI.1.1. Let $\mathfrak{T}$ be a hyperplane of $\mathfrak{S}$ which is singular for one of the maps $\mathrm{Res}_{\mathfrak{S}}^{G} E(\phi,\pi)$ where $\phi \in P_{(M,\mathfrak{P})}$; $\mathfrak{T}$ belongs to $S_{\mathfrak{X}}$. We suppose that $\mathfrak{T}$ intersects the set $\{\pi \in \mathfrak{S} | \mathrm{Re}\,\pi = o(\mathfrak{S})\}$: then $\mathfrak{T}$ is singular for at least one of the functions:*

$$\pi \in \mathfrak{S} \mapsto \mathrm{Res}_{\mathfrak{S}}^{G}\langle M(\tau^{-1}, -\tau\overline{\pi})\phi'(-\tau\overline{\pi}), \phi(\pi)\rangle$$

where $\tau \in W$ satisfies $\tau\pi = -\overline{\pi}$ for every $\pi \in \mathfrak{S}$ of real part $o(\mathfrak{S})$. In particular τ is an element of the Weyl group of $M_{\mathfrak{S}}$ and, defining $M_{\mathfrak{T}}$ as in VI.1.1, we have:

$$M_{\mathfrak{S}} = M_{\mathfrak{T}}.$$

Remark *$\mathfrak{T}$ is a singular hyperplane for one of the operators $\pi \mapsto \mathrm{Res}_{\mathfrak{S}}^{G} E(\phi, \pi)$ if and only if it is singular for one of the functions $\pi \mapsto \mathrm{Res}_{\mathfrak{S}}^{G} A(\phi', \phi)(\pi)$, where $\phi' \in P_{\mathfrak{X}}$ and $\phi \in P_{(M,\mathfrak{P})}$.*

This remark is an immediate consequence of the calculation:

$$\langle \theta_{\phi'}, \mathrm{Res}_{\mathfrak{S}}^{G} E(\phi, \pi)\rangle = \mathrm{Res}_{\mathfrak{S}}^{G} A(\phi', \phi)(\pi)$$

given in V.3.7, lines preceding (1).

We first prove the first assertion. Fix a finite set of K-types such that one of the maps $\mathrm{Res}_{\mathfrak{S}}^{G} E(\phi, \pi)$ with $\phi \in P_{(M,\mathfrak{P})}^{\mathfrak{F}}$, admits $\mathfrak{T}$ as singular hyperplane. We denote by $\mathfrak{C}$ the element of $[S_{\mathfrak{X}}]$ containing $\mathfrak{S}$ and we fix a general point π_0 of $\{\pi \in \mathfrak{T} | \mathrm{Re}\,\pi = o(\mathfrak{S})\}$, general in the following sense:

> $w\pi_0 = \pi_0 \Rightarrow w\pi = \pi$ for every $\pi \in \mathfrak{T}$, and every hyperplane of $\mathfrak{S}$ belonging to $S_{\mathfrak{X}}^{\mathfrak{F}}$ and containing π_0 coincides with $\mathfrak{T}$.

We fix an element v of $\mathrm{Re}\,\mathfrak{S}^0$ not belonging to $\mathrm{Re}\,\mathfrak{T}^0$. We also fix an open neighborhood U of 0 in $\mathrm{Im}\,\mathfrak{S}^0 / \mathrm{Fix}_{X_M^G} \mathfrak{P}$ and we suppose that $U \cap (\mathrm{Im}\,\mathfrak{T}^0) / \mathrm{Fix}_{X_M^G} \mathfrak{P}$ is a relatively compact neighborhood of 0 in $\mathrm{Im}\,\mathfrak{T}^0 / \mathrm{Fix}_{X_M^G} \mathfrak{P}$. For every $n \in \mathbb{N}$, we set:

$$U_n := \bigcup_{y \in [1/2n, 1/n]} (U \cap (\mathrm{Im}\,\mathfrak{T}^0) / \mathrm{Fix}_{X_M^G} \mathfrak{P}) + \mathrm{i}\, y v.$$

We note that for large n, $\pi_0 = U_n$ does not intersect $\mathfrak{T}$ but we suppose, for large n, that $\pi_0 + U_n$ is contained in $\pi_0 + U$ (this means that U contains a small parallelepiped). Fix n_0 such that this property occurs for $n > n_0$. Let us show that if U is sufficiently small and n sufficiently large, we have:

(1) $$\forall \mathfrak{S}' \in \mathfrak{C} - \{\mathfrak{S}\}, \mathfrak{S}' \cap (\pi_0 + U_n) = \emptyset,$$

(2) $$\forall w \in \mathrm{Norm}\,\mathfrak{S} - \mathrm{Fix}\,\mathfrak{S}, w(\pi_0 + U_n) \cap (\pi_0 + U_n) = \emptyset.$$

(1) is very easy to obtain: we can replace this condition by the stronger condition:

(1)′ $$\forall \mathfrak{T}' \in S_{\mathfrak{X}}^{\mathfrak{F}}, \mathfrak{T}' \subsetneqq \mathfrak{S}, \quad \mathfrak{T}' \cap (\pi_0 + U_n) = \emptyset.$$

Now, up to taking U sufficiently small, the second property of π_0 ensures that for $\mathfrak{X}'$ as above:

$$\mathfrak{X}' \cap (\pi_0 + U) \neq \emptyset \Rightarrow \mathfrak{X}' \subset \mathfrak{X}.$$

If n is large enough for $\pi_0 + U_n$ to be contained in $\pi_0 + U$, the property (1)$'$ is realised. Suppose this is the case and suppose moreover that U is so small that $w(\pi_0 + U) \cap (\pi_0 + U) \neq \emptyset$ for $w \in \mathrm{Norm}\,\mathfrak{S}$ only if $w\pi_0 = \pi_0$, i.e. by the generality of π_0, only if $w \in \mathrm{Fix}\,\mathfrak{X}$. Let $\pi' \in w(\pi_0 + U_n) \cap (\pi_0 + U_n)$; we write $\pi' = \pi_1 + \mathrm{i}\,y v$ with $\pi_1 \in \pi_0 + (U \cap \mathrm{Im}\,\mathfrak{X}^0 / \mathrm{Fix}_{X_M^G}\,\mathfrak{P})$ and $y \in [1/2n, 1/n]$. Then

$$w\pi' = \pi_1 + \mathrm{i}\,y\,w(v).$$

Since w normalises $\mathfrak{S}$ and thus $\mathrm{Re}\,\mathfrak{S}^0$, there exists $v_1 \in \mathrm{Re}\,\mathfrak{X}^0$ and $\alpha \in \mathbb{R}$ such that $w(v) = \alpha v + v_1$. As w fixes the elements of $\mathfrak{X}$ and thus those of $\mathrm{Re}\,\mathfrak{X}^0$, we have for every $k \in \mathbb{N}$:

$$w^k v = \begin{cases} \alpha^k v + \big((1-\alpha^k)/(1-\alpha)\big)v_1, & \text{if } \alpha \neq 1 \\ v + k v_1 & \text{if } \alpha = 1. \end{cases}$$

As w is of finite order, we necessarily have either $\alpha = -1$ or $\alpha = 1$ and $v_1 = 0$. The second hypothesis means that w fixes $\mathfrak{S}$ and the first is excluded since $\pi_1 + \mathrm{i}\,y v_1 - \mathrm{i}\,y v \notin \pi_0 + U_n$ whenever n is large. Thus we obtain (2).

We will need the following sublemma:

Sublemma *Let $\mathfrak{S} \in \mathfrak{C}$. We set:*

$$\mathring{\mathfrak{S}} = \{\pi \in \mathfrak{S} | \mathrm{Re}\,\pi = o(\mathfrak{S})\} - \bigcup_{\mathfrak{S}' \subsetneq \mathfrak{S},\ \mathfrak{S}' \in S_{\mathfrak{X}}^{\mathfrak{F}}} \mathfrak{S}'.$$

Let $\pi_1 \in \mathring{\mathfrak{S}}$ and $\phi \in P_{(M,\mathfrak{P})}^{\mathfrak{F}}$; then $\mathrm{Res}_{\mathfrak{S}}^G E(\phi, \pi)$ is holomorphic at $\pi = \pi_1$ and $\mathrm{Res}_{\mathfrak{S}}^G E(\phi, \pi_1) \in A_{\mathfrak{C},\pi_1}$, in the notation of V.3.2. Moreover the square of the norm of $\mathrm{Res}_{\mathfrak{S}}^G E(\phi, \pi_1)$ in $A_{\mathfrak{C},\pi_1}$ is:

$$\text{(3)}\qquad \begin{aligned} &|\mathrm{Norm}\,\mathfrak{S}| / |\{\tau \in \mathrm{Norm}\,\mathfrak{S} | \tau\pi_1 = \pi_1\}| \\ &\mathrm{Res}_{\mathfrak{S}}^G \langle \textstyle\sum M(\tau^{-1}, -\tau\overline{\pi}_1)\phi(-\tau\overline{\pi}_1), \phi(\pi_1)\rangle, \end{aligned}$$

where the sum runs over the set $\{\tau | \tau\mathfrak{S} = -\overline{\mathfrak{S}}$ and $\tau\pi_1 = -\overline{\pi}_1\}$. In particular if $\pi_1 \in \pi_0 + U_n$ the sum only runs over $\{\tau | \tau\mathfrak{S} = -\overline{\mathfrak{S}}$ and $\tau\pi = -\overline{\pi}, \forall \pi \in \mathfrak{S}$, of real part $o(\mathfrak{S})\}$. Let $\phi' \in P_{\mathfrak{X}}$. Then in $A_{\mathfrak{C},\pi_1}$ with the notation of V.3.2(4), we have:

$$\text{(4)}\qquad \begin{aligned} &\langle e_{\mathfrak{C}}(\phi', \pi_1), \mathrm{Res}_{\mathfrak{S}}^G E(\phi, \pi_1)\rangle \\ &= |\{\tau \in \mathrm{Norm}\,\mathfrak{S} | \tau\pi_1 = \pi_1\}|^{-1}\, \mathrm{Res}_{\mathfrak{S}}^G A(\phi', \phi)(\pi_1). \end{aligned}$$

We know $\mathrm{Res}_{\mathfrak{S}}^{G} E(\phi, \pi)$ is holomorphic at π_1 since, *a priori*, the singularities of this map are on hyperplanes of $\mathfrak{S}$ belonging to $S_{\mathfrak{X}}$; however none of these hyperplanes contains π_1. Let $\phi' \in P_{(M,\mathfrak{P})}^{\mathfrak{F}}$ be an element vanishing to a high order at every point of $\{\{\mathrm{Stab}(M,\mathfrak{P}).\pi_1\} - \{\pi_1\}\}$ and such that $\phi - \phi'$ vanishes to a high order at π_1. Note that

$$\left(\mathrm{Res}_{w\mathfrak{S}}^{G} A(\phi', \phi')(\pi)\right)_{\pi'=w\pi_1} \text{ and } \left(\mathrm{Res}_{w\mathfrak{S}}^{G} E(\phi', \pi')\right)_{\pi'=w\pi_1}$$

are zero if $w\pi_1 \neq \pi_1$. We set

$$y := |\{w \in \mathrm{Stab}(M,\mathfrak{P}), w\pi_1 = \pi_1\}|,$$

and we have:

$$\mathrm{Res}_{\mathfrak{S}}^{G} E(\phi, \pi_1) = y^{-1} \sum_{w \in \mathrm{Stab}(M,\mathfrak{P})} (\mathrm{Res}_{w\mathfrak{S}}^{G} E)(\phi', w\pi_1) = |\mathrm{Norm}\,\mathfrak{S}| y^{-1} e_{\mathfrak{C}}(\phi', \pi_1),$$

with the notation of V.3.2(4)(i). It is thus an element of $A_{\mathfrak{C},\pi_1}$. On the other hand in $A_{\mathfrak{C},\pi_1}$ we have:

$$\begin{aligned} \|\mathrm{Res}_{\mathfrak{S}}^{G} E(\phi, \pi_1)\|^2 &= |\mathrm{Norm}\,\mathfrak{S}|^2 y^{-2} \|e_{\mathfrak{C}}(\phi', \pi_1)\|^2 \\ &= |\mathrm{Norm}\,\mathfrak{S}| y^{-2} \sum_{\substack{w \in \mathrm{Stab}(M,\mathfrak{P}) \\ w\pi_1 = \pi_1}} (\mathrm{Res}_{w\mathfrak{S}}^{G} A(\phi', \phi'))(w\pi_1) \\ &= |\mathrm{Norm}\,\mathfrak{S}| y^{-1} (\mathrm{Res}_{\mathfrak{S}}^{G} A(\phi', \phi'))(\pi_1); \end{aligned}$$

and the desired result can be obtained by expanding $(\mathrm{Res}_{\mathfrak{S}}^{G} A(\phi', \phi'))$. If $\pi_1 \in \pi_0 + U_n$, we obtain the desired properties thanks to (1) and (2). We prove (4) similarly.

Let $\tilde{\ell}$ be a non-trivial linear form on $\mathrm{Re}\, X_M$, defined over $\mathbb{Q}$, invariant under translation by $(\mathrm{Re}\,\mathfrak{S}^0)^{\perp}$ (i.e. zero on $\mathrm{Re}\,\mathfrak{S}^{0\perp}$) and zero on $\mathrm{Re}\,\mathfrak{T}^0$; $\tilde{\ell}$ can be extended to a linear form on $\mathfrak{a}_M^*$ (see I.1.4). We define the function ℓ on $\mathfrak{P}$ as follows:

If k is a number field: We identify $\mathfrak{P}$ and $X_M^{\mathbf{G}}$ by fixing ρ_0 to be the element of $\mathfrak{P}$ trivial on $A_M^{\mathbf{G}}$ and which identifies $\pi \in \mathfrak{P}$ with $\lambda_\pi \in X_M^{\mathbf{G}}$ if $\pi \simeq \lambda_\pi \otimes \rho_0$. We then set:

$$\ell(\pi) := \tilde{\ell}(\lambda_\pi - \lambda_{\pi_0}) e^{\langle \lambda_\pi, \lambda_\pi \rangle}.$$

If k is a function field: We proceed as in V.1.2, third example. There exists an integer N such that the map $\mu \in \mathfrak{a}_M^* \mapsto q^{N\tilde{\ell}(\mu)}$ passes to the quotient by the kernel of the composed map

$$\mathfrak{a}_M^* \xrightarrow{\kappa} X_M^{\mathbf{G}} \longrightarrow X_M^{\mathbf{G}} / \mathrm{Fix}_{X_M^{\mathbf{G}}} \mathfrak{P}$$

and thus naturally defines a function on $X_M^{\mathbf{G}} / \mathrm{Fix}_{X_M^{\mathbf{G}}} \mathfrak{P}$. We set:

$$\ell(\pi) = (q^{N\tilde{\ell}(\lambda_\pi)} - 1)$$

where λ_π is any element of $X_M^{\mathbf{G}}$ satisfying $\pi \simeq \lambda_\pi \otimes \pi_0$.

We note that we have:

$$\ell(\pi) = 0, \forall \pi \in \mathfrak{T}, \tag{5}$$

$$\forall \epsilon \in \operatorname{Re}\mathfrak{S}^0 - \operatorname{Re}\mathfrak{T}^0, (d\ell(\pi_0 + t\epsilon)/dt)_{t=0} \neq 0, \tag{6}$$

$$\forall \epsilon \in (\operatorname{Re}\mathfrak{S}^0)^\perp, \pi \in \mathfrak{S} \mapsto (d(\ell(\pi + t\epsilon)/dt)_{t=0} \text{ is identically zero.} \tag{7}$$

Let $\phi', \phi \in P_{\mathfrak{X}}$; *a priori* $\mathfrak{T}$ is the only possible singular hyperplane for the function $\operatorname{Res}_{\mathfrak{S}}^{\mathfrak{G}} A(\phi', \phi)(\pi)$ passing through π_0 and it is indeed singular, by hypothesis, for suitably chosen ϕ' and ϕ. It is a consequence of this and of (5) and (6) that there exists an integer $r \geq 1$ such that for $\phi', \phi \in P_{\mathfrak{X}}$, the function $\operatorname{Res}_{\mathfrak{S}}^{\mathfrak{G}} A(\phi', \ell^r \phi)$ is holomorphic at a general point of $\mathfrak{T}$ and its restriction to $\mathfrak{T}$ is non-zero for ϕ' and ϕ general in $P_{\mathfrak{X}}$. Up to changing π_0, we can suppose that there exists $\phi'_0, \phi_0 \in P_{\mathfrak{X}}$ such that we have:

$$\operatorname{Res}_{\mathfrak{S}}^{\mathfrak{G}} A(\phi'_0, \ell^r \phi_0)(\pi_0) \neq 0.$$

Here we fix T large and T' very large with respect to T. We return to the notation of V.3.11. Thanks to the sublemma, for $n \in \mathbb{N}$ large, we define an element F_n of $\operatorname{Hilb}_{\mathfrak{C},T}^{\mathfrak{F}}$, where $\mathfrak{F}$ is a suitable set of **K**-types, by setting:

$$F_n(\pi) = \begin{cases} 0 & \text{if } \pi \notin \cup_{w \in \operatorname{Norm}\mathfrak{S}} w(\pi_0 + U_n), \\ (\operatorname{Res}_{\mathfrak{S}}^{\mathfrak{G}} E)(n\ell^r \phi_0, w^{-1}\pi) & \text{if } \pi \in w(\pi_0 + U_n) \end{cases}$$

where $w \in \operatorname{Norm}\mathfrak{S}$. We defined $F_{\phi'_0}$ in V.3.11. We will show that we have:

$$\lim_{n \to \infty} \langle F_{\phi'_0}, F_n \rangle \text{ exists and is non-zero.} \tag{8}$$

We have:

$$\langle F_{\phi'_0}, F_n \rangle = |\operatorname{Norm}\mathfrak{S}| \int_{\pi \in \pi_0 + U_n} \langle e_{\mathfrak{C}}(\phi'_0, \pi), \operatorname{Res}_{\mathfrak{S}}^{\mathfrak{G}} E(n\ell^r \phi_0, \pi) \rangle \, d_{\mathfrak{S}}\pi,$$

and with (4) which can be applied at every point of $\pi_0 + U_n$ thanks to (1), this is equal to:

$$|\operatorname{Norm}\mathfrak{S}| / |\operatorname{Fix}\mathfrak{S}| \int_{\pi \in \pi_0 + U_n} \operatorname{Res}_{\mathfrak{S}}^{\mathfrak{G}} A(\phi'_0, n\ell^r \phi_0)(\pi) \, d_{\mathfrak{S}}\pi.$$

Up to shrinking U, by the choice of r, the function to be integrated is holomorphic in a neighborhood of every point of $\pi_0 + (U \cap \operatorname{Im}\mathfrak{T}^0)$ and for a suitable measure on $\pi_0 + (U \cap \operatorname{Im} \mathfrak{T}^0)$, we have:

$$\begin{aligned}
&\lim_{n\to\infty} \langle F_{\phi_0'}, F_n \rangle \\
&= |\operatorname{Norm} \mathfrak{S}| / |\operatorname{Fix} \mathfrak{S}| \big(\lim_{n\to\infty} n(1/n - 1/2n)\big) \\
&\qquad \int_{\pi\in\pi_0+(U\cap \operatorname{Im}\mathfrak{T}^0)} \operatorname{Res}_{\mathfrak{S}}^{G} A(\phi_0', \ell^r \phi_0)(\pi)\, d\pi \\
&= |\operatorname{Norm} \mathfrak{S}|/2 |\operatorname{Fix} \mathfrak{S}| \int_{\pi\in\pi_0+(U\cap \operatorname{Im}\mathfrak{T}^0)} \operatorname{Res}_{\mathfrak{S}}^{G} A(\phi_0', \ell^r \phi_0)(\pi)\, d\pi.
\end{aligned}$$

Since the function we are integrating is non-zero and continuous at π_0 (up to shrinking U again if necessary), we can suppose that the integral is non-zero. This gives the result stated in (8).

We will now suppose that $\mathfrak{T}$ is not a singular hyperplane for any of the operators $\operatorname{Res}_{\mathfrak{S}}^{G} M(\tau^{-1}, -\tau\overline{\pi})$ where τ satisfies $\tau\pi = -\overline{\pi}$ for every $\pi \in o(\mathfrak{S}) + \operatorname{Im} \mathfrak{S}$. We show that then we have:

$$\lim_{n\to\infty} \|F_n\| \text{ exists but equals } 0.$$

With (8), this will give the desired contradiction, since we have the general inequality:

$$|\langle F_{\phi_0'}, F_n \rangle| \le \|F_{\phi_0'}\| \, \|F_n\|.$$

We calculate $\|F_n\|^2$ using (3):

$$\begin{aligned}
&\|F_n\|^2 = |\operatorname{Norm} \mathfrak{S}| / |\operatorname{Fix} \mathfrak{S}| \\
&\int_{\pi\in\pi_0+U_n} \sum_{\tau} \operatorname{Res}_{\mathfrak{S}}^{G} \langle M(\tau^{-1}, -\tau\overline{\pi}(n\ell^r \phi_0)(-\tau\overline{\pi}), (n\ell^r \phi_0)(\pi) \rangle \, d_{\mathfrak{S}}\pi,
\end{aligned}$$

where τ runs through the same set as above. Thanks to (7), for every $\pi \in \pi_0 + U_n$, we have:

$$\begin{aligned}
&\operatorname{Res}_{\mathfrak{S}}^{G} \langle M(\tau^{-1}, -\tau\overline{\pi})(\ell^r \phi_0)(-\tau\overline{\pi}), \ell^r \phi_0(\pi) \rangle \\
&= |\ell(\pi)|^{2r} \operatorname{Res}_{\mathfrak{S}}^{G} \langle M(\tau^{-1}, -\tau\overline{\pi})\phi_0(-\tau\overline{\pi}), \phi_0(\pi) \rangle.
\end{aligned}$$

The hypothesis on $\mathfrak{T}$ ensures that

$$\operatorname{Res}_{\mathfrak{S}}^{G} \langle M(\tau^{-1}, -\tau\overline{\pi})\phi_0(-\tau\overline{\pi}), \phi_0(\pi) \rangle$$

is holomorphic at π_0 and on all of $\pi_0 + U$ if U is small enough. For every $\pi \in \pi_0 + (U \cap \operatorname{Im} \mathfrak{T}^0) + i\, y\, v$, $\ell(\pi)$ is a function of y only; we denote it by

$\ell(y)$. We then obtain:

$$\lim_{n\to\infty} \|F_n\| =$$

$$|\mathrm{Norm}\,\mathfrak{S}|/|\,\mathrm{Fix}\,\mathfrak{S}|\ \lim_{n\to\infty}\left(n^2\int_{1/2n}^{1/n}\ell(y)^{2r}\,\mathrm{d}y\right)\times$$

$$\int_{\pi\in\pi_0+(U\cap \mathrm{Im}\,\mathfrak{T}^0)} \sum_{\tau} \mathrm{Res}_{\mathfrak{S}}^{\mathbf{G}}\langle M(\tau^{-1},-\tau\overline{\pi})\phi_0(-\tau\overline{\pi}),\phi_0(\pi)\rangle\ \mathrm{d}_{\mathfrak{S}}\pi,$$

where τ runs through the set described above. It remains to show that the limit:

$$\lim_{n\to\infty}\left(n^2\int_{1/2n}^{1/n}|\ell(y)|^{2r}\right)\mathrm{d}y$$

is zero.

If k is a number field, we have $\ell(y) = \mathrm{i}\,y\tilde{\ell}(v)$, and the result is clear since $r \geq 1$.

If k is a function field, we have $\ell(y) = (q^{\mathrm{i}Ny\tilde{\ell}(v)} - 1)$ whence $|\ell(y)| \leq c\,y$ for a well chosen constant c, and the result is still clear. This gives the first assertion of the proposition.

Let us prove the second: since $\mathfrak{T} \subset \mathfrak{S}$, we have $M_{\mathfrak{S}} \subset M_{\mathfrak{T}}$. Fix τ such that $\mathfrak{T}$ is a singular hyperplane for the operator $M(\tau^{-1},-\tau\overline{\pi})$ where τ satisfies $\tau\pi = -\overline{\pi}$, for every $\pi \in o(\mathfrak{S}) + \mathrm{Im}\,\mathfrak{S}$. It is clear that this last property implies that τ is the identity on $\mathrm{Im}\,\mathfrak{S}^0$ and thus on $\mathfrak{S}^0$. Thus $\tau M_{\mathfrak{S}} = M_{\mathfrak{S}}$ and τ acts trivially on $X_{M_{\mathfrak{S}}}^{\mathbf{G}}$. This implies that τ is an element of the Weyl group of $M_{\mathfrak{S}}$. In particular the singular hyperplanes of $M(\tau^{-1},-\tau\overline{\pi})$ are stable under translation by $X_{M_{\mathfrak{S}}}^{\mathbf{G}}$ and those of $\mathrm{Res}_{\mathfrak{S}}^{\mathbf{G}} M(\tau^{-1},-\tau\overline{\pi})$ are stable under translation by $X_{M_{\mathfrak{S}}}^{\mathbf{G}} \cap \mathfrak{S}^0$. If $M_{\mathfrak{T}}$ strictly contained $M_{\mathfrak{S}}, \mathfrak{T}^0$ could not contain $\mathrm{Re}\,\mathfrak{S}^0$ and we would have:

$$\mathfrak{S} = \mathfrak{T} + (X_{M_{\mathfrak{S}}}^{\mathbf{G}} \cap \mathfrak{S}^0).$$

Thus $\mathfrak{S}$ would be singular for $\mathrm{Res}_{\mathfrak{S}}^{\mathbf{G}} M(\tau^{-1},-\tau\overline{\pi})$ which is absurd. This gives the proposition.

VI.1.3. Corollary

Let $\mathfrak{C} \in [S_{\mathfrak{X}}]$; $\mathfrak{S} \in \mathfrak{C}$ and let L be a standard Levi containing $M_{\mathfrak{S}}$ (i.e. $\mathfrak{S}$ is stable under translation by $X_L^{\mathbf{G}}$). We denote by W_L the Weyl group of

L. Then for every $\phi \in P_{(M,\mathfrak{P})}$, *the map:*

$$\sum_{\sigma \in W_L \cap \mathrm{Stab}(M,\mathfrak{P})} (\mathrm{Res}^{\mathbf{G}}_{\sigma\mathfrak{S}} E)(\phi, \sigma\pi)$$

is holomorphic at every point of $o(\mathfrak{S}) + \mathrm{Im}\,\mathfrak{S}$. *This is also true of the function:*

$$\sum_{\sigma \in W_L \cap \mathrm{Stab}(M,\mathfrak{S})} (\mathrm{Res}^{\mathbf{G}}_{\sigma\mathfrak{S}} A(\phi', \phi))(\sigma\pi),$$

for every $\phi' \in P_{\mathfrak{X}}$.

We know *a priori* that the singularities of this operator are on hyperplanes of $\mathfrak{S}$ belonging to $S_{\mathfrak{X}}$. Indeed, fix $\phi \in P_{\mathfrak{X}}$ and let $\mathfrak{T}$ be a hyperplane of $\mathfrak{S}$ belonging to $S_{\mathfrak{X}}$. We suppose that $\mathfrak{T}$ meets $o(\mathfrak{S}) + \mathrm{Im}\,\mathfrak{S}$ and that $\mathfrak{T}$ is indeed a singular hyperplane. Let π_0 be a general point of $\mathfrak{T} \cap (o(\mathfrak{S}) + \mathrm{Im}\,\mathfrak{S})$; we require that π_0 not belong to any other hyperplane of $\mathfrak{S}$ belonging to $S_{\mathfrak{X}}$. This hypothesis implies that for every $\sigma \in \mathrm{Stab}(M, \mathfrak{P})$, $\mathfrak{T}$ is the only possible singular hyperplane passing through π_0 for the operator $(\mathrm{Res}^{\mathbf{G}}_{\sigma\mathfrak{S}} E)(\phi, \sigma\pi)$. We pick up the notation $\ell(\pi)$ of the proof of VI.1.2 and we choose $r \in \mathbb{N}_{>0}$ such that:

(1) $$\pi \in \mathfrak{S} \mapsto \ell(\pi)^r \sum_{\sigma \in \mathrm{Stab}(M,\mathfrak{S}) \cap W_L} (\mathrm{Res}^{\mathbf{G}}_{\sigma\mathfrak{S}} E)(\phi, \sigma\pi)$$

is holomorphic at π_0 and non-zero. By V.3.2.(4)(i), the function:

(2) $$\ell(\pi)^r \sum_{\substack{\sigma \in \mathrm{Stab}(M,\mathfrak{P}) \\ \sigma \notin (\mathrm{Stab}(M,\mathfrak{P}) \cap W_L)}} (\mathrm{Res}^{\mathbf{G}}_{\sigma\mathfrak{S}} E)(\phi, \sigma\pi)$$

is also defined at π_0 and on it takes the negative of the value (1); in particular it is non-zero. Let $\tilde{\phi} \in P_{\mathfrak{X}}$ be such that $\tilde{\phi} - \phi$ vanishes to a very high order at every point of $\{(\mathrm{Stab}(M, \mathfrak{P}) \cap W_L)\pi_0\}$ and $\tilde{\phi}$ vanishes to a high order at every point of $\{\mathrm{Stab}(M, \mathfrak{P})\pi_0\} - \{(\mathrm{Stab}(M, \mathfrak{P}) \cap W_L)\pi_0\}$. The generality of π_0 ensures that for $\sigma \in \mathrm{Stab}(M, \mathfrak{P})$ we have $\sigma\pi_0 = \pi_0$ only if the restriction of σ to $\mathfrak{T}^0$ is the identity. As $\mathfrak{T}^0$ contains $\mathrm{Re}\,X^{\mathbf{G}}_L$ this property implies that σ is an element of W_L. Thus:

$$\begin{aligned}&\{\big(\mathrm{Stab}(M, \mathfrak{P}) - (\mathrm{Stab}(M, \mathfrak{P}) \cap W_L)\big)\pi_0\} \\ &\quad = \{\mathrm{Stab}(M, \mathfrak{P})\pi_0\} - \{(\mathrm{Stab}(M, \mathfrak{P}) \cap W_L)\pi_0\}.\end{aligned}$$

In (2), we can thus replace ϕ by $\tilde{\phi}$ without changing the value of the function at π_0. The vanishing hypotheses on $\tilde{\phi}$ then ensure that (2) is zero, which is a contradiction.

We obtain the second assertion of the proposition by calculating the scalar product of $\theta_{\phi'}$ against the function $(\mathrm{Res}^{\mathbf{G}}_{\sigma\mathfrak{S}} E)(\phi, \sigma\pi)$ (see V.3.7, just before (1)).

VI.1.4 Reduction to the Levi subgroup

VI.1.4.(a) Preliminary remarks

Let $\mathfrak{X}$ be as before and $(M, \mathfrak{P}) \in \mathfrak{X}$. We fix a standard Levi subgroup L of G containing M. At the start of Chapter VI we defined $\mathfrak{P}_L$; $(M, \mathfrak{P}_L)$ is a cuspidal datum in the sense of II.1.1 for the algebraic group L and its covering $\mathbf{L}$. We define $S_{(M,\mathfrak{P}_L)}$ in a similar way to $S_{(M,\mathfrak{P})}$; see V.1.1. We denote by $\mathfrak{X}_L$ the equivalence class of $(M, \mathfrak{P}_L)$ and we then define $[S_{\mathfrak{X}_L}]$. We have:

(1) $$\mathfrak{P} = \mathfrak{P}_L + X_L^{\mathbf{G}}.$$

Lemma *Let $\mathfrak{S}_L \in S_{(M,\mathfrak{P}_L)}$. Let $\mathfrak{a}_M^{G*}$ act by translation on $X_M^{\mathbf{G}}$ via κ; see I.1.4(a). Then:*

(i) $\mathfrak{S} := \mathfrak{S}_L + (\operatorname{Re} X_L^{\mathbf{G}} \otimes_{\mathbb{R}} \mathbb{C})$ *is an element of $S_{(M,\mathfrak{P})}$ such that* $\operatorname{Re}\mathfrak{S} = \operatorname{Re}\mathfrak{S}_L + \operatorname{Re} X_L^{\mathbf{G}}$ *and $o(\mathfrak{S}) = o(\mathfrak{S}_L)$.*

(ii) *$\mathfrak{S}_L$ is an irreducible component of $\mathfrak{S} \cap \mathfrak{P}_L$. If k is a number field, we even have the equality $\mathfrak{S}_L = \mathfrak{S} \cap \mathfrak{P}_L$; if k is a function field two irreducible components of $\mathfrak{S} \cap \mathfrak{P}_L$ can be deduced from each other by translation by an element of $X_L^{\mathbf{L}}$.*

(iii) *$M_{\mathfrak{S}} \subset L$ and $M_{\mathfrak{S}} = M_{\mathfrak{S}_L}$, where $M_{\mathfrak{S}_L}$ is defined in an analoguous way to $M_{\mathfrak{S}}$.*

(i) To check that $\mathfrak{S} \in S_{(M,\mathfrak{P})}$, it suffices, by the definitions, to check it when $\mathfrak{S}_L$ is a singular hyperplane of $\mathfrak{P}_L$ for one of the intertwining operators $\mathrm{o}M(w^{-1}, w\pi)$, for the group L, where $w \in W_L(M)$ (the analogue of $W(M)$ in the Weyl group of L) and $\pi \in \mathfrak{P}_L$. In this case $\mathfrak{S}$ is a singular hyperplane of $\mathfrak{P}$ for the operator $M(w^{-1}, w\pi)$ considered for the group G; see IV.4.5. The remainder of (i) is evident taking into account that the orthogonal complement of $\operatorname{Re} X_L^{\mathbf{G}}$ in $\operatorname{Re} X_M^{\mathbf{G}}$ is $\operatorname{Re} X_M^{\mathbf{L}}$.

(ii) We note that $\mathfrak{S} \cap \mathfrak{P}_L$ is a union of affine spaces of $\mathfrak{P}_L$ with vector part defined on $\mathbb{R}$ (see V.1.1) and is irreducible if k is a number field. As $\mathfrak{S}_L$ is contained in $\mathfrak{S} \cap \mathfrak{P}_L$, to prove the first assertion of (ii) it suffices to show that $\operatorname{Re}(\mathfrak{S} \cap \mathfrak{P}_L)$ is equal to $\operatorname{Re}\mathfrak{S}_L$. We fix $\pi_L \in \mathfrak{S}_L$ and we have:

$$\pi \in \mathfrak{S} \Leftrightarrow \exists \lambda \in \mathfrak{S}^0 \text{ such that } \pi \simeq \pi_L \otimes \lambda.$$

Thus

$$\operatorname{Re}\mathfrak{S} \cap \operatorname{Re} X_M^{\mathbf{L}} = \operatorname{Re}(\mathfrak{S} \cap \mathfrak{P}_L).$$

Now we have the decomposition:

$$\operatorname{Re} X_M^{\mathbf{G}} = \operatorname{Re} X_M^{\mathbf{L}} \oplus \operatorname{Re} X_L^{\mathbf{G}}.$$

Thus we obtain $\operatorname{Re}\mathfrak{S} \cap \operatorname{Re} X_M^{\mathbf{L}} = \operatorname{Re}\mathfrak{S}_L$ and the desired assertion. It remains to study the non-connectedness of $\mathfrak{S} \cap \mathfrak{P}_L$ in the case where k is a function field: we want to show that $\mathfrak{S} + X_L^{\mathbf{L}}$ contains this intersection. We have the exact sequence:

$$0 \to X_L^{\mathbf{L}} \to X_M^{\mathbf{L}} \to X_M^{\mathbf{G}}/X_L^{\mathbf{G}} \to 0.$$

This shows that taking the intersection with $X_M^{\mathbf{L}}$ induces a bijection between the subspaces of $X_M^{\mathbf{G}}$ stable under $X_L^{\mathbf{G}}$ and those of $X_M^{\mathbf{L}}$ stable under $X_L^{\mathbf{L}}$; the inverse map can be obtained by saturating by $X_L^{\mathbf{G}}$. Now $(\mathfrak{S}_L^0 + X_L^{\mathbf{L}}) + X_L^{\mathbf{G}}$ has real part $\mathrm{Re}\,\mathfrak{S}^0$ and thus contains $\mathfrak{S}^0 + X_L^{\mathbf{G}}$, which implies the equality. Thus:

$$(\mathfrak{S}^0 + X_L^{\mathbf{G}}) \cap X_M^{\mathbf{L}} = (\mathfrak{S}_L^0 + X_L^{\mathbf{L}})$$

and

(1) $$(\mathfrak{S} + X_L^{\mathbf{G}}) \cap \mathfrak{P}_L = (\mathfrak{S}_L + X_L^{\mathbf{L}}).$$

In other words, we obtain (ii).

(iii) The inclusion of $M_{\mathfrak{S}}$ in L results from the definitions and the equality $M_{\mathfrak{S}} = M_{\mathfrak{S}_L}$ which comes from the calculation of real parts done in (i).

Corollary *Let $\mathfrak{S}_L \in S_{(M,\mathfrak{P}_L)}$ and $\mathrm{Res}_{\mathfrak{S}_L,L}$ be a residue datum for $\mathfrak{S}_L$ into $\mathfrak{P}_L$; then $\mathrm{Res}_{\mathfrak{S}_L,L}$ naturally defines a residue datum for $\mathfrak{S}$ into $\mathfrak{P}$.*

This results from (i) above and from the definitions.

VI.1.4.(b) Statement

We fix for every $\mathfrak{S}_L \in [S_{\mathfrak{X}_L}]$ a residue datum, denoted by $\mathrm{Res}^L_{\mathfrak{S}_L}$, satisfying V.3.13(a) for L. We note (see VI.1.4(a)(i) above) that for every positive real number T:

(2) $$\{\pi \in \mathfrak{S}_{L,\leq T} | \mathrm{Re}\,\pi = o(\mathfrak{S})\} \neq \emptyset \text{ if and only if } \{\pi \in \mathfrak{S}_{\leq T} | \mathrm{Re}\,\pi = o(\mathfrak{S})\} \neq \emptyset.$$

Proposition *For every $\mathfrak{S}_L \in [S_{\mathfrak{X}_L}]$, the operator $\mathrm{Res}^G_{\mathfrak{S}} - \mathrm{Res}^L_{\mathfrak{S}_L}$ kills all functions of the type $A(\phi', \phi)$ for $\phi', \phi \in P_{\mathfrak{X}}$.*

We denote by $(M, \mathfrak{P}_L)$ the cuspidal datum of $\mathfrak{S}_L$, i.e. $\mathfrak{S}_L \in S_{(M,\mathfrak{P}_L)}$ and $\mathfrak{S} \in S_{(M,\mathfrak{P})}$. We are then led to study the scalar products $\langle \theta_{\phi'}, \theta_\phi \rangle$ for $\phi' \in P_{\mathfrak{X}}$ and $\phi \in P_{(M,\mathfrak{P})}$. We fix $\lambda_0 \in \mathrm{Re}\,X_M^{\mathbf{G}}$ very positive and satisfying the following additional property:

Decompose $\lambda_0 = \lambda_0^L + \mu_0$ with $\lambda_0^L \in \mathrm{Re}\,X_M^{\mathbf{L}}$ and $\mu_0 \in \mathrm{Re}\,X_L^{\mathbf{G}}$. Then, for every root α in $R^+(T_M, G) - R^+(T_M, L)$, we have:

(3) $$\langle \mu_0, \check{\alpha} \rangle \gg \langle \lambda_0^L, \check{\alpha} \rangle.$$

For $\phi' \in P_{\mathfrak{X}}, \phi \in P_{(M,\mathfrak{P})}$, we define a meromorphic function on $\mathfrak{P}$ by setting:

$$W_L(M) := W_L \cap W(M),$$
$$A_L(\phi', \phi)(\pi) := \sum_{\sigma \in W_L(M)} \langle M(\sigma^{-1}, -\sigma\overline{\pi})\phi'(-\sigma\overline{\pi}), \phi(\pi)\rangle.$$

We generalise this definition to the case where ϕ' is only a meromorphic function on $\cup_{\sigma \in W_L(M)} \sigma\mathfrak{P}$. Let $\tau^L \in W$ be minimal in its right class modulo W_L; for $\phi' \in P_{\mathfrak{X}}$, we define a function ϕ'_{τ^L}, meromorphic on $\cup_{\sigma \in W_L(M)} \sigma\mathfrak{P}$, by setting for every $(M', \mathfrak{P}')$ conjugate to $(M, \mathfrak{P})$ by an element of $W_L(M)$ and for every $\pi \in \mathfrak{P}'$:

$$\phi'_{\tau^L}(\pi) = \begin{cases} 0, & \text{if } \tau^L \notin W(M') \\ M((\tau^L)^{-1}, \tau^L\pi)\phi'(\tau^L\pi), & \text{otherwise.} \end{cases}$$

We will show that we have:

(4) $$A(\phi', \phi) = \sum_{\tau^L, \text{ minimal in } \tau^L W_L} A_L(\phi'_{\tau^L}, \phi).$$

Let $\tau \in W(M)$. We decompose τ as follows:

$$\tau = \tau^L \tau_L$$

where $\tau_L \in W_L$ and τ^L is the element of minimal length in its right class modulo W_L. We know (see I.1.9) that:

(5) $$\tau_L \in W_L(M) \text{ and } \tau^L \in W(\tau_L(M)).$$

We calculate, for every $\pi \in \mathfrak{P}$:

$$\begin{aligned} &A(\phi', \phi)(\pi) \\ &= \sum_{\tau \in W(M)} \langle M(\tau^{-1}, -\tau\overline{\pi})\phi'(-\tau\overline{\pi}), \phi(\pi)\rangle \\ &= \sum_{\tau \in W(M)} \langle M(\tau_L^{-1}, -\tau_L\overline{\pi})M((\tau^L)^{-1}, -\tau^L\tau_L\overline{\pi})\phi'(-\tau^L\tau_L\overline{\pi}), \phi(\pi)\rangle \\ &= \sum_{\tau \in W(M)} \langle M(\tau_L^{-1}, -\tau_L\overline{\pi})\phi'_{\tau^L}(-\tau_L\overline{\pi}), \phi(\pi)\rangle \\ &= \sum_{\tau_L \in W(M)} \left(\sum_{\substack{\tau^L \in W(\tau_L(M)) \\ \text{minimal in } \tau^L W_L}} \langle M(\tau_L^{-1}, -\tau_L\overline{\pi})\phi'_{\tau^L}(-\tau_L\overline{\pi}), \phi(\pi)\rangle \right) \end{aligned}$$

and adding some terms which are equal to zero, this is still equal to:

$$\sum_{\tau_L \in W_L(M)} \sum_{\tau^L \text{ minimal in } \tau^L W_L} \langle M(\tau_L^{-1}, -\tau_L\overline{\pi})\phi'_{\tau^L}(-\tau_L\overline{\pi}), \phi(\pi)\rangle.$$

We thus obtain (4).

Let $\tau^L \in W$ be minimal in its right class modulo W_L. In II.2.2, we proved that for every $\phi' \in P_{\mathfrak{X}}$ the restriction of the function ϕ'_{τ^L} to the set

$\mu + \cup_{(M',\mathfrak{P}'_L)\in\mathfrak{X}_L}\mathfrak{P}'_L$, where μ is an element of $X_L^{\mathbf{G}}$ of real part μ_0 (see above), is a suitable element of $P^R_{\mathfrak{X}_L}$ for R. On the other hand the restriction of ϕ to $\mu + \mathfrak{P}_L$ is an element of $P_{(M,\mathfrak{P}_L)}$ for every $\mu \in X_L^{\mathbf{G}}$. Fix $\mu \in X_L^{\mathbf{G}}$ of real part μ_0; for every large positive real number T_0, we can generalise V.3 (where we replace G by L) to the case where the fixed central character, denoted by ξ in V.3, is not unitary, as here where it is equal to $\mu + \xi_L$ (we multiply by the function $\mu \circ m_P$, where P is the standard parabolic Levi subgroup L). We obtain (recalling that $\phi \in P_{(M,\mathfrak{P})}$):

$$\int_{\substack{\pi\in\mu+\mathfrak{P}_L\\ \mathrm{Re}\,\pi=\lambda_0^L+\mu_0}} A_L(\phi'_{\tau^L},\phi)(\pi)\,d\pi =_{T_0} \sum_{\mathfrak{C}_L\in[S_{\mathfrak{X}_L}]} |\mathrm{Norm}_{W_L(M)}\mathfrak{S}_{\mathfrak{C}_L,L}|^{-1}$$

$$\int_{\substack{\pi\in\mu+(\mathfrak{S}_{\mathfrak{C}_L},L)_{\le T_0}\\ \mathrm{Re}\,\pi=o(\mathfrak{S}_{\mathfrak{C}_L},L)+\mu_0}} \sum_{w\in\mathrm{Stab}_{W_L}(M,\mathfrak{P}_L)} \mathrm{Res}^L_{w(\mathfrak{S}_{\mathfrak{C}_L},L)} A_L(\phi'_{\tau^L},\phi)(w\pi)\,d_{(\mathfrak{S}_{\mathfrak{C}_L},L)}\pi,$$

where $\mathfrak{S}_{(\mathfrak{C}_L,L)}$ is a fixed element of $\mathfrak{C}_L \cap S_{(M,\mathfrak{P}_L)}$. We sum over τ^L using (4), which allows us to replace A_L by A and ϕ'_{τ^L} by ϕ'. Then we integrate over $\mathrm{Im}\,X_L^{\mathbf{G}}/X_L^{\mathbf{L}}$. We separate the number field case from the function field case. We first suppose that k is a number field and set $T := T_0 - \|\mu_0\|^2$. We obtain:

$$\int_{\substack{\pi\in\mathfrak{P}_{\le T}\\ \mathrm{Re}\,\pi=\lambda_0}} A(\phi',\phi)(\pi)\,d\pi =_T \sum_{\mathfrak{C}_L\in[S_{\mathfrak{X}_L}]} |\mathrm{Norm}_{W_L(M)}\mathfrak{S}_{\mathfrak{C}_L,L}|^{-1}$$

(6)

$$\int_{\substack{\pi\in\mathfrak{S}_{\mathfrak{C}_L,\le T}\\ \mathrm{Re}\,\pi=o(\mathfrak{S}_{\mathfrak{C}_L})+\mu_0}} \left(\sum_{w\in\mathrm{Stab}_{W_L}(M,\mathfrak{P}_L)} (\mathrm{Res}^L_{w\mathfrak{S}_{\mathfrak{C}_L},L} A(\phi',\phi))(w\pi)\right) d_{\mathfrak{S}_{\mathfrak{C}_L}}\pi,$$

where $\mathfrak{S}_{\mathfrak{C}_L} := \mathfrak{S}_{\mathfrak{C}_L,L} + \mathrm{Re}\,\mathfrak{S}_L^{\mathbf{G}} \otimes_{\mathbf{R}} \mathbf{C}$.

Recall from VI.1.4(a)(ii) that $\mathfrak{S}_{\mathfrak{C}_L,L}$ is equal to $\mathfrak{S}_{\mathfrak{C}_L} \cap \mathfrak{P}_L$. Thus writing $\mathrm{Res}^L_{w\mathfrak{S}_{\mathfrak{C}_L}}$ instead of $\mathrm{Res}^L_{w\mathfrak{S}_{\mathfrak{C}_L,L}}$ does not lead to any confusion; it is a residue datum for $w\mathfrak{S}_{\mathfrak{C}_L}$. On the other hand:

$$|\mathrm{Norm}_{W_L(M)}\mathfrak{S}_{\mathfrak{C}_L,L}|^{-1} = |\mathrm{Norm}_{W_L(M)}\mathfrak{S}_{\mathfrak{C}_L}|^{-1}.$$

We denote by $[S']$ the set of conjugacy classes under

$$\mathrm{Stab}_{W_L(M)}(M,\mathfrak{P}_L)$$

among the elements of $S_{(M,\mathfrak{P})}$ stable under $\mathrm{Re}\,X_L^{\mathbf{G}}$; here, since k is a number field, this means stable under $X_L^{\mathbf{G}}$. For every $\mathfrak{S}'$ belonging to one of the elements of $[S']$, we set $\mathrm{Res}^L_{\mathfrak{S}'} = 0$ if $(\mathfrak{S}' \cap \mathfrak{P}_L) \notin \mathrm{Sing}^L$. For every element $\mathfrak{C}'$ of $[S']$, we fix $\mathfrak{S}_{\mathfrak{C}'} \in \mathfrak{C}'$ and the above equality can be rewritten:

$$(6)_n \quad \int_{\substack{\pi\in\mathfrak{P}_{\leq T}\\ \mathrm{Re}\,\pi=\lambda_0}} A(\phi',\phi)(\pi)\,\mathrm{d}\pi =_T \sum_{\mathfrak{C}'\in[S']} |\mathrm{Norm}_{W_L(M)}\mathfrak{S}_{\mathfrak{C}'}|^{-1} \int_{\substack{\pi\in\mathfrak{S}_{\mathfrak{C}',\leq T}\\ \mathrm{Re}\,\pi=o(\mathfrak{S}_{\mathfrak{C}'})+\mu_0}} \left(\sum_{w\in\mathrm{Stab}_{W_L(M)}(M,\mathfrak{P}_L)} (\mathrm{Res}^L_{w\mathfrak{S}_{\mathfrak{C}'}} A(\phi',\phi))(w\pi)\right) \mathrm{d}_{\mathfrak{S}_{\mathfrak{C}'}}\pi.$$

Now suppose that k is a function field. To be able to integrate on $\mathrm{Im}\, X_L^{\mathbf{G}}/X_L^{\mathbf{L}}$ we must start by regrouping the elements of $[S_{\mathfrak{X}_L}]$ under the action of $X_L^{\mathbf{L}}$ by translation; for $\mathfrak{C}_L \in [S_{\mathfrak{X}_L}]$ and for $\nu \in X_L^{\mathbf{L}}$, we define another element of $[S_{\mathfrak{X}_L}]$, denoted by $\mathfrak{C}_L \otimes \nu$, by setting:

$$\mathfrak{C}_L \otimes \nu := \{\mathfrak{S}_L \otimes \nu, \ \text{ where } \mathfrak{S}_L \in \mathfrak{C}_L\}.$$

It is a consequence of Remark V.3.13(c) applied to L instead of G that:

$$\mathfrak{C}_L \in \mathrm{Sing}^L \Leftrightarrow \mathfrak{C}_L \otimes \nu \in \mathrm{Sing}^L.$$

Obviously, we can have $\mathfrak{C}_L \otimes \nu = \mathfrak{C}_L$; we denote by V the subgroup of $X_L^{\mathbf{L}}$ satisfying this equality and by V' the subgroup of V consisting of elements ν such that

$$\mathfrak{S}_{\mathfrak{C}_L,L} \otimes \nu = \mathfrak{S}_{\mathfrak{C}_L,L}.$$

Set $\mathfrak{S}_{\tilde{\mathfrak{C}}_L,L} := \mathfrak{S}_{\mathfrak{C}_L,L} + X_L^{\mathbf{L}}$ and note that:

$$|V/V'| = |\mathrm{Norm}_{W_L(M)}\mathfrak{S}_{\tilde{\mathfrak{C}}_L,L}|/|\mathrm{Norm}_{W_L(M)}\mathfrak{S}_{\mathfrak{C}_L,L}|.$$

Again using V.3.13(c), we see that the function:

$$\sum_{w\in\mathrm{Stab}_{W_L(M)}(M,\mathfrak{P}_L)} (\mathrm{Res}^L_{w(\mathfrak{S}_{\mathfrak{C}_L,L})} A(\phi',\phi))(w\pi)\ \mathrm{d}_{(\mathfrak{S}_{\mathfrak{C}_L,L})}\pi$$

is invariant under V. We thus have:

$$|\mathrm{Norm}_{W_L(M)}\mathfrak{S}_{\mathfrak{C}_L,L}|^{-1} \times \int_{\substack{\pi\in\mu+(\mathfrak{S}_{\mathfrak{C}_L,L})\\ \mathrm{Re}\,\pi=o(\mathfrak{S}_{\mathfrak{C}_L,L})+\mu_0}} \sum_{w\in\mathrm{Stab}_{W_L(M)}(M,\mathfrak{P}_L)} \mathrm{Res}^L_{w(\mathfrak{S}_{\mathfrak{C}_L,L})} A(\phi',\phi)(w\pi)\ \mathrm{d}_{(\mathfrak{S}_{\mathfrak{C}_L,L})}\pi$$

$$= |\mathrm{Norm}_{W_L(M)}\mathfrak{S}_{\tilde{\mathfrak{C}}_L,L}|^{-1} \times \int_{\substack{\pi\in\mu+(\mathfrak{S}_{\mathfrak{C}_L,L})+V\\ \mathrm{Re}\,\pi=o(\mathfrak{S}_{\mathfrak{C}_L,L})+\mu_0}} \sum_{w\in\mathrm{Stab}_{W_L(M)}(M,\mathfrak{P}_L)} \mathrm{Res}^L_{w(\mathfrak{S}_{\mathfrak{C}_L,L})} A(\phi',\phi)(w\pi)\ \mathrm{d}_{(\mathfrak{S}_{\mathfrak{C}_L,L})}\pi.$$

We denote by $\tilde{\mathfrak{C}}_L$ the union of equivalence classes which can be deduced from $\mathfrak{C}_L$ by translation by an element of $X_L^{\mathbf{L}}$; the group V introduced above is the same for every element of $\tilde{\mathfrak{C}}_L$ and the construction just done for $\mathfrak{S}_{\mathfrak{C}_L,L}$ can be done for every element $\mathfrak{S}_{\mathfrak{C}'_L}$ contained in $\mathfrak{S}_{\tilde{\mathfrak{C}}_L,L}$ representing one of these classes. The residue data which occur at the origin of the real part do not depend on the chosen class, and summing over the contribution of all these elements comes down to replacing the domain of integration of the above integral by the domain $\mu + \mathfrak{S}_{\tilde{\mathfrak{C}}_L,L}$, which is to say that the contribution of conjugacy classes contained in $\tilde{\mathfrak{C}}_L$ is, with the obvious notation:

$$|\mathrm{Norm}_{W_L(M)}\mathfrak{S}_{\tilde{\mathfrak{C}}_L,L}|^{-1} \times \int_{\substack{\pi\in\mu+\mathfrak{S}_{\tilde{\mathfrak{C}}_L,L}\\ \mathrm{Re}\,\pi=o(\mathfrak{S}_{\tilde{\mathfrak{C}}_L,L})+\mu_0}} \sum_{w\in\mathrm{Stab}_{W_L(M)}(M,\mathfrak{P}_L)} (\mathrm{Res}^L_{w(\mathfrak{S}_{\tilde{\mathfrak{C}}_L,L})}A(\phi',\phi))(w\pi)\ \mathrm{d}_{\mathfrak{S}_{\tilde{\mathfrak{C}}_L,L}}\pi.$$

We can now integrate over $\mathrm{Im}\,X_L^{\mathbf{G}}/X_L^{\mathbf{L}}$ by setting $\tilde{\mathfrak{S}}_{\tilde{\mathfrak{C}}_L} := \mathfrak{S}_{\tilde{\mathfrak{C}}_L,L} + X_L^{\mathbf{G}}$. We obtain:

$$|\mathrm{Norm}_{W_L(M)}\mathfrak{S}_{\tilde{\mathfrak{C}}_L,L}|^{-1} \times \int_{\substack{\pi\in\tilde{\mathfrak{S}}_{\tilde{\mathfrak{C}}_L}\\ \mathrm{Re}\,\pi=o(\mathfrak{S}_{\tilde{\mathfrak{C}}_L})+\mu_0}} \sum_{w\in\mathrm{Stab}_{W_L(M)}(M,\mathfrak{P}_L)} (\mathrm{Res}^L_{w(\mathfrak{S}_{\tilde{\mathfrak{C}}_L,L})}A(\phi',\phi))(w\pi)\ \mathrm{d}_{\tilde{\mathfrak{S}}_{\tilde{\mathfrak{C}}_L,L}}\pi.$$

We must now redecompose along the connected components of $\tilde{\mathfrak{S}}_{\tilde{\mathfrak{C}}_L}$ and regroup those which are conjugate by an element of

$$\mathrm{Stab}_{W_L(M)}(M,\mathfrak{P}_L).$$

By VI.1.4 (a), if $\mathfrak{S}_{\tilde{\mathfrak{C}}_L}$ is one of these connected components, it satisfies:

$$\mathfrak{S}_{\tilde{\mathfrak{C}}_L} + X_L^{\mathbf{L}} = \tilde{\mathfrak{S}}_{\tilde{\mathfrak{C}}_L},$$

and the procedure is thus exactly the inverse of what was just explained. We find that the above expression is equal to:

$$\sum_{\mathfrak{C}} \int_{\substack{\pi\in\mathfrak{S}_{\mathfrak{C}'}\\ \mathrm{Re}\,\pi=o(\mathfrak{S}_{\mathfrak{C}'})+\mu_0}} |\mathrm{Norm}_{W_L(M)}\mathfrak{S}_{\mathfrak{C}'}|^{-1} \times \sum_{w\in\mathrm{Stab}_{W_L(M)}(M,\mathfrak{P}_L)} (\mathrm{Res}^L_{w\mathfrak{S}_{\mathfrak{C}'}}A(\phi',\phi))(w\pi)\ \mathrm{d}_{\mathfrak{S}_{\mathfrak{C}'}}\pi,$$

where $\mathfrak{C}'$ runs through the set of conjugacy classes under $\mathrm{Stab}_{W_L(M)}(M,\mathfrak{P}_L)$ of components of $\tilde{\mathfrak{S}}_{\tilde{\mathfrak{C}}_L}$. Finally, we obtain the exact analogue of $(6)_n$:

$$(6)_f \quad \int_{\substack{\pi\in\mathfrak{P} \\ \mathrm{Re}\,\pi=\lambda_0}} A(\phi',\phi)(\pi)\,\mathrm{d}\pi = \sum_{\mathfrak{C}'\in[S']} |\mathrm{Norm}_{W_L(M)}\mathfrak{S}_{\mathfrak{C}'}|^{-1} \times \int_{\substack{\pi\in\mathfrak{S}_{\mathfrak{C}'} \\ \mathrm{Re}\,\pi=o(\mathfrak{S}_{\mathfrak{C}'})+\mu_0}} \left(\sum_{w\in\mathrm{Stab}_{W_L(M)}(M,\mathfrak{P}_L)} (\mathrm{Res}^L_{w\mathfrak{S}_{\mathfrak{C}'}} A(\phi',\phi))(w\pi) \right) \mathrm{d}_{\mathfrak{S}_{\mathfrak{C}'}}\pi.$$

Let us return to the case of arbitrary k. We note for what follows that:

$$\mathrm{Stab}_{W_L(M)}(M,\mathfrak{P}_L) = \mathrm{Stab}_{W_L(M)}(M,\mathfrak{P}).$$

In what follows, we require that ϕ vanish to a very high order on the elements of $S_{(M,\mathfrak{P})}$ which are not stable under translation by $\mathrm{Re}\,X^{\mathbf{G}}_L$. Let $\mathfrak{S}_L \in S_{(M,\mathfrak{P}_L)}$. We define $\mathfrak{S}$ as above, saturating by $\mathrm{Re}\,X^{\mathbf{G}}_L$, and let $\mathfrak{S}' \in S_{(M,\mathfrak{P})}$ be such that $\mathrm{Re}\,\mathfrak{S}'$ properly intersects the segment of $\mathrm{Re}\,\mathfrak{S}$ joining $o(\mathfrak{S})+\mu_0$ to $o(\mathfrak{S})$. Then $\mathfrak{S}'$ is not stable under $\mathrm{Re}\,X^{\mathbf{G}}_L$ since $\mu_0 \in \mathrm{Re}\,X^{\mathbf{G}}_L$ and all the points of the segment are of the form $o(\mathfrak{S})+t\mu_0$ with $t\in\mathbb{R}$. As in V.3.1, for every $\mathfrak{S}\in S_{(M,\mathfrak{P})}$, we fix a general point near $o(\mathfrak{S})$, denoted by $z(\mathfrak{S})$. An immediate application of the residue theorem gives, using $(6)_n$ and $(6)_f$:

$$(7) \quad \langle\theta_{\phi'},\theta_\phi\rangle =_T \sum_{\mathfrak{C}'\in[S']} |\mathrm{Norm}_{W_L(M)}\mathfrak{S}_{\mathfrak{C}'}|^{-1} \times \int_{\substack{\pi\in\mathfrak{S}_{\mathfrak{C}',\le T} \\ \mathrm{Re}\,\pi=z(\mathfrak{S}_{\mathfrak{C}'})}} \sum_{w\in\mathrm{Stab}_{W_L(M)}(M,\mathfrak{P})} (\mathrm{Res}^L_{w\mathfrak{S}_{\mathfrak{C}'}} A(\phi',\phi))(w\pi)\,\mathrm{d}_{\mathfrak{S}_{\mathfrak{C}'}}\pi.$$

We also have:

$$(8) \quad \langle\theta_{\phi'},\theta_\phi\rangle =_T \sum_{\mathfrak{C}\in[S_{(M,\mathfrak{P})}]} |\mathrm{Norm}\,\mathfrak{S}_{\mathfrak{C}}|^{-1} \times \int_{\substack{\pi\in\mathfrak{S}_{\mathfrak{C},\le T} \\ \mathrm{Re}\,\pi=o(\mathfrak{S}_{\mathfrak{C}})}} \sum_{w\in\mathrm{Stab}(M,\mathfrak{P})} (\mathrm{Res}^G_{w\mathfrak{S}_{\mathfrak{C}}} A(\phi',\phi))(w\pi)\,\mathrm{d}_{\mathfrak{S}_{\mathfrak{C}}}\pi.$$

Given the vanishing properties of ϕ, we can replace the sum over $\mathrm{Stab}(M,\mathfrak{P})$ by the sum over

$$W'_{\mathfrak{C}} := \{w\in\mathrm{Stab}(M,\mathfrak{P}) | w\mathfrak{S}_{\mathfrak{C}} \text{ is stable under } \mathrm{Re}\,X^{\mathbf{G}}_L\}.$$

Let $w\in W'_{\mathfrak{C}}$; we saw in VI.1.3 that the function:

$$\sum_{\sigma\in\mathrm{Stab}_{W_L(M)}(M,\mathfrak{P})} (\mathrm{Res}^G_{\sigma w\mathfrak{S}_{\mathfrak{C}}} A(\phi',\phi))(\sigma w\pi)$$

is holomorphic on $\{\pi\in\mathfrak{S}_{\mathfrak{C}} | \mathrm{Re}\,\pi = o(\mathfrak{S}_{\mathfrak{C}})\}$. With an obvious change of variables, for $w \in W'_{\mathfrak{C}}$:

$$\int_{\substack{\pi\in\mathfrak{S}_{\mathfrak{C},\le T}\\ \mathrm{Re}\,\pi=o(\mathfrak{S}_{\mathfrak{C}})}} \sum_{\sigma\in \mathrm{Stab}\, W_L(M)(M,\mathfrak{P})} (\mathrm{Res}^{L}_{\sigma w\mathfrak{S}_{\mathfrak{C}_L,L}} A(\phi',\phi))(\sigma w\pi)\ \mathrm{d}_{\mathfrak{S}}\pi$$

depends only on the right class of w modulo $\mathrm{Norm}\,\mathfrak{S}_{\mathfrak{C}}$. To write (8) using only the conjugacy classes under the action of W_L, we identify

$$\mathrm{Stab}\, W_{L(M)}((M,\mathfrak{P}))\backslash W'_{\mathfrak{C}}/\mathrm{Norm}\,\mathfrak{S}_{\mathfrak{C}}$$

with the set of conjugacy classes of $\mathrm{Stab}\, W_{L(M)}((M,\mathfrak{P}))$ in the set of elements of $\mathfrak{C}$ stable under translation by $\mathrm{Re}\,X_L^{\mathrm{G}}$, and we obtain for $\mathfrak{C}$ fixed as above:

$$\sum_{\mathfrak{C}'} x_{\mathfrak{C}'} \int_{\substack{\pi\in\mathfrak{S}_{\mathfrak{C}',\le T}\\ \mathrm{Re}\,\pi=o(\mathfrak{S}_{\mathfrak{C}'})}} \sum_{w\in \mathrm{Stab}\, W_L(M)(M,\mathfrak{P})} (\mathrm{Res}^{\mathrm{G}}_{w\mathfrak{S}_{\mathfrak{C}'}} A(\phi',\phi))(w\pi)\ \mathrm{d}_{\mathfrak{S}_{\mathfrak{C}'}}\pi,$$

where $\mathfrak{C}'$ runs through a set of representatives for the conjugacy classes described above and where, fixing $w^\bullet$ such that $w^\bullet\mathfrak{C} = \mathfrak{C}'$,

$$x_{\mathfrak{C}'}^{-1} := |\mathrm{Norm}\,\mathfrak{S}_{\mathfrak{C}} \cap (w^\bullet)^{-1}\mathrm{Stab}\, W_{L(M)}(M,\mathfrak{P})w^\bullet| = |\mathrm{Norm}\, W_{L(M)}\mathfrak{S}_{\mathfrak{C}'}|.$$

Thus (8) can be rewritten:

$$\langle\theta_{\phi'},\theta_\phi\rangle =_T \sum_{\mathfrak{C}'\in[S']} |\mathrm{Norm}\, W_{L(M)}\mathfrak{S}_{\mathfrak{C}'}|^{-1} \times$$

(9)
$$\int_{\substack{\pi\in\mathfrak{S}_{\mathfrak{C}',\le T}\\ \mathrm{Re}\,\pi=o(\mathfrak{S}_{\mathfrak{C}'})}} \sum_{w\in \mathrm{Stab}\, W_L(M)(M,\mathfrak{P})} (\mathrm{Res}^{\mathrm{G}}_{w\mathfrak{S}_{\mathfrak{C}'}} A(\phi',\phi))(w\pi)\ \mathrm{d}_{\mathfrak{S}_{\mathfrak{C}'}}\pi,$$

with the notation already introduced. Up to $=_T$, we can replace $o(\mathfrak{S}_{\mathfrak{C}'})$ by $z(\mathfrak{S}_{\mathfrak{C}'})$. The equality $=_T$ of (7) and (9) will give the proposition; we put a total order on the set of elements of $[S']$ satisfying the same properties as those used in V.3.3. Let $\mathfrak{C}'$ be a maximal element for this order, if one exists, such that there exists $\mathfrak{S} \in \mathfrak{C}'$, and $\phi',\phi \in P_{\mathfrak{X}}$ as above satisfying:

$$(\mathrm{Res}^{L}_{\mathfrak{S}_L} - \mathrm{Res}^{\mathrm{G}}_{\mathfrak{S}})A(\phi',\phi) \not\equiv 0.$$

As in V.3.6, we see that we can keep the property of non-vanishing while requiring that ϕ be zero to a high order on all the elements $\mathfrak{C}''$ of $[S']$ satisfying $\mathfrak{C}'' < \mathfrak{C}'$. These vanishing properties of ϕ and the maximality of $\mathfrak{C}'$ imply:

(10)
$$\int_{\substack{\pi\in\mathfrak{S}_{\le T}\\ \mathrm{Re}\,\pi=z(\mathfrak{S})}} (\mathrm{Res}^{L}_{\mathfrak{S}} - \mathrm{Res}^{\mathrm{G}}_{\mathfrak{S}})A(\phi',\phi)(\pi)\ \mathrm{d}_{\mathfrak{S}}\pi =_T 0$$

while:

(11)
$$(\mathrm{Res}^{L}_{\mathfrak{S}} - \mathrm{Res}^{\mathrm{G}}_{\mathfrak{S}})A(\phi',\phi) \not\equiv 0.$$

If k is a function field, the contradiction results from the technical Lemma V.3.6.

If k is a number field, we use still V.3.6 to impose the following properties on ϕ: for every f function on $\mathfrak{P}$ having polynomial growth:

(12) $(\mathrm{Res}^{L}_{\mathfrak{S}} - \mathrm{Res}^{\mathbf{G}}_{\mathfrak{S}})A(\phi', f\phi) = f_{|\mathfrak{S}}(\mathrm{Res}^{L}_{\mathfrak{S}} - \mathrm{Res}^{\mathbf{G}}_{\mathfrak{S}})A(\phi', \phi)$,

(13) $(\mathrm{Res}^{L}_{\mathfrak{S}} - \mathrm{Res}^{\mathbf{G}}_{\mathfrak{S}})A(\phi', \phi)$ is holomorphic on $\{\pi \in \mathfrak{S}_{\leq T} | \mathrm{Re}\,\pi = o(\mathfrak{S})\}$,

(14) $$\int_{\substack{\pi\in\mathfrak{S}_{\leq T}\\ \mathrm{Re}\,\pi=o(\mathfrak{S})}} (\mathrm{Res}^{L}_{\mathfrak{S}} - \mathrm{Res}^{\mathbf{G}}_{\mathfrak{S}})A(\phi', \phi)(\pi)\; d_{\mathfrak{S}}\pi \neq 0.$$

(13) and (10) imply (see the residue theorem):

$$\int_{\substack{\pi\in\mathfrak{S}_{\leq T}\\ \mathrm{Re}\,\pi=o(\mathfrak{S})}} (\mathrm{Res}^{L}_{\mathfrak{S}} - \mathrm{Res}^{\mathbf{G}}_{\mathfrak{S}})A(\phi', \phi)(\pi)\; d_{\mathfrak{S}}\pi =_T 0$$

and (12) implies that this is really an equality (see the proof of V.2.9). This gives a contradiction with (14).

VI.1.5. Functional equation for residues of Eisenstein series
We fix a cuspidal datum $(M, \mathfrak{P})$ as above; let $w \in W(M)$ and $\mathfrak{S}$ be an element of $S_{(M,\mathfrak{P})}$. Then w takes a residue datum $\mathrm{Res}_{\mathfrak{S}}$ for $\mathfrak{S}$ in $\mathfrak{P}$ into a residue datum $w\mathrm{Res}_{\mathfrak{S}}$ for $w\mathfrak{S}$ in $w\mathfrak{P}$ and we have:

Proposition *Let $w, \mathfrak{S}$ be as above and suppose $\mathfrak{S} \in \mathrm{Sing}^G$ and $w \in W(M_{\mathfrak{S}})$. Then:*

(i) $\forall \phi' \in P_{\mathfrak{X}}, \phi \in P_{w(M,\mathfrak{P})}$, $(w\mathrm{Res}^{\mathbf{G}}_{\mathfrak{S}})A(\phi', \phi) \equiv \mathrm{Res}^{G}_{w\mathfrak{S}}A(\phi', \phi)$.

This can still be expressed by:

$$\forall \phi \in P_{w(M,\mathfrak{P})},\quad (w\mathrm{Res}^{\mathbf{G}}_{\mathfrak{S}})E(\phi, \pi) \equiv \mathrm{Res}^{G}_{w\mathfrak{S}}E(\phi, \pi),$$

the equality of meromorphic functions on $w\mathfrak{S}$.

(ii) $\forall \phi \in P_{(M,\mathfrak{P})}$, $\mathrm{Res}^{G}_{\mathfrak{S}}E(\phi, \pi) \equiv \left(\mathrm{Res}^{G}_{w\mathfrak{S}}E(M(w, w^{-1}\pi')\phi, \pi')\right)_{\pi'=w\pi}$.

We set $L := M_{\mathfrak{S}}$ and $L' = wM_{\mathfrak{S}}w^{-1}$; they are standard Levis of G which may be equal. We fix ξ_L as in the introduction of VI and set $\xi_{Le} = w\xi_L$ which in the function field case can be different from ξ_L even if $Le = L$ and $w\mathfrak{P} = \mathfrak{P}$. We have then defined $\mathfrak{P}_L, \mathfrak{X}_L$ and $(w\mathfrak{P})_{L'}, \mathfrak{X}_{L'}$, which satisfies $(w\mathfrak{P})_{L'} = w\mathfrak{P}_L$. It is clear that w takes the decomposition of the scalar product of pseudo-Eisenstein series for L associated with $\mathfrak{X}_L$ to the decomposition of the scalar product of pseudo-Eisenstein series for L' and $\xi_{L'}$. This shows that in V.3.13(a), for L', we can take for

$\mathrm{Res}_{w\mathfrak{S}_L}^{L'}$ the residue datum $w\mathrm{Res}_{\mathfrak{S}_L}^{L}$. Taking into account VI.1.4(b), this immediately gives (i).

(ii) The functional equation for the Eisenstein series (see IV.1.10) gives:

$$E(\phi, \pi) = E(M(w, \pi)\phi, w\pi),$$

this equality depending meromorphically on $\pi \in \mathfrak{P}$. Thus:

$$\left(\mathrm{Res}_{w\mathfrak{S}}^{G} E(M(w, w^{-1}\pi')\phi, \pi')\right)_{\pi'=w\pi} = (w^{-1}\mathrm{Res}_{w\mathfrak{S}}^{G})(E(\phi, \pi)).$$

By (i), this is still equal to $\mathrm{Res}_{\mathfrak{S}}^{G} E(\phi, \pi)$, i.e. (ii).

VI.1.6. Positivity property of the origin of singular planes

Fix $(M, \mathfrak{P}) \in \mathfrak{X}$.

VI.1.6.(a) A lemma on root systems

Lemma *Let M be a standard Levi subgroup of G and V a vector subspace of* $\mathrm{Re}\, X_M^{\mathbf{G}}$ *defined by equations of the type $\langle \lambda, \check{\alpha} \rangle = 0$ where $\alpha \in R(T_M, G)$. Then:*

(i) *$V^{\perp}$ is generated as a $\mathbb{R}$-vector space by the elements of $V^{\perp} \cap R^{+}(T_M, G)$.*

(ii) *$V^{\perp} \cap R(T_M, G)$ is the projection onto* $\mathrm{Re}\, X_M^{\mathbf{G}}$ *of a sub-root system of $R(T_0, G)$.*

(iii) *Let $\lambda \in V^{\perp}$. Suppose that $\langle \lambda, \check{\alpha} \rangle > 0$, for every element α of $V^{\perp} \cap R^{+}(T_M, G)$. Then λ belongs to the interior of the cone of $V^{\perp}$ generated by the elements of $V^{\perp} \cap R^{+}(T_M, G)$.*

Let $\alpha \in R(T_M, G)$ be such that $\langle \lambda, \check{\alpha} \rangle = 0$ for every $\lambda \in V$. Then by definition of the scalar product on $\mathrm{Re}\, X_M^{\mathbf{G}}$, the element α is in $V^{\perp}$. Thus, by the form of V, we know that the codimension of V is the dimension of the vector space generated by $V^{\perp} \cap R^{+}(T_M, G)$. Now this codimension is also the dimension of the orthogonal complement of V, which gives (i).

(ii) Taking into account the inclusion of $\mathrm{Re}\, X_M^{\mathbf{G}}$ in $\mathrm{Re}\, X_{M_0}^{\mathbf{G}}$, we consider V as a subspace of $\mathrm{Re}\, X_{M_0}^{\mathbf{G}}$. It is defined by the equations $\langle \lambda, \check{\alpha} \rangle = 0$ where now α is either an element of $R(T_0, M)$, or projects onto an element of $V^{\perp} \cap R(T_M, G)$. The orthogonal complement of V in $\mathrm{Re}\, X_{M_0}^{\mathbf{G}}$ contains $\mathrm{Re}\, X_{M_0}^{\mathbf{M}}$ and projects modulo $\mathrm{Re}\, X_{M_0}^{\mathbf{G}}$ onto the orthogonal complement of V in $\mathrm{Re}\, X_M^{\mathbf{G}}$. To prove (ii), we can thus suppose that $M = M_0$, which we will do,and show that $\Phi := V^{\perp} \cap R(T_0, G)$ is a root system. By ([S], nos. 1, 3), there are three conditions to check:

Φ is finite, not containing 0;

$\forall\alpha,\beta \in \Phi, \langle\alpha,\check{\beta}\rangle \in \mathbf{Z}$;
s_α, the symmetry associated with α, leaves Φ stable.

These properties are more or less immediate. For the last one we note that we have $s_\alpha(\beta) \in R(T_0, G)$, and $s_\alpha(\beta) = \beta - \langle\beta, \check{\alpha}\rangle\alpha$ belongs like β and α to $V^\perp$.

(iii) To prove (iii), we begin by replacing λ by $\lambda + \epsilon$ where $\epsilon \in \mathrm{Re}\, X_{M_0}^{\mathbf{M}}$ is in the positive cone and near 0, supposing that $M = M_0$. In the proof of (ii) we saw that $V^\perp \cap R(T_0, G)$ is a root system, in which case (iii) becomes a classical property of root systems (a direct consequence of Langlands' geometric lemma, see [L] Lemma 2.4.) We obtain (iii) by projecting onto $\mathrm{Re}\, X_M^{\mathbf{G}}$.

VI.1.6(b) Lemma

*Fix a finite set $\mathfrak{F}$ of **K**-types, and large real numbers T and T' with $T' \gg T$. There exists a subset of $S_{(M,\mathfrak{P})}^{\mathfrak{F}}$, denoted by S', whose elements $\mathfrak{S}$ are equipped with a residue datum, denoted by $\mathrm{Res}_{\mathfrak{S}}^1$ and with a point $z'(\mathfrak{S})$ belonging to $\mathrm{Re}\,\mathfrak{S}$ satisfying:*

(i) *$\forall \mathfrak{S} \in S', o(\mathfrak{S})$ belongs to the interior of the positive cone generated by the elements of $(\mathrm{Re}\,\mathfrak{S}^0)^\perp \cap R^+(T_M, G) =:^+ (\mathrm{Re}\,\mathfrak{S}^0)^\perp$.*

(ii) *$z'(\mathfrak{S})$ is very near $o(\mathfrak{S})$ but T'-general (see V.1.5) and also satisfies (i)*

(iii) *$\forall \phi' \in P_{\mathfrak{X}}^{\mathfrak{F}}, \forall \phi \in P_{(M,\mathfrak{P})}^{\mathfrak{F}}$*

$$\langle \theta_{\phi'}, \theta_\phi \rangle =_T \sum_{\mathfrak{S}\in S'} \int_{\pi\in\mathfrak{S}_{\leq T}|\mathrm{Re}\,\pi = z'(\mathfrak{S})} \mathrm{Res}_{\mathfrak{S}}^1 A(\phi', \phi)(\pi)\, \mathrm{d}_{\mathfrak{S}}\pi.$$

We adopt the notation $^+(\mathrm{Re}\,\mathfrak{S}^0)^\perp$ for every element $\mathfrak{S}$ of $S_{(M,\mathfrak{P})}$. We proceed as in the proof of the first step of V.2.3, checking step by step that it suffices to run through a set of very particular paths. We reuse the proof of V.2.3 but without fixing the point $z(\mathfrak{S})$ beforehand; its role is played by $z'(\mathfrak{S})$ which is also constructed in the descending induction. More precisely, in the description of the construction by induction, we add to V.2.3(2) and (3) the following properties (the notation is that of V.2.3):

(*) Every $y \in P(\mathfrak{T})$, such that $\mathrm{Res}_{\mathfrak{T}}^y \not\equiv 0$ on the set of functions $A(\phi', \phi)$, belongs to the interior of the convex envelope generated by the closure of $^+(\mathrm{Re}\,\mathfrak{T}^0)^\perp$ and by $R^+(T_M, G)$;

(**) For $\mathfrak{T} \in S(d-1)$, such that there exists $y \in P(\mathfrak{T})$ with $\mathrm{Res}_{\mathfrak{T}}^y \not\equiv 0$ as in (*), $o(\mathfrak{T}) \in {}^+(\mathrm{Re}\,\mathfrak{T}^0)^\perp$.

We fix $z'(\mathfrak{T})$, T'-general but very near $o(\mathfrak{T})$, also satisfying $(*)$. Let us show what we must add to the proof of V.2.3 to obtain these additional properties.
We begin by showing that $(*)$ implies $(**)$: fix $y \in P(\mathfrak{T})$ satisfying $(*)$. Then there exists Y in the closure of ${}^{+}(\operatorname{Re}\mathfrak{T}^0)^{\perp}$ and Y' belonging to the positive Weyl chamber such that

$$y = Y + Y'.$$

Since $\operatorname{Re}\mathfrak{T}$ is an affine space containing y and $o(\mathfrak{T})$, there exists $\mu \in \operatorname{Re}\mathfrak{T}^0$ such that:

$$o(\mathfrak{T}) = y + \mu.$$

Since Y is in the closure of ${}^{+}(\operatorname{Re}\mathfrak{T}^0)^{\perp}$, it suffices to prove that $Y' + \mu$ is in ${}^{+}(\operatorname{Re}\mathfrak{T}^0)^{\perp}$. By VI.1.6(a)(iii) applied to $V = \operatorname{Re}\mathfrak{T}^0$ and $\lambda = Y' + \mu$, it suffices to show that $\langle Y' + \mu, \check{\alpha}\rangle > 0$ for every $\alpha \in (\operatorname{Re}\mathfrak{T}^0)^{\perp} \cap R^{+}(T_M, G)$. Now, this is equal to $\langle Y', \check{\alpha}\rangle$ and the property of Y' gives the result. This proves $(**)$.

Let us return to the proof by induction. To begin with, if k is a number field we must define $z'(\mathfrak{P})$ and if k is a function field, the point z' for an irreducible component of $\mathfrak{P}$ (all these components have the same real part and we take the same point). We simply require that this point be T'-general, very near 0 and in the interior of the positive Weyl chamber (for $R^{+}(T_M, G)$). The starting point λ_0 is also in the interior of this cone and to apply the residue theorem (V.1.5), we fix a path in the interior of this cone, as we may. Thus the elements $y(\mathfrak{T})$ which occur obviously satisfy $(*)$ Now let d be arbitrary. We must consider $\mathfrak{S} \in S(d)$ with the induction hypothesis indicating that every $y \in P(\mathfrak{S})$, thus in particular the $z'(\mathfrak{S})$ which was constructed, satisfies $(*)$for $\mathfrak{S}$. We apply the residue theorem to a path of $\operatorname{Re}\mathfrak{S}$ joining one of the points of $P(\mathfrak{S})$ to $z'(\mathfrak{S})$, which we choose to be very near the segment joining these points: in particular every point of this path must satisfy $(*)$ for $\mathfrak{S}$. The only thing to remark is that for every hyperplane $\mathfrak{T}$ of $\mathfrak{S}$ belonging to $S_{\mathfrak{T}}$, $(*)$ for $\mathfrak{S}$ implies $(*)$ for $\mathfrak{T}$: for this it suffices to check that the closure of ${}^{+}(\operatorname{Re}\mathfrak{T}^0)^{\perp}$ contains that of ${}^{+}(\operatorname{Re}\mathfrak{S}^0)^{\perp}$. This is an immediate consequence of the inclusion $\operatorname{Re}\mathfrak{T}^0 \subset \operatorname{Re}\mathfrak{S}^0$ and of the definitions. This concludes the proof.

VI.1.6(c) Proposition
Let $\mathfrak{S} \in S_{(M,\mathfrak{P})}$ be such that the operator $\operatorname{Res}^{G}_{\mathfrak{S}}$ is not identically zero on the functions $A(\phi', \phi)$ (i.e. $\mathfrak{S} \in \operatorname{Sing}^{G}$); then $o(\mathfrak{S})$ is in the interior of the positive cone generated by the elements of $R^{+}(T_M, M_{\mathfrak{S}})$.

For every $\mathfrak{S} \in S_{(M,\mathfrak{P})}$, we fix a point $z(\mathfrak{S})$ from the set of these points satisfying V.2.2(*). Observe that the points $z'(\mathfrak{S})$ constructed in the preceding paragraph unfortunately do not work in general. For the descending induction on d going from $\dim \operatorname{Re} X_M^{\mathbf{G}}$ to -1, we will prove that there exists a subset of $\bigcup_{d' \le d} S(d')$, denoted by S'_d, and for every $\mathfrak{S} \in S'_d$ a residue datum $\operatorname{Res}_{\mathfrak{S},d}$ and a point $z'(\mathfrak{S})$ of $\operatorname{Re}\mathfrak{S}$ satisfying:

$$\langle \theta_{\phi'}, \theta_\phi \rangle =_T \sum_{\mathfrak{S} \in \operatorname{Sing}^G \cap (\bigcup_{d'>d} S(d'))} |\operatorname{Norm} \mathfrak{S}|^{-1} \times$$

(1) $$\int_{\substack{\pi \in \mathfrak{S}_{\le T} \\ \operatorname{Re}\pi = z(\mathfrak{S})}} \sum_{w \in \operatorname{Norm}\mathfrak{S}} (\operatorname{Res}_{\mathfrak{S}}^G A(\phi', \phi))(w\pi)\, \mathrm{d}_{\mathfrak{S}}\pi$$

$$+ \sum_{\mathfrak{S} \in S'_d} \int_{\substack{\pi \in \mathfrak{S}_{\le T} \\ \operatorname{Re}\pi = z'(\mathfrak{S})}} \operatorname{Res}_{\mathfrak{S},d} A(\phi', \phi)(\pi)\, \mathrm{d}_{\mathfrak{S}}\pi.$$

(2) for every $\mathfrak{S} \in (\operatorname{Sing}^G \cap \bigcup_{d'>d} S(d')) \cup S'_d$, $o(\mathfrak{S})$ satisfies the condition of the statement and $z'(\mathfrak{S})$ is T'-general but very near $o(\mathfrak{S})$.

We first note that for $\mathfrak{S} \in S_{(M,\mathfrak{P})}$, defining ${}^+(\operatorname{Re}\mathfrak{S}^0)^\perp$ as in (b), we have:

(3) ${}^+(\operatorname{Re}\mathfrak{S}^0)^\perp$ is contained in the interior of the positive cone generated by the elements of $R^+(T_M, M_{\mathfrak{S}})$.

Every element of $R^+(T_M, G) \cap (\operatorname{Re}\mathfrak{S}^0)^\perp$ is an element of $R^+(T_M, M_{\mathfrak{S}})$; it thus suffices to check that every element of $R^+(T_M, M_{\mathfrak{S}})$ which is a projection of a simple reduced root of $R^+(T_0, M_{\mathfrak{S}})$ occurs in the decomposition of at least one element of $R^+(T_M, G) \cap (\operatorname{Re}\mathfrak{S}^0)^\perp$. But this is the definition of $M_{\mathfrak{S}}$.

Fix $S', \operatorname{Res}_{\mathfrak{S}}^1$ and $z'(\mathfrak{S})$ with the properties of (b). For $d = \dim \operatorname{Re} X_M^{\mathbf{G}}$, we set $S'_d = S'$ and $\operatorname{Res}_{\mathfrak{S},d} = \operatorname{Res}_{\mathfrak{S}}^1$. It is a consequence of (b) and (3) that (1) and (2) are satisfied for this value of d. When we have proved the assertion for d we prove it for $d-1$: we first check that for every $\mathfrak{S} \in S(d)$, we can take $\operatorname{Res}_{\mathfrak{S}}^G = \operatorname{Res}_{\mathfrak{S},d}$ if $\mathfrak{S} \in S(d) \cap S'_d$ and $\operatorname{Res}_{\mathfrak{S}}^G = 0$ if $\mathfrak{S} \in S(d) - S'_d$. This is an immediate application of V.2.4 and V.2.5. Now fix $\mathfrak{S} \in S(d) \cap S'_d$. To start with we must replace $z'(\mathfrak{S})$ by $z(\mathfrak{S})$ and then replace the function $\pi \mapsto \operatorname{Res}_{\mathfrak{S}}^G A(\phi', \phi)(\pi)$, which, we recall, is equal to $\operatorname{Res}_{\mathfrak{S},d} A(\phi', \phi)$, by $\pi \mapsto |\operatorname{Norm}\mathfrak{S}|^{-1} \sum_{w \in \operatorname{Norm}\mathfrak{S}} (\operatorname{Res}_{\mathfrak{S}}^G A(\phi', \phi))(w\pi)$ and show that (1) and (2) remain true. Now these operations are performed using V.2.5 and V.2.6, where the only elements which actually occur are $\mathfrak{T}$ of $S_{(M,\mathfrak{P})}$ of the form $\mathfrak{T} = \bigcap_{\mathfrak{H} \in J} \mathfrak{H}$ where J is a set of hyperplanes of $\mathfrak{S}$, contained in $S_{(M,\mathfrak{P})}$, intersecting $\{\pi \in \mathfrak{S}_{\le T} | \operatorname{Re}\pi = o(\mathfrak{S})\}$ and singular for at least one of the functions $\pi \in \mathfrak{S} \mapsto \operatorname{Res}_{\mathfrak{S}}^G A(\phi', \phi)(\pi)$. For every $\mathfrak{H} \in J$, we have $M_{\mathfrak{S}} = M_{\mathfrak{H}}$ by VI.1.2. Since $\mathfrak{T} = \cap_{\mathfrak{H} \in J} \mathfrak{H}$, $\mathfrak{T}$ is, like every $\mathfrak{H} \in J$, stable under translation by $\mathrm{Re}\,X^G_{M_{\mathfrak{S}}}$, so $M_{\mathfrak{T}}$ contains $M_{\mathfrak{S}}$. As $\mathfrak{T} \subset \mathfrak{S}$, we have the opposite inclusion, so $M_{\mathfrak{T}} = M_{\mathfrak{S}}$. Moreover $o(\mathfrak{T}) = o(\mathfrak{S})$ and (1) and (2) remain true for $d-1$, which concludes the proof.

VI.1.6.(d) Corollary

Let $\mathfrak{C}$ be an element of $[S_{\mathfrak{X}}]$ intersecting Sing^G. We use the notation $L^2_{\mathfrak{C}}$ of V.3.14. Either $L^2_{\mathfrak{C}}$ is contained in the discrete spectrum or there exists $\mathfrak{S} \in \mathfrak{C} \cap \mathrm{Sing}^G$ such that $M_{\mathfrak{S}} \subsetneq G$. In the first case every element of $\mathfrak{C}$ is reduced to a point which satisfies the positivity conditions of Proposition III.3.1.

Suppose that for every $\mathfrak{S} \in \mathfrak{C}$, we have $M_{\mathfrak{S}} = G$. We fix $\mathfrak{S} \in \mathfrak{C} \cap \mathrm{Sing}^G$ and we denote by $(M, \mathfrak{P})$ the cuspidal datum attached to $\mathfrak{S}$. We fix $\phi \in P_{(M,\mathfrak{P})}$ and a general point π_0 of $\{\pi \in \mathfrak{S} | \mathrm{Re}\,\pi = o(\mathfrak{S})\}$ such that $\mathrm{Res}^G_{\mathfrak{S}} E(\phi, \pi)$ is holomorphic and non-zero in π_0. We will show that $\mathrm{Res}^G_{\mathfrak{S}} E(\phi, \pi_0)$ is an automorphic form on $G(k)\backslash \mathbf{G}$, square integrable modulo the centre. For this, we calculate its cuspidal support; see III.3. By V.3.16, it is contained in the set of elements of the form $w\pi_0$ where $w \in W(M)$ satisfies $-w\overline{\mathfrak{S}} \in \mathrm{Sing}^G \cap \mathfrak{C}$. Thus, for such a w,

$$\mathrm{Re}\, w\,\pi_0 = -o(-w\overline{\mathfrak{S}}).$$

By assumption $M_{-w\overline{\mathfrak{S}}} = G$, and by VI.1.6(c), since $-w\overline{\mathfrak{S}} \in \mathrm{Sing}^G$, the point $-o(-w\mathfrak{S})$ satisfies the negativity condition needed in I.4.11. By III.3.1, the imaginary parts of the cuspidal exponents are in a finite set, which implies that $\mathrm{Im}\ \mathfrak{S}$ is a finite set. This is then also true for $\mathfrak{S}^0$ and, by connectedness, $\mathfrak{S}$ is reduced to a point. This proves the lemma.

Remark *The two cases of the lemma are mutually exclusive.*

Indeed, it is a consequence of V.3.13(a)(iii) that $L^2_{\mathfrak{C}}$ is in the discrete spectrum if and only if the elements of $\mathfrak{C}$ are reduced to a point. It thus suffices to check that if $\mathfrak{S} \in S_{\mathfrak{X}}$ is reduced to a point then $M_{\mathfrak{S}} = G$; this is trivially a consequence of the fact that for all proper standard Levis L of G, the dimension of $\mathrm{Re}\,X^G_L$ is strictly positive.

VI.1.7. Association classes of standard Levis attached to an element of $[S_{\mathfrak{X}}]$

Proposition *Let $\mathfrak{C} \in [S_{\mathfrak{X}}]$ intersect Sing^G; then there exists a unique association class of standard Levis of G, denoted by $L(\mathfrak{C})$, such that for every $L \in L(\mathfrak{C})$ there exists $\mathfrak{S} \in \mathfrak{C} \cap \mathrm{Sing}^G$ satisfying:*

$$\mathrm{Re}\,\mathfrak{S}^0 = \mathrm{Re}\,X^G_L.$$

The proof is by induction on the semi-simple rank of G: by VI.1.6(d), if for every $\mathfrak{S} \in \mathfrak{C}$, we also have $\operatorname{Re}\mathfrak{S}^0 = 0$, every element of $\mathfrak{C}$ is then reduced to a point. The proposition is then trivially true. Suppose that this is not the case. Then there exists $\mathfrak{S} \in \mathfrak{C} \cap \operatorname{Sing}^G$ such that $M_{\mathfrak{S}} \neq G$. We fix $\mathfrak{S}$ with this property and we denote by $(M, \mathfrak{P})$ the cuspidal datum of $\mathfrak{S}$. We set $L' := M_{\mathfrak{S}}$. We define $\mathfrak{P}_{L'}$ after choice of a character $\xi_{L'}$ as in the introduction to VI. We fix an irreducible component $\mathfrak{S}_{L'}$ of $\mathfrak{S} \cap \mathfrak{P}_{L'}$ and we write $\mathfrak{C}_{L'}$ for the equivalence class generated by $\mathfrak{S}_{L'}$. We know (see VI.1.4(b)) that $\mathfrak{C}_{L'}$ intersects $\operatorname{Sing}^{L'}$. We apply the induction hypothesis to L': it implies the existence of a standard Levi L of L', such that

$$\exists\ \mathfrak{S}'_{L'} \in \operatorname{Sing}^{L'} \cap \mathfrak{C}_{L'} \text{ such that } \operatorname{Re}\mathfrak{S}'^{0}_{L'} = \operatorname{Re}X_L^{L'}$$

We denote by w an element of the Weyl group of L' satisfying:

$$w\mathfrak{S}_{L'} = \mathfrak{S}'_L.$$

We have $w\mathfrak{S} \in \mathfrak{C}$ and $w\operatorname{Re}\mathfrak{S}^0 = w\operatorname{Re}\mathfrak{S}^0_{L'} + \operatorname{Re}X^G_{L'} = \operatorname{Re}X^G_L$. Moreover, by VI.1.4, since $\mathfrak{S}'_{L'}$ is an element of $\operatorname{Sing}^{L'}$, we have $w\mathfrak{S} \in \operatorname{Sing}^G$. Let us check that the association class of L corresponds to the conditions of the proposition; it remains to prove that for any standard Levi L'' of G associated with L, there exists $\mathfrak{S}'' \in \mathfrak{C} \cap \operatorname{Sing}^G$ such that $\operatorname{Re}\mathfrak{S}''^0 = \operatorname{Re}X^G_{L''}$. Fix L'' and an element τ of W such that:

$$\tau L = L''$$

and τ is of minimal length in its right class modulo the Weyl group of L and in its left class modulo that of L''; in particular $\tau \in W(L)$. Let $\mathfrak{T} \in \mathfrak{C} \cap \operatorname{Sing}^G$ be such that $M_{\mathfrak{T}} \subset L$; then $\tau \in W(M_{\mathfrak{T}})$ and by VI.1.5, $\tau\mathfrak{T} \in \operatorname{Sing}^G$. The same argument can be applied to τ^{-1}: exchanging the roles of L and L'', we see that τ takes $\{\mathfrak{T} \in \mathfrak{C} \cap \operatorname{Sing}^G | M_{\mathfrak{T}} \subset L\}$ onto $\{\mathfrak{T} \in \mathfrak{C} \cap \operatorname{Sing}^G | M_{\mathfrak{T}} \subset L''\}$. Thus $\tau w\mathfrak{S} \in \mathfrak{C} \cap \operatorname{Sing}^G$, and obviously $\operatorname{Re}\tau w\mathfrak{S}^0 = \tau\operatorname{Re} X^G_L = X^G_{L''}$. This proves the assertion. The uniqueness of the association class is clear.

VI.1.8. 'Reality' of the origin of elements of Sing^G

Corollary *Suppose that k is a number field. Let $\mathfrak{S} \in \operatorname{Sing}^G$; let $(M, \mathfrak{P})$ be the cuspidal datum of $\mathfrak{S}$ and π_0 the element of $\mathfrak{P}$ trivial on A^G_M. Then $\mathfrak{S}$ contains the element $o(\mathfrak{S}) \otimes \pi_0$. Fix a finite set $\mathfrak{F}$ of* **K**-*types. Then without hypotheses on k, the set of elements of $\operatorname{Sing}^{G\,\mathfrak{F}}$ which have the same cuspidal datum is finite.*

If k is a function field, the finiteness of $\operatorname{Sing}^{G\,\mathfrak{F}}$ results from the local finiteness of $S_{\mathfrak{X}}^{\mathfrak{F}}$ and from the fact that there exists $R \in \mathbb{R}_+^*$ such that every element of Sing^G intersects the compact set:

$$\bigcup_{w \in W(M)} w\{\pi \in \mathfrak{P} |\ \|\mathrm{Re}\,\pi\| \leq R\}.$$

Suppose that k is a number field. Let $\mathfrak{S}, (M, \mathfrak{P})$ and π_0 be as in the statement. Let $\mathfrak{C}$ be the conjugacy class of $\mathfrak{S}$. Let us show that $\mathfrak{S}$ contains the element $o(\mathfrak{S}) \otimes \pi_0$. Up to conjugating, we can still suppose, using VI.1.7, that:

(1) $$\mathrm{Re}\,\mathfrak{S}^0 = \mathrm{Re}\,X_{M_{\mathfrak{S}}}^{\mathbf{G}}, \quad i.e. \quad M_{\mathfrak{S}} \in L(\mathfrak{C}).$$

Suppose first that $L(\mathfrak{C}) = \{G\}$, i.e. that every element of $\mathfrak{C}$ is reduced to a point. Let $\mathfrak{S} \in \mathfrak{C}$. Let $\pi_{\mathfrak{S}}$ be the element of $\mathfrak{S}$; we must show that $\pi_{\mathfrak{S}} = o(\mathfrak{S}) \otimes \pi_0$. But this assertion is equivalent to proving that the restriction of the central character of $\pi_{\mathfrak{S}}$ to $A_M^{\mathbf{G}}$ has positive real values. This results first from VI.1.6(d), which shows that $L_{\mathfrak{C}}^2$ is the closure of the vector space consisting of the automorphic forms $\mathrm{Res}_{\mathfrak{S}}^{G} E(\phi, \pi)$ and that these are square integrable modulo the centre; next from V.3.17, which shows that the cuspidal support of these automorphic forms consists of representations $w\pi_{\mathfrak{S}}$ with $w \in W(M)$; and finally from III.3.1, which shows that for such w, the restriction of the central character of $w\pi_{\mathfrak{S}}$ to $A_{wMw^{-1}}^{\mathbf{G}}$ has positive real values.

Suppose now that $L(\mathfrak{C}) \neq \{G\}$ i.e. $M_{\mathfrak{S}} \neq G$; then we set $M_{\mathfrak{S}} = L$. We use the notation ξ_L of the introduction of VI and we set:

$$\mathfrak{S}_L = \mathfrak{S} \cap \mathfrak{P}_L.$$

We know that $\mathfrak{S}_L \in \mathrm{Sing}^L$, see VI.1.4(b), and obviously, by (1), that $\mathfrak{S}_L$ is reduced to a point. We apply VI.1.6(d) in L to the equivalence class of $\mathfrak{S}_L$; we must first note that $\pi_0 \in \mathfrak{P}_L$ since ξ_L is trivial on $A_L^{\mathbf{G}}$. Thanks to this, $\mathfrak{S}_L$ (which is reduced to a point) coincides with $\pi_0 \otimes o(\mathfrak{S}_L)$. Now from VI.1.4(a) $o(\mathfrak{S}_L) = o(\mathfrak{S})$. This gives the desired result:

$$o(\mathfrak{S}) \otimes \pi_0 \in \mathfrak{S}.$$

Moreover, since $\mathfrak{S} \in \mathrm{Sing}^G$, we have $\|o(\mathfrak{S})\| < R$ where R is as above and the finiteness of the set of elements of $\mathrm{Sing}^{G\mathfrak{F}}$ having the same cuspidal datum is a consequence, as in the function field case, of the local finiteness of $S_{\mathfrak{X}}^{\mathfrak{F}}$.

VI.1.9 Discrete parameter

Definition

We fix an equivalence class $\mathfrak{X}$ of cuspidal data. A discrete parameter for $\mathfrak{X}$ is a pair (L, δ) where L is a standard Levi of G and where δ is a subrepresentation (in general of infinite length) of the discrete spectrum of $\mathbf{L}$ having a central character ξ_L extending ξ, trivial on $A_L^{\mathbf{G}}$, and such that δ is of the form:

$$\delta = L^2(L(k)\backslash\mathbf{L})_{\mathfrak{C}_L} \subset L^2(L(k)\backslash\mathbf{L})_{\mathfrak{X}_L}$$

where $\mathfrak{X}_L$ is an equivalence class of cuspidal data (relative to L) 'contained' in $\mathfrak{X}$ and where $\mathfrak{C}_L$ is an element of $[S_{\mathfrak{X}_L}]$ consisting of points. We say that $\mathfrak{C}_L$ is the singular class attached to δ; we note that $\mathfrak{C}_L$ is uniquely determined by δ and we sometimes replace (L, δ) by $(L, \mathfrak{C}_L)$. We define the association class of (L, δ), denoted by $[(L, \delta)]$, to be the set of pairs $(wLw^{-1}, w\delta)$ where w runs through $W(L)$.

Let (L, δ) be a discrete parameter and let $\mathfrak{C} \in [S_{\mathfrak{X}}]$; we say that (L, δ) is a discrete parameter for $\mathfrak{C}$ if, $\mathfrak{C}_L$ being the singular class attached to δ, we have:

> for every element $\mathfrak{S}_L$ of $\mathfrak{C}_L$, $\mathfrak{C}$ contains a connected component of $\mathfrak{S}_L + X_L^{\mathbf{G}}$. If k is a number field, this means that $\mathfrak{S}_L + X_L^{\mathbf{G}}$ is an element of $\mathfrak{C}$ and $\mathfrak{C}$ is uniquely determined by $\mathfrak{C}_L$. If k is a function field the situation is more complicated, in particular $\mathfrak{C}$ is no longer determined by $\mathfrak{C}_L$ and we denote by $\tilde{\mathfrak{C}}$ the set of conjugacy classes admitting a discrete parameter which is also one for $\mathfrak{C}$. The proposition below clarifies this situation.

Proposition *Every element $\mathfrak{C}$ of $[S_{\mathfrak{X}}]$ intersecting* Sing^G *admits a discrete parameter and every element of the association class of a discrete parameter for $\mathfrak{C}$ is still one. If k is a number field, $\mathfrak{C}$ admits a discrete parameter unique up to association. Suppose that k is a function field and let $(L, \mathfrak{C}_L)$ be a discrete parameter for $\mathfrak{C}$; fix $\mathfrak{S}_L \in \mathfrak{C}_L$. Every discrete parameter for $\mathfrak{C}$ belongs to the association class of $(L, \mathfrak{C}'_L)$ where $\mathfrak{C}'_L$ is the conjugacy class of an element $\mathfrak{S}'_L$ of the form $\mathfrak{S}_L + v$, where v is an element of $X_L^{\mathbf{G}}$ trivial on $A_L^{\mathbf{G}}$. There are only a finite number of these. Every element of $\tilde{\mathfrak{C}}$ (see above for the definition) is the conjugacy class of one of the connected components of $\mathfrak{S}_L + X_L^{\mathbf{G}}$.*

The existence of a discrete parameter (under the hypotheses of the statement) is a consequence of VI.1.7. The fact that every element of the association class of a discrete parameter is still a discrete parameter results from VI.1.4(c). Let us prove the assertions relative to the uniqueness of the association class of discrete parameters. Let $(L, \mathfrak{C}_L)$ and $(L', \mathfrak{C}_{L'})$ be discrete parameters for $\mathfrak{C}$. We denote by $\mathfrak{X}_L$ and $\mathfrak{X}'_{L'}$ the equivalence classes of relative cuspidal data. As in the proof of VI.1.6, we show that L and L' are conjugate, so we can suppose them equal. We also check that we can find elements $\mathfrak{S}_L$ and $\mathfrak{S}'_L$ of $\mathfrak{C}_L$ and $\mathfrak{C}'_L$ respectively having cuspidal data $(M, \mathfrak{P}_L)$ and $(M, \mathfrak{P}'_L)$, (i.e. the Levi M is common to the two data) such that $\mathfrak{P}_L + X_L^{\mathbf{G}} = \mathfrak{P}'_L + X_L^{\mathbf{G}}$. We denote by ξ_L and $\xi_{L'}$ the characters of $Z_{\mathbf{L}}$ determined by $\mathfrak{S}_L$ and $\mathfrak{S}'_L$ respectively; by hypothesis, they are trivial on $A_L^{\mathbf{G}}$. Suppose first that k is a number field. Here ξ_L and ξ'_L coincide and thus so do $\mathfrak{P}_L$ and $\mathfrak{P}'_L$. We define:

$$\mathfrak{S} := \mathfrak{S}_L + X_L^{\mathbf{G}},$$

$$\mathfrak{S}' := \mathfrak{S}'_L + X_L^{\mathbf{G}}.$$

These are elements of $\mathfrak{C}$ which satisfy, see VI.1.4(a)(ii):

$$\mathfrak{S}_L = \mathfrak{S} \cap \mathfrak{P}_L,$$

$$\mathfrak{S}'_L = \mathfrak{S}' \cap \mathfrak{P}_L.$$

Let $w \in \mathrm{Stab}_{W(M)}(M, \mathfrak{P})$ conjugate $\mathfrak{S}$ and $\mathfrak{S}'$. It is a consequence of what precedes that w normalises L, so also ξ_L and $\mathfrak{P}_L$. Thus w sends $\mathfrak{S}_L$ onto $\mathfrak{S}'_L$, which is the desired uniqueness. Suppose now that k is a function field. In this case, we define $\mathfrak{S}$ and $\mathfrak{S}'$ to be elements of $\mathfrak{C}$ which are connected components of $\mathfrak{S}_L + X_L^{\mathbf{G}}$ and $\mathfrak{S}'_L + X_L^{\mathbf{G}}$ respectively. As above, w conjugates $\mathfrak{S}$ and $\mathfrak{S}'$; conjugating by w, we come back to the case where ξ_L and ξ'_L (defined as above) coincide. $\mathfrak{P}_L$ is then well-defined and it only remains to apply VI.1.4(a)(ii) to see that $\mathfrak{S}_L$ differs from $\mathfrak{S}'_L$ by translation by an element of $X_L^{\mathbf{L}}$. The description of $\tilde{\mathfrak{C}}$ can be deduced from this.

VI.2. Spectral decomposition

VI.2.1. The most general statement (cf [A3])
To start with, we do not make any hypotheses on k. Let L be a standard Levi of G; we denote by $L^2(L(k)\backslash\mathbf{L})_{d,\xi}$ the part of the discrete spectrum of L (modulo the centre) whose central character is trivial on $A_L^{\mathbf{G}}$ and whose restriction to the centre of $\mathbf{G}$ is the character ξ. We denote by $A(L(k)\backslash\mathbf{L})_{d,\xi}$ the intersection of this space with the space of automorphic forms for L. Writing U_L for the radical unipotent of the standard parabolic of Levi L, we define:

$$A(U_L(\mathbb{A})L(k)\backslash\mathbf{G})_{d,\xi} := \{f \in A(U_L(\mathbb{A})L(k)\backslash\mathbf{G}) | \forall k \in \mathbf{K},\ f_k \in A(L(k)\backslash\mathbf{L})_{d,\xi}\};$$

see I.2.13 for the notation f_k. We denote by $P_{L,d}^R$ the space of holomorphic functions on $\{\mu \in X_L^{\mathbf{G}} | \|\mathrm{Re}\,\mu\| < R\}$ with values in $A(U_L(\mathbb{A})L(k)\backslash\mathbf{G})_{d,\xi}$ satisfying the growth condition of II.1.4. We have:

Theorem *Let L be a standard Levi of G.*

(i) *Let $\phi \in P^{R}_{L,d}$ and let $\mu \in X^{\mathbf{G}}_{L}$. The Eisenstein series $E(\phi,\mu)$ (or more precisely $E^G_L(\phi,\mu)$) defined by a series which converges for very positive* $\operatorname{Re}\mu$ *(see II.1.5) can be continued to a meromorphic operator on μ holomorphic on* $\operatorname{Im} X^{\mathbf{G}}_{L}$. *This is also true for the intertwining operator $M(t^{-1},t\mu)$ for $t \in W$ of minimal length in its right class modulo W_L such that tLt^{-1} is still a standard Levi. We have the functional equation:*

$$E^G_L(\phi,\mu) = E^G_{tLt^{-1}}(M(t,\mu)\phi, t\mu).$$

(ii)
$$\lim_{T\to\infty} \int_{\substack{\mu\in \operatorname{Im} X^{\mathbf{G}}_{L} \\ \|\mu\|\le T}} E(\phi,\mu)\, d\mu$$

exists in the L^2 i.e. defines an element of $L^2(G(k)\backslash\mathbf{G})_\xi$. We denote by $L^2(G(k)\backslash\mathbf{G})_{L,\xi}$ the closed subspace $L^2(G(k)\backslash\mathbf{G})$ generated by these elements. Let L' be another standard Levi of G; suppose that L and L' are conjugate. Then $L^2(G(k)\backslash\mathbf{G})_{L,\xi} = L^2(G(k)\backslash\mathbf{G})_{L',\xi}$. We denote this space by $L^2(G(k)\backslash\mathbf{G})_{[L],\xi}$ where [L] is the association class of L.

(iii) *We have the orthogonal decomposition:*

$$L^2(G(k)\backslash\mathbf{G})_\xi = \oplus_{[L]} L^2(G(k)\backslash\mathbf{G})_{[L],\xi}$$

where [L] runs through the set of association classes of standard Levis of G.

(iv) *Suppose that k is a number field. Let W^L be the set of elements of W of minimal length modulo W_L and*

$$\operatorname{Stab}_{W^L} L = \{w \in W^L | wLw^{-1} = L\}.$$

Then $L^2(G(k)\backslash\mathbf{G})_{d,\xi}$ is isometric to the following Hilbert space:

$$\left\{\begin{array}{l} \text{Measurable functions } F \text{ on } \operatorname{Im} X^{\mathbf{G}}_{L} \text{ with values in} \\ \operatorname{ind}^{\mathbf{K}}_{\mathbf{K}\cap\mathbf{L}} L^2(L(k)\backslash\mathbf{L})_{d,\xi} \text{ such that :} \\ (1)\quad M(t^{-1},t\mu)F(t\mu) = F(\mu), \\ \qquad \text{almost everywhere, for every } t \in \operatorname{Stab}_{W^L} L \\ (2)\quad \lim_{T\to\infty} \displaystyle\int_{\substack{\mu\in \operatorname{Im} X^{\mathbf{G}}_{L} \\ \|\mu\|\le T}} |\operatorname{Stab}_{W^L} L|^{-1} \|F(\mu)\|^2\, d\mu < +\infty, \end{array}\right\}$$

where the norm is that of $\operatorname{ind}^{\mathbf{K}}_{\mathbf{K}\cap\mathbf{L}} L^2(L(k)\backslash\mathbf{L})_{d,\xi}$. The norm of this Hilbert space is given by (2).

VI.2.2. A more technical but more precise statement

Let $\mathfrak{X}$ be an equivalence class of cuspidal data and let $\mathfrak{C}$ be an equivalence class in $S_{\mathfrak{X}}$ intersecting Sing^G with which, in the corollary to V.3.12, we associated the subspace $L^2(G(k)\backslash \mathbf{G})_\xi$. Let (L,δ) be a discrete parameter for $\mathfrak{C}$. We denote by $\mathfrak{C}_L$ the singular class attached to δ and we recall that δ determines $\mathfrak{C}_L$. We define in a natural way:

$$\mathrm{Stab}_{W^L}(L,\delta) := \{w \in W(L) \text{ such that } wLw^{-1} = L \text{ and } w\mathfrak{C}_L = \mathfrak{C}_L\}.$$

By the preceding discussion, $w\mathfrak{C}_L = \mathfrak{C}_L$ is equivalent to $w\delta = \delta$. In the function field case, we have already defined $\tilde{\mathfrak{C}}$ (see VI.1.9); we still need to define:

$$\mathrm{Stab}'_{W^L}(L,\delta) := \{w \in W(L) \text{ such that } wLw^{-1} = L \text{ and } (L, w\mathfrak{C}_L) \text{ is a discrete parameter for } \mathfrak{C}\}.$$

Note that if π_L is an element of $\mathfrak{C}_L$, we have:

(1) $$\mathrm{Stab}_{W^L}(L,\delta) = \{t \in \mathrm{Stab}_{W^L} L \text{ such that } \exists\, w_L \in W_L(M), \text{ with } t\pi_L = w_L\pi_L\}.$$

The inclusion of the left-hand term in the right-hand one is a consequence of the definitions. Let us show the inclusion in the other direction: let $t \in \mathrm{Stab}_{W^L}(L)$ and let $w_L \in W_L(M)$ be such that $t\pi_L = w_L\pi_L$. In particular,

$$tMt^{-1} = w_L M w_L^{-1}.$$

For every $w' \in W_L(M)$, we thus have $tw't^{-1}w_L M(tw't^{-1}w_L)^{-1} = tw'M(tw')^{-1}$. Now $w'Mw'^{-1}$ is a standard Levi of L and $tw'M(tw')^{-1}$ is thus also one. In other words, since $tw't^{-1}w_L \in W_L$, we have:

$$tw't^{-1}w_L \in W_L(M).$$

Thus

$$tw'\pi_L = (tw't^{-1})w_L\pi_L \in \mathfrak{C}_L$$

i.e. $t\mathfrak{C}_L = \mathfrak{C}_L$.

We show similarly that:

(2) $$\mathrm{Stab}'_{W^L}(L,\delta) = \{t \in \mathrm{Stab}_{W^L} L \text{ such that } \exists w_L \in W_L(M), \text{ with } t\pi_L + X_L^{\mathbf{G}} = w_L\pi_L + X_L^{\mathbf{G}}\}.$$

Let $t \in \mathrm{Stab}'_{W^L}(L,\delta)$; then t acts on $X_L^{\mathbf{G}}/\mathrm{Fix}_{X_L^L}\delta$ in the following twisted fashion: let $\mu \in X_L^{\mathbf{G}}/\mathrm{Fix}_{X_L^L}\delta$; then $t.\mu = t\mu + \nu$ where ν is the element of $X_L^{\mathbf{G}}/\mathrm{Fix}_{X_L^L}\delta$ defined by $t\delta = w_L\delta \otimes \nu$, in the above notation.

Theorem *Let $\mathfrak{X}$ and (L,δ) be as above.*

(i) *We suppose that k is a number field. $L^2(G(k)\backslash \mathbf{G})_{\mathfrak{C}}$ is isomorphic to the* *following Hilbert space, denoted by* $H_{\mathfrak{C}}$:

$$\left\{\begin{array}{l} \textit{Measurable functions } F \textit{ on } \operatorname{Im} X_L^{\mathbf{G}} \textit{ with values in} \\ \operatorname{ind}_{\mathbf{K}\cap\mathbf{L}}^{\mathbf{K}}\delta \textit{ such that :} \\ \forall t \in \operatorname{Stab}_{W^L}(L,\delta),\ M(t^{-1},t\mu)F(t\mu) = F(\mu) \\ \lim\limits_{T\to\infty} \displaystyle\int\limits_{\substack{\mu\in\operatorname{Im}X_L^{\mathbf{G}}\\ \|\mu\|\leq T}} |\operatorname{Stab}_{W^L}(L,\delta)|^{-1}\|F(\mu)\|^2\, d\mu < +\infty. \end{array}\right\}$$

For the definition of $M(t^{-1},t\mu)$, *we refer the reader to VI.2.1(i).*

(ii) *We still suppose that* k *is a number field. Let* ϕ *be a function on* $X_L^{\mathbf{G}}$ *with values in* $A(U_L(\mathbb{A})L(k)\backslash\mathbf{G})_\delta$, *holomorphic in a neighborhood of every point of* $\operatorname{Im} X_L^{\mathbf{G}}$ *and rapidly decreasing on this set. Then*

$$e(\phi) := \lim_{T\to\infty} \int\limits_{\substack{\mu\in\operatorname{Im}X_L^{\mathbf{G}}\\ \|\mu\|\leq T}} E(\phi,\mu)\, d\mu$$

exists in the L^2 *sense and defines an element of* $L^2(G(k)\backslash\mathbf{G})_{\mathfrak{C}}$. *The space* $L^2(G(k)\backslash\mathbf{G})_{\mathfrak{C}}$ *is the closure of the vector space generated by the set of these functions.*

Suppose k *is a function field. We set:*

$$L^2(G(k)\backslash\mathbf{G})_{\mathfrak{F}} := \oplus_{\mathfrak{C}\in\mathfrak{F}} L^2(G(k)\backslash\mathbf{G})_{\mathfrak{C}}.$$

We have the exact analogues of (i) and (ii) above, i.e.

(iii) $L^2(G(k)\backslash\mathbf{G})_{\mathfrak{F}}$ *is isomorphic to the Hilbert space* $H_{\mathfrak{F}}$ *defined as follows:*

$$\left\{\begin{array}{l} \textit{Measurable functions } F \textit{ on } \operatorname{Im} X_L^{\mathbf{G}}/\operatorname{Fix}_{X_L^{\mathbf{L}}}\delta \textit{ with values in} \\ \operatorname{ind}_{\mathbf{K}\cap\mathbf{L}}^{\mathbf{K}}\delta \textit{ such that :} \\ \forall t \in \operatorname{Stab}'_{W^L}(L,\delta),\ M(t^{-1},t.\mu)F(t.\mu) = F(\mu) \\ \displaystyle\int\limits_{\mu\in\operatorname{Im}X_L^{\mathbf{G}}/\operatorname{Fix}_{X_L^{\mathbf{L}}}\delta} |\operatorname{Stab}'_{W^L}(L,\delta)|^{-1}\|F(\mu)\|^2\, d\mu < +\infty. \end{array}\right\}$$

(iv) *Let* ϕ *be a function on* $X_L^{\mathbf{G}}$ *with values in* $A(U_L(\mathbb{A})L(k)\backslash\mathbf{G})_\delta$, *invariant under* $\operatorname{Fix}_{X_L^{\mathbf{L}}}\delta$ *(see I.1.2), holomorphic in a neighborhood of every point of* $\operatorname{Im} X_L^{\mathbf{G}}$. *Then:*

$$e(\phi) := \int\limits_{\mu\in\operatorname{Im}X_L^{\mathbf{G}}/\operatorname{Fix}_{X_L^{\mathbf{L}}}\delta} E(\phi,\mu)\, d\mu$$

defines an element of $L^2(G(k)\backslash\mathbf{G})_{\mathfrak{F}}$. *The space* $L^2(G(k)\backslash\mathbf{G})_{\mathfrak{F}}$ *is the closure of the vector space generated by the set of these functions.*

VI.2.3. Proof of VI.2.1 (i)

Fix $\mathfrak{X}$ and a discrete parameter (L,δ) for $\mathfrak{X}$. Let $\mathfrak{C}_L$ be the singular class attached to δ (see VI.1.9). Fix $\mathfrak{S}_L \in \mathfrak{C}_L$ and let $(M,\mathfrak{P})$ be its cuspidal datum; as $\mathfrak{S}_L$ consists of a unique representation, we denote this representation by π_L. We also write $A(L(k)\backslash\mathbf{L})_{\mathfrak{C}_L}$ for the space of automorphic forms for $\mathbf{L}$ contained in $L^2(L(k)\backslash\mathbf{L})_{\mathfrak{C}_L}$, and we define:

$$A(U_L(\mathbb{A})L(k)\backslash\mathbf{G})_{\mathfrak{C}_L} :=$$
$$\{f \in A(U_L(\mathbb{A})L(k)\backslash\mathbf{G}) | \forall k \in \mathbf{K},\ f_k \in A(L(k)\backslash\mathbf{L})_{\mathfrak{C}_L}\}.$$

For every $\mu \in X_L^{\mathbf{G}}$, we denote by

$$A(U_L(\mathbb{A})L(k)\backslash\mathbf{G})_{\mathfrak{C}_{L,\mu}}$$

the translation of

$$A(U_L(\mathbb{A})L(k)\backslash\mathbf{G})_{\mathfrak{C}_L}$$

by the function $\mu \circ m_{P_L}$ where P_L is the standard parabolic of Levi L. Using V.3.17 and setting $\pi_\mu := \pi_L \otimes \mu$ for every $\mu \in X_L^{\mathbf{G}}$, we see that $A(U_L(\mathbb{A})L(k)\backslash\mathbf{G})_{\mathfrak{C}_{L,\mu}}$ is generated by the elements

$$\sum_{w \in W_L(M)} \operatorname{Res}^L_{w\pi_L} E^L(\phi, w\pi_\mu), \tag{1}$$

or equivalently

$$\operatorname{Res}^L_{w\pi_L} E^L(\phi, w\pi_\mu),$$

where E^L is the Eisenstein series for L and ϕ runs through

$$\bigcup_{w \in W_L(M)} P_{w(M,\mathfrak{P})}.$$

We define $P^R_{(L,\delta)}$ in a similar way to $P^R_{d,\xi}$. Let $f \in P^R_{(L,\delta)}$: the finiteness properties of f under the action of $\mathbf{K}$ and (1) ensure that there exists $\phi \in P^R_{\mathfrak{X}}$ such that:

$$f(\mu) = \sum_{w \in W_L(M)} \operatorname{Res}^L_{w\pi_L} E^L(\phi, w\pi_\mu). \tag{2}$$

For $\operatorname{Re}\mu$ very positive, we defined in II.1.5 the Eisenstein series from $\mathbf{L}$ to $\mathbf{G}$, which we here denote by $E_L^G(\mu)$, by a convergent series. Let $\phi \in P^R_{\mathfrak{X}}$ (R large) and $w \in W_L(M)$; we set $\mathfrak{S} := \mathfrak{S}_L + \operatorname{Re} X_L^{\mathbf{G}} \otimes_{\mathbf{R}} \mathbf{C}$ and check that for $\operatorname{Re}\mu$ very positive, we have:

$$\operatorname{Res}^G_{w\mathfrak{S}} E(\phi, w\pi_\mu) = E_L^G(\mu)(\operatorname{Res}^L_{w\pi_L} E^L(\phi, w\pi_\mu)). \tag{3}$$

We know that $\operatorname{Res}^G_{w\mathfrak{S}} E(\phi, w\pi_\mu) = \operatorname{Res}^L_{w\pi_L} E(\phi, w\pi_\mu)$, see VI.1.4 (c). We must thus prove that we have:

$$E_L^G(\mu)\operatorname{Res}^L_{w\pi_L} E^L(\phi, w\pi_\mu) = \operatorname{Res}^L_{w\pi_L}(E_L^G(\mu)E^L(\phi, w\pi_\mu)).$$

This equality is true for $\operatorname{Re}\mu \gg 0$ using the definition of $E_L^G(\mu)$ as an absolutely and uniformly convergent series on every compact set of $\mathbf{G}$. The left-hand side of (3) admits a meromorphic continuation: thus this is also true of the right-hand side and (3) becomes an equality of meromorphic functions.

As a corollary of (3) we obtain the meromorphic continuation of the Eisenstein series $E_L^G(f, \mu)$. It is holomorphic on the unitary axis because of VI.1.3. We show similarly the continuation of the intertwining operators. The intertwining operators are holomorphic on the unitary axis since these operators are unitary there, and defined on an open dense set.

We now establish the functional equation: let $t \in W(L)$. We must prove the following formula:

$$E_{tLt^{-1}}^G(t\mu)(M(t,\mu)f(\mu)) = E_L^G(\mu)f(\mu). \tag{4}$$

We first check that for t as above we have:

$$\begin{aligned} &\forall w \in W_L(M),\ \forall \phi \in P_{\mathfrak{X}}, \\ &M(t,\mu)\left(\mathrm{Res}^L_{w\pi_L} E^L(\phi,\pi')\right)_{\pi'=w\pi_\mu} = \\ &\qquad \left(\mathrm{Res}^{tLt^{-1}}_{tw\pi_L} E^{tLt^{-1}}(M(t,\mu)\phi,\pi')\right)_{\pi'=tw\pi_\mu}. \end{aligned} \tag{5}$$

Fix v, v' very positive in $\mathrm{Re}\,X^{\mathbf{L}}_{wMw^{-1}}$ and $\mathrm{Re}\,X^{\mathbf{G}}_L$ respectively: we easily obtain the equality:

$$M(t,\mu+v')E^L(\phi, w\pi_{\mu+v'}+v) = E^{tLt^{-1}}(M(t,\mu+v')\phi, tw\pi_{\mu+v'}+tv),$$

by writing down the definitions; it comes to exchanging a sum and an integral. We can now set $v' = 0$ and apply the operator $\mathrm{Res}^L_{w\pi_L}$ at $\pi' = w\pi_\mu$; on the right-hand side, this comes to applying the operator $t\mathrm{Res}^L_{w\pi_L}$ to the point $tw\pi_\mu$, or even, by VI.1.6(c), the operator $\mathrm{Res}^{tLt^{-1}}_{tw\pi_L}$ to the same point. Thus

$$\mathrm{Res}^L_{w\pi_L} M(t,\mu)E^L(\phi, w\pi_\mu) = \mathrm{Res}^{tLt^{-1}}_{tw\pi_L} E^{tLt^{-1}}(M(t,\mu)\phi, tw\pi_\mu).$$

To prove (5), it remains to check that we have:

$$\mathrm{Res}^L_{w\pi_L} M(t,\mu) = M(t,\mu)\mathrm{Res}^L_{w\pi_L}.$$

This can be proved by meromorphic continuation by translating μ by an element of $X^{\mathbf{G}}_L$ of the very positive cone and exchanging the derivative and the integral; we use the fact that $w \in W_L$. We apply $E^{\mathbf{G}}_{tLt^{-1}}(t\mu)$ to the two sides of (5); the right-hand side becomes (see (3))

$$\mathrm{Res}^{tLt^{-1}}_{tw\pi_L} E(M(t,\mu)\phi, tw\pi_\mu).$$

Using VI.1.4(c), then the functional equation VI.1.5, we obtain:

$$\begin{aligned} \mathrm{Res}^{tLt^{-1}}_{tw\pi_L} E(M(t,\mu)\phi, tw\pi_\mu) &= \mathrm{Res}^G_{tw\mathfrak{S}} E(M(t,\mu)\phi, tw\pi_\mu) \\ &= \mathrm{Res}^G_{w\mathfrak{S}} E(\phi, w\pi_\mu). \end{aligned}$$

Thus:

$$E^G_{tLt^{-1}}(t\mu)M(t,\mu)\mathrm{Res}^L_{w\pi_L}E^L(\phi,w\pi_\mu) = \mathrm{Res}^G_{w\mathfrak{S}}E(\phi,w\pi_\mu).$$

Summing over w, we obtain (4).

We now establish the key lemma (which is Lemma 7.4 of Langlands, [L]).

VI.2.4. Description of $A_{\mathfrak{C},\pi}$ (for general π)

Using the notation of V.3.14, the following holds.

Lemma *Let $\mathfrak{C} \in S_{\mathfrak{X}}$ and let L be a standard Levi belonging to $L(\mathfrak{C})$. Fix $\mathfrak{S} \in \mathfrak{C}$ such that $M_{\mathfrak{S}} = L$, and an element π_L of $\mathfrak{S}$ the restriction of whose central character to $A_L^{\mathbf{G}}$ is trivial (such an element exists). We note that $\pi_L \in \mathrm{Sing}^L$ and we denote by $\mathfrak{C}_L$ the singular class defined by it. Let μ be a general point of $\mathrm{Re}\,X_L^{\mathbf{G}} \otimes_{\mathbf{R}} \mathbf{C} \cap \mathrm{Im}\,X_L^{\mathbf{G}}$; see below for the definition of general point. We set $\pi_\mu = \pi_L \otimes \mu$; it is an element of $\mathfrak{S}$.*

(i) *$\forall \phi \in P_{\mathfrak{X}}$, we set:*

$$e^L(\phi,\pi_\mu) := |\mathrm{Norm}_{W_L(M)}\pi_L|^{-1} \sum_{w\in W_L(M)} \mathrm{Res}^L_{w\pi_L}E^L(\phi,w\pi_\mu).$$

It is an element of $A(U_L(\mathbb{A})L(k)\backslash\mathbf{G})_{\mathfrak{C}_L,\mu}$, in the notation of VI.2.3, and the set of these elements generates $A(U_L(\mathbb{A})L(k)\backslash\mathbf{G})_{\mathfrak{C}_L,\pi_\mu}$. We have:

$$E_L^G(\mu)e^L(\phi,\pi_\mu) \in A_{\mathfrak{C},\pi_\mu}.$$

(ii) *$E_L^G(\mu)$ is, up to a constant, an isometry of*

$$A(U_L(\mathbb{A})L(k)\backslash\mathbf{G})_{\mathfrak{C}_L,\mu}$$

on $A_{\mathfrak{C},\pi_\mu}$; in particular it is bijective.

We suppose that $L \neq G$ and we say that μ is a general point if no conjugate π_μ is in the cuspidal support of a square integrable automorphic form (see III.4.1) and if π_μ does not belong to any hyperplane of $\mathfrak{S}$ contained in $S_{\mathfrak{X}}$ (to be strictly correct, we should fix a finite set $\mathfrak{F}$ of **K**-types and put $\mathfrak{F}$ in the exponent everywhere). In particular, $\mathfrak{S}$ is the only element of $\mathfrak{C}$ containing π_μ.

The first part of (i) follows from VI.2.3(1). We saw in VI.2.3(3) that

(1) $$E_L^R(\mu)\mathrm{Res}^L_{w\pi_L}E^L(\phi,w\pi_\mu) = \mathrm{Res}^G_{w\mathfrak{S}}E(\phi,w\pi_\mu).$$

Then (i) results from the sublemma of VI.1.2, where we saw that

$$\mathrm{Res}^G_{w\mathfrak{S}}E(\phi,w\pi_\mu) \in A_{\mathfrak{C},\pi_\mu}.$$

Let us prove (ii). Note first that V.3.17 shows that

$$A(U_L(\mathbb{A})L(k)\backslash\mathbf{G})_{\mathfrak{C}_L,\mu}$$

is generated by the functions $\mathrm{Res}^L_{w\pi_L} E^L(\phi, w\pi_\mu)$, where ϕ runs through $P_{\mathfrak{X}}$ and w runs through $W_L(M)$.

Thus to prove surjectivity of $E^G_L(\mu)$, it will suffice to prove that:

(2) $A_{\mathfrak{C},\pi_\mu}$ is the vector space generated by the set of functions $\mathrm{Res}^G_{w\mathfrak{S}} E(\phi, w\pi_\mu)$ where $\phi \in P_{\mathfrak{X}}$ and $w \in W_L(M)$.

But first let us prove injectivity of $E^G_L(\mu)$ by showing that it preserves the scalar product up to a scalar. For every $\phi, \phi' \in P_{\mathfrak{X}}$, we must calculate the scalar product in $A(U_L(\mathbb{A})L(k)\backslash \mathbf{G})_{\mathfrak{C}_L,\mu}$:

$$\langle e^L(\phi', \pi_\mu), e^L(\phi, \pi_\mu)\rangle.$$

By the definition and V.3.2(5)(i) applied to **L** instead of **G**, we see that

$$\langle e^L(\phi', \pi_\mu), e^L(\phi, \pi_\mu)\rangle = \sum_{w\in W_L(M)} |\mathrm{Norm}_{W_L(M)}\pi_L|^{-1} (\mathrm{Res}^L_{w\pi_L} A^L(\phi', \phi))(w\pi_\mu)$$

where

$$A^L(\phi', \phi)(w\pi_\mu) := \sum_{\sigma\in W_L(wMw^{-1})} \langle M(\sigma^{-1}, -\sigma w\overline{\pi}_\mu)\phi'(-\sigma w\overline{\pi}_\mu), \phi(w\pi_\mu)\rangle.$$

A calculation identical to that done in VI.1.2(4) shows that:

$$\begin{aligned} &\forall\phi, \phi' \in P_{\mathfrak{X}},\ \forall\sigma, w \in W(M),\\ (3)\quad &\langle \mathrm{Res}^G_{\sigma\mathfrak{S}} E(\phi', \sigma\pi_\mu), \mathrm{Res}^G_{w\mathfrak{S}} E(\phi, w\pi_\mu)\rangle_{A_{\mathfrak{C},\pi_\mu}}\\ &= y\mathrm{Res}^G_{w\mathfrak{S}} \sum_\tau \langle M(\tau^{-1}, -\tau w\overline{\pi}_\mu)\phi'(-\tau w\overline{\pi}_\mu), \phi(w\pi_\mu)\rangle, \end{aligned}$$

where $y := |\mathrm{Norm}\ \mathfrak{S}|\,|\,\mathrm{Fix}\ \mathfrak{S}|^{-1}$ and where the sum over τ only runs over the elements of the set

$$\{\tau \in W(wM) | -\tau w\overline{\mathfrak{S}} = \sigma\mathfrak{S},\ \tau w_{|\mathrm{Re}\,\mathfrak{S}^0} = \sigma_{|\mathrm{Re}\,\mathfrak{S}^0}\}.$$

We must describe this set more precisely when σ, w are elements of $W_L(M)$. Under this hypothesis, the restrictions of w and σ to $\mathrm{Re}\,\mathfrak{S}^0$ coincide with the identity and $\mathrm{Re}\,\mathfrak{S}^0 = \mathrm{Re}\,X^{\mathbf{G}}_L$. Thus $\tau L\tau^{-1} = L$ and $\tau_{|\mathrm{Re}\,\mathfrak{S}^0}$ is then also the identity; this implies that $\tau \in W_L$. Still using $\mathrm{Res}^G_{w\mathfrak{S}}$ '=' $\mathrm{Res}^L_{w\pi_L}$ (see VI.1.4 for the meaning of this equality), we obtain for every $\phi', \phi \in P_{\mathfrak{X}}$ and for every $w \in W_L(M)$:

$$\begin{aligned} &\langle \sum_{\sigma\in W_L(M)} \mathrm{Res}^G_{\sigma\mathfrak{S}} E(\phi', \sigma\pi_\mu), \mathrm{Res}^G_{w\mathfrak{S}} E(\phi, w\pi_\mu)\rangle\\ &= |\mathrm{Norm}_{W_L}\mathfrak{S}|y\\ &\mathrm{Res}^L_{w\pi_L}\langle \sum_{\tau|\tau w\in W_L(M)} M(\tau^{-1}, -\tau w\overline{\pi}_\mu)\phi'(-\tau w\overline{\pi}_\mu), \phi(w\pi_\mu)\rangle\\ &= y(\mathrm{Res}^L_{w\pi_L} A^L(\phi', \phi))(w\pi_\mu). \end{aligned}$$

Now Fix $\mathfrak{S} = \text{Norm}_{W_L(M)}\pi_L$. Thus $E_L^G(\mu)$ is an isometry up to a factor $(\text{Norm}\,\mathfrak{S}/\text{Norm}_{W_L(M)}\pi_L)^{1/2}$.

It remains to prove (2). Let $w \in W(M)$. We first show that

$$\text{Res}_{w\mathfrak{S}}^G E(\phi, w\pi_\mu)$$

belongs to the vector space generated by the elements $\text{Res}_{\tau\mathfrak{S}}^G E(\phi, \tau\pi_\mu)$ such that $M_{\tau\mathfrak{S}}$ is a Levi conjugate of L. We prove this assertion by induction on the semi-simple rank of G.

First step

Fix $w \in W(M)$ and $\phi \in P_{\mathfrak{X}}$. Let us show that $\text{Res}_{w\mathfrak{S}}^G E(\phi, w\pi_\mu)$ is in the vector subspace of $A_{\mathfrak{C},\pi_\mu}$ generated by the elements $\text{Res}_{\sigma\mathfrak{S}}^G E(\phi', \sigma\pi_\mu)$ where $\phi' \in P_{\mathfrak{X}}$ and σ is an element of $W(M)$ such that $M_{\sigma\mathfrak{S}} \neq G$. Using the finiteness of $\dim A_{\mathfrak{C},\pi_\mu}^{\mathfrak{F}}$, which results from finiteness theorems of automorphic forms, it suffices to prove that the orthogonal complement of this subspace is zero, i.e. to prove that if $\phi \in P_{\mathfrak{X}}$ satisfies

$$\langle \text{Res}_{\sigma\mathfrak{S}}^G E(\phi', \sigma\pi_\mu), \sum_{w \in W(M)} \text{Res}_{w\mathfrak{S}}^G E(\phi, w\pi_\mu)\rangle = 0$$

for every $\phi' \in P_{\mathfrak{X}}$ and for every $\sigma \in W(M)$ such that $M_{\sigma\mathfrak{S}} \neq G$, then $\sum_{w \in W(M)} \text{Res}_{w\mathfrak{S}}^G E(\phi, w\pi_\mu) = 0$. Let ϕ satisfy this hypothesis. By VI.1.2 (4), the hypothesis can be rewritten:

$$\text{Res}_{\sigma\mathfrak{S}}^G A(\phi, \phi')(\sigma\pi_\mu) = 0,$$

for every σ as above.

By V.3.16, the cuspidal support of $\sum_{w \in W(M)} \text{Res}_{w\mathfrak{S}}^G E(\phi, w\pi_\mu)$ is in the set $\{\sigma\pi_\mu | -\sigma\overline{\mathfrak{S}} \in \text{Sing}^G\}$. Using the hypothesis of orthogonality and the precise calculation done in V.3.16, we see that this set is contained in the set $\{\sigma\pi_\mu$ where $M_{\sigma\mathfrak{S}} = G\}$. Clearly $M_{\sigma\mathfrak{S}} = M_{-\sigma\overline{\mathfrak{S}}}$ and we conclude as at the end of VI.1.6(d) that

$$\sum_{w \in W(M)} \text{Res}_{w\mathfrak{S}}^G E(\phi, w\pi_\mu)$$

is square integrable; this contradicts the choice of π_μ.

Second step

Fix $w \in W(M)$ and $\phi \in P_{\mathfrak{X}}$; suppose that $M_{w\mathfrak{S}} \neq G$. Let us show that the automorphic form $\text{Res}_{w\mathfrak{S}}^G E(\phi, w\pi_\mu)$ belongs to the vector space generated by the automorphic forms $\text{Res}_{\sigma\mathfrak{S}}^G E(\phi', \sigma\pi_\mu)$ where $\phi' \in P_{\mathfrak{X}}$ and $\sigma \in W(M)$ satisfies $M_{\sigma\mathfrak{S}} \in L(\mathfrak{C})$.

Set $L' = M_{w\mathfrak{S}}$; it is a proper standard Levi of G. Let $\xi_{L'}$ be the character of $Z_{L'}$ defined by $w\pi_L$; this allows us to define, as in the intoduction to Chapter VI, the subset $(w\mathfrak{P})_{L'}$ of $w\mathfrak{P}$. Let $(w\mathfrak{S})_{L'}$ be the irreducible component of $w\mathfrak{S} \cap \mathfrak{P}_{L'}$ containing $w\pi_L$. We define an

Eisenstein series from $\mathbf{L}'$ to $\mathbf{G}$ similarly to $E_L^G(\mu)$ denoted simply by $E_{L'}^G$, satisfying the analogue of (1):

$$E_{L'}^G(\operatorname{Res}_{w\mathfrak{S}_{L'}}^{L'} E^{L'}(\phi, w\pi_\mu)) = \operatorname{Res}_{w\mathfrak{S}}^G E(\phi, w\pi_\mu) \tag{4}$$

for every $\phi \in P_{\mathfrak{X}}$. (Here $E^{L'}$ is the Eisenstein series for $\mathbf{L}'$.) By VI.1.4:

$$\operatorname{Res}_{w\mathfrak{S}}^G E^{L'}(\phi, w\pi_\mu) = \operatorname{Res}_{(w\mathfrak{S})_{L'}}^{L'} E^{L'}(\phi, w\pi_\mu).$$

We denote by $\mathfrak{X}_{L'}$ the equivalence class of cuspidal data for L' containing $(wMw^{-1}, (w\mathfrak{P})_{L'})$ and by $\mathfrak{C}_{L'}$ the equivalence class of $(w\mathfrak{S})_{L'}$. By definition $L(\mathfrak{C})$ contains $L(\mathfrak{C}_{L'})$ and the induction hypothesis ensures that there exist elements $\sigma_1, \ldots, \sigma_R \in W_{L'}(wMw^{-1})$ and $\phi_1, \ldots, \phi_R \in P_{\mathfrak{X}}$ such that:

(5) $\operatorname{Res}_{(w\mathfrak{S})_{L'}}^{L'} E^{L'}(\phi, w\pi_\mu)$ belongs to the vector space generated by the elements $\operatorname{Res}_{\sigma_i(w\mathfrak{S})_{L'}}^{L'} E^{L'}(\phi_i, \sigma_i w\pi_\mu), i \in [1, R]$,

and

$$M_i := M_{\sigma_i(w\mathfrak{S})_{L'}} \in L(\mathfrak{C}_{L'}),\ i \in [1, R]. \tag{6}$$

Now:

$$\sigma_i(w\mathfrak{S})_{L'} + \operatorname{Re} X_{L'}^{\mathbf{G}} \otimes_{\mathbf{R}} \mathbf{C} = \sigma_i((w\mathfrak{S})_{L'} + \operatorname{Re} X_{L'}^{\mathbf{G}} \otimes_{\mathbf{R}} \mathbf{C}) = \sigma_i w\mathfrak{S},$$

by VI.1.4, so

$$\operatorname{Res}_{\sigma_i(w\mathfrak{S})_{L'}}^{L'} E^{L'}(\phi_i, \sigma_i w\pi_\mu) = \operatorname{Res}_{\sigma_i w\mathfrak{S}}^G E^{L'}(\phi_i, \sigma_i w\pi_\mu)$$

and with (6)

$$\operatorname{Re} \sigma_i w\mathfrak{S}^0 = \operatorname{Re} X_{M_i}^{\mathbf{G}}.$$

We replace w by $\sigma_i w$ in (4) and we apply $E_{L'}^G$ to (5); this gives the desired assertion.

To conclude the proof of (2), it remains to prove the following assertion:

Let $w \in W(M)$ be such that $\operatorname{Re} w\mathfrak{S}^0 = \operatorname{Re} X_{M_{w\mathfrak{S}}}^{\mathbf{G}}$, *i.e.* $M_{w\mathfrak{S}} \in L(\mathfrak{C})$, *and let $\phi \in P_{\mathfrak{X}}$; then there exists $\phi \in P_{\mathfrak{X}}$, $w_L \in W_L(M)$ such that:*

$$\operatorname{Res}_{w\mathfrak{S}}^G E(\phi, w\pi_\mu) = \operatorname{Res}_{w_L\mathfrak{S}}^G E(\phi', w_L\pi_\mu). \tag{8}$$

We have $wLw^{-1} = M_{w\mathfrak{S}}$. Decompose w into $w^L w_L$, where $w_L \in W_L$ and w^L is the element of minimal length in its right class modulo W_L. By I.1.9, $w_L \in W_L(M)$ and $w^L \in W(w_L M w_L^{-1})$. By VI.1.5:

$$\operatorname{Res}_{w\mathfrak{S}}^G E(\phi, w\pi_\mu) = \operatorname{Res}_{w_L\mathfrak{S}}^G E\left(M((w^L)^{-1}, w\pi_\mu)\phi, w_L\pi_\mu\right).$$

We fix $\phi' \in P_{\mathfrak{X}}$ such that ϕ' and $M((w^L)^{-1}, w\pi_\mu)\phi$ coincide to a high order at every point of the orbit of π_μ and we still have:

$$\operatorname{Res}_{w_L\mathfrak{S}}^G E(M((w^L)^{-1}, w\pi_\mu)\phi, w_L\pi_\mu) = \operatorname{Res}_{w_L\mathfrak{S}}^G E(\phi', w_L\pi_\mu).$$

(It is an equality at the point $w_L\pi_\mu$ and only at this point.) This concludes the proof of (2).

VI.2.5. Proof of the spectral decomposition of VI.2.2 (number field case) We first suppose that k is a number field and postpone the function field case to VI.2.7. We fix a finite set $\mathfrak{F}$ of **K**-types and we also fix $T \in \mathbb{R}_+^*$; we define $H_{\mathfrak{C},T}^{\mathfrak{F}}$ in a similar way to $H_{\mathfrak{C}}$ with $\mathrm{Im}\, X_L^{\mathbf{G}}$ replaced by:

$$\{\mu \in \mathrm{Im}\, X_L^{\mathbf{G}} |\ \|\mu\|^2 < T + \|o(\mathfrak{S})\|^2\},$$

and restricting to linear combinations of elements on which K acts via an element of $\mathfrak{F}$. Let us prove VI.2.2. We recall the notation $\mathrm{Hilb}_{\mathfrak{C},T}^{\mathfrak{F}}$ of V.3.11. We construct an isometry j of $H_{\mathfrak{C},T}^{\mathfrak{F}}$ onto $\mathrm{Hilb}_{\mathfrak{C},T}^{\mathfrak{F}}$: with

$$F \in H_{\mathfrak{C},T}^{\mathfrak{F}}$$

we associate

$$j(F) := E_L^G \circ F,$$

i.e.

$$j(F)(\pi_\mu) = E_L^G(\mu) F(\mu).$$

We note that the map $\mu \in \mathrm{Im}\, X_L^{\mathbf{G}} \mapsto \pi_\mu \in \mathfrak{S}$ induces an isomorphism between:

$$\{\mu \in \mathrm{Im}\, X_L^{\mathbf{G}} |\ \|\mu\|^2 < T + \|o(\mathfrak{S})\|^2\} \text{ and } \mathfrak{S}_{\leq T}.$$

Taking VI.2.4 into account, the only thing to check is thus the compatibility of j with the invariance properties; more precisely we must show the following equivalence:

$$\begin{aligned} &F(t\mu) = M(t,\mu)F(\mu),\ \forall t \in \mathrm{Stab}_{W^L}(L,\delta) \\ (1) \qquad &\Leftrightarrow (jF)(\sigma\pi_\mu) = (jF)(\pi_\mu),\ \forall \sigma \in \mathrm{Norm}\ \mathfrak{S}. \end{aligned}$$

We first establish the relation between Norm $\mathfrak{S}$ and $\mathrm{Stab}_{W^L}(L,\delta)$. Let us show that:

(2) $\quad \forall t \in \mathrm{Stab}_{W^L}(L,\delta)\ \exists\, w_L \in W_L(M)$, such that $tw_L \in \mathrm{Norm}\,\mathfrak{S}$,

(3) $\quad \forall \sigma \in \mathrm{Norm}\,\mathfrak{S},\ \exists\, t \in \mathrm{Stab}_{W^L}(L,\delta)$ such that $t^{-1}\sigma \in W_L(M)$.

Let us prove (2). Let $t \in \mathrm{Stab}_{W^L}(L,\delta)$ and let $w_L \in W_L(M)$ be such that $t\pi_L = w_L\pi_L$. Then $t(t^{-1}w^{-1}t) = w_L^{-1}t$ stabilises π_L and normalises $\mathrm{Re}\, X_L^{\mathbf{G}}$, i.e. is an element of $\mathrm{Norm}\,\mathfrak{S}$. It is easy to see that $t^{-1}w_L^{-1}t \in W_L(M)$ which gives (2). Let us prove (3). Let $\sigma \in \mathrm{Norm}\,\mathfrak{S}$: write $\sigma = tw_L$ with t of minimal length in σW_L and $w_L \in W_L$. We already saw that $w_L \in W_L(M)$. We also have $\sigma L \sigma^{-1} = L$ since σ normalises $\mathrm{Re}\, X_L^{\mathbf{G}}$ and thus $tLt^{-1} = L$ and $t \in \mathrm{Stab}_{W^L} L$. We also have $\sigma\pi_L = \pi_L$ since k is a number field, so by VI.2.1(1), $t^{-1} \in \mathrm{Stab}_{W^L}(L,\delta)$, and as $\mathrm{Stab}_{W^L}(L,\delta)$ is a group, we have $t \in \mathrm{Stab}_{W^L}(L,\delta)$ hence (3).

Taking (2) and (3) into account, we see that to prove (1) it suffices to prove the following result:

i*Let* $t \in \mathrm{Stab}_{W^L}(L, \delta), \sigma \in \mathrm{Norm}\,\mathfrak{S}$ *and* $w_L \in W_L(M)$. *Suppose that* $\sigma = tw_L$. *Then for every measurable function* F *on* $\mathrm{i}\,\mathrm{Re}\,X_L^{\mathbf{G}}$:

(4) $M(t, \mu)F(\mu) = F(t\mu) \Leftrightarrow (jF)(\sigma\pi_\mu) = (jF)(\pi_\mu)$, almost everywhere.

To prove this we need the formula proved in VI.2.3(4) (here $tLt^{-1} = L$), for F as above:

$$(5) \qquad E_L^G(t\mu)(M(t, \mu)F(\mu)) = E_L^G(\mu)F(\mu).$$

Using the fact that E_L^G is injective, we see (at least generically, see VI.2.4(ii)) that (4) is equivalent to:

$$E_L^G(\mu)F(\mu) = E_L^G(t\mu)F(t\mu) \Leftrightarrow (jF)(\sigma\pi_\mu) = (jF)(\pi_\mu).$$

But the second term can be rewritten as $E_L^G(\sigma\mu)F(\sigma\mu) = E_L^G(\mu)F(\mu)$. The equivalence is trivially a consequence of the fact that $w_L\mu = \mu$ for every $\mu \in X_L^{\mathbf{G}}$.

This concludes the proof of VI.2.2(i).

We now prove VI.2.2(ii). Let ϕ be as in the statement: we suppose in addition that it is a linear combination of elements on which K acts via one of the K-types contained in $\mathfrak{F}$. As in VI.2.2 (1) and (2), we see that there exists:

$$\phi' \in \cup_{w \in W_L(M)} P_{w(M,\mathfrak{P})}^{\mathfrak{F}}$$

such that

$$\phi(\mu) = \sum_{w \in W_L(M)} \mathrm{Res}_{w\mathfrak{S}_L}^L E^L(\phi', w\pi_\mu).$$

We first check that the function:

$$F_\phi : \mu \mapsto E(\phi, \mu)$$

defines an element of $\mathrm{Hilb}_{\mathfrak{C}}^{\mathfrak{F}}$. We have

$$E(\phi, \mu) = \sum_{w \in W_L(M)} \mathrm{Res}_{w\mathfrak{S}}^G E(\phi', w\pi_\mu).$$

By VI.1.2, this function depends holomorphically on μ. For every T, it thus defines an element of $\mathrm{Hilb}_{\mathfrak{C},T}^{\mathfrak{F}}$; we pass to the limit by using the decreasing properties of ϕ. We show that $e(\phi) = j(F_\phi)$ by calculating the scalar product against every pseudo-Eisenstein series as we did in V.3.7. To prove that the vector space generated by the $e(\phi)$ is dense, we proceed by showing that its orthogonal complement is zero. As in V.3.11 (see the beginning of the proof) we must prove that for almost every μ the space $A_{\mathfrak{C},\pi_\mu}^{\mathfrak{F}}$ coincides with the image of $E_L^G(\mu)$, which is VI.2.3(ii). This concludes the proof.

VI.2.6. Proof of the spectral decomposition of VI.2.1 (number field case) We proved (i) in VI.2.2, as well as the first part of (ii): the second part is a consequence of the functional equation.

(iii) Taking VI.2.2(ii) into account, we have:

$$L^2(G(k)\backslash\mathbf{G})_{[L],\xi} = \bigoplus_{\mathfrak{X}} \bigoplus_{\mathfrak{C}} L^2(G(k)\backslash\mathbf{G})_{\mathfrak{C}}$$

where $\mathfrak{X}$ runs through the set of equivalence classes of cuspidal data and $\mathfrak{C}$ runs through the set of elements of $[S_{\mathfrak{X}}]$ such that $L(\mathfrak{C}) = [L]$. This gives (iii).

(iv) This results from the following reformulation of VI.2.2(i): let $\mathfrak{C} \in [S_{\mathfrak{X}}]$ and let (L, δ) be a discrete parameter for $\mathfrak{C}$. Set:

$$L^2(L(k)\backslash\mathbf{L})_{\mathfrak{C}} = \oplus_{\delta'} L^2(L(k)\backslash\mathbf{L})_{\delta'}$$

where δ' runs through the set of subrepresentations of the discrete spectrum of $\mathbf{L}$ such that (L, δ') is a discrete parameter for $\mathfrak{C}$, i.e. δ' runs through the set $w\delta$ where $w \in W^L$ satisfies $wLw^{-1} = L$. Then the Hilbert space described in VI.2.2(i) is isomorphic to the following Hilbert space:

$$\left\{\begin{array}{l} \textit{Measurable functions } F' \textit{ on } \mathrm{Im}\, X_L^{\mathbf{G}} \textit{ with values in} \\ \mathrm{ind}_{\mathbf{K}\cap\mathbf{L}}^{\mathbf{K}} L^2(L(k)\backslash\mathbf{L})_{\mathfrak{C}} \textit{ such that :} \\ M(t^{-1}, t\mu)F'(t\mu) = F'(\mu),\ \forall t \in \mathrm{Stab}_{W^L} L \\ \lim\limits_{T\to\infty} \displaystyle\int_{\substack{\mu\in \mathrm{Im}\, X_L^{\mathbf{G}} \\ \|\mu\|\le T}} \|F'(\mu)\|^2 < +\infty. \end{array}\right\}$$

Indeed, with F as in VI.2.1(1), we associate F' defined by:

(1)
$$F'(\mu) := \sum_{t\in \mathrm{Stab}_{W^L} L} M(t^{-1}, t\mu)F(t\mu)$$

and with F' as above we associate F defined by:

$$F(\mu) := \mathrm{proj}_{\mathrm{ind}_{L\cap\mathbf{K}}^{\mathbf{K}} L^2(L(k)\backslash\mathbf{L})_\delta} F'(\mu).$$

We denote this projection by proj_δ. Now,

$$M(t^{-1}, t\mu)F(t\mu) \in \mathrm{ind}_{L\cap\mathbf{K}}^{\mathbf{K}} L^2(L(k)\backslash\mathbf{L})_{t\delta}$$

almost everywhere, whence, almost everywhere

$$\mathrm{proj}_\delta M(t^{-1}, t\mu)F(t\mu) = \begin{cases} 0 & \text{if } t\delta \neq \delta, \\ F(\mu), & \text{otherwise.} \end{cases}$$

Thus if F and F' are related by (1), $\mathrm{proj}_\delta F'(\mu) = F(\mu)$ up to a (non-zero) scalar for every μ and t as above. This implies that $F' = 0$. So (iv) is clear.

VI.2.7. The function field case

We suppose that k is a function field; in the statement of VI.2.2, we defined $L^2(G(k)\backslash \mathbf{G})_{\tilde{\mathfrak{C}}}$. Set:

$$\tilde{\mathfrak{S}} := \mathfrak{S} \otimes X_L^{\mathbf{G}} = \pi_L \otimes X_L^{\mathbf{G}}.$$

We note that the elements of $\tilde{\mathfrak{C}}$ are exactly the conjugacy classes of connected components of $\tilde{\mathfrak{S}}$. By a method analogous to that of VI.2.6, we check that the spectral decomposition of

$$L^2(G(k)\backslash \mathbf{G})_{\mathfrak{C}}$$

given in V.3.11 can be generalised to a spectral decomposition of $L^2(G(k)\backslash \mathbf{G})_{\tilde{\mathfrak{C}}}$ simply by replacing $\mathfrak{S}$ by $\tilde{\mathfrak{S}}$ and thus $\operatorname{Norm}\mathfrak{S}$ by $\operatorname{Norm}\tilde{\mathfrak{S}}$ (we note that $o(\mathfrak{S}) = o(\mathfrak{S}')$ for every connected component $\mathfrak{S}'$ of $\tilde{\mathfrak{S}}$). We can then repeat word for word the proofs of VI.2.5 and VI.2.6, simply replacing $\operatorname{Stab}_{W^L}(L, \delta)$ by $\operatorname{Stab}'_{W^L}(L, \delta)$ (see VI.1.9 for the definition) and $\operatorname{Norm}\mathfrak{S}$ by $\operatorname{Norm}\tilde{\mathfrak{S}}$; we have the exact analogue of VI.2.5(2),...,(5) and the remaining part of the argument can be used unchanged.

APPENDIX I

Lifting of Unipotent Subgroups into a Central Extension

This appendix was entirely communicated to us by Deligne, who was inspired by [S].

We recall the situation: G is a reductive algebraic group (connected, but this plays no role here) defined over the global field k. We denote by $\mathbf{G}$ a central extension of $G(\mathbb{A})$ by a finite group $\mathbf{N}$. We thus have the exact sequence:

$$1 \to \mathbf{N} \to \mathbf{G} \overset{\mathrm{pr}}{\to} G(\mathbb{A}) \to 1.$$

We suppose that there exists a section of $G(k)$ into $\mathbf{G}$. We note that $G(\mathbb{A})$ acts by conjugation on $\mathbf{G}$ via the formula:

$$g \in G(\mathbb{A}), \mathbf{g} \in \mathbf{G} \mapsto \tilde{g}\mathbf{g}\tilde{g}^{-1},$$

where $\tilde{g}$ is any element of $\mathrm{pr}^{-1}(g)$. The goal of this appendix is to prove the following:

Proposition (a) *Let P be a parabolic subgroup of G defined over k and let k' be a k-subalgebra of $\mathbb{A}$. Let U be the radical unipotent of P. There exists a unique $P(k)$-equivariant homomorphism $s_{k'}$ of $U(k')$ into $\mathbf{G}$ which is a section of* pr. *This homomorphism is $P(k')$-equivariant and its restriction to $U(k)$ is the restriction to $U(k)$ of the fixed lifting of $G(k)$ in $\mathbf{G}$. Moreover if $k' = \mathbb{A}$, then $s_{\mathbb{A}}$ is continuous.*

We note immediately that this proposition is easy in characteristic 0, in which case we have an even more precise result:

Proposition *Suppose k is of characteristic 0 and let k' be a k-algebra* *contained in* $\mathbb{A}$. *Let* U *be a unipotent subgroup of* G. *There exists a unique homomorphism* $s_{k'}$ *of* $U(k')$ *into* $\mathbf{G}$ *which is a section of* pr. *Moreover if* $k' = \mathbb{A}$, *then* $s_{\mathbb{A}}$ *is continuous.*

Let N be the order of $\mathbf{N}$. The map $u \in U(k') \mapsto u^N$ is an isomorphism; let $u \in U(k')$ and let u' be the element of $U(k')$ such that $u'^N = u$. We also fix a lifting $\tilde{u}'$ of u' in $\mathbf{G}$. We necessarily have:

$$s_{k'}(u) = \tilde{u}'^N,$$

and the right-hand side is independent of the chosen lifting. This allows us to define $s_{k'}$ uniquely; it is clear that $s_{k'}$ is the inverse of the restriction of pr to the subset $\tilde{U}$ of $\mathbf{G}$ consisting of elements of the form $\tilde{u}^N$ where $\mathrm{pr}(u) \in U(k')$. But we must still prove that $s_{k'}$ is a group homomorphism and is continuous if $k' = \mathbb{A}$, i.e. that $\tilde{U}$ is a subgroup of $\mathbf{G}$ which is open in $\mathrm{pr}^{-1}(U(\mathbb{A}))$ if $k' = \mathbb{A}$. This is clear if U is abelian, and in the general case it can be proved by induction using a filtration of U by normal subgroups, the quotient of two successive terms of this filtration being commutative.

Let us return to the first proposition. Let L be a Levi subgroup of P, defined over k, and let T be its central maximal split torus. We will show:

Proposition (b) *There exists a unique* $T(k)$*-equivariant homomorphism* $s_{k'}$ *of* $U(k')$ *into* $\mathbf{G}$ *which lifts* pr. *Moreover if* $k' = \mathbb{A}$ *then* $s_{\mathbb{A}}$ *is continuous.*

Let us suppose for the moment that this result is known. The uniqueness of $s_{k'}$ ensures that $s_{k'}$ is $L(k')$-equivariant and as this homomorphism respects the group structure, it is also $U(k')$-equivariant. The uniqueness of s_k ensures that s_k coincides on the one hand with the restriction of $s_{k'}$ to $U(k)$ and on the other with the restriction to $U(k)$ of the fixed section of $G(k)$ in $\mathbf{G}$. This proves the proposition.

We now prove the proposition which was temporarily supposed known. It can be deduced from the following lemma:

Lemma *Let* X, Y *be* k*-vector spaces and let* μ_i, μ_j *be rational characters of* T *defined over* k; *we suppose that there exists no* $(a, b) \in \mathbb{N}^2 - \{(0,0)\}$ *such that* $\mu_i^a = \mu_j^{-b}$. *The sub* $k \otimes_{\mathbf{Z}} k$*-module of* $X \otimes_{\mathbf{Z}} Y$ *generated by the set of elements:*

$$\mu_i(t)x \otimes \mu_j(t)y - x \otimes y,\ t \in T(k),\ x \in X,\ y \in Y,$$

coincides with $X \otimes_{\mathbf{Z}} Y$.

We easily reduce to the case where $X \simeq k \simeq Y$, and thus to the problem of proving that the ideal of the ring $k \otimes_{\mathbf{Z}} k$ generated by the elements $\mu_i(t) \otimes \mu_j(t) - 1$, where t runs through $T(k)$, is not proper. We can replace

k by a suitably chosen subfield of k. The lemma is clear if k is a field of characteristic 0, since in this case $\mathbb{Q} \otimes_{\mathbb{Z}} \mathbb{Q} \simeq \mathbb{Q}$ is a field. We thus can suppose that k contains $\mathbb{F}_p(T)$ and even that $k = \mathbb{F}_p(T)$. If μ_i and μ_j are linearly independent characters of T, the lemma is also clear. We can thus suppose that there exist $a, b \in \mathbb{N}$, not both zero, such that the subgroup of $k^* \times k^*$ which is the image of $T(k)$ under (μ_i, μ_j) is the set of elements (λ^a, λ^b) where λ runs through k^*. Thus the ideal we consider is the ideal I of $k \otimes_{\mathbb{Z}} k$ generated by the set of elements $\lambda^a \otimes \lambda^b - 1$ where λ runs through k^*. Note that $k \otimes_{\mathbb{Z}} k \simeq \mathbb{F}_p(T) \otimes_{\mathbb{F}_p} \mathbb{F}_p(T)$ is isomorphic to the $\mathbb{F}_p$-algebra of polynomials in two variables U, V, localised at the set of elements which are products of an element of $\mathbb{F}_p[U] - \{0\}$ with an element of $\mathbb{F}_p[V] - \{0\}$. We fix an algebraic closure $\overline{\mathbb{F}}_p$ of $\mathbb{F}_p$ and a non-trivial algebraically closed extension K of $\overline{\mathbb{F}}_p$. We denote by I_K the ideal of the K-algebra $K[U]_{\{\mathbb{F}_p[U]-0\}} \otimes_K K[V]_{\{\mathbb{F}_p[V]-0\}}$ obtained by extension of the base field of $\mathbb{F}_p$ to K. Evidently I is proper if and only if I_K is, and I_K is generated by the set of elements $g(U)^a g(V)^b$ where $g(T)$ runs through $\mathbb{F}_p[T]$. Now for each proper ideal of $K[U]_{\{\mathbb{F}_p[U]-0\}} \otimes_K K[V]_{\{\mathbb{F}_p[V]-0\}}$, there exists at least one point u, v of $K \times K$, with u and v transcendent over $\mathbb{F}_p$, at which all the elements of the ideal vanish. Fix such u, v transcendent over $\mathbb{F}_p$: we show that there exists at least one polynomial $g(T)$ with coefficients in $\mathbb{F}_p$ such that $g(u)^a g(v)^b - 1$ is not zero, which will give a contradiction. Let m be the smallest common multiple of a and b and write $aa' = m$ and $bb' = m$. Set $x = u^a$. We first consider $g = T$; if the desired conclusion is not immediate with this polynomial, we have:

$$x = v^{-b}.$$

We next consider the polynomials g of the form $f(T^m)$. Since x is transcendent over $\mathbb{F}_p$, the equality

$$f(x^{a'})^a f(x^{-b'})^b = 1,$$

cannot hold for every polynomial $f(T)$. This gives the lemma.

Let us return to the existence and the uniqueness of $s_{k'}$. Let $g, g' \in G(k')$; we denote by $c(g, g')$ the element of $\mathbf{G}$ which equals $\tilde{g}\tilde{g}'\tilde{g}^{-1}\tilde{g}'^{-1}$, where $\tilde{g}$ and $\tilde{g}'$ are any elements of $\mathrm{pr}^{-1}(g)$ and $\mathrm{pr}^{-1}(g')$) respectively. We fix a filtration of U by normal subgroups, stable under T:

$$U = U_0 \supset \cdots \supset U_i \supset \cdots \supset U_d = \{1\}.$$

We suppose (as we may) that for every $i \in [1, d]$, there exists a T-equivariant nilpotent subgroup of U_i, denoted by V_i, such that the natural map of V_{i-1} in U_{i-1}/U_i is an isomorphism and such that the action of T on V_i is given by a character, μ_i; this means that we have implicitly identified V_i with a vector space in a T-equivariant way. Obviously, we suppose that everything is defined over k. By the structure of U, we know that μ_i is non-trivial for every i and that there exists no i, j such that $\mu_i = \mu_j^{-1}$. Let $i \in [0, d]$ and denote by $\tilde{U}_i$ the group generated by the set of elements $c(t, u)$ where t runs through $T(k)$ and u runs through $U_i(k')$. We will prove by decreasing induction on $i \in [0, d]$ that the restriction of pr to $\tilde{U}_i$ is an isomorphism on $U_i(k')$, whose inverse we denote by s_i. As $\tilde{U}_i$ is clearly stable under conjugation by $T(k)$ and as pr is $T(k)$ equivariant, s_i will also be $T(k)$-equivariant. The morphism s_0 corresponds to the conditions imposed on $s_{k'}$ in the statement; this proves the existence. The uniqueness results from the fact that if $s_{k'}$ has the desired properties, $s_{k'}(U(k'))$ necessarily contains $\tilde{U}_0$, and by what was just said, must coincide with $\tilde{U}_0$. Let us prove the desired properties of $\tilde{U}_i$ for fixed i in $[0, d]$. These properties are clear if $i = d$; we thus suppose them proved down to $i+1$ and prove them for i. We fix $t_i \in T(k)$ such that $\mu_i(t_i) - 1 \neq 0$ and we denote by $\tilde{V}_i$ the set of elements $c(t_i, u)$ where u runs through V_i. It is clear that $\tilde{V}_i$ is $T(k)$-invariant. We will show:

(1) $\tilde{V}_i$ is a group,
(2) $\tilde{V}_i$ normalises $\tilde{U}_{i+1}$,
(3) $\operatorname{pr} \tilde{V}_i = V_i$,
(4) $\tilde{V}_i \cap \mathbf{N} = 1$.

Thus, thanks to (1) and (2), the subset of $\mathbf{G}$ consisting of products $\tilde{v}\tilde{u}$, where $\tilde{v}$ runs through $\tilde{V}_i$ and $\tilde{u}$ runs through $\tilde{U}_{i+1}$, is a subgroup of $\mathbf{G}$. Thanks to (3), it contains $\tilde{U}_i$ and thus coincides with $\tilde{U}_i$. We easily deduce from (3) and (4) the desired properties of $\tilde{U}_i$. We can now prove (1) to (4). Assertions (3) and (4) are immediate consequences of the properties of t_i. We remark that to prove (1), it suffices to prove that $\tilde{V}_i$ consists of elements which commute with each other: indeed, let $u, v \in V_i$; with the help of (3) we fix a representative $\tilde{v} \in \tilde{V}_i$ of v and a representative $\tilde{u} \in \mathbf{G}$ of u such that $\tilde{u}^{-1} \in \tilde{V}_i$. We use the fact that $\tilde{u}^{-1}$ commutes with $c(t_i, v)$ to obtain:

$$c(t_i, u)c(t_i, v) = (t_i \tilde{u} t_i^{-1})c(t_i, v)\tilde{u}^{-1} = t_i \tilde{u} t_i^{-1} t_i \tilde{v} t_i^{-1} \tilde{v}^{-1} \tilde{u}^{-1} = c(t_i, uv).$$

An analogous argument proves that $c(t_i, u)^{-1} \in \tilde{V}_i$ for every $u \in V_i$. We will prove by descending induction on $j \in [i, d]$ that:

$$\forall u \in V_i, \forall v \in V_j, \quad c(u, v) \in \tilde{U}_{j+1}.$$

This will prove (1) and also (2). We denote by $\bar{c}$ the map deduced from the commutator c of $V_i \times V_j$ in $\tilde{U}_{j+1}\mathbf{N}/\tilde{U}_{j+1} \simeq \mathbf{N}$. We check that $\bar{c}$ is biadditive and thus defines a morphism of $V_i \otimes_{\mathbf{Z}} V_j$ in $\mathbf{N}$. This morphism is $T(k)$-equivariant; its triviality is a consequence of the lemma. This concludes the proof of existence and uniqueness of $s_{k'}$. It remains to prove continuity of $s_{\mathbb{A}}$; for this it is equivalent to prove that $s_{\mathbb{A}}(U(\mathbb{A}))$ is open in $\mathrm{pr}^{-1}(U(\mathbb{A}))$, or that there exists a neighborhood X of the unit element in $V_i(\mathbb{A})$ over which the covering $\mathrm{pr}^{-1}(U_i(\mathbb{A})) \to U(\mathbb{A})$ can be trivialised, such that $s_{\mathbb{A}}(X)$ is contained in the sheet of $\mathrm{pr}^{-1}(X)$ which contains the unit element $1_{\mathbf{G}}$ of $\mathbf{G}$; we denote this sheet by $\tilde{X}$. Up to restricting X, we suppose that for every $i \in [0, d]$, $X \cap V_i(\mathbb{A})$ is a neighborhood of the unit element in $V_i(\mathbb{A})$ and that multiplication induces an isomorphism:

$$X \cap V_1(\mathbb{A}) \times \cdots \times X \cap V_i(\mathbb{A}) \times \cdots \times X \cap V_d(\mathbb{A}) \to X.$$

Up to restricting X again, we suppose that no product of d elements of $\tilde{X}$ belongs to

$$\bigcup_{n \in \mathbf{N} - \{1_{\mathbf{G}}\}} \tilde{X}n.$$

It thus suffices to show that for every $i \in [0, d]$, $s_{\mathbb{A}}(X \cap V_i(\mathbb{A}))$ is contained in $\tilde{X}$. For $i \in [0, d]$, we recall the notation t_i introduced above and up to restricting X yet again, we suppose that there exists a neighborhood X' of the unit element in $U(\mathbb{A})$ such that for every $i \in [0, d]$, and for every $v \in V_i(\mathbb{A}) \cap X$ there exists $v' \in X' \cap V_i(\mathbb{A})$ such that $v = t_i v' t_i^{-1} v'^{-1}$. Up to restricting once more, we also suppose that for every $i \in [0, d]$ the open set $t_i \mathrm{pr}^{-1}(X') t_i^{-1} \mathrm{pr}^{-1}(X'^{-1})$ of $\mathrm{pr}^{-1}(U(\mathbb{A}))$ does not intersect $\cup_{n \in \mathbf{N} - \{1_{\mathbf{G}}\}} \tilde{X}n$. Let $i \in [0, d]$; all these hypotheses imply that:

$$s_{\mathbb{A}}(X \cap V_i(\mathbb{A})) \subset t_i \mathrm{pr}^{-1}(X') t_i^{-1} \mathrm{pr}^{-1}(X'^{-1}) \cap \mathrm{pr}^{-1}(X) \subset \tilde{X}.$$

This concludes the proof.

The above proof can obviously also be applied to the case of finite central extensions of $G(F)$ where F is a local field, and even more generally, a field of characteristic 0 or non-algebraic over the prime field.

APPENDIX II

Automorphic Forms and Eisenstein series over a Function Field

1. Derivatives of Eisenstein series

1.1. Some reminders

When M is a standard Levi subgroup of G, we set with no further explanations P to be the standard parabolic subgroup of Levi M and U to be the unipotent radical of P.

Let M be a standard Levi subgroup of G and $\pi \in \Pi_0(\mathbf{M})$ a (class of) irreducible automorphic cuspidal representations of $\mathbf{M}$. We denote by $A_0(M(k)\backslash\mathbf{M})_\pi$ the isotypic subspace of type π of $A_0(M(k)\backslash\mathbf{M})$ and by $A_0(U(\mathbb{A})M(k)\backslash\mathbf{G})_\pi$ the space of $\varphi \in A_0(U(\mathbb{A})M(k)\backslash\mathbf{G})$ such that for every $k \in \mathbf{K}$, $\varphi_k \in A_0(M(k)\backslash\mathbf{M})_\pi$. Recall that $\varphi_k : M(k)\backslash\mathbf{M} \to \mathbb{C}$ is defined by

$$\varphi_k(m) = m^{-\rho_P}\varphi(mk)$$

for every $m \in \mathbf{M}$.

The orbit of π under X_M is $\{\pi \otimes \lambda; \lambda \in X_M\}$. This orbit is a complex analytic variety, for it is locally isomorphic to X_M.

For $\psi \in A_0(U(\mathbb{A})M(k)\backslash\mathbf{G})_\pi$ and $\lambda \in X_M$, we define

$$\lambda \circ \psi \in A_0(U(\mathbb{A})M(k)\backslash\mathbf{G})_{\pi\otimes\lambda}$$

by

$$(\lambda \circ \psi)(g) = m_P(g)^\lambda \psi(g).$$

We denote by $-\pi$ the contragredient of π; there exists a natural bilinear product on

$$A_0(U(\mathbb{A})M(k)\backslash \mathbf{G})_\pi \times A_0(U(\mathbb{A})M(k)\backslash \mathbf{G})_{-\pi}$$

denoted by (,). We emphasize the fact that we use a bilinear and not a sesquilinear product throughout this appendix.

1.2. Eisenstein series

Let M be a standard Levi subgroup, π an irreducible cuspidal representation of $\mathbf{M}$, $\mathfrak{P}$ its orbit and $\mathfrak{F}$ a finite set of $\mathbf{K}$-types, i.e. of isomorphism classes of irreducible representations of $\mathbf{K}$. Denote by $P_{\mathrm{hol},\pi}^{\mathfrak{F}}$ the space of functions ϕ such that

(i) there exists a neighborhood V_ϕ of π in $\mathfrak{P}$ such that ϕ is defined on V; for $\pi' \in V_\phi$, we have

$$\phi(\pi') \in A_0(U(\mathbb{A})M(k)\backslash \mathbf{G})_{\pi'};$$

ϕ is smooth and $\mathbf{K}$-finite, the group acting in the obvious way on the functions satisfying (i); the $\mathbf{K}$-invariant space generated by ϕ can be decomposed into irreducible subspaces whose isomorphism classes belong to $\mathfrak{F}$;

(iii) for every $\psi \in A_0(U(\mathbb{A})M(k)\backslash \mathbf{G})_{-\pi}$, the function

$$\{\lambda \in X_M ; \pi \otimes \lambda \in V_\phi\} \to \mathbb{C}$$

$$\lambda | \mapsto (\phi(\pi \otimes \lambda), (-\lambda) \circ \psi)$$

is holomorphic.

We consider these functions as germs, i.e. we identify ϕ and ϕ' if they coincide in a neighborhood of π.

For $\phi \in P_{\mathrm{hol},\pi}^{\mathfrak{F}}$ and $\pi' \in V_\phi$, we can define an Eisenstein series $E(\phi(\pi'), \pi')$ to be an automorphic form on $G(k)\backslash \mathbf{G}$. More exactly the function $\pi' \mapsto E(\phi(\pi'), \pi')$ is meromorphic on V. We know that we can find a holomorphic function d, not identically zero on a neighborhood of π, such that for every $\phi \in P_{\mathrm{hol},\pi}^{\mathfrak{F}}$ and every $g \in \mathbf{G}$, the function

$$\pi' \longmapsto d(\pi')E(\phi(\pi'), \pi')(g)$$

is holomorphic on a neighborhood of π. Denote by $I_{\mathrm{hol},\pi}^{\mathfrak{F}}$ the set of germs of functions d satisfying this condition. It is an ideal of the local ring ϑ_π of germs of holomorphic functions at the point π.

1.3. Derivations

Let M be a standard Levi subgroup. Recall that $\mathrm{Re}\, X_M$ is a real vector space by virtue of the isomorphism $\mathrm{Re}\, X_M \simeq \mathrm{Re}\ \mathfrak{a}_M^*$. The space $\mathrm{Sym}(\mathrm{Re}\, X_M) \otimes_{\mathbb{R}} \mathbb{C}$ can be interpreted as a space of derivations on X_M:

with $\mu \in \mathrm{Re}X_M$, we associate the derivation ∂_μ defined by

$$\partial_\mu f(\lambda) = \frac{\mathrm{d}}{\mathrm{d}t} f(\lambda + t\mu)|_{t=0}$$

for every function f on X_M. Similarly, if $\mathfrak{P}$ is the orbit of a representation π of $\mathbf{M}$, $\mathrm{Sym}(\mathrm{Re}X_M) \otimes \mathbb{C}$ can be interpreted as a space of derivations on $\mathfrak{P}$. We denote by $\mathrm{Der}(X_M)$ and $\mathrm{Der}(\mathfrak{P})$, respectively, the spaces of derivations defined in this way. They are holomorphic derivations, i.e. they preserve the space of holomorphic functions.

1.4. Derivatives of Eisenstein series

Let M be a standard Levi subgroup, π an irreducible cuspidal representation of $\mathbf{M}$, $\mathfrak{P}$ its orbit, $\mathfrak{F}$ a finite set of $\mathbf{K}$-types, $\phi \in P^{\mathfrak{F}}_{\mathrm{hol},\pi}$, $d \in I^{\mathfrak{F}}_{\mathrm{hol},\pi}$ and $D \in \mathrm{Der}(\mathfrak{P})$. We define the automorphic form on $G(k)\backslash\mathbf{G}$:

$$D[d(\pi')E(\phi(\pi'), \pi')]_{\pi'=\pi},$$

see IV.1.9. The derivative of an Eisenstein series is an element of the space of automorphic forms generated by these functions as we let M, π, $\mathfrak{F}, \phi, d$ and D vary.

Conjecture *Every automorphic form is the derivative of an Eisenstein series.*†

The goal of this appendix is to prove this conjecture when k is a function field.

1.5. Function fields

We suppose from now on that k is a function field. We can then make the preceding constructions algebraic. Let M be a standard Levi subgroup, π an irreducible cuspidal representation of $\mathbf{M}$, $\mathfrak{F}$ its orbit. The sets X_M and $\mathfrak{F}$ are equipped with structures of algebraic varieties; a polynomial on X_M is a linear combination of functions of the form $\lambda \mapsto m^\lambda$ where m is any fixed element of $\mathbf{M}$; a polynomial on $\mathfrak{P}$ is a function Q such that the function on $X_M : \lambda \mapsto Q(\pi \otimes \lambda)$ is a polynomial. We denote by $\mathbb{C}[X_M]$ and $\mathbb{C}[\mathfrak{P}]$, the space of polynomials on X_M and $\mathfrak{P}$, respectively. We denote by $\mathbb{C}[\mathfrak{P}]_\pi$ the localisation of $\mathbb{C}[\mathfrak{P}]$ at π.

Remark As $X_M \simeq \mathbb{C}^n/(2\pi\mathrm{i}/\log q)L \simeq \mathbb{C}^{*n}$, for a certain integer n and a certain lattice L of $\mathbb{Q}^n$, the ring $\mathbb{C}[X_M]$ is isomorphic to a ring $\mathbb{C}[T_1, T_1^{-1}, \ldots, T_n, T_n^{-1}]$. The same is true of $\mathbb{C}[\mathfrak{P}]$.

† This conjecture was recently proved by Franke in his paper 'Harmonic analysis in weighted L^2 spaces'.

Let $\mathfrak{F}$ be a finite set of $\mathbf{K}$-types. Denote by $P^{\mathfrak{F}}_{(M,\mathfrak{P})}$ the space of functions ϕ such that

(i) ϕ is defined on $\mathfrak{P}$ and for every $\pi' \in \mathfrak{P}$,

$$\phi(\pi') \in A_0(U(\mathbb{A})M(k)\backslash\mathbf{G})_{\pi'};$$

the same as in 1.2;

(iii) for every $\psi \in A_0(U(\mathbb{A})M(k)\backslash\mathbf{G})_{-\pi}$, the function

$$X_M \to \mathbb{C}$$

$$\lambda \mapsto (\phi(\pi \otimes \lambda), (-\lambda) \circ \psi)$$

is a polynomial.

Set

$$P^{\mathfrak{F}}_{\pi} = P^{\mathfrak{F}}_{(M,\mathfrak{P})} \otimes_{\mathbb{C}[\mathfrak{P}]} \mathbb{C}[\mathfrak{P}]_{\pi}.$$

We know that for every $\phi \in P^{\mathfrak{F}}_{\pi}$ and every $g \in \mathbf{G}$, the function $\pi' \mapsto E(\phi(\pi'), \pi')(g)$ is a rational function on $\mathfrak{P}$; see IV.1.12. More precisely, there exists $d \in \mathbb{C}[\mathfrak{P}]_{\pi}$ such that for every $\phi \in P^{\mathfrak{F}}_{\pi}$ and every $g \in \mathbf{G}$, the function

$$\pi' \mapsto d(\pi')E(\phi(\pi'), \pi')(g)$$

belongs to $\mathbb{C}[\mathfrak{P}]_{\pi}$, i.e. has no singularity at π. Let $I^{\mathfrak{F}}_{\pi}$ be the ideal of d satisfying this condition. For $\phi \in P^{\mathfrak{F}}_{\pi}$, $d \in I^{\mathfrak{F}}_{\pi}$, $D \in \mathrm{Der}(\mathfrak{P})$, we define as in 1.2 the automorphic form

$$D[d(\pi')E(\phi(\pi'), \pi')]_{\pi'=\pi}.$$

In fact these automorphic forms generate the same space as in 1.2. This is an easy consequence of the fact that the natural inclusion $\mathbb{C}[\mathfrak{P}]_{\pi} \to \vartheta_{\pi}$ becomes an isomorphism after completion.

2. Relations between the constant terms of automorphic forms

2.1. Notation

Let φ be an automorphic form on $G(k)\backslash\mathbf{G}$. For every standard parabolic subgroup $P = MU$ of G, we have defined the constant term φ_P of φ along P, its cuspidal component $\varphi_P^{\mathrm{cusp}}$ and the cuspidal support $\Pi_0(\mathbf{M}, \varphi)$; see I.3.5. Fix a (finite) set of cuspidal data

$$D(M, \varphi) \subset \mathbb{C}[\mathrm{Re}\,\mathfrak{a}_M] \times \Pi_0(\mathbf{M}) \times A_0(U(\mathbb{A})M(k)\backslash\mathbf{G})$$

such that

(i) if $(Q, \pi, \psi) \in D(M, \varphi)$, then $\psi \in A_0(U(\mathbb{A})M(k)\backslash\mathbf{G})_{\pi}$;

(ii) for every $g \in \mathbf{G}$, we have the equality

$$\varphi_P^{\text{cusp}}(g) = \sum_{(Q,\pi,\psi)\in D(M,\varphi)} Q(\log_M m_P(g))\psi(g);$$

(iii) $\Pi_0(\mathbf{M},\varphi)$ is the set of π which occur in some triple $(Q,\pi,\psi) \in D(M,\varphi)$.

We fix a finite set $\mathfrak{F}$ of $\mathbf{K}$-types, stable under passage to the contragredient, such that for every standard Levi M and every $(Q,\pi,\psi) \in D(M,\varphi)$, the representation of $\mathbf{K}$ in the $\mathbf{K}$-invariant space generated by φ can be decomposed into irreducible components whose classes belong to $\mathfrak{F}$.

Denote by Y the set of pairs $(M,\mathfrak{P})$, where M is a standard Levi and $\mathfrak{P}$ is the orbit under X_M of an irreducible cuspidal representation of $\mathbf{M}$. Set

$$P^{\mathfrak{F}} = \bigoplus_{(M,\mathfrak{P})\in Y} P^{\mathfrak{F}}_{(M,\mathfrak{P})}.$$

For $\phi = (\phi_{M,\mathfrak{P}})_{(M,\mathfrak{P})\in Y}$, M a standard Levi and $\pi \in \Pi_0(\mathbf{M})$, we set $\phi(\pi) = \phi_{M,\mathfrak{P}}(\pi)$, where $\mathfrak{P}$ is the orbit of π.

For $(M,\mathfrak{P}) \in Y$, we defined a function $\theta_{\phi_{M,\mathfrak{P}}}$ on $G(k)\backslash\mathbf{G}$, called a pseudo-Eisenstein series, by

$$\theta_{\phi_{M,\mathfrak{P}}}(g) = \int_{\pi\in\mathfrak{P},\ \mathrm{Re}\,\pi=\lambda_0} E(\phi_{M,\mathfrak{P}}(\pi),\pi)(g)\,d\pi,$$

λ_0 being a 'sufficiently positive' fixed point in $\mathrm{Re}\,X_M$. For $\phi = (\phi_{M,\mathfrak{P}}) \in P^{\mathfrak{F}}$, we set

$$\theta_\phi = \sum_{(M,\mathfrak{P})\in Y} \theta_{\phi_{M,\mathfrak{P}}}.$$

The field k being a function field, θ_ϕ has compact support on $G(k)\backslash\mathbf{G}$.

We have the isomorphism $\mathbb{C}[\mathrm{Re}\,\mathfrak{a}_M] \simeq \mathrm{Sym}(\mathrm{Re}\ X_M) \otimes_{\mathbf{R}} \mathbb{C}$. We can thus associate a derivation ∂_Q on X_M with an element Q of $\mathbb{C}[\mathrm{Re}\,\mathfrak{a}_M]$, see 1.3.

Recall that $\mathbf{G}$, $\mathbf{M}$, X_M and so on were equipped with Haar measures. They are invariant under conjugation, i.e. if M and M' are two standard Levi subgroups and if $g \in G(k)$ conjugates M to M', the measures on $\mathbf{M}'$, $X_{M'}$ and so on are the images of measures on $\mathbf{M}$, X_M and so on under the conjugation defined by g.

2.2. Calculation of a scalar product

Lemma *For every $(M,\mathfrak{P}) \in Y$, there exists a constant $c_{M,\mathfrak{P}} > 0$, depending only on the Haar measures, and such that $c_{M,\mathfrak{P}} = c_{M',\mathfrak{P}'}$ if $(M,\mathfrak{P})$ and*

$(M', \mathfrak{P}')$ are conjugate, such that for every $\phi \in P^{\mathfrak{F}}$, we have the equality

$$\int_{G(k)\backslash \mathbf{G}} \varphi(g)\theta_\phi(g)\, dg = \sum_M \sum_{(Q,\pi,\psi)\in D(M,\varphi)} c_{M,\mathfrak{P}} \partial_Q (\lambda \circ \psi, \phi(-(\pi \otimes \lambda)))\,|_{\lambda=0},$$

where $\mathfrak{P}$ is the orbit of π.

We fix constants as above and we define $I(\varphi, \phi)$ to be the right-hand side of the above equality.

Proof Set $\phi = (\phi_{M,\mathfrak{P}})_{(M,\mathfrak{P})\in Y}$, and denote by $J(\varphi, \phi)$ the integral of the left-hand side. We have

$$\begin{aligned} J(\varphi,\phi) &= \sum_{(M,\mathfrak{P})\in Y} \int_{G(k)\backslash \mathbf{G}} \varphi(g)\theta_{\phi_{M,\mathfrak{P}}}(g)\, dg \\ &= \sum_{(M,\mathfrak{P})\in Y} \int_{U(\mathbb{A})M(k)\backslash \mathbf{G}} \varphi_P(g)\varepsilon F\phi_{M,\mathfrak{P}}(g)\, dg, \end{aligned}$$

where by definition

$$\varepsilon F\phi_{M,\mathfrak{P}}(g) = \int_{\pi\in\mathfrak{P}, \mathrm{Re}\,\pi=\lambda_0} \phi_{M,\mathfrak{P}}(\pi)(g)\, d\pi.$$

Thus

$$\begin{aligned} J(\varphi,\phi) &= \sum_{(M,\mathfrak{P})\in Y} \int_{U(\mathbb{A})M(k)\backslash \mathbf{G}} \varphi_P^{\mathrm{cusp}}(g)\varepsilon F\phi_{M,\mathfrak{P}}(g)\, dg \\ &= \sum_{(M,\mathfrak{P})\in Y} \sum_{(Q,\pi,\psi)\in D(M,\varphi)} \int_{U(\mathbb{A})M(k)\backslash \mathbf{G}} Q(\log_M m_P(g))\psi(g)\varepsilon F\phi_{M,\mathfrak{P}}(g)\, dg. \end{aligned}$$

For $(M, \mathfrak{P}) \in Y$, $(Q, \pi, \psi) \in D(M, \varphi)$ and $\lambda \in X_M$, set

$$j(\psi, \phi_{M,\mathfrak{P}}, \lambda) = \int_{U(\mathbb{A})M(k)\backslash \mathbf{G}} (\lambda \circ \psi)(g)\varepsilon F\phi_{M,\mathfrak{P}}(g)\, dg.$$

Using the equalities

$$(\lambda \circ \psi)(g) = m_p(g)^\lambda \psi(g),$$

$$Q(\log_M m_P(g)) = \partial_Q(m_P(g)^\lambda)|_{\lambda=0},$$

we obtain

$$(1) \qquad J(\varphi,\phi) = \sum_{(M,\mathfrak{P})\in Y} \sum_{(Q,\pi,\psi)\in D(M,\varphi)} \partial_Q j(\psi, \phi_{M,\mathfrak{P}}, \lambda)|_{\lambda=0}.$$

It remains to calculate $j(\psi, \phi_{M,\mathfrak{P}}, \lambda)$ for ψ and $\phi_{M,\mathfrak{P}}$ fixed. Set $Z^1_{\mathbf{M}} =$

$Z_{\mathbf{M}} \cap \mathbf{M}^1$. We have

$$j(\psi, \phi_{M,\mathfrak{P}}, \lambda) = \int_{Z_{\mathbf{M}} U(\mathbb{A}) M(k) \backslash \mathbf{G}} \int_{Z^1_{\mathbf{M}} \backslash Z_{\mathbf{M}}} \int_{(Z_M(k) \cap Z_{\mathbf{M}}) \backslash Z^1_{\mathbf{M}}} (\lambda \circ \psi)(g)\, z^{\lambda} \chi_{\pi}(zz_1)$$
$$\int_{\pi' \in \mathfrak{P},\ \mathrm{Re}\,\pi' = \lambda_0} \chi_{\pi'}(zz_1) \phi_{M,\mathfrak{P}}(\pi')(g)\, d\pi'\, dz_1\, dz\, dg.$$

The integral in $z_1 \in (Z_M(k) \cap Z_{\mathbf{M}}) \backslash Z^1_{\mathbf{M}}$ is zero unless $(\chi_\pi + \chi_{\pi'})|_{Z^1_{\mathbf{M}}} = 1$ for $\pi' \in \mathfrak{P}$ (note that $\chi_{\pi'|Z^1_{\mathbf{M}}}$ does not depend on the choice of $\pi' \in \mathfrak{P}$). Suppose this condition is satisfied. Then the double integral in $z \in Z^1_{\mathbf{M}} \backslash Z_{\mathbf{M}}$ and $\pi' \in \mathfrak{P}$ can be simplified by Fourier-Mellin inversion. We obtain

$$j(\psi, \phi_{M,\mathfrak{P}}, \lambda) = c \int_{Z_{\mathbf{M}} U(\mathbb{A}) M(k) \backslash \mathbf{G}} (\lambda \circ \psi)(g) \sum_{\pi' \in \mathfrak{P}; \chi_{\pi'} = -\chi_\pi - \lambda} \phi_{M,\mathfrak{P}}(\pi')(g)\, dg$$

where $c > 0$ depends only on the Haar measures on $Z_{\mathbf{M}}$ and $\mathfrak{P}$. The remaining integral is easy to calculate and we finally obtain

$$j(\psi, \phi_{M,\mathfrak{P}}, \lambda) = \begin{cases} 0 & \text{if } -\pi \notin \mathfrak{P}, \\ c\,(\lambda \circ \psi, \phi_{M,\mathfrak{P}}(-(\pi \otimes \lambda))) & \text{if } -\pi \in \mathfrak{P} \end{cases}.$$

Plugging this value into the expression (1), we obtain the desired equality. By construction, $c_{M,\mathfrak{P}}$ depends only on the Haar measures. As these are invariant under conjugation, we have $c_{M,\mathfrak{P}} = c_{M',\mathfrak{P}'}$ if $(M, \mathfrak{P})$ and $(M', \mathfrak{P}')$ are conjugate. □

2.3. Vanishing of a pseudo-Eisenstein series

For a Levi M and an irreducible cuspidal representation π of $\mathbf{M}$, denote by $A_0(U(\mathbb{A})M(k)\backslash \mathbf{G})^{\mathfrak{F}}_\pi$ the subspace of elements of $A_0(U(\mathbb{A})M(k)\backslash \mathbf{G})_\pi$ such that the $\mathbf{K}$-invariant subspace generated by them decomposes into irreducible subspaces whose isomorphism classes belong to $\mathfrak{F}$.

Let M and M' be two standard Levi subgroups. Recall that we denote by $W(M, M')$ the set of elements w of the Weyl group W of G such that $wMw^{-1} = M'$ and w is minimal in its right class modulo the Weyl group of M. Let w be an element of $W(M, M')$ and π an irreducible cuspidal representation of $\mathbf{M}$. We know how to define an intertwining operator

$$M(w, \pi) : A_0(U(\mathbb{A})M(k)\backslash \mathbf{G})^{\mathfrak{F}}_\pi \to A_0(U'(\mathbb{A})M'(k)\backslash \mathbf{G})^{\mathfrak{F}}_{w\pi}.$$

The base field being a function field, this operator is a rational function. More precisely, let $\mathfrak{P}$ be the orbit of π; there exists a polynomial d on $\mathfrak{P}$, not identically zero, such that

(i) for all $\psi \in A_0(U(A)M(k)\backslash\mathbf{G})^{\mathfrak{F}}_{\pi}$, $\psi' \in A_0(U'(\mathbb{A})M'(k)\backslash\mathbf{G})_{-w\pi}$, the function on X_M given by:

$$\lambda \mapsto d(\pi\otimes\lambda)(M(w,\pi\otimes\lambda)(\lambda\circ\psi),(-w\lambda)\circ\psi')$$

is a polynomial; see Proposition IV.1.12.

For every $(M,\mathfrak{P}) \in Y$, fix a polynomial $d_{M,\mathfrak{P}}$ on $\mathfrak{P}$, not identically zero, satisfying (i) for every M', every $w \in W(M,M')$ and every $\pi \in \mathfrak{P}$ (or equivalently, for a fixed π in $\mathfrak{P}$) and also satisfying

(ii) for every M', $w \in W(M,M')$, $\pi \in \mathfrak{P}$, $d_{M',w\mathfrak{P}}(w\pi) = d_{M,\mathfrak{P}}(\pi)$;

(iii) for every $\pi \in \mathfrak{P}$, $d_{M,-\mathfrak{P}}(-\pi) = d_{M,\mathfrak{P}}(\pi)$, where $-\mathfrak{P}$ is the orbit of $-\pi$.

In this situation, we set

$$N(w,\pi) = d_{M,\mathfrak{P}}(\pi)M(w,\pi)$$

where $\mathfrak{P}$ is the orbit of π.

Lemma *Let $\phi \in P^{\mathfrak{F}}$. Suppose that for every M and every $\pi \in \Pi_0(\mathbf{M})$, we have the equality*

$$\sum_{M'}\sum_{w\in W(M,M')} N(w^{-1},w\pi)\phi(w\pi) = 0$$

in $A_0(M(k)U(\mathbb{A})\backslash\mathbf{G})_\pi$. Then $\theta_\phi = 0$.

Proof Obviously the hypothesis of the statement is equivalent to the same hypothesis where $N(w^{-1},w\pi)$ is replaced by $M(w^{-1},w\pi)$. But then the lemma is simply a consequence of the scalar product formula which in our situation can be written

$$\int_{G(k)\backslash\mathbf{G}} \theta_\phi(g)\overline{\theta_\phi}(g)\,dg =$$

$$\sum_{(M,\mathfrak{P})\in Y} c'_{M,\mathfrak{P}} \int_{\pi\in\mathfrak{P},\ \mathrm{Re}\,\pi=\lambda_{\mathfrak{P}}} \sum_{M'}\sum_{w\in W(M,M')} (M(w^{-1},-w\pi)\phi(-w\pi),\overline{\phi}(\overline{\pi}))\,d\pi$$

where $c'_{M,\mathfrak{P}}$ is a constant > 0 depending only on the Haar measures. □

Corollary *Under these hypotheses, $I(\varphi,\phi) = 0$.*

This is an immediate consequence of Lemma 2.2 and the preceding lemma.

2.4. The global sections generate the local sections

Let M be a standard Levi subgroup. Let S_M be the local ring $\mathbb{C}[X_M]_0$ and $\hat{S}$ its completion. The latter is a formal series ring. In particular it is a filtered ring: for an integer N, we can define the subspace $\hat{S}_M^{\geq N}$ of elements which vanish to the order at least N. Let $\pi \in \Pi_0(\mathbf{M})$ and $\mathfrak{P}$ be its orbit. We set

$$S_{M,\pi} = A_0(M(k)U(\mathbb{A})\backslash\mathbf{G})_\pi^{\mathfrak{F}} \otimes_{\mathbb{C}} S_M,$$
$$\hat{S}_{M,\pi} = A_0(M(k)U(\mathbb{A})\backslash\mathbf{G})_\pi^{\mathfrak{F}} \otimes_{\mathbb{C}} \hat{S}_M,$$
$$\hat{S}_{M,\pi}^{\geq N} = A_0(M(k)U(\mathbb{A})\backslash\mathbf{G})_\pi^{\mathfrak{F}} \otimes_{\mathbb{C}} \hat{S}_M^{\geq N}.$$

In particular if $\phi = (\phi_{M,\mathfrak{P}})_{(M,\mathfrak{P})\in Y} \in P^{\mathfrak{F}}$, the function on X_M given by:

$$\lambda \mapsto \lambda \circ \phi_{M,\mathfrak{P}}(\pi \otimes (-\lambda))$$

can be identified with an element denoted by $s_{M,\pi}\phi$ of $S_{M,\pi}$ or of $\hat{S}_{M,\pi}$.

Let M' be another Levi subgroup and $w \in W(M, M')$. Then w naturally defines maps $S_{M'} \to S_M$, $\hat{S}_{M'} \to \hat{S}_M$ Similarly, we define an operator

$$sN(w,\pi) : S_{M,\pi} \to S_{M',w\pi}.$$

by

$$(sN(w,\pi)(\psi \otimes f))(\lambda) = \lambda \circ N\left(w, \pi \otimes (-w^{-1}\lambda)\right)(-w^{-1}\lambda \circ \psi)f(w^{-1}\lambda)$$

for $\psi \in A_0(M(k)U(\mathbb{A})\backslash\mathbf{G})_\pi^{\mathfrak{F}}$, $f \in S_M$, $\lambda \in X_{M'}$. This operator can be extended to an operator also denoted by $sN(w,\pi) : \hat{S}_{M,\pi} \to \hat{S}_{M',w\pi}$.

The key point of our proof is the following proposition.

Proposition *Let Z be a finite set of pairs (M,π), where M is a standard Levi and $\pi \in \Pi_0(\mathbf{M})$; for every $(M,\pi) \in Z$, let $\hat{\phi}_{M,\pi} \in \hat{S}_{M,\pi}$. Let N be an integer. Suppose that Z is stable under conjugation (i.e. if $(M,\pi) \in Z$, if $w \in W(M,M')$, then $(M', w\pi) \in Z$), and that for every $(M,\pi) \in Z$, we have the equality*

$$\sum_{M'} \sum_{w\in W(M,M')} sN(w^{-1}, w\pi)\hat{\phi}_{M',w\pi} = 0$$

in $\hat{S}_{M,\pi}$. Then there exists $\phi \in P^{\mathfrak{F}}$ such that

(i) *for every M, every $\pi \in \Pi_0(\mathbf{M})$, we have the equality*

$$\sum_{M'} \sum_{w\in W(M,M')} N(w^{-1}, w\pi)\phi(w\pi) = 0$$

in $A_0(M(k)U(\mathbb{A})\backslash\mathbf{G})_\pi^{\mathfrak{F}}$;

(ii) *for every $(M,\pi) \in Z$, we have the congruence*

$$s_{M,\pi}\phi - \hat{\phi}_{M,\pi} \in \hat{S}_{M,\pi}^{\geq N}.$$

We refer to Section 3 for the (purely algebraic) proof of this proposition.

2.5. A linear system

For the set Z, we take the set of pairs (M,π) such that there exists a pair (M',π') conjugate to (M,π), with $\pi' \in \Pi_0(\mathbf{M}',\varphi)$. Fix an integer N strictly greater than the degree of every polynomial Q occurring in a cuspidal datum of φ. The formula which defines $I(\varphi,\phi)$ for $\phi \in P^{\mathfrak{F}}$ uses only the expansions to the order $< N$ of ϕ at the points $-\pi$, for π occurring in a pair $(M,\pi) \in Z$. Set

$$\hat{S}_Z = \otimes_{(M,\pi)\in Z}\hat{S}_{M,-\pi},\ \hat{S}_Z^{\geq N} = \otimes_{(M,\pi)\in Z}\hat{S}_{M,-\pi}^{\geq N}$$

and define a map

$$s : P^{\mathfrak{F}} \to \hat{S}_Z$$

$$\phi \mapsto s\phi = (s_{M,-\pi}\phi)_{(M,\pi)\in Z}.$$

We then see that there exists a linear form $\hat{I}(\varphi,.)$ on $\hat{S}_Z$, zero on $\hat{S}_Z^{\geq N}$, such that $\hat{I}(\varphi,s\phi) = I(\varphi,\phi)$ for every $\phi \in P^{\mathfrak{F}}$.

Lemma *Let $\hat{\phi} = (\hat{\phi}_{M,-\pi})_{(M,\pi)\in Z}$ be an element of $\hat{S}_Z$. Suppose that for every $(M,\pi) \in Z$, we have*

$$\sum_{M'}\ \sum_{w\in W(M,M')} sN(w^{-1},-w\pi)\hat{\phi}_{M',-w\pi} = 0$$

in $\hat{S}_{M,-\pi}$. Then $\hat{I}(\varphi,\hat{\phi}) = 0$.

Proof We apply Proposition 2.4 to the set $Z' = \{(M,-\pi);\ (M,\pi) \in Z\}$, to the integer N and to $\hat{\phi}_{M,-\pi}$. We deduce from this the existence of an element ϕ of $P^{\mathfrak{F}}$ satisfying conditions (i) and (ii) of the proposition. By (i) and Corollary 2.3, we have $I(\varphi,\phi) = 0$. By (ii) and the fact that $\hat{I}(\varphi,.)$ kills $\hat{S}_Z^{\geq N}$, we have

$$\hat{I}(\varphi,\hat{\phi}) = \hat{I}(\varphi,s\phi) = I(\varphi,\phi).$$

This gives the result. □

2.6. Resolution of the system

The algebra $\mathrm{Der}(X_M)$ acts on $\mathbb{C}[X_M]$. Its action can be extended to $\mathbb{C}[X_M]_0$ then to $\hat{S}_M$ by continuity. On the other hand we can evaluate an element of $\hat{S}_M$ at 0 (which comes down to considering the map $\hat{S}_M \to \hat{S}_M/\hat{S}_M^{\geq 1} \simeq \mathbb{C}$). For $D \in \mathrm{Der}(X_M)$ and $f \in \hat{S}_M$, we denote by $D(f)|_0$ the value of $D(f)$ at 0. This construction defines an embedding $\mathrm{Der}(X_M) \hookrightarrow \hat{S}_M^*$ where $\hat{S}_M^*$ is the dual of $\hat{S}_M$ as a complex vector space. Let $\pi \in \Pi_0(\mathbf{M})$. Then $A_0(M(k)U(\mathbb{A})\backslash\mathbf{G})_\pi^{\mathfrak{F}}$ is the dual of $A_0(M(k)U(\mathbb{A})\backslash\mathbf{G})_{-\pi}^{\mathfrak{F}}$. We obtain an embedding

$$A_0(M(k)U(\mathbb{A})\backslash\mathbf{G})_\pi^{\mathfrak{F}} \otimes \mathrm{Der}(X_M) \hookrightarrow (\hat{S}_{M,-\pi})^*.$$

We adopt the following notation: for $\psi \in A_0(M(k)U(\mathbb{A})\backslash\mathbf{G})_\pi^{\mathfrak{F}}$, $D \in$

$\mathrm{Der}(X_M)$ and $\phi_{M,-\pi} \in \hat{S}_{M,-\pi}$, the value of $\psi \otimes D$ applied to $\phi_{M,-\pi}$ is denoted by

$$D(\psi, \phi_{M,-\pi})|_0.$$

Lemma *For every $(M, \pi) \in Z$, there exists a finite set*

$$\Delta(M, \pi) \subset A_0(M(k)U(\mathbb{A})\backslash \mathbf{G})_\pi^{\mathfrak{F}} \times \mathrm{Der}(X_M)$$

such that for every $\hat{\phi} = (\hat{\phi}_{M,-\pi})_{(M,\pi)\in Z} \in \hat{S}_Z$, we have the equality

$$\hat{I}(\varphi, \hat{\phi}) = \sum_{(M,\pi)\in Z} \sum_{(\psi,D)\in\Delta(M,\pi)} D\left(\psi, \sum_{M'} \sum_{w\in W(M,M')} sN(w^{-1}, -w\pi)\hat{\phi}_{M'-w\pi}\right)|_0.$$

Proof Denote by R the subspace of $\hat{\phi} = (\hat{\phi}_{M,-\pi})_{(M,\pi)\in Z} \in \hat{S}_Z$ such that for every $(M, \pi) \in Z$, we have

$$\sum_{M'} \sum_{w\in W(M,M')} sN(w^{-1}, -w\pi)\hat{\phi}_{M,-w\pi} = 0.$$

Define a map

$$j : \bigoplus_{(M,\pi)\in Z} (A_0(M(k)U(\mathbb{A})\backslash \mathbf{G})_\pi^{\mathfrak{F}} \otimes_{\mathbb{C}} \mathrm{Der}(X_M)) \to \hat{S}_Z^*$$

as follows: for

$$x = \left(\sum_{(\psi,D)\in\Delta(M,\pi)} \psi \otimes D\right)_{(M,\pi)\in Z}$$

in the domain, where the $\Delta(M, \pi)$ are finite sets, and

$$\hat{\phi} = (\hat{\phi}_{M,-\pi})_{(M,\pi)\in Z} \in \hat{S}_Z,$$

set

$$(j(x), \hat{\phi}) = \sum_{(M,\pi)\in Z} \sum_{(\psi,D)\in\Delta(M,\pi)} D\left(\psi, \sum_{M'} \sum_{w\in W(M,M')} sN(w^{-1}, -w\pi)\hat{\phi}_{M,-w\pi}\right)|_0.$$

Let V be the image of j.

Note that if $f \in \hat{S}_M$ satisfies $D(f)|_0 = 0$ for every $D \in \mathrm{Der}(X_M)$, then $f = 0$. We deduce from this that R is exactly the subspace of elements killed by V.

For every integer n, denote by V^n the subspace of elements of V which kill $\hat{S}_Z^{\geq n}$, and R^n the subspace of elements of $\hat{S}_Z$ which are killed by V^n. It is clear that we have the inclusion

$$\hat{S}_Z^{\geq N} + R \subset R^n.$$

Let us show that in fact we have the equality

$$\hat{S}_Z^{\geq N} + R = R^n. \tag{1}$$

For this, we must show that we have

$$\text{for all } n' \geq n,\ \hat{S}_Z^{\geq N} + R^{n'} = R^n. \tag{2}$$

So let $n, n' \in \mathbb{N}$ with $n' \geq n$. Set $\tilde{S}_Z = \hat{S}_Z / \hat{S}_Z^{\geq n'}$ and let $\tilde{S}_Z^{\geq n}$, $\tilde{R}^{n'}$ and $\tilde{R}^n$ be the images of $\hat{S}_Z^{\geq n}$, $R^{n'}$ and R^n in $\tilde{S}_Z$. As these last spaces are invariant under $\hat{S}_Z^{\geq n'}$, the equality (2) is equivalent to

$$\tilde{S}_Z^{\geq n} + \tilde{R}^{n'} = \tilde{R}^n. \tag{3}$$

The elements of $V^{n'}$ and V^n which kill $\hat{S}_Z^{\geq n'}$ naturally define elements of the dual $\tilde{S}_Z^*$. Denote by $\tilde{V}^{n'}$ and $\tilde{V}^n$, respectively, the subspaces consisting of these elements. The following properties are consequences of the definitions:

(i) $\tilde{R}^{n'}$ and $\tilde{R}^n$, are the subspaces of elements of $\tilde{S}_Z$ killed by $\tilde{V}^{n'}$ and $\tilde{V}^n$ respectively;

(ii) $\tilde{V}^n$ is the subspace of elements of $\tilde{V}^{n'}$ which kill $\tilde{S}_Z^{\geq n}$.

The equality (3) is an easy consequence of these properties and of the biduality theorem, which states that if E is a finite-dimensional vector space, $E' \subset E$ a subspace and $E'^{\perp}$ its annihilator in E^*, then $E' = (E'^{\perp})^{\perp}$; note that to apply this theorem $\tilde{S}_Z$ must be finite-dimensional.

This proves (3) and (2). Now let $\hat{\phi} \in R^n$. Thanks to (2), we can construct by induction a sequence $(\hat{\phi}_{n'})_{n' \geq n}$ such that

$$\begin{cases} \text{(i) } \hat{\phi}_{n'} \in R^{n'}; \\ \text{(ii) } \hat{\phi}_{n'+1} - \hat{\phi}_{n'} \in \hat{S}_Z^{\geq n'}; \\ \text{(iii) } \hat{\phi}_n = \phi. \end{cases} \tag{4}$$

Recall that for every M, $\hat{S}_M$ is equipped with a topology for which it is complete. The space $\hat{S}_Z$ is thus itself equipped with a topology for which it is complete. Condition (4)(ii) above implies that $(\hat{\phi}_{n'})_{n' \geq n}$ is a Cauchy sequence. It thus converges to a limit denoted by $\hat{\phi}_\infty$, for which we have

$$\hat{\phi}_\infty - \hat{\phi}_{n'} \in \hat{S}_Z^{\geq n'} \text{ for every } n' \geq n. \tag{5}$$

Together with (4)(i), this relation implies that $\hat{\phi}_\infty$ is killed by $V^{n'}$ for every $n' \geq n$. Now by construction

$$V = \bigcup_{n' \in \mathbb{N}} V^{n'} = \bigcup_{n' \geq n} V^{n'}.$$

Thus $\hat{\phi}_\infty$ is killed by V, which implies that $\hat{\phi}_\infty \in R$ as we already noted.

The relation (5) for $n' = n$ shows that $\hat{\phi}_n \in \hat{S}_Z^{\geq n} + R$. Hence the inclusion $R^n \subset \hat{S}_Z^{\geq n} + R$, and the equality (1).

Introducing as above a space $\tilde{S}_Z$ equal this time to $\hat{S}_Z/\hat{S}_Z^{\geq n}$ and using the biduality theorem in the space $\tilde{S}_Z$, it is a consequence of (1) that V^n is exactly the subspace of elements of $\hat{S}_Z^*$ which kill $\hat{S}_Z^{\geq n} + R$. It is a consequence of its construction that the linear form $\hat{I}(\varphi,.)$ kills $\hat{S}_Z^{\geq N}$. It kills R by Lemma 2.5. Thus it belongs to V^N, *a fortiori* to V. The definition of V then implies the statement of the lemma. □

2.7. Expression of an automorphic form as the derivative of an Eisenstein series

For every $(M, \pi) \in Z$, fix a finite set $\Delta(M, \pi)$ such that the conditions of the preceding lemma are satisfied. Fix an integer N' strictly greater than the order of the derivatives appearing in the formula of the lemma. If $(M, \pi) \in Z$, $(\psi, D) \in \Delta(M, \pi)$ and $f \in \hat{S}_M^{\geq N'}$, we thus have $D(f)|_0 = 0$. For the same data, we can find $\phi_\psi \in P^{\mathfrak{F}}_{(M,\mathfrak{P})}$, where $\mathfrak{P}$ is the orbit of π, such that

$$s^+\phi_\psi - \psi \in \hat{S}_{M,\pi}^{\geq N'}$$

where we denote by $s^+\phi_\psi$ the element of $\hat{S}_{M,\pi}$ which can be deduced from the function

$$\lambda \mapsto (-\lambda) \circ \phi_\psi(\pi \otimes \lambda).$$

We fix such an element ϕ_ψ. Note that the function $d_{M,\mathfrak{P}}$ which we fixed in 2.3 belongs to $I_\pi^{\mathfrak{F}}$ (see 1.5): if $\phi \in P_\pi^{\mathfrak{F}}$, the function

$$\pi' \mapsto d_{M,\mathfrak{P}}(\pi')E(\phi(\pi'), \pi')$$

has no singularities in π, as its constant terms (which can be calculated using the operators $M(w, \pi')$, see below) have none. We can then consider the derivative of the Eisenstein series

$$E_{M,\pi,\psi,D} := D[d_{M,\mathfrak{P}}(\pi')E(\phi_\psi(\pi'), \pi')]_{\pi'=\pi}.$$

Lemma *We have the equality*

$$\varphi = \sum_{(M,\pi)\in Z} c_{M,\mathfrak{P}}^{-1} \sum_{(\psi,D)\in\Delta(M,\pi)} E_{M,\pi,\psi,D}$$

(where $\mathfrak{P}$ is the orbit of π).

Proof Let E be the right-hand side of the equality in the statement. It suffices (see Proposition I.3.4) to prove that for every standard parabolic $P = MU$ of G, we have the equality $\varphi_P^{\text{cusp}} = E_P^{\text{cusp}}$. An expression for

φ_P^{cusp} is given by hypothesis; see 2.1. Calculation of the constant terms of Eisenstein series leads to the equality

$$E_P^{\text{cusp}} = \sum_{(M',\pi')\in Z} c_{M',\mathfrak{P}'}^{-1} \sum_{(\psi',D')\in\Delta(M',\pi')} D'[d_{M',\mathfrak{P}'}(\pi'') \sum_{w\in W(M',M)} M(w,\pi'')\phi_{\psi'}(\pi'')]_{\pi''=\pi'}.$$

It is clear that the cuspidal support of E_P^{cusp} is contained in $\{\pi; (M,\pi)\in Z\}$. This is also true for φ_P^{cusp} by the definition of Z. We can thus fix $\pi \in \Pi_0(\mathbf{M})$ such that $(M,\pi)\in Z$ and define the components $\varphi_{P,\pi}^{\text{cusp}}$ and $E_{P,\pi}^{\text{cusp}}$ of φ_P^{cusp} and E_P^{cusp} supported by π in the obvious way: then we must show that $\varphi_{P,\pi}^{\text{cusp}} = E_{P,\pi}^{\text{cusp}}$. Fix $g\in\mathbf{G}$. We have

$$\varphi_{P,\pi}^{\text{cusp}}(g) = \sum Q(\log_M m_P(g))\psi(g),$$

the sum being over the pairs (Q,ψ) such that $(Q,\pi,\psi)\in D(M,\varphi)$;

(1)
$$E_{P,\pi}^{\text{cusp}}(g) = \sum_{M'} \sum_{w\in W(M',M)} \sum_{(\psi',D')\in\Delta(M',w^{-1}\pi)} c_{M',w^{-1}\mathfrak{P}}^{-1} D'[d_{M',w^{-1}\mathfrak{P}}(\pi') (M(w,\pi')\phi_{\psi'}(\pi'))(g)]_{\pi'=w^{-1}\pi},$$

where $\mathfrak{P}$ is the orbit of π.

As $\psi \mapsto \psi(g)$ is a linear form on $A_0(M(k)U(\mathbb{A})\backslash\mathbf{G})_\pi^{\mathfrak{F}}$, whose dual is $A_0(M(k)U(\mathbb{A})\backslash\mathbf{G})_{-\pi}^{\mathfrak{F}}$, there exists $\psi_g \in A_0(M(k)U(\mathbb{A})\backslash\mathbf{G})_{-\pi}^{\mathfrak{F}}$ such that

$$\psi(g) = (\psi,\psi_g)$$

for every $\psi \in A_0(M(k)U(\mathbb{A})\backslash\mathbf{G})_\pi^{\mathfrak{F}}$. Fix such an element ψ_g. Define an element $\hat{\phi} = (\hat{\phi}_{M',-\pi'})_{(M',\pi')\in Z} \in \hat{S}_Z$ by

$$\hat{\phi}_{M',-\pi'} = 0, \text{ if } (M',\pi') \neq (M,\pi);$$
$$\hat{\phi}_{M,-\pi} = \psi_g \otimes m_p(g)^\bullet,$$

where we write $m_p(g)^\bullet$ for the natural image in $\hat{S}_M$ of the polynomial on X_M given by $\lambda \mapsto m_P(g)^\lambda$. It is then a consequence of the definitions that

$$\hat{I}(\varphi,\hat{\phi}) = \sum c_{M,\mathfrak{P}}\, \partial_Q(\lambda\circ\psi, m_p(g)^\lambda(-\lambda)\circ\psi_g)|_{\lambda=0},$$

the sum being over the pairs (Q,ψ) such that $(Q,\pi,\psi)\in D(M,\varphi)$,

(2)
$$\begin{aligned}\hat{I}(\varphi,\hat{\phi}) &= c_{M,\mathfrak{P}} \sum \partial_Q(m_P(g)^\lambda)|_{\lambda=0}(\psi,\psi_g),\\ &= c_{M,\mathfrak{P}} \sum Q(\log_M m_P(g))\psi(g),\\ &= c_{M,\mathfrak{P}}\varphi_{P,\pi}^{\text{cusp}}(g).\end{aligned}$$

Let us use Lemma 2.6. We obtain another expression for $\hat{I}(\varphi, \hat{\phi})$:
(3)
$$\hat{I}(\varphi,\hat{\phi}) = \sum_{M'} \sum_{w\in W(M',M)} \sum_{(\psi',D')\in\Delta(M',w^{-1}\pi)} D'(\psi', sN(w^{-1},-\pi)\hat{\phi}_{M,-\pi})|_0.$$
Let x denote the summand. We can replace ψ' by $s\phi_{\psi'}$ in x. Then
$$x = D'[d_{M,-\mathfrak{P}}(-(\pi\otimes w\lambda'))\, m_P(g)^{w\lambda'}((-\lambda')\circ\phi_{\psi'}(w^{-1}\pi\otimes\lambda'),$$
$$\lambda'\circ M(w^{-1},-(\pi\otimes w\lambda'))((-w\lambda')\circ\psi_g))]_{\lambda'=0}.$$
We now use relations (ii) and (iii) of 2.3 and the adjunction formula for the intertwining operators. We obtain
$$\begin{aligned} x &= D'[d_{M',w^{-1}\mathfrak{P}}(w^{-1}\pi\otimes\lambda')m_P(g)^{w\lambda'}((-w\lambda')\circ \\ &\qquad M(w,w^{-1}\pi\otimes\lambda')\phi_{\psi'}(w^{-1}\pi\otimes\lambda'),\psi_g)]_{\lambda'=0}, \\ &= D'[d_{M',w^{-1}\mathfrak{P}}(w^{-1}\pi\otimes\lambda')m_P(g)^{w\lambda'}((-w\lambda')\circ \\ &\qquad M(w,w^{-1}\pi\otimes\lambda')\phi_{\psi'}(w^{-1}\pi\otimes\lambda'))(g)]_{\lambda'=0}, \\ &= D'[d_{M',w^{-1}\mathfrak{P}}(w^{-1}\pi\otimes\lambda')(M(w,w^{-1}\pi\otimes\lambda') \\ &\qquad \phi_{\psi'}(w^{-1}\pi\otimes\lambda'))(g)]_{\lambda'=0}, \end{aligned}$$
or equivalently,
$$x = D'[d_{M',w^{-1}\mathfrak{P}}(\pi'')(M(w,\pi'')\phi_{\psi'}(\pi''))(g)]_{\pi''=w^{-1}\pi}.$$
This calculation, the relations (1) and (3), and the fact that the constants $c_{M',\mathfrak{P}'}$ are invariant under conjugation lead to the equality
$$\hat{I}(\varphi,\hat{\phi}) = c_{M,\mathfrak{P}} E^{\text{cusp}}_{P,\pi}(g).$$
Comparing this with (2), we obtain the equality
$$\varphi^{\text{cusp}}_{P,\pi}(g) = E^{\text{cusp}}_{P,\pi}(g)$$
which is what we wanted to prove. This concludes the proof. □

2.8. Theorem

Every automorphic form on $G(k)\backslash\mathbf{G}$ is the derivative of an Eisenstein series (see 1.4).

As the automorphic form fixed in 2.1 was arbitrary, the theorem is a corollary of Lemma 2.7.

2.9. Remarks

(1) It is clear that the constructions in Section 1, the theorem and its proof admit variations when we fix a character ξ of $Z_{\mathbf{G}}$ and restrict our attention to the automorphic forms and Eisenstein series on which $Z_{\mathbf{G}}$ acts via ξ. We must of course then restrict ourselves in 1.3 to the

derivatives coming from $\mathrm{Sym}(\mathrm{Re}\, X_M^{\mathbf{G}}) \otimes_{\mathbf{R}} \mathbf{C}$, where $X_M^{\mathbf{G}}$ is the subgroup of elements of X_M which are trivial on $Z_{\mathbf{G}}$.

(2) It is a consequence of the theorem that the Eisenstein series constructed from (non-cuspidal) automorphic forms admit meromorphic continuations and functional equations. This result was recently obtained by Bernstein; Bernstein's result is also valid when k is a number field).

3. Proof of Proposition 2.4

3.1. Transformation of the problem

Decompose the set Z of Proposition 2.4 into a disjoint sum of sets Z^i stable under conjugation, such that this decomposition is maximal. With notation as in the proposition, suppose that we have constructed for every i an element ϕ^i satisfying (i) and (ii) for every pair $(M,\pi) \in Z^i$. For every pair $(M,\mathfrak{P})$, we can find a function $f^i_{M,\mathfrak{P}} \in \mathbf{C}[\mathfrak{P}]$ such that

(i) $f^i_{M,\mathfrak{P}}$ vanishes to the order at least N at every point π such that $(M,\pi) \in Z - Z^i$;

(ii) $f^i_{M,\mathfrak{P}} - 1$ vanishes to the order at least N at every point π such that $(M,\pi) \in Z^i$;

(iii) for $(M,\mathfrak{P})$, $(M',\mathfrak{P}') \in Y$, $w \in W(M,M')$ such that $w\mathfrak{P} = \mathfrak{P}'$, and for $\pi \in \mathfrak{P}$, we have

$$f^i_{M,\mathfrak{P}}(\pi) = f^i_{M',\mathfrak{P}'}(w\pi).$$

Then the element $\phi \in P^{\mathfrak{F}}$ defined by

$$\phi_{M,\mathfrak{P}} = \sum_i f^i_{M,\mathfrak{P}} \phi^i_{M,\mathfrak{P}}$$

still satisfies relation (i) of 2.4, and now satisfies (ii) as well for every pair $(M,\pi) \in Z$. This proves the proposition for Z.

This reasoning brings us back to the case where $Z = Z^i$. We thus suppose from now on that Z satisfies the condition:

(1) if (M,π), $(M',\pi') \in Z$, there exists $w \in W(M,M')$ such that $M' = wMw^{-1}$, $\pi' = w\pi$.

Then fix $(M,\pi) \in Z$ and set $W(M) = \bigcup_{M'} W(M,M')$. For $w \in W(M)$, define

$$A_w = A_0(M'(k)U'(\mathbb{A})\backslash \mathbf{G})^{\mathfrak{F}}_{w\pi} \bigotimes_{\mathbf{C}} \mathbf{C}[X_M],$$

where $M' = wMw^{-1}$. For $w, w' \in W(M)$, we define

$$N_{w,w'} : A_{w'} \to A_w$$

by

$$N_{w,w'}(\psi \otimes f)(\lambda) = (w\lambda) \circ N(ww'^{-1}, w'\pi \otimes (-w'\lambda))((-w'\lambda) \circ \psi)f(\lambda)$$

for $\psi \in A_0(M'(k)U'(\mathbb{A})\backslash \mathbf{G})^{\mathfrak{F}}_{w'\pi}$, $f \in \mathbb{C}[X_M]$, $\lambda \in X_M$. Set

$$A_Z = \bigoplus_{w \in W(M)} A_w.$$

This is a free finitely-generated $\mathbb{C}[X_M]$-module. The matrix $(N_{w,w'})_{w,w' \in W(M)}$ defines a $C[X_M]$-linear endomorphism of A_Z, denoted by N_Z. Let R be its kernel.

Set $\hat{A}_Z = A_Z \otimes_{\mathbb{C}[X_M]} \hat{S}_M$. The map N_Z can be extended by continuity to a $\hat{S}_M$-linear endomorphism of $\hat{A}_Z$. Let $\hat{R}$ be its kernel. The crucial point is that $\hat{S}_M$ is flat over $C[X_M]$ ([B2] Chapter III.3.4, Theorem 3). Thus

(2) $$\hat{R} = R \bigotimes_{\mathbb{C}[X_M]} \hat{S}_M.$$

3.2. Calculation in the completion

An element w of $W(M, M')$ naturally defines a map also denoted by $w : \mathbb{C}[X_M] \to \mathbb{C}[X_{M'}]$, which can be extended by continuity to a map $w : \hat{S}_M \to \hat{S}_{M'}$. Tensoring with the identity on $A_0(M'(k)U'(\mathbb{A})\backslash \mathbf{G})^{\mathfrak{F}}_{w\pi}$, we deduce maps

$$w : A_w \to A_0(M'(k)U'(\mathbb{A})\backslash \mathbf{G})^{\mathfrak{F}}_{w\pi} \bigotimes_{\mathbb{C}} \mathbb{C}[X_{M'}]$$

and

$$w : \hat{A}_w \to \hat{S}_{M,\pi}.$$

In the situation of Proposition 2.4, define an element

$$\hat{\phi} = (\hat{\phi}_w)_{w \in W(M)} \in \hat{A}_Z$$

by

$$\hat{\phi}_w = w^{-1}\hat{\phi}_{wMw^{-1},w\pi}.$$

Lemma *The element $\hat{\phi}$ belongs to $\hat{R}$.*

Proof Let $w \in W(M)$. We must show that

(1) $$\sum_{w' \in W(M)} N_{w,w'}\hat{\phi}_{w'} = 0.$$

Set $M'' = wMw^{-1}$, $\pi'' = w\pi$. We have $W(M) = \{w''w;\ w'' \in W(M'')\}$ and (1) is equivalent to

(2) $$\sum_{w'' \in W(M'')} N_{w,w''w}\hat{\phi}_{w''w} = 0.$$

Fix $w'' \in W(M'')$ and set $M' = w''M''w''^{-1}$, $\pi' = w''\pi''$. Let $(\phi^n_{M',\pi'})_{n\in\mathbf{N}}$ be a sequence of elements of $S_{M',\pi'}$ having limit equal to $\hat{\phi}_{M',\pi'}$. It is a consequence of the definitions that $N_{w,w''w}\hat{\phi}_{w''w}$ is the limit in $\hat{A}_w$ of the sequence consisting of the natural images in $\hat{A}_w$ of the elements $\psi^n \in A_w$ defined by

$$\psi^n(\lambda) = (w\lambda) \circ N\left(w''^{-1}w''\pi'' \otimes (-w''w\lambda)\right)\left((-w''w\lambda) \circ \phi^n_{M',\pi'}(w''w\lambda)\right)$$

for $\lambda \in X_M$. On the other hand $sN(w''^{-1}, w''\pi'')\hat{\phi}_{M',\pi'}$ is the limit in $\hat{S}_{M'',\pi''}$ of the sequence consisting of the natural images in $\hat{S}_{M'',\pi''}$ of the elements $\psi''^n \in S_{M'',\pi''}$ defined by

$$\psi''^n(\lambda'') = \lambda'' \circ N(w''^{-1}, w''\pi'' \otimes (-w''\lambda''))((-w''\lambda'') \circ \phi^n_{M',\pi'}(w''\lambda''))$$

for $\lambda'' \in X_{M''}$. We deduce from these formulae the equality

$$N_{w,w''w}\hat{\phi}_{w''w} = w^{-1}[sN(w''^{-1}, -w''\pi'')\hat{\phi}_{M',w''\pi''}].$$

But then (2) is equivalent to

$$w^{-1}\left[\sum_{w''\in W(M'')} sN(w''^{-1}, -w''\pi'')\hat{\phi}_{M',w''\pi''}\right] = 0$$

which is a consequence of the hypothesis of Proposition 2.4 applied to the pair (M'', π'').

3.3. End of the proof

By the preceding lemma and relation (2) of 3.1, we can find $\phi = (\phi_w)_{w\in W(M)} \in R$ such that $\phi - \hat{\phi} \in \hat{A}^{\geq N}_M$, where we of course set

$$\hat{A}^{\geq N}_Z = A_Z \bigotimes_{\mathbb{C}[X_M]} \hat{S}^{\geq N}_M.$$

Choose such a ϕ. Also choose an element $Q \in \mathbb{C}[X_M]$ such that

(1)(i) $Q - 1$ vanishes to the order at least N at $\lambda = 0$;

(ii) Let $(w, \lambda) \in W(M, M) \times X_M$ be such that $w(\pi \otimes \lambda) = \pi$ and $\lambda \neq 0$; then Q vanishes to the order at least N at λ.

For a standard Levi M' and an orbit $\mathfrak{P}'$ of irreducible cuspidal representations of $\mathbf{M}'$, define $\phi_{M',\mathfrak{P}'}$ by

$$\phi_{M',\mathfrak{P}'}(\pi') = c^{-1} \sum_{w\in W(M,M')} \sum_{\lambda\in X_M; w(\pi\otimes\lambda)=\pi'} (w\lambda) \circ \phi_w(-\lambda)Q(-\lambda)$$

for $\pi' \in \mathfrak{P}'$, where

$$c = |\{w \in W(M, M);\ w\pi = \pi\}|.$$

We have $\phi_{M',\mathfrak{P}'} = 0$ for almost every pair $(M', \mathfrak{P}')$; we can thus define $\phi \in P^{\mathfrak{F}}$ by $\phi = (\phi_{M',\mathfrak{P}'})_{(M',\mathfrak{P}')\in Y}$. We will show that this element satisfies the conditions of Proposition 2.4.

(i) Let M'' be a standard Levi and $\pi'' \in \Pi_0(\mathbf{M}'')$. Clearly relation (i) of 2.4 for the pair (M'', π'') is true if there exist no elements $w \in W(M, M'')$ and $\lambda \in X_M$ such that $w(\pi \otimes \lambda) = \pi''$. Suppose thus that such elements exist, and denote by $\mathscr{W}$ the set of these pairs (w, λ). For $(w, \lambda) \in \mathscr{W}$, evaluate

$$x_{w,\lambda} := \left(\sum_{w' \in W(M)} N_{w,w'} \phi_{w'} Q \right) (-\lambda).$$

We have

$$x_{w,\lambda} = (-w\lambda) \circ \sum_{w' \in W(M)} N(ww'^{-1}, w'(\pi \otimes \lambda))(w'\lambda \circ \phi_{w'}(-\lambda))Q(-\lambda).$$

Set $w' = w''w$, with $w'' \in W(M'')$. We obtain

$$x_{w,\lambda} = (-w\lambda) \circ \sum_{w'' \in W(M'')} N(w''^{-1}, w''\pi'')(w''w\lambda \circ \phi_{w''w}(-\lambda))Q(-\lambda).$$

Then

$$\sum_{(w,\lambda) \in \mathscr{W}} w\lambda \circ x_{w,\lambda} =$$

$$\sum_{w'' \in W(M'')} N(w''^{-1}, w''\pi'') \left[\sum_{(w,\lambda) \in \mathscr{W}} w''w\lambda \circ \phi_{w''w}(-\lambda)\, Q(-\lambda) \right].$$

When w'' is fixed and (w, λ) describes $\mathscr{W}$, the pair $(w''w, \lambda)$ describes the set of $(w', \lambda') \in W(M, w''M''w''^{-1}) \times X_M$ such that $w'(\pi \otimes \lambda) = w''\pi''$. The term between brackets is thus equal to $c\phi(w''\pi'')$ and we obtain

$$\sum_{(w,\lambda) \in \mathscr{W}} w\lambda \circ x_{w,\lambda} = c \sum_{w'' \in W(M'')} N(w''^{-1}, w''\pi'')\phi(w''\pi'').$$

It remains to recall that ϕ (and thus also ϕQ) belongs to R, so $x_{w,\lambda} = 0$ for every $(w, \lambda) \in \mathscr{W}$. The right-hand side of the above equality is thus zero, which is exactly assertion (i) of 2.4 for the pair (M'', π'').

(ii) Let $(M', \pi') \in Z$. Now let $\mathscr{W}$ be the set of pairs $(w, \lambda) \in W(M, M') \times X_M$ such that $w(\pi \otimes \lambda) = \pi'$, and for $\lambda' \in X_{M'}$, let $\mathscr{W}(\lambda')$ be the set of (w, λ) such that $w(\pi \otimes \lambda) = \pi' \otimes (-\lambda')$. We have the equality

$$\mathscr{W}(\lambda') = \{(w, \lambda - w^{-1}\lambda');\ (w, \lambda) \in \mathscr{W}\}.$$

By definition, we have the equality

$$\begin{aligned} s_{M',\pi'}\phi(\lambda') &= \lambda' \circ \phi_{M',\mathfrak{P}'}(\pi' \otimes (-\lambda)) \\ &= c^{-1} \sum_{(w,\lambda) \in \mathscr{W}(\lambda')} (\lambda' + w\lambda) \circ \phi_w(-\lambda)\, Q(-\lambda) \\ &= c^{-1} \sum_{(w,\lambda) \in \mathscr{W}} w\lambda \circ \phi_w(-\lambda + w^{-1}\lambda')\, Q(-\lambda + w^{-1}\lambda'). \end{aligned}$$

But let $(w, \lambda) \in \mathscr{W}$. As $(M', \pi') \in Z$, π' is conjugate to π. Thus $\pi \otimes \lambda$ is conjugate to π. By the properties (1) of the element Q, the function $\lambda' \mapsto Q(-\lambda + w^{-1}\lambda')$ belongs to $\mathbb{C}[X_{M'}] \cap \hat{S}_{M'}^{\geq N}$ if $\lambda \neq 0$, and is congruent to 1 modulo $\mathbb{C}[X_{M'}] \cap \hat{S}_{M'}^{\geq N}$ if $\lambda = 0$. Note that

$$\{(w, \lambda) \in \mathscr{W};\ \lambda = 0\} \simeq \{w \in W(M, M');\ w\pi = \pi'\}.$$

Set

$$\phi'(\lambda') = c^{-1} \sum_{w \in W(M,M'); w\pi = \pi'} \phi_w(w^{-1}\lambda').$$

We obtain

$$\text{(2)} \qquad s_{M',\pi'}\phi - \phi' \in \hat{S}_{M',\pi'}^{\geq N}.$$

But by definition

$$\phi' = c^{-1} \sum_{w \in W(M,M'); w\pi = \pi'} w\phi_w.$$

As $\phi - \hat{\phi} \in \hat{A}_Z^{\geq N}$, we have $w\phi_w - w\hat{\phi}_w \in \hat{S}_{wMw^{-1},w\pi}^{\geq N}$ for every $w \in W(M)$. On the other hand

$$w\hat{\phi}_w = \hat{\phi}_{wMw^{-1},w\pi}.$$

We obtain

$$\text{(3)} \qquad \phi' - c^{-1}|\{w \in W(M, M'); w\pi = \pi'\}|\hat{\phi}_{M',\pi'} \in \hat{S}_{M',\pi'}^{\geq N}.$$

Noting that $c = |\{w \in W(M, M'); w\pi = \pi'\}|$, (2) and (3) imply relation (ii) of Proposition 2.4 for the pair (M', π'), so the proof of this proposition is complete.

APPENDIX III

On the Discrete Spectrum of G_2

1. The result

Let k be a number field and G the split simple group over k of type G_2. There is only one such group: it is both simply connected and adjoint. Set $\mathbf{G} = G(\mathbb{A})$. Denote by $\{M_0, 1\}$ the equivalence class of the pair consisting of the split torus M_0 and of the trivial representation of $\mathbf{M}_0$. Denote by $L^2_{\{M_0,1\},\mathrm{disc}}$ the direct sum of irreducible subspaces of $L^2_{\{M_0,1\}}$. Langlands determined the subspace of $\mathbf{K}$-invariant vectors of $L^2_{\{M_0,1\},\mathrm{disc}}$. It is of dimension 2. Besides the constants, it contains an element whose cuspidal exponents are short roots. We are interested here in what happens when we suppress the hypothesis of invariance under $\mathbf{K}$. A complete study shows that $L^2_{\{M_0,1\},\mathrm{disc}}$ decomposes into two subspaces. The first is of dimension 1 and is reduced to the constants. The $\mathbf{K}$-finite elements of the other all have short roots as cuspidal exponents. We propose to determine the representation of the group $\mathbf{G}$ in this last space. A complete study would necessitate a local study at the archimedean places which has not been done. We will study the space V consisting of $\mathbf{K}$-finite elements of $L^2_{\{M_0,1\},\mathrm{disc}}$ whose cuspidal exponents are short roots and which are invariant under $\mathbf{K}_\infty$. The group $\mathbf{G}_f$ operates on V. Denote by Σ the set of finite places of k.

Proposition *For every place $v \in \Sigma$, there exist two distinct irreducible representations V_v and V'_v of $G(k_v)$ such that the following properties are satisfied: the representation V_v possesses a non-zero vector invariant under*

$\mathbf{K}_v$. *For every subfinite set* $F \subset \Sigma$, *set*

$$V^F = (\bigotimes_{v\in F} V'_v) \otimes (\bigotimes_{v\in\Sigma-F} V_v).$$

Set $\mathfrak{F} = \{F;\ F \subset \Sigma,\ F$ *finite*, $\mathrm{card}(F) \neq 1\}$. *Then we have the isomorphism*

$$V \simeq \bigoplus_{F\in\mathfrak{F}} V^F.$$

Remarks

(a) The condition $\mathrm{card}(F) \neq 1$ is extremely surprising (unrealistic?). We can ask ourselves if for $\mathrm{card}(F) = 1$, the representation V^F does not appear in a subspace of the space of cuspidal automorphic forms on **G**. (b) We could study the function field case. The situation for the elements whose cuspidal exponents are short roots would remain unchanged. But elements with non-real cuspidal exponents also appear: $-\beta_2+(i\pi/\log q)\beta_4$, $-\beta_4+(i\pi/\log q)\beta_6$, $-\beta_4+(i\pi/\log q)\beta_2$, $-\beta_4 \pm (2i\pi/3\log\ q)\beta_3$ where q is the number of elements in the constant field of k. It would obviously also be interesting to study these elements.

2. Explicit calculations for G_2

Let us recall some facts concerning $\mathfrak{a}^*_{M_0}$. The set of roots > 0 has 6 elements, which we denote by $\beta_1, \ldots, \beta_6$, corresponding to the figure below.

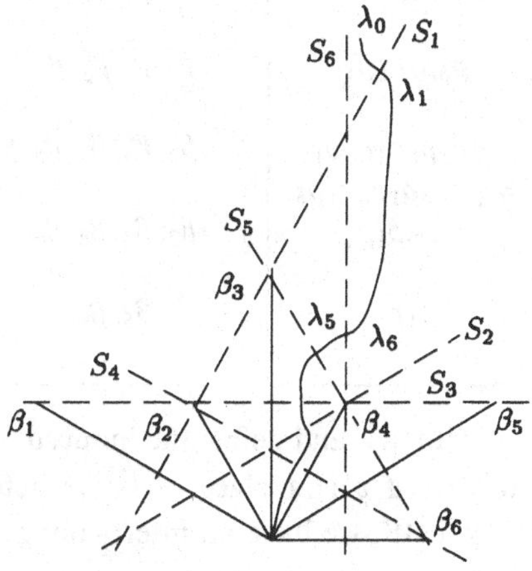

The simple roots are β_1 which is long and β_6 which is short. The products $\langle\check{\beta}_i, \beta_j\rangle$ are given in the following table

	β_1	β_2	β_3	β_4	β_5	β_6
$\check{\beta}_1$	2	1	1	0	-1	-1
$\check{\beta}_2$	3	2	3	1	0	-1
$\check{\beta}_3$	1	1	2	1	1	0
$\check{\beta}_4$	0	1	3	2	3	1
$\check{\beta}_5$	-1	0	1	1	2	1
$\check{\beta}_6$	-3	-1	0	1	3	2

We introduce coordinates in $\mathfrak{a}^*_{M_0}$ by setting $\lambda = x\beta_3 + y\beta_4$ to be a general element, where $x, y \in \mathbb{C}$.

The Weyl group W has 12 elements. It contains the symmetries ρ_i, $i = 1, \ldots, 6$ associated with the roots β_i and the rotations $\sigma(\theta)$ of angle $\theta \in (\pi/3)\mathbb{Z}/2\pi\mathbb{Z}$. The simple symmetries are ρ_1 and ρ_6 and we have the following table.

$w \in W$	decomposition	$\{\beta > 0; w\beta < 0\}$	$w\lambda$
1			x, y
ρ_1	ρ_1	β_1	$-x, 3x + y$
ρ_2	$\rho_1\rho_6\rho_1$	$\beta_1, \beta_2, \beta_3$	$-2x - y,$ $3x + 2y$
ρ_3	$\rho_1\rho_6\rho_1\rho_6\rho_1$	$\beta_1, \beta_2, \beta_3, \beta_4, \beta_5$	$-x - y, y$
ρ_4	$\rho_6\rho_1\rho_6\rho_1\rho_6$	$\beta_2, \beta_3, \beta_4, \beta_5, \beta_6$	$x,$ $-3x - y$
ρ_5	$\rho_6\rho_1\rho_6$	$\beta_4, \beta_5, \beta_6$	$2x + y,$ $-3x - 2y$
ρ_6	ρ_6	β_6	$x + y, -y$
$\sigma(\pi/3)$	$\rho_6\rho_1$	β_1, β_2	$2x + y,$ $-3x - y$
$\sigma(2\pi/3)$	$\rho_6\rho_1\rho_6\rho_1$	$\beta_1, \beta_2.\beta_3, \beta_4$	$x + y,$ $-3x - 2y$
$\sigma(\pi)$	$\rho_6\rho_1\rho_6\rho_1\rho_6\rho_1$ $= \rho_1\rho_6\rho_1\rho_6\rho_1\rho_6$	$\beta_1, \beta_2, \beta_3, \beta_4, \beta_5, \beta_6$	$-x, -y$
$\sigma(4\pi/3)$	$\rho_1\rho_6\rho_1\rho_6$	$\beta_3, \beta_4, \beta_5, \beta_6$	$-2x - y,$ $3x + y$
$\sigma(5\pi/3)$	$\rho_1\rho_6$	β_5, β_6	$-x - y,$ $3x + 2y$

We identify X_{M_0} with $\mathfrak{a}^*_{M_0}$ and define the induced representation $I(\lambda) = \mathrm{Ind}_{\mathbf{P}_0}^{\mathbf{G}} \lambda$ of $\mathbf{G}$. Similarly if v is a place of k, we define a representation $I_v(\lambda)$ of $G(k_v)$. For $w \in W$, we have an intertwining operator

$$M(w, \lambda) : I(\lambda) \to I(w\lambda).$$

If w is a simple symmetry, it reduces immediately to an operator relative to SL_2, which is well-known. Define a meromorphic function r on $\mathbb{C}$ by

$$r(z) = \zeta(z)(\varepsilon(z)\zeta(z+1))^{-1}$$

where ζ is the zeta function of the field k and ε its ε factor. For $j \in \{1, 6\}$, we can write

$$M(\rho_j, \lambda) = r(\langle\check{\beta}_j, \lambda\rangle)N(\rho_j, \lambda)$$

where $N(\rho_j, \lambda)$ is a normalised operator. This last decomposes into a product over all places v of k of operators

$$N_v(\rho_j, \lambda) : I_v(\lambda) \longrightarrow I_v(\rho_j\lambda)$$

normalised such that they act via the identity on the $\mathbf{K}_v$-invariant vectors. Let us identify the spaces of induced representations with spaces of functions on $\mathbf{K}$ or $\mathbf{K}_v$, which are independent of λ. The operators $N(\rho_j, \lambda)$ or $N_v(\rho_j, \lambda)$ depend only on $\langle\check{\beta}_j, \lambda\rangle$, i.e. for $z \in \mathbb{C}$, there exist operators $N_j(z)$ or $N_{j,v}(z)$ such that

$$N(\rho_j, \lambda) = N_j(\langle\check{\beta}_j, \lambda\rangle), \quad N_v(\rho_j, \lambda) = N_{j,v}(\langle\check{\beta}_j, \lambda\rangle).$$

The holomorphic properties of the functions introduced above are well-known. ζ has two simple poles at 0 and 1 and is holomorphic everywhere else; its zeros are in the band $0 < \mathrm{Re}\, z < 1$. We have the functional equation $\epsilon(z)\zeta(1-z) = \zeta(z)$. We deduce that r has a pole at $z = 1$, a zero at $z = -1$ and that $r(0) = -1$. The other poles of r are in the band $-1 < \mathrm{Re}\, z < 0$. We will need the expansions

$$r(z) = c(z-1)^{-1} + 0(1)$$

in a neighborhood of $z = 1$ and

$$r(z) = -1 + c'z + 0(z^2)$$

in a neighborhood of $z = 0$. If v is a finite place of k, denote by q_v the number of elements in the residue field of k_v. For $j \in \{1, 6\}$, $N_{j,v}(z)$ has simple poles at the points $z \in -1 + (2i\pi/\log\ q_v)\mathbb{Z}$ and is holomorphic everywhere else. We are not concerned about whether or not $N_v(\rho_j, \lambda)$ is holomorphic when v is archimedean since we restrict ourselves to the $\mathbf{K}_\infty$-invariant elements on which this operator acts via the identity.

Let $w \in W$ and $w = \rho^1 \ldots \rho^n$ with $\rho^i \in \{\rho_1, \rho_6\}$ be a reduced decomposition. Denote by $\beta^i \in \{\beta_1, \beta_6\}$ the root associated with ρ^i. We have the equality

$$M(w, \lambda) = M(\rho^1, \rho^2 \ldots \rho^n\lambda)M(\rho^2, \rho^3 \ldots \rho^n\lambda) \ldots M(\rho^n, \lambda).$$

Set

$$r(w, \lambda) = \prod_{i=1}^{n} r(\langle\check{\beta}^i, \rho^{i+1} \ldots \rho^n\lambda\rangle),$$

$$N(w, \lambda) = N(\rho, \rho \ldots \rho\lambda) \ldots N(\rho, \lambda).$$

We have the equality

$$M(w,\lambda) = r(w,\lambda)N(w,\lambda).$$

Note that we can write

(1) $$r(w,\lambda) = \prod_{\beta>0, w\beta<0} r(\langle\check{\beta},\lambda\rangle)$$

and that there exists a function $j : \{\beta;\ \beta > 0, w\beta < 0\} \to \{1,6\}$ such that

(2) $$N(w,\lambda) = \prod_{\beta>0, w\beta<0} N_{j(\beta)}(\langle\check{\beta},\lambda\rangle)$$

the product being taken in a certain order. Note that we have the equalities

$$r(w^{-1}, -w\lambda) = r(w,\lambda), \quad r(w,\overline{\lambda}) = \overline{r(w,\lambda)}.$$

3. Reminder of the method

The space V is obtained by the general method of which we recall the outline. First recall the definition of residue functions. Let v be a non-zero element of $\mathfrak{v}\mathfrak{a}_{M_0}$ and $c \in \mathbb{R}$. The equation $\langle v,\lambda\rangle = c$, where λ is a variable in $X_{M_0} \simeq \mathfrak{a}^*_{M_0}$, defines an affine hyperplane S of X_{M_0}. Let S° be the vector hyperplane defined by $\langle v,\lambda\rangle = 0$, $\operatorname{Im} S^\circ = S^\circ \cap \operatorname{Im} X_{M_0}$, $\operatorname{Re} S = S \cap \operatorname{Re} X_{M_0}$. The origin $o(S)$ of S is the unique vector of S which is orthogonal to $S^\circ \cap \operatorname{Re} X_{M_0}$. Fix another element v' of $\operatorname{Re}\mathfrak{a}_{M_0}$ not colinear with v. Then $\{v,v'\}$ is a coordinate system on X_{M_0}. Let f be a meromorphic function $f : X_{M_0} \to \mathbb{C}$. We define a meromorphic function $\operatorname{Res}_S f : S \to \mathbb{C}$. In the neighborhood of a general point $\lambda_0 \in S$, we consider f as a function of variables $\langle v,\lambda\rangle$, $\langle v',\lambda\rangle$ and $\operatorname{Res}_S f(\lambda_0)$ is the usual residue with respect to the variable $\langle v,\lambda\rangle$. This depends on the choices of v and v', but the final formulae in our calculation do not. We deduce from v a measure $d_S\lambda$ on $\operatorname{Re} S$: choose v' as above such that $d\lambda = d(\langle v,\lambda\rangle)d(\langle v',\lambda\rangle)$, let v'_S be the restriction of v' to Re S, and for $\lambda \in$ Re S, set $d_S\lambda = d(\langle v'_S,\lambda\rangle)$. From this measure we deduce a measure on every set of the form λ_0+ Im S°. Set $P = P^{\mathbf{K}_\infty}_{\{M_0,1\}}$, i.e. P is the space of holomorphic and Paley-Wiener functions φ defined on X_{M_0} which associate with $\lambda \in X_{M_0}$ a $\mathbf{K}_\infty$-invariant element $\varphi(\lambda) \in I(\lambda)$. Let $\varphi, \varphi' \in P$, $\lambda \in X_{M_0}$, and set

(3) $$A(\varphi',\varphi;\lambda) = \sum_{w\in W} (M(w^{-1}, -w\overline{\lambda})\varphi'(-w\overline{\lambda}), \varphi(\lambda)),$$

the product being the natural sesquilinear form on $I(-\overline{\lambda}) \times I(\lambda)$. On every variety of the form $\lambda_0 + \operatorname{Im} X_{M_0}$, let $d\lambda$ be the measure deduced from the differential form dx dy. For an element $\lambda_0 \in \operatorname{Re} X_{M_0}$ sufficiently > 0, the measures being suitably normalised, we have the equality

$$(\theta_{\varphi'}, \theta_{\varphi}) = (2\pi)^{-2} \int_{\lambda_0 + \operatorname{Im} X_{M_0}} A(\varphi', \varphi; \lambda)\, d\lambda.$$

We shift the domain of integration along a path in $\operatorname{Re} X_{M_0}$ joining λ_0 to 0. We choose it according to the figure in Section 2. Using the properties described in Section 2, we see that the singular hyperplanes which are crossed are the S_j, $j = 1, \ldots, 6$ defined by the equations $\langle \check{\beta}_j, \lambda \rangle = 1$. Let λ_j be the intersection of the path with S_j and let $o(S_j)$ be the origin of S_j. The contribution of S_j is

$$(2\pi)^{-1} \int_{\lambda_j + \operatorname{Im} S_j^{\circ}} \operatorname{Res}_{S_j} A(\varphi', \varphi; \lambda)\, d_{S_j} \lambda.$$

We begin again, now shifting the domain of integration along the segment joining λ_j to $o(S_j)$ in $\operatorname{Re} S_j$. Singularities then appear which form the discrete spectrum. Using the properties of Section 2, we check that the functions to be integrated are holomorphic at the extremities of the segments which appear. In any case, considering the figure, these extremities could not create functions whose cuspidal exponents were short roots. The figure shows that the above procedure creates three contributions to the space V:

$$\operatorname{Res}_{\beta_2} \operatorname{Res}_{S_1} A(\varphi', \varphi; \lambda)$$

$$\operatorname{Res}_{\beta_4} \operatorname{Res}_{S_6} A(\varphi', \varphi; \lambda)$$

$$\operatorname{Res}_{\beta_4} \operatorname{Res}_{S_5} A(\varphi', \varphi; \lambda).$$

4. Calculation of a residue

We begin by calculating $\operatorname{Res}_{\beta_2} \operatorname{Res}_{S_1} A(\varphi', \varphi; \lambda)$. The line S_1 is defined by the equation $x = 1$. The point β_2 has coordinates $(1, -1)$. At the point λ_1, the coordinate y is much greater than 0. The term we seek can thus be obtained by calculating the residue of $A(\varphi', \varphi; \lambda)$ with respect to the variable $x - 1$, then the residue of the term obtained with respect to the variable $y + 1$. The residue with respect to $x - 1$ is easy to calculate: in formulae (1) and (2), only the term $r(\langle \check{\beta}_1, \lambda \rangle)$ creates a singularity. In formula (3), the terms corresponding to w with $w\beta_1 > 0$ thus have no pole, those for which $w\beta_1 < 0$ have a simple pole. For $w \in W$, set

$$M^1(w^{-1}, -w\lambda) = \left[\prod_{\beta > 0, w\beta < 0, \beta \neq \beta_1} r(\langle \check{\beta}, \lambda \rangle) \right] N(w^{-1}, -w\lambda).$$

We obtain

$$\mathrm{Res}_{S_1} A(\varphi',\varphi;\lambda) = c \sum_{w;w\beta_1<0} (M^1(w^{-1},-w\bar{\lambda})\varphi'(-w\bar{\lambda}),\varphi(\lambda)).$$

For $\lambda = \beta_3 + y\beta_4 \in S_1$, we calculate

$$\langle\check{\beta}_2,\lambda\rangle = 3+y,\ \ \langle\check{\beta}_3,\lambda\rangle = 2+y,\ \ \langle\check{\beta}_4,\lambda\rangle = 3+2y,$$
$$\langle\check{\beta}_5,\lambda\rangle = 1+y,\ \langle\check{\beta}_6,\lambda\rangle = y.$$

The singularities of $M^1(w^{-1},-w\lambda)$ at $y=-1$ come from the terms $r(\langle\check{\beta}_3,\lambda\rangle)$, $r(\langle\check{\beta}_4,\lambda\rangle)$, $N_i(\langle\check{\beta}_6,\lambda\rangle)$. Let us show that that these singularities cannot exist. They occur only for w such that $w\beta_1<0$, $w\beta_6<0$, i.e. for $w=\sigma(\pi)=\rho_5\rho_2$, so $N(\sigma(\pi),\lambda) = N(\rho_5,\rho_2\lambda)N(\rho_2,\lambda)$. The singularity is in the first factor. Let us consider the local operator $N_v(\rho_5,\rho_2\lambda)$ for a finite place v, at a point $\lambda = x\beta_3 + y\beta_4$. We have

$$N_v(\rho_5,\rho_2\lambda) = N_{6,v}(y)N_{1,v}(x+y)N_{6,v}(3x+2y).$$

The operator $(y+1)N_v(\rho_5,\rho_2\lambda)$ is holomorphic in a neighborhood of β_2. On the line $x+y=0$, we have $N_{1,v}(x+y)=$id, $N_{6,v}(3x+2y) = N_{6,v}(-y) = N_{6,v}(y)^{-1}$, whence $(y+1)N_v(\rho_5,\rho_2\lambda) = (y+1)$id. Thus there exists an operator $P(x,y)$, holomorphic in a neighborhood of β_2, such that

$$N_v(\rho_5,\rho_2\lambda) = \mathrm{id}\ +\ \frac{x+y}{y+1}P(x,y).$$

When we restrict to S_1, we obtain

$$N_v(\rho_5,\rho_2\lambda) = \mathrm{id}\ +\ P(1,y)$$

which is thus holomorphic at β_2. Let E_v be the operator defined in this way for $\lambda=\beta_2$ and E be the product of the E_v.

The singularities thus come from functions r and occur for $w=\rho_2$ or $\sigma(\pi)$ for which the pole is simple (note that $r(\ \langle\check{\beta}_6,\lambda\rangle)$ creates a zero at β_2) and for $w=\rho_3$ or $\sigma(2\pi/3)$ for which the pole is double.

The contribution of ρ_2 to the residue at β_2 can be calculated immediately: it is

(4) $$c^2r(2)\ (N(\rho_2,\beta_2)\varphi'(\beta_2),\varphi(\beta_2)).$$

To calculate the contribution of $\sigma(\pi)$, note that $r(y)r(1+y)r(3+2y)$ equals $1/2$ when $y=-1$. We obtain

(5) $$(1/2)c^2r(2)\ (EN(\rho_2,\beta_2)\varphi'(\beta_2),\varphi(\beta_2)).$$

Let us regroup the contributions of ρ_3 and $\sigma(2\pi/3)$ to $\mathrm{Res}_{S_1}A(\varphi',\varphi;\lambda)$. As $\rho_3^{-1} = \sigma(4\pi/3)\rho_1$, we obtain

$$cr(3+y)r(2+y)r(3+2y)\ (N(\sigma(4\pi/3),-\sigma(2\pi/3)\bar{\lambda})$$
$$[\varphi'(-\sigma(2\pi/3)\bar{\lambda}) + r(\bar{y}+1)N_1(\bar{y}+1)\varphi'(-\rho_3\bar{\lambda})],\ \varphi(\lambda)).$$

As $\rho_3\beta_2 = \sigma(2\pi/3)\beta_2 = -\beta_4$ and $r(0)N_1(0) = -\text{id}$, the term between brackets is zero when $y = -1$. Let $p\varphi'(\beta_4)(y+1) + 0((y+1)^2)$ denote its expansion to the order 1. The residue of the above expression is then

$$(1/2)c^3r(2)\ \left(N(\sigma(4\pi/3), \beta_4)p\varphi'(\beta_4), \varphi(\beta_2)\right)$$

which can be written

$$(6) \qquad (1/2)c^3r(2)\ \left(N(\rho_2, \beta_2)N(\rho_6, \beta_4)p\varphi'(\beta_4), \varphi(\beta_2)\right).$$

We easily see that

$$p\varphi'(\beta_4) = D_1\varphi'(\beta_4) + c'\varphi'(\beta_4) - N_1'(0)\varphi'(\beta_4),$$

where N_1' is the derivative of N_1 and

$$D_1\varphi'(\lambda) = \frac{\mathrm{d}}{\mathrm{d}t}\varphi'(\lambda - t\beta_1)|_{t=0}, \text{ i.e. } D_1\varphi' = \left(-2\frac{\partial}{\partial x} + 3\frac{\partial}{\partial y}\right)\varphi'.$$

The term $\operatorname{Res}_{\beta_2}\operatorname{Res}_{S_1}A(\varphi', \varphi; \lambda)$ is the sum of the expressions (4), (5) and (6).

5. Calculation continued

For the calculation, it is best to regroup the two residues at β_4. The coordinates of β_4 are (0,1). The operators $N(w^{-1}, -w\lambda)$ are holomorphic at β_4. The singularities come from the terms $r(\langle\check{\beta}_j, \lambda\rangle)$ for $j = 2, 3, 5, 6$. We deduce from this that if we set

$$Q(\lambda) = \prod_{j=2,3,5,6}(\langle\check{\beta}_j, \lambda\rangle - 1) = (3x+y-1)(2x+y-1)(x+y-1)(y-1),$$

$$F(\lambda) = Q(\lambda)A(\varphi', \varphi; \lambda),$$

the function F is holomorphic at the point β_4. For every function f holomorphic at the point β_4, we introduce its series expansion

$$f(\lambda) = \sum_{m,n\in\mathbb{N}} a_{mn}(f)x^m(y-1)^n.$$

The term $\operatorname{Res}_{\beta_4}\operatorname{Res}_{S_6}A(\varphi', \varphi; \lambda)$ is obtained by calculating the residue of $A(\varphi', \varphi; \lambda)$ with respect to the variable $y-1$, then the residue of the term thus obtained with respect to x. We easily calculate

$$\operatorname{Res}_{\beta_4}\operatorname{Res}_{S_6}A(\varphi', \varphi; \lambda) = \frac{1}{6}a_{20}(F).$$

The term $\operatorname{Res}_{\beta_4}\operatorname{Res}_{S_5}A(\varphi', \varphi; \lambda)$ is obtained by calculating the residue of $A(\varphi', \varphi; \lambda)$ with respect to the variable $x+y-1$, then the residue of the term thus obtained with respect to the variable x restricted to S_5. We calculate

$$\operatorname{Res}_{S_5}A(\varphi', \varphi; \lambda) = \frac{-1}{2}x^{-3}\sum_{m,n\in\mathbb{N}} a_{mn}(F)(-1)^n x^{m+n},$$

$$\mathrm{Res}_{\beta_4}\,\mathrm{Res}_{S_5} A(\varphi',\varphi;\lambda) = -\frac{1}{2}a_{20}(F) + \frac{1}{2}a_{11}(F) - \frac{1}{2}a_{02}(F).$$

The sum of the residues at β_4 is thus equal to $\ell(F)$, where for every function f holomorphic at β_4, we set

$$\ell(f) = -\frac{1}{3}a_{20}(f) + \frac{1}{2}a_{11}(f) - \frac{1}{2}a_{02}(f).$$

A simple calculation shows that for every f, $\ell(f\circ\rho_1) = \ell(f)$. For $w \in W$, set

$$F[w](\lambda) = Q(\lambda)(M(w^{-1},-w\overline{\lambda})\varphi'(-w\overline{\lambda}),\varphi(\lambda)).$$

Set $W' = \{w \in W;\ w\beta_1 > 0\}$ and for $w \in W$, $H[w] = F[w]+F[w\rho_1]\circ\rho_1$. We have the equalities

$$\begin{aligned}\ell(F) = \sum_{w\in W}\ell(F[w]) &= \sum_{w\in W'}(\ell(F[w]) + \ell(F[w\rho_1]))\\ &= \sum_{w\in W'}(\ell(F[w]) + \ell(F[w\rho_1]\circ\rho_1))\\ &= \sum_{w\in W'}\ell(H[w]).\end{aligned}$$

Noting that $Q\circ\rho_1 = Q$, we calculate

$$\begin{aligned}H[w](\lambda) = Q(\lambda)\big[&(M(w^{-1},-w\overline{\lambda})\varphi'(-w\overline{\lambda}),\varphi(\lambda))\\ &+ (M(\rho_1 w^{-1},-w\overline{\lambda})\varphi'(-w\overline{\lambda}),\varphi(\rho_1\lambda))\big].\end{aligned}$$

Using the adjunction relation for $M(\rho_1,\overline{\lambda})$, we obtain

$$H[w](\lambda) = Q(\lambda)(M(w^{-1},-w\overline{\lambda})\varphi'(-w\overline{\lambda}),\tilde{\varphi}(\lambda)),$$

where $\tilde{\varphi}(\lambda) = \varphi(\lambda)+M(\rho_1,\rho_1\lambda)\varphi(\rho_1\lambda)$. As $\rho_1\beta_4 = \beta_4$ and $M(\rho_1,\beta_4) = -\mathrm{id}$, we have $\tilde{\varphi}(\beta_4) = 0$. We easily calculate the expansion of $\tilde{\varphi}$ to the order 1:

(7) $\tilde{\varphi}(\lambda) = \left(-D_1\varphi(\beta_4) - c'\varphi(\beta_4) + N_1'(0)\varphi(\beta_4)\right)x + \ldots = -p\varphi(\beta_4)x + \ldots$

If w does not invert at least three elements of the set $\{\beta_i;\ i = 2,3,5,6\}$, the operator $Q(\lambda)M(w^{-1},-w\lambda)$ vanishes to the order at least 2 at β_4; thus $H[w]$ vanishes to the order at least 3 at β_4 and $\ell(H[w]) = 0$. There are only two elements $w \in W'$ which invert at least three elements, namely $\sigma(4\pi/3)$ and ρ_4.

The element $w = \sigma(4\pi/3)$ does not invert β_2, so

$$Q(\lambda)M(\sigma(4\pi/3)^{-1},-\sigma(4\pi/3)\lambda)$$

is divisible by $3x+y-1$. We calculate the leading term more precisely:

$$\begin{aligned}&Q(\lambda)M\left(\sigma(4\pi/3)^{-1},-\sigma(4\pi/3)\lambda\right)\varphi'(-\sigma(4\pi/3)\lambda)\\ &\quad= c^3r(2)(3x+y-1)N\left(\sigma(2\pi/3),\beta_2\right)\varphi'(\beta_2)+\ldots\end{aligned}$$

Using (7), we obtain

$$\ell(H[\sigma(4\pi/3)]) = \frac{1}{2}c^3r(2)\left(N\left(\sigma(2\pi/3),\beta_2\right)\varphi'(\beta_2),p\varphi(\beta_4)\right)$$

which can be written, using the adjunction property and the relation $\sigma(2\pi/3) = \rho_6\rho_2$, as

$$(8) \qquad \ell(H[\sigma(4\pi/3)]) = \frac{1}{2}c^3 r(2)\left(N(\rho_2, \beta_2)\varphi'(\beta_2), N(\rho_6, \beta_4)p\varphi(\beta_4)\right).$$

For $w = \rho_4$, we recall that by construction, $H[w] = H[w\rho_1] \circ \rho_1$, thus $\ell(H[w]) = \frac{1}{2}\ell(H[w] + H[w\rho_1])$. On the other hand $\rho_4\varphi_1 = \sigma(\pi) = \rho_1\rho_4$, which gives

$$\begin{aligned} H[\rho_4](\lambda) + H[\rho_4\varphi_1](\lambda) &= Q(\lambda)\left[\left(M(\rho_4, -\rho_4\bar{\lambda})\varphi'(-\rho_4\bar{\lambda}), \tilde{\varphi}(\lambda)\right) + \right. \\ &\qquad \left.\left(M(\rho_4\rho_1, -\rho_1\rho_4\bar{\lambda})\varphi'(-\rho_1\rho_4\bar{\lambda}), \tilde{\varphi}(\lambda)\right)\right] \\ &= Q(\lambda)\left(M(\rho_4, -\rho_4\bar{\lambda})\tilde{\varphi}'(-\rho_4\bar{\lambda}), \tilde{\varphi}(\lambda)\right). \end{aligned}$$

We deduce from (7) the expansion

$$\tilde{\varphi}'(-\rho_4\lambda) = p\varphi'(\beta_4)x + \ldots$$

whence

$$H[\rho_4](\lambda) + H[\rho_4\rho_1](\lambda) = -c^4 r(2)\left(N(\rho_4, \beta_4)p\varphi'(\beta_4), p\varphi(\beta_4)\right)x^2 + \ldots$$

and

$$\ell(H[\rho_4]) = \frac{1}{6}c^4 r(2)\left(N(\rho_4, \beta_4)p\varphi'(\beta_4), p\varphi(\beta_4)\right)$$

which can also be written, using the adjunction property

$$(9) \quad \ell(H[\rho_4]) = \frac{1}{6}c^4 r(2)\left(N(\rho_2, \beta_2)N(\rho_6, \beta_4)p\varphi'(\beta_4), N(\rho_6, \beta_4)p\varphi(\beta_4)\right).$$

The sum of the residues at β_4 is the sum of (8) and (9).

6. Expression of the scalar product

Let $I(\beta_2)_f$ and $I(\beta_4)_f$ be the subspaces of $\mathbf{K}_\infty$-invariant vectors in $I(\beta_2)$ and $I(\beta_4)$, respectively. The map

$$P \to I(\beta_2)_f$$
$$\varphi \mapsto \varphi(\beta_2)$$

is a surjective intertwiner between representations of $\mathbf{G}$. The map

$$P \to I(\beta_4)_f$$
$$\varphi \mapsto p\varphi(\beta_4)$$

is also a surjective intertwiner: for φ of the form

$$\varphi(\lambda) = -(1/2)\psi(\lambda)x$$

we have $p\varphi(\beta_4) = \psi(\beta_4)$, whence the surjectivity; by definition the map is a limit of intertwining operators, thus it is an intertwiner. The operator $N(\rho_6, \beta_4) : I(\beta_4)_f \to I(\beta_2)_f$ is an intertwiner but is not surjective. Denote its image by J. Define a Hermitian form on $I(\beta_2)_f \oplus J$ by the following formula, where $f, f' \in I(\beta_2)_f$, $g, g' \in J$:

$$(f' \oplus g', f \oplus g) = (N(\rho_2, \beta_2)f', f) + (1/2)(EN(\rho_2, \beta_2)f', f) + (1/2)(N(\rho_2, \beta_2)g', f) + (1/2)(N(\rho_2, \beta_2)f', g) + (1/6)(N(\rho_2, \beta_2)g', g).$$

Denote by Ker the kernel of this Hermitian form. Fix a root $r(2)^{1/2}$. The formulae of Sections 4 and 5 show that the map

$$\varphi \longmapsto (f = cr(2)^{1/2}\varphi(\beta_2),\ g = c^2 r(2)^{1/2} N(\rho_6, \beta_4)p\varphi(\beta_4))$$

realises, via passage to the quotient, an isomorphism between the Hermitian space V and the Hermitian space $(I(\beta_2)_f \oplus J)/\mathrm{Ker}$.

7. The Hecke algebra

Fix a finite place v and denote by q the number of elements of the residue field of k_v. We want to study the representations $I_v(\lambda)$ and their intertwining operators. Recall that the category of representations all of whose irreducible subquotients are also subquotients of a representation $I_v(\lambda)$ for some suitable λ is equivalent to the category of representations of the Hecke algebra H which we will now describe using [Ro], §§1 and 2 (Rogawski himself uses results of Bernstein). H is the $\mathbb{C}$-algebra generated by four generators T_1, T_6, X_1, X_6, with the relations

$$(T_i + 1)(T_i - q) = 0 \text{ for } i \in \{1, 6\};$$

$$(T_1 T_6)^3 = (T_6 T_1)^3;$$

$$X_i X_j = X_j X_i,\ T_i X_j = X_j T_i, \text{ for } i, j \in \{1, 6\},\ i \neq j;$$

$$X_1 T_1 = T_1 X_6 X_1^{-1} + (q-1)X_1,$$

$$X_6 T_6 = T_6 X_1^3 X_6^{-1} + (q-1)X_6.$$

Let $\mathfrak{X}$ be the commutative subalgebra generated by X_1 and X_6, H_0 the finite-dimensional subalgebra generated by T_1 and T_6. For $w \in W$, we define $T(w) \in H_0$: if $w = \rho_1\rho_6\ldots$ (or $\rho_6\rho_1\ldots$) is a reduced decomposition, we set $T(w) = T_1 T_6 \ldots$ (or $T_6 T_1 \ldots$). The set $\{T(w);\ w \in W\}$ is a basis of H_0. We set $\underline{1} = T(1)$. It is the unit element of H_0.

For $\lambda \in \mathfrak{a}^*_{M_0}$, we define an H-module $H(\lambda)$: as a vector space $H(\lambda) = H_0$; the algebra H_0 acts on it by left multiplication; to define the action of $\mathfrak{X}$, it suffices to define $X_i\underline{1}$. For $i = 1, 6$; we set

$$X_1\underline{1} = q^{2x+1}\underline{1},\ X_6\underline{1} = q^{3x+2y}\underline{1};$$

if $\check{\varpi}_1, \check{\varpi}_6$ are the fundamental coweights, the formula is $X_i\underline{1} = q^{\langle\check{\varpi}_i, \lambda\rangle}\underline{1}$. The module $H(\lambda)$ corresponds to the induced one $I_v(\lambda)$ by equivalence of categories. For $i \in \{1,6\}$ and $z \in \mathbb{C}$, define the element of H_0

$$n_i(z) = (1-q^z)(1-q^{z+1})^{-1}T_i + (1-q)(1-q^{z+1})^{-1}\underline{1}.$$

For $w \in W$ and $\lambda \in \mathfrak{a}^*_{M_0}$, define $n(w,\lambda) \in H_0$ as follows: let $w = \rho^1 \ldots \rho^m$ be a reduced decomposition, with $\rho^j \in \{\rho_1, \rho_6\}$; for $j = 1,\ldots,m$, set $\beta^j = \beta_1$ or β_6 and $n^j = n_1$ or n_6 according to whether $\rho^j = \rho_1$ or ρ_6; then

$$n(w,\lambda) = n^m(\langle\check{\beta}^m,\lambda\rangle)n^{m-1}(\langle\check{\beta}^{m-1},\rho^m\lambda\rangle)\ldots n^1(\langle\check{\beta}^1,\rho^2\ldots\rho^m\lambda\rangle).$$

Then the operator $N_v(w,\lambda) : I_v(\lambda) \to I_v(w\lambda)$ can be interpreted as right multiplication by $n(w,\lambda)$ which sends $H(\lambda) \simeq H_0$ in $H(w\lambda) \simeq H_0$.

For general λ, the eigenvectors for the action of $\mathfrak{X}$ form a basis of $H(\lambda)$. Indeed $n(w^{-1}, w\lambda)$, considered as an element of $H(\lambda)$, is the image of $\underline{1} \in H(w\lambda)$ under the above intertwining operator. It is thus an eigenvector of eigenvalue $w\lambda$ and $\{n(w^{-1},w\lambda);\ w \in W\}$ is a basis of eigenvectors.

Basis of $H(\lambda)$, for $\lambda = x\beta_3 + y\beta_4$, λ general	eigenvalue
$\underline{1}$	λ
$n_1(-x)$	$\rho_1\lambda$
$n_6(-3x-y)n_1(-x)$	$\sigma(\pi/3)\lambda$
$n_1(-2x-y)n_6(-3x-y)n_1(-x)$	$\rho_2\lambda$
$n_6(-3x-2y)n_1(-2x-y)\times$ $n_6(-3x-y)n_1(-x)$	$\sigma(2\pi/3)\lambda$
$n_1(-x-y)n_6(-3x-2y)\times$ $n_1(-2x-y)n_6(-3x-y)n_1(-x)$	$\rho_3\lambda$
$n_6(-y)n_1(-x-y)n_6(-3x-2y)\times$ $n_1(-2x-y)n_6(-3x-y)n_1(-x)$	$\sigma(\pi)\lambda$
$n_6(-y)$	$\rho_6\lambda$
$n_1(-x-y)n_6(-y)$	$\sigma(5\pi/3)\lambda$
$n_6(-3x-2y)n_1(-x-y)n_6(-y)$	$\rho_5\lambda$
$n_1(-2x-y)n_6(-3x-2y)\times$ $n_1(-x-y)n_6(-y)$	$\sigma(4\pi/3)\lambda$
$n_6(-3x-y)n_1(-2x-y)\times$ $n_6(-3x-2y)n_1(-x-y)n_6(-y)$	$\rho_4\lambda$

Table 1

When λ is not regular or when the terms $n(w^{-1}, -w\lambda)$ have singularities, this construction cannot be applied. In the table below we give a basis of $H(\beta_2)$ obtained as the limit of a basis of $H(\lambda)$ for λ regular. For $i \in \{1,6\}$, we set $v_i = (1-q^{-1})T_i + 1 - q$. We denote by $n(\rho_5$, "$\rho_5\beta_2$")

and $n(\rho_5, \text{“}-\beta_2\text{”})$, the values at $\lambda = \beta_2$ of the functions $n(\rho_5, \rho_5\lambda)$ and $n(\rho_5, -\lambda)$, respectively, restricted to S_1. A reasoning analogous to the one in §4 shows that these values are well-defined although the functions on all of $\mathfrak{a}^*_{M_0}$ are not holomorphic at β_2. The fact that the vectors described form a basis of $H(\beta_2)$ is a consequence of the fact that the matrix which expresses them in the basis $\{T(w);\ w \in W\}$ is triangular.

Basis of $H(\beta_2)$	eigenvalue
$\underline{1}$	β_2
ν_1	β_6
$n_6(-2)\nu_1$	$-\beta_6$
$\nu_1 n_6(-2)\nu_1$	$-\beta_2$
$\nu_6\nu_1 n_6(-2)\nu_1$	$-\beta_4$
$T_1\nu_6\nu_1 n_6(-2)\nu_1$	not eigen
$n(\rho_5, \text{“}-\beta_2\text{”})\nu_1 n_6(-2)\nu_1$	$-\beta_2$
$n_6(1)$	β_4
$T_1 n_6(1)$	not eigen
$n(\rho_5, \text{“}\rho_5\beta_2\text{”})$	β_2
$\nu_1 n(\rho_5, \text{“}\rho_5\beta_2\text{”})$	β_6
$n_6(-2)\nu_1 n(\rho_5, \text{“}\rho_5\beta_2\text{”})$	$-\beta_6$

Table 2

8. Intertwining operators

The image of the map $N_v(\rho_2, \beta_2) : I_v(\beta_2) \to I_v(-\beta_2)$ corresponds to the image $\mathrm{Im}(N)$ of the map $N : H(\beta_2) \to H(-\beta_2)$ given by right multiplication by $n(\rho_2, \beta_2)$. We have $n(\rho_2, \beta_2) = n_1(1)n_6(2)n_1(1)$. Note that $\nu_1 n_1(1) = 0$. Thus N kills the vectors of Table 2 which end with ν_1. On the other hand $n(\rho_5, \text{“}\rho_5\beta_2\text{”})n(\rho_2, \beta_2)$ is the value at β_2 of the restriction to S_1 of $n(\rho_5\rho_2, \rho_5\lambda)$. Now $\rho_5\rho_2 = \rho_2\rho_5 = \sigma(\pi)$, whence

$$n(\rho_5\rho_2, \rho_5\lambda) = n(\rho_2, \rho_5\lambda)n(\rho_5, -\lambda),$$

and

$$\begin{aligned} n(\rho_5, \text{“}\rho_5\beta_2\text{”})n(\rho_2, \beta_2) &= n(\rho_2, \beta_2)n(\rho_5, \text{“}-\beta_2\text{”}), \\ &= n_1(1)n_6(2)n_1(1)n(\rho_5, \text{“}-\beta_2\text{”}). \end{aligned}$$

The image under N of the last two vectors of Table 2 is the term above multiplied on the left by ν_1 and $n_6(-2)\nu_1$, respectively. Since $\nu_1 n_1(1) = 0$, it is actually zero. Finally $\mathrm{Im}(N)$ is generated by the following vectors in $H(-\beta_2)$:

$$e_1 = n(\rho_2, \beta_2),\ e_2 = n_6(1)n(\rho_2, \beta_2),$$

$$e_3 = T_1 n_6(1) n(\rho_2, \beta_2),\ e_4 = n(\rho_5, ``\rho_5\beta_2") n(\rho_2, \beta_2).$$

We check that they are linearly independent.

Denote by $n(\rho_5, ``\rho_2\beta_2")$ and $n(\rho_5, ``\beta_2")$, the values at $\lambda = \beta_2$ of the functions $n(\rho_5, \rho_2\lambda)$ and $n(\rho_5, \lambda)$, respectively, restricted to S_1. The endomorphism E_v of $I_v(-\beta_2)$ can be interpreted as the endomorphism of $H(-\beta_2)$, also denoted by E_v, which is given by right multiplication by $n(\rho_5, ``\rho_2\beta_2")$. The term $n(\rho_2, \beta_2) n(\rho_5, ``\rho_2\beta_2")$ is the value at β_2 of the restriction to S_1 of $n(\rho_2\rho_5, \lambda)$. We have

$$n(\rho_2\rho_5, \lambda) = n(\rho_5\rho_2, \lambda) = n(\rho_5, \lambda) n(\rho_2, \rho_5\lambda),$$

whence

(10) $$n(\rho_2, \beta_2) n(\rho_5, ``\rho_2\beta_2") = n(\rho_5, ``\beta_2") n(\rho_2, \beta_2).$$

We see as in Section 4 that there exists a term $p(\lambda) \in H(\lambda)$, depending holomorphically on λ in a neighborhood of β_2, such that

$$n(\rho_5, \lambda) = \underline{1} + \frac{x+y}{y+1} p(\lambda).$$

We then have

$$n(\rho_5, \rho_5\lambda) = \underline{1} + \frac{x+y}{3x+2y-1} p(\rho_5\lambda),$$

whence

$$n(\rho_5, ``\beta_2") = \underline{1} + p(\beta_2),\ n(\rho_5, ``\rho_5\beta_2") = \underline{1} + \frac{1}{2} p(\beta_2),$$

(11) $$n(\rho_5, ``\beta_2") = 2n(\rho_5, ``\rho_5\beta_2") - \underline{1},$$

and

(12) $$n(\rho_2, \beta_2) n(\rho_5, ``\rho_2\beta_2") = 2n(\rho_5, ``\rho_5\beta_2") n(\rho_2, \beta_2) - n(\rho_2, \beta_2).$$

In other words $E_v(e_1) = 2e_4 - e_1$.

Let us consider $n_6(1) n(\rho_5, ``\rho_5\beta_2") n(\rho_2, \beta_2)$. It is the value in β_2 of the restriction to S_1 of $n(\rho_6, \rho_1\rho_6\lambda) n(\rho_5, \rho_5\lambda) n(\rho_2, \lambda)$, i.e. of $n(\rho_3, \rho_1\rho_6\lambda)$. This last term is holomorphic at β_2 and its value there is

$$n_1(0) n_6(1) n(\rho_2, \beta_2),$$

which can be written $n_6(1) n(\rho_2, \beta_2)$ since $n_1(0) = \underline{1}$. By (12), we see that

(13) $$n_6(1) n(\rho_2, \beta_2) n(\rho_5, ``\rho_2\beta_2") = n_6(1) n(\rho_2, \beta_2),$$

i.e. $E_v(e_2) = e_2$. For left multiplication by T_1, we also have $E_v(e_3) = e_3$.

Let us consider $n(\rho_5, ``\rho_5\beta_2")^2$. By (11), it is equal to

$$\frac{1}{2} n(\rho_5, ``\beta_2") n(\rho_5, ``\rho_5\beta_2") + \frac{1}{2} n(\rho_5, ``\rho_5\beta_2").$$

Now $n(\rho_5, ``\beta_2") n(\rho_5, ``\rho_5\beta_2")$ is the value at β_2 of the restriction to S_1 of $n(\rho_5, \lambda)\ n(\rho_5, \rho_5\lambda)$, i.e. of $\underline{1}$. Thus,

$$n(\rho_5, ``\rho_5\beta_2")^2 = \underline{1}/2 + \frac{1}{2} n(\rho_5, ``\rho_5\beta_2").$$

By (12), we see that

$$n(\rho_5, \text{“}\rho_5\beta_2\text{”})n(\rho_2, \beta_2)n(\rho_5, \text{“}\rho_2\beta_2\text{”}) = n(\rho_2, \beta_2),$$

i.e. $E_v(e_4) = e_1$. Set $f_1 = 2e_4 + e_1,\ f_2 = e_2,\ f_3 = e_3,\ f_4 = e_4 - e_1$. We see that these elements form a basis of $\mathrm{Im}(N)$ and are eigenvectors of E_v. The associated eigenvalues are 1 for f_1, f_2, f_3 and -2 for f_4.

Note that relation (13) can be written

(14) $$E_v N_v(\rho_2, \beta_2) N_v(\rho_6, \beta_4) = N_v(\rho_2, \beta_2) N_v(\rho_6, \beta_4).$$

By adjunction, (13) becomes

$$n(\rho_5, \text{“}\beta_2\text{”})n(\rho_2, \beta_2)n_6(1) = n(\rho_2, \beta_2)n_6(1),$$

whence, by (10):

$$n(\rho_2, \beta_2)n(\rho_5, \text{“}\rho_2\beta_2\text{”})n_6(1) = n(\rho_2, \beta_2)n_6(1).$$

This can be written

(15) $$N_v(\rho_6, -\beta_2) E_v N_v(\rho_2, \beta_2) = N_v(\rho_6, -\beta_2) N_v(\rho_2, \beta_2).$$

Denote by $\mathrm{Im}(N)^\circ$ the subspace of $H(-\beta_2)$ generated by f_1, f_2, f_3 and $\mathrm{Im}(N)' = \mathbb{C} f_4$. These spaces are sub$H$-modules of $H(-\beta_2)$. We calculate quite easily the actions of generators of H on these spaces. We obtain

$$T_6 f_4 = -f_4,\quad T_1 f_4 = X_1 f_4 = X_6 f_4 = q f_4;$$

in the basis f_1, f_2, f_3 of $\mathrm{Im}(N)^\circ$:

$$T_1 = \begin{bmatrix} q & 0 & 0 \\ 0 & 0 & q \\ 0 & 1 & q-1 \end{bmatrix} \qquad T_6 = \begin{bmatrix} -1 & 0 & -q \\ 3(q+1) & q & 3q \\ 0 & 0 & q \end{bmatrix}$$

$$X_1 = \begin{bmatrix} q & 0 & 0 \\ 0 & q & q(q-1) \\ 0 & 0 & q \end{bmatrix} \qquad X_6 = \begin{bmatrix} q & 0 & 0 \\ 0 & q^2 & 0 \\ 0 & 0 & q^2 \end{bmatrix}$$

We see that $\mathrm{Im}(N)^\circ$ is irreducible and possesses a vector on which T_1 and T_6 act via multiplication by q, namely

$$-q f_1 + (q+1)(f_2 + f_3).$$

Let V_v and V'_v, be the subspaces of $I_v(-\beta_2)$ which correspond to $\mathrm{Im}(N)^\circ$ and $\mathrm{Im}(N)'$, respectively. These subspaces are irreducible and V_v possesses a vector which is invariant under $\mathbf{K}_v$. The operator E_v acts via the identity on V_v and via multiplication by -2 on V'_v.

9. The final calculation

Recalling the definition of J (see Section 6) and using formulae (14) and (15), we can write the Hermitian form on $I(\beta_2)_f \oplus J$ as follows:

$$(f' \oplus g', f \oplus g) = ((1 + E/2)N(\rho_2, \beta_2)(f' + g'/3), f + g/3).$$

Then $(I(\beta_2)_f \oplus J)/\mathrm{Ker}$ can be identified with the image of $I(\beta_2)_f$ under the operator $(1 + E/2)\ N(\rho_2, \beta_2)$. By the results of Section 8, $N(\rho_2, \beta_2)I(\beta_2)_f$ is isomorphic as a $\mathbf{G}_f$-module to the restricted tensor product:

$$\bigotimes_{v \in \Sigma} (V_v \oplus V'_v),$$

and even, with the notation of Section 1, to $\oplus V^F$, where the sum is over the finite subsets F of Σ. The operator $1 + E/2$ acts on a subspace V^F by multiplication by $1 + (-2)^{\mathrm{card}(F)}2^{-1}$. This term is zero if and only if $\mathrm{card}(F) = 1$. We deduce from this that

$$(I(\beta_2)_f \oplus J)/\mathrm{Ker} \simeq \bigoplus_{F \in \mathfrak{F}} V^F,$$

which gives the proposition by the result of Section 6.

APPENDIX IV

Non-Connected Groups

We will briefly indicate the modifications necessary to the theory when the group G is no longer supposed connected.

1. Definitions

Let k still be a global field and G a group defined over k. We suppose G linear, i.e. that we have an embedding $i'_G \hookrightarrow GL_n$, defined over k, whose image is closed. Denote by G^0 the connected component of the identity of G. We suppose that G^0 is reductive. We define $i_G : G \hookrightarrow GL_{2n}$ as in I.1.1. There still exists a finite set S of places of k, containing the archimedean places, such that the image of i_G is defined and smooth on $\mathfrak{o}^S$.

Lemma *For almost all places $v \notin S$, we have the equality $G(k_v) = G^0(k_v)G(\mathfrak{o}_v)$.*

Proof Recall (see [Bor] §1.3) that every connected component of G is defined over a separable extension of k. Let k_s be a separable closure of k and $\mathfrak{c}$ the finite set of connected components of G. The group $\mathrm{Gal}(k_s/k)$ acts on $\mathfrak{c}$. For $C \in \mathfrak{c}$, let k_C be the field of definition of C, i.e. $\mathrm{Gal}(k_s/k_C)$ is the stabiliser of C in $\mathrm{Gal}(k_s/k)$. Up to increasing S, we can suppose that every component C is defined over $\mathfrak{o}_C^S$, where $\mathfrak{o}_C^S$ is the ring of S-integers of k_C. For every valuation v of k, fix a lifting $\overline{v}$ of v to k_s. If k' is a separable extension of k, we also use $\overline{v}$ to denote the restriction of $\bar{v}$ to k'. Finally, let $\mathfrak{c}(\bar{v})$ be the subset of $C \in \mathfrak{c}$ such that $k_{C,\bar{v}} = k_v$. For $v \notin S$, we have the isomorphisms

$$G(k_v) \simeq \bigcup_{C \in \mathfrak{c}(\bar{v})} C(k_{C,\bar{v}}),$$

$$G(\mathfrak{o}_v) \simeq \bigcup_{C \in \mathfrak{c}(\bar{v})} C(\mathfrak{o}_{C,\bar{v}}).$$

It thus suffices to prove the following assertion: let $C \in \mathfrak{c}$ and let S_C be the set of places of k_C whose restriction to k belongs to S; then for almost every place v of k_C, $v \notin S_C$, the set $C(\mathfrak{o}_{C,v})$ is non-empty. Fix C. Up to increasing S_C, we can suppose that there exists a group $G^0_{\mathfrak{o}}$, a variety $C_{\mathfrak{o}}$ and an action of $G^0_{\mathfrak{o}}$ on $C_{\mathfrak{o}}$, all defined over $\mathfrak{o}_C^{S_C}$, such that the group G^0, considered as a group over k_C, the variety C and the action of G^0 on C come from the preceding objects by extension of scalars of $\mathfrak{o}_C^{S_C}$ to k_C. For every finite place v of k_C, let f_v be the residue field of $\mathfrak{o}_{C,v}$. Up to making S_C bigger, we can suppose that for $v \notin S_C$, the residual varieties $G^0_{\mathfrak{o}} \times_{\mathfrak{o}_{C,v}} f_v$, $C_{\mathfrak{o}} \times_{\mathfrak{o}_{C,v}} f_v$ are connected. Then for $v \notin S_C$, $C_{\mathfrak{o}} \times_{\mathfrak{o}_{C,v}} f_v$ is the principal homogeneous space of a connected group. Such a space has a point over f_v. Thus $C_{\mathfrak{o}} \times_{\mathfrak{o}_{C,v}} f_v(f_v) \neq \emptyset$. By a Henselian lemma, $C_{\mathfrak{o}}(\mathfrak{o}_{C,v})$ is also non-empty. Now $C_{\mathfrak{o}}(\mathfrak{o}_{C,v}) = C(\mathfrak{o}_{C,v})$. This concludes the proof. □

2. Levi subgroups

Fix a parabolic subgroup $P^0_{\min}$ of G^0, defined over k and minimal (it will be the P_0 of Chapter I) and a Levi subgroup $M^0_{\min}$ of $P^0_{\min}$, defined over k. Let $P^0 = M^0U$ be a standard parabolic subgroup of G^0, defined over k, and T_{M^0} the largest split torus in the centre of M^0. Let M be the centraliser of T_{M^0} in G. It is a subgroup defined over k. Then $M \cap G^0 = M^0$ and M normalises M^0 and U. Set $P = MU$. By definition, we call a group P or M constructed in this way a standard parabolic, or Levi subgroup, respectively, of G.

Remarks

(a) In general, G does not appear as one of its own standard parabolic subgroups.

(b) With the obvious notation, if $M_1^0 \subset M_2^0$, then $M_1 \subset M_2$.

Let M^0 and M be as above, k' an extension of k, $m \in M(k')$ and $\chi \in \mathrm{Rat}(M^0)$. Let us show that we have:

(1) $\qquad$ for all $m^0 \in M^0(k')$, $\chi(m^{-1}m^0m) = \chi(m^0)$.

Proof Let $\overline{k}$ be an algebraic closure of k. As $M(k') \subset M(\overline{k})M^0(k'\overline{k})$, we can suppose $k' = \overline{k}$. Denote by $\mathrm{Rat}_{\overline{k}}M^0$ the group of algebraic characters of M^0 defined over $\overline{k}$. The group of automorphisms of $\overline{k}/k$, denoted by Γ, operates on $\mathrm{Rat}_{\overline{k}}M^0$. We know that every algebraic character of M^0 is defined over a separable extension of k. Thus $\mathrm{Rat}(M^0) = (\mathrm{Rat}_{\overline{k}}M^0)^\Gamma$, the subgroup of Γ-invariants. On the other hand $M(\overline{k})$ operates on $\mathrm{Rat}_{\overline{k}}M^0$. The action factors through the finite quotient $\mathfrak{m} = M(\overline{k})/M^0(\overline{k})$. We must show that

(2) $$(\mathrm{Rat}_{\overline{k}}M^0)^\Gamma \subset (\mathrm{Rat}_{\overline{k}}M^0)^{\mathfrak{m}}.$$

Let us decompose $V = (\mathrm{Rat}_{\overline{k}}M^0)\bigotimes_{\mathbf{Z}}\mathbb{C}$ into a sum of $\mathfrak{m}$-isotypical subspaces

$$V = \bigoplus V_\tau,$$

where τ runs through the set of irreducible representations of $\mathfrak{m}$. For $\gamma \in \Gamma$, we have $\gamma V_\tau = V_{\gamma\tau}$, where $(\gamma\tau)(m) = \tau(\gamma^{-1}m)$. In particular V_1 and $\oplus_{\tau\neq 1}V_\tau$ are stable under Γ. Thus $V^\Gamma = V_1^\Gamma \oplus (\oplus_{\tau\neq 1}V_\tau)^\Gamma$. Let us consider the restriction

$$r : V \to (\mathrm{Rat}_{\overline{k}}T_{M^0})\bigotimes_{\mathbf{Z}}\mathbb{C}.$$

It is $\mathfrak{m} \times \Gamma$-equivariant for the natural actions of $\mathfrak{m}$ and Γ. Now, these actions are trivial on the image space. Thus $\oplus_{\tau\neq 1}V_\tau \subset \mathrm{Ker}(r)$. Now, we know that r induces an isomorphism of V^Γ onto $(\mathrm{Rat}_{\overline{k}}T_{M^0}) \otimes_{\mathbf{Z}} \mathbb{C}$. Thus $(\oplus_{\tau\neq 1}V_\tau)^\Gamma = \{0\}$, then $V^\Gamma = V_1^\Gamma \subset V^{\mathfrak{m}}$. Intersecting with the lattice $\mathrm{Rat}_{\overline{k}}M^0$, we obtain (2). □

Let $\chi \in \mathrm{Rat}(M^0)$. We can define a character $|\chi| : M^0(\mathbb{A}) \to \mathbb{R}_+^\times$. Let us show that:

(3) there exists a unique character, still denoted by $|\chi| : M(\mathbb{A}) \to \mathbb{R}_+$ which extends the character $|\chi|$ of $M^0(\mathbb{A})$.

Proof It suffices to prove, for every place v, the analogous assertion obtained by replacing $\mathbb{A}$ by k_v. Fix a section (of sets) $s : M(k_v)/M^0(k_v) \to M(k_v)$. For $m_1, m_2 \in M(k_v)/M^0(k_v)$, set

$$c(m_1, m_2) = |\chi|(s(m_1)s(m_2)s(m_1m_2)^{-1}).$$

We see using (1) that c is a 2-cocycle with values in $\mathbb{R}_+^\times$, equipped with the trivial action of $M(k_v)/M^0(k_v)$. But $H^2(M(k_v)/M^0(k_v), \mathbb{R}_+^\times) = 0$ since $M(k_v)/M^0(k_v)$ is finite and $\mathbb{R}_+^\times$ is uniquely divisible. Thus there exists a map $c' : M(k_v) \to \mathbb{R}_+^\times$ such that

$$c(m_1, m_2) = c'(m_1)c'(m_2)c'(m_1m_2)^{-1}$$

for all $m_1, m_2 \in M(k_v)/M^0(k_v)$. Now let $m \in M(k_v)$. We can write

$m = s(m_1)m^0$ in a unique way, with $m_1 \in M(k_v)/M^0(k_v)$, $m^0 \in M^0(k_v)$. Set $|\chi|(m) = c'(m_1)|\chi|(m^0)$. This defines a map $|\chi| : M(k_v) \to \mathbb{R}_+^\times$ which, as is easily seen, solves our problem. The uniqueness is equally clear. □

As in I.1.4, we can define groups M^1, X_M, the space $\mathfrak{a}_M$ (actually equal to $\mathfrak{a}_{M^0}$), and so on. We have an injection

$$M^{0,1}\backslash M^0(\mathbb{A}) \to M^1\backslash M(\mathbb{A})$$

with finite cokernel and a surjection

$$X_{M^0} \longleftarrow X_M$$

with finite kernel. If k is a number field, they are bijections.

3. The Weyl group

Let $T_{\min}$ be the largest split subtorus in the centre of $M^0_{\min}$ and $\mathrm{Norm}_{G(k)} T_{\min}$ its normaliser $G(k)$. We set

$$W = \mathrm{Norm}_{G(k)} T_{\min}/M^0_{\min}(k).$$

We write W for the Weyl group of G. It contains the Weyl group W^0 of G^0 as normal subgroup. Denote by $\mathrm{Norm}_{G(k)}(T_{\min}, P^0_{\min})$ the subgroup of elements of $G(k)$ which normalise both $T_{\min}$ and $P^0_{\min}$. Set

$$B = \mathrm{Norm}_{G(k)}(T_{\min}, P^0_{\min})/M^0_{\min}(k).$$

It is a subgroup of W and we have the decomposition into a semi-direct product

$$W \simeq W^0 \ltimes B. \tag{1}$$

Indeed, no element of W^0 normalises $P^0_{\min}$ except the identity, so $B \cap W^0 = \{1\}$. To show that $W = W^0 B$, it suffices to show that

$$G(k) = G^0(k)\mathrm{Norm}_{G(k)}(T_{\min}, P^0_{\min}).$$

Now, if $g \in G(k)$, $\mathrm{ad}(g)T_{\min} \subset \mathrm{ad}(g)P^0_{\min}$ is a maximal split torus defined over k, and a minimal parabolic subgroup defined over k, respectively. Such a pair is conjugate in $G^0(k)$ to $(T_{\min}, P^0_{\min})$. That is, there exists $g^0 \in G^0(k)$ such that $g^0 g \in \mathrm{Norm}_{G(k)}(T_{\min}, P^0_{\min})$, which proves our assertion.

The group W normalises $T_{\min}$ and thus also $M^0_{\min}$ and its associated Levi $M_{\min}$ of G. It acts on $\mathrm{Rat}(M^0_{\min})$, $\mathfrak{a}_{M^0}$ etc. ... In general this action is not faithful: its kernel is precisely the Weyl group $W_{M_{\min}}$ of the Levi $M_{\min}$.

Let M^0 be a standard Levi of G^0 and M its associated Levi. We set

$$W_M = W \cap (M(k)/M^0_{\min}(k)).$$

It is the analogue of W for M. Also set

$$W_M^0 = W^0 \cap (M^0(k)/M_{\min}^0(k)), \ B_M = B \cap (M(k)/M_{\min}^0(k)).$$

We have the decomposition

$$W_M \simeq W_M^0 \ltimes B_M.$$

Indeed, it is the analogue of (1) for the group M, at least if we show that B_M is exactly the analogue of B, i.e. that we have the equality

$$\mathrm{Norm}_{G(k)}(T_{\min}, P_{\min}^0)\ M(k) = \mathrm{Norm}_{M(k)}(T_{\min}, P_{\min}^0 M^0).$$

But if U is the unipotent radical of the standard parabolic subgroup of G^0 of Levi M^0, we have already noted that M normalises U. As $P_{\min}^0 = (P_{\min}^0 \cap M^0)U$, it is the same as for an element of $M(k)$ of normaliser $P_{\min}^0$ or $P_{\min}^0 \cap M^0$.

Let $P = MU$ and $P' = M'U'$ be two standard parabolic subgroups of G. Set $P^0 = P \cap G^0$, $P'^0 = P' \cap G^0$. Thanks to the properties described above, the well-known Bruhat decomposition for $G^0(k)$ can be generalised to the decompositions into disjoint unions

$$G(k) = \bigcup_{w \in W_{M'}^0 \backslash W / W_M^0} P'^0(k)\dot{w}P^0(k) = \bigcup_{w \in W_{M'} \backslash W / W_M} P'(k)\dot{w}P(k),$$

where $\dot{w}$ is a representative of w.

4. The compact subgroup

Let v be a place of k and K_v a subgroup of $G(k_v)$. Consider the conditions

(i) K_v is a maximal compact subgroup of $G(k_v)$;

(ii) $G(k_v) = P_{\min}^0(k_v)K_v$.

If v is archimedean, there exists K_v satisfying these conditions. Choose such a K_v. Lemma 1 and the analogous properties for the group G^0 show that there exists a finite set S_0 of finite places of k such that if v is a finite place and $v \notin S_0$, the group $K_v = G(\mathfrak{o}_v)$ satisfies these conditions. Fix such a S_0. For a finite place $v \notin S_0$, set $K_v = G(\mathfrak{o}_v)$.

For $v \in S_0$, it is not always possible to find K_v satisfying conditions (i) and (ii). We then fix a maximal compact subgroup K_v of $G^0(k_v)$ satisfying the analogue of (ii) for the group G^0. We choose a (finite) set D_v of representatives of the set of double cosets

$$P_{\min}^0(k_v)\backslash G(k_v)/K_v.$$

We set $K = \prod_v K_v$, $D = \prod_v D_v$, where $D_v = \{1\}$ if $v \notin S_0$. We have the decomposition into a disjoint union

$$G(A) = \bigcup_{d \in D} P_{\min}^0(\mathbb{A})dK.$$

Let $P = MU$ be a standard parabolic subgroup of G. We must fix a compact subgroup of $M(\mathbb{A})$ analogous to the subgroup K of $G(\mathbb{A})$. For every place v, we choose a maximal compact subgroup of $M(k_v)$ (or $M^0(k_v)$ if $v \in S_0$) which contains the image of $P(k_v) \cap K_v$ under the map $P(k_v) \to P(k_v)/U(k_v) \simeq M(k_v)$.

5. Coverings, centres

Let $\mathbf{G}$ be a topological group which is a finite central covering of $G(\mathbb{A})$. We suppose that $G(k)$ lifts to $\mathbf{G}$. We denote by $\mathbf{K}$ the inverse image of K in $\mathbf{G}$ and we fix a subset, again denoted by D, of $\mathbf{G}$, which projects bijectively onto the subset D of $G(\mathbb{A})$ fixed in Section 4.

Let $P = MU$ be a standard parabolic subgroup of G. We denote by $\mathbf{M}$ and $\mathbf{M}^1$ the inverse images of $M(\mathbb{A})$ and M^1 in $\mathbf{G}$. We define as in the connected case the map

$$\log_M : \mathbf{M}^1 \backslash \mathbf{M} \to \mathrm{Re}\, \mathfrak{a}_M.$$

There is a problem with the centre. The algebraic group Z_G is well-defined, but $Z_G(\mathbb{A})$ is not in general the centre of $G(\mathbb{A})$. It is, however, if $G(k)$ intersects every connected component of G. We will denote by $Z_{\mathbf{G}}$ the intersection of the centre of $\mathbf{G}$ with the inverse image of $Z_G(\mathbb{A})$.

In the same way that the group M^1 was deduced from a Levi M, we deduce a group Z_G^1 from Z_G. Set $Z_{\mathbf{G}}^1 = Z_{\mathbf{G}} \cap \mathrm{pr}^{-1}(Z_G^1)$. Then Lemma I.1.5 remains true.

Proof We show as in I.1.5 that the subgroups $Z_G(k)Z_{\mathbf{G}}^1$ and $G(k)Z_{\mathbf{G}}^1$ are closed in $\mathbf{G}$ and that the quotient $Z_G(k) \cap Z_{\mathbf{G}} \backslash Z_{\mathbf{G}}^1$ is compact. Set $Z'_G = Z_G \cap G^0$, $Z'_{\mathbf{G}} = Z_{\mathbf{G}} \cap \mathrm{pr}^{-1}(Z'_G(\mathbb{A}))$. As $Z'_G(k)$ is of finite index in $Z_G(k)$, the quotient $Z'_G(k) \cap Z_{\mathbf{G}} \backslash Z_{\mathbf{G}}^1$ is also compact. *A fortiori* $Z'_{\mathbf{G}} \cap Z_{\mathbf{G}}^1 \backslash Z_{\mathbf{G}}^1$. It is clear that $Z'_{\mathbf{G}} Z_{\mathbf{G}}^1$ is of finite index in $Z_{\mathbf{G}}$. Thus $Z'_{\mathbf{G}} \backslash Z_{\mathbf{G}}$ is compact. The end of the proof of I.1.5 shows that $Z'_G(k)Z'_{\mathbf{G}}$ and $G^0(k)Z'_{\mathbf{G}}$ are closed subgroups of $\mathbf{G}$. As the product of a closed subgroup and a compact subgroup is closed, $Z'_G(k)Z_{\mathbf{G}}$ and $G^0(k)Z_{\mathbf{G}}$ are also closed, thus so are $Z_G(k)Z_{\mathbf{G}}$ and $G(k)Z_{\mathbf{G}}$. □

As in Section 2, denote by $\bar{k}$ an algebraic closure of k, Γ the group of automorphisms of $\bar{k}/k$ and $\mathfrak{m} = G(\bar{k})/G^0(\bar{k})$. The group $(\mathrm{Rat}_{\bar{k}} G^0)^{\mathfrak{m}}$ is invariant under Γ. Consider the space $\mathrm{Re}\, \mathfrak{a}_G^* = ((\mathrm{Rat}_{\bar{k}} G^0)^{\mathfrak{m}})^{\Gamma} \otimes_{\mathbb{Z}} \mathbb{R}$. As $((\mathrm{Rat}_{\bar{k}} G^0)^{\mathfrak{m}})^{\Gamma} \subset \mathrm{Rat}(G^0)$, we have $\mathrm{Re}\, \mathfrak{a}_G^* \subset \mathrm{Re}\, \mathfrak{a}_{G^0}^*$. From an element $x \in \mathrm{Re}\, \mathfrak{a}_G^*$, we deduce a character of $\mathbf{G}^0$ with values in $\mathbb{R}_+^\times$. As x is invariant under $\mathfrak{m}$, we show as in Section 2 that this character can be extended to a character of $\mathbf{G}$ with values in $\mathbb{R}_+^\times$, denoted by ξ_x. It is trivial on $G(k)$ since it is trivial on $G^0(k)$ and $G^0(k)$ is of finite index in $G(k)$. Set $\mathrm{Re}\, X_G = \{\xi_x;\ x \in \mathrm{Re}\,\mathfrak{a}_G^*\}$. We have the following property:

(*) the restriction map of $\mathbf{G}$ to $Z_\mathbf{G}$ defines an isomorphism between $\mathrm{Re}\, X_G$ and the group of characters of $Z_\mathbf{G}$ which are trivial on $Z_G(k) \cap Z_\mathbf{G}$, with values in $\mathbb{R}_+^\times$.

Proof Thanks to Lemma I.1.5, we see that this last group can be identified, via the projection $\mathrm{pr} : \mathbf{G} \to G(\mathbb{A})$, with the group of characters of $Z_G^1 \backslash Z_G(\mathbb{A})$ having values in $\mathbb{R}_+^\times$. This group can itself be identified with $\mathrm{Rat}(Z_G^0) \otimes_\mathbb{Z} \mathbb{R}$. Now, we have the equality

$$\mathrm{Rat}(Z_G^0) = (\mathrm{Rat}_{\bar{k}} Z_G^0)^\Gamma,$$

and injective maps with finite cokernel

$$(\mathrm{Rat}_{\bar{k}} Z_{G^0}^0) \to \mathrm{Rat}_{\bar{k}} Z_G^0,$$

$$\mathrm{Rat}_{\bar{k}} G^0 \to \mathrm{Rat}_{\bar{k}} Z_{G^0}^0.$$

We thus deduce the isomorphism

$$\mathrm{Re}\,\mathfrak{a}_G^* \simeq \mathrm{Rat}(Z_G^0) \bigotimes_\mathbb{Z} \mathbb{R},$$

which concludes the proof. □

We denote by $\mathrm{Re}\,\mathfrak{a}_G$ the subspace of $\mathrm{Re}\,\mathfrak{a}_{M_{\min}}$ generated by the image of the map

$$Z_\mathbf{G} \longrightarrow \mathbf{M}_{\min}^1 \backslash \mathbf{M}_{\min} \overset{\log_{M_{\min}}}{\longrightarrow} \mathrm{Re}\,\mathfrak{a}_{M_{\min}}.$$

Note that in general it is different from $\mathrm{Re}\,\mathfrak{a}_{G^0}$. We denote by $\mathrm{Re}\,\mathfrak{a}_{M_{\min}}^G$ the annihilator in $\mathrm{Re}\,\mathfrak{a}_{M_{\min}}$ of $\mathrm{Re}\,\mathfrak{a}_G^*$ ($\subset \mathrm{Re}\,\mathfrak{a}_{G^0}^* \subset \mathrm{Re}\,\mathfrak{a}_{M_{\min}}^*$). These spaces are invariant under W and we have the decomposition

$$\mathrm{Re}\,\mathfrak{a}_{M_{\min}} = \mathrm{Re}\,\mathfrak{a}_G \oplus \mathrm{Re}\,\mathfrak{a}_{M_{\min}}^G.$$

We equip $\mathrm{Re}\,\mathfrak{a}_{M_{\min}}$ with a scalar product invariant under W and for which the above decomposition is orthogonal.

6. Automorphic forms

Let M be a standard Levi subgroup of G and $M^0 = M \cap G^0$. In I.2.1 we defined a subgroup $A_{\mathbf{M}^0}$ of $Z_{\mathbf{M}^0}$. Set $A_\mathbf{M} = A_{\mathbf{M}^0} \cap Z_\mathbf{M}$. As $\mathrm{pr}(A_{\mathbf{M}^0}) \subset T_{M^0}(\mathbb{A})$ and M commutes with T_{M^0}, we easily see that $A_\mathbf{M}$ is of finite index in $A_{\mathbf{M}^0}$. The index is 1 if k is a number field. We define $A_\mathbf{M}^G$ as in I.2.1.

As in the connected case, we define the notion of a function with moderate growth on $\mathbf{G}$. In fact a function $f : \mathbf{G} \to \mathbb{C}$ has moderate growth if and only if there exists a moderately increasing function f^0 : $\mathbf{G}^0 \to \mathbb{R}_+$ such that for all $g \in \mathbf{G}^0$, $k \in D\mathbf{K}$, $|f(gk)| \leq f^0(g)$. If k is a number field, denote by $\mathfrak{z}^0$ the centre of the enveloping algebra of the Lie algebra of $\mathbf{G}_\infty$ and by $\mathfrak{z}$ the subalgebra of elements invariant under $\mathbf{G}_\infty$. If k is a function field, denote by $\mathfrak{z}^0$ the Bernstein centre of $\mathbf{G}^0_{v_0}$ and $\mathfrak{z}$ the subalgebra of elements invariant under $\mathbf{G}_{v_0}$. In both cases, $\mathfrak{z}^0$ is a finitely-generated $\mathfrak{z}$-module. Let $\phi : G(k)\backslash\mathbf{G} \to \mathbb{C}$ be a function. We say that ϕ is automorphic if it satisfies conditions (i) to (iv) of I.2.17. It is equivalent to require that ϕ be smooth, $\mathbf{K}$-finite and that for every $k \in D\mathbf{K}$, the function $\phi^0_k : G^0(k)\backslash\mathbf{G}^0 \to \mathbb{C}$ defined by

$$\phi^0_k : g \mapsto \phi(gk)$$

be automorphic.

Let $P = MU$ be a standard parabolic subgroup of G. For every measurable and locally L^1 function ϕ on $U(k)\backslash\mathbf{G}$, we define as in the connected case its constant term ϕ_P which is a measurable and locally L^1 function on $U(\mathbb{A})\backslash\mathbf{G}$. Let ϕ be an automorphic form on $G(k)\backslash\mathbf{G}$. We say that ϕ is cuspidal if for every standard parabolic subgroup $P \subsetneq G$, we have $\phi_P = 0$. This definition masks a trick. Suppose G satisfies the condition

(∗) G commutes with T_{G^0}, which is the largest split torus in the centre of G^0.

Then ϕ is cuspidal if and only if for every $k \in D\mathbf{K}$, the function ϕ^0_k is cuspidal. Suppose on the contrary that G does not satisfy the condition (∗). Denote by G' the commutator of T_{G^0}. According to our definitions, $G' \subsetneq G$ is a standard parabolic subgroup of G. As $\phi_{G'} = \phi$, the only cuspidal form on $G(k)\backslash\mathbf{G}$ is then the function which is identically zero.

We adopt the same notations $A(G(k)\backslash\mathbf{G})$ and so on, as in the connected case. Let $P = MU$ be a standard parabolic subgroup of G. We define similarly the notions of automorphic form and of cuspidal automorphic form on $U(\mathbb{A})M(k)\backslash\mathbf{G}$. Note that by definition M is a group satisfying the condition (∗).

Suppose that G satisfies the condition (∗). Then the quotient $Z_\mathbf{G}(Z_{G^0}(k) \cap Z_{\mathbf{G}^0})\backslash Z_{\mathbf{G}^0}$ is compact. The following properties are then a consequence of the analogous properties for the group G^0:

$Z_\mathbf{G} G(k)\backslash\mathbf{G}$ is of finite measure;

let ξ be a unitary character of $Z_\mathbf{G}$; then every element of $A_0(G(k)\backslash\mathbf{G})_\xi$ is square integrable modulo $Z_\mathbf{G}$;

under the same hypothesis, $A_0(G(k)\backslash\mathbf{G})_\xi$ is dense in $L^2_0(G(k)\backslash\mathbf{G})_\xi$;

let ξ be a character of $Z_{\mathbf{G}}$; there exists a map

$$A(G(k)\backslash\mathbf{G})_{\xi} \to A_0(G(k)\backslash\mathbf{G})_{\xi}$$

$$\phi| \mapsto \phi^{\mathrm{cusp}}$$

analogous to that of I.2.18.

Suppose as before that G satisfies the condition $(*)$. Let $I \subset \mathfrak{z}$ be an ideal of finite codimension and σ an irreducible representation of $\mathbf{K}$. Denote by $A_0(G(k)\backslash\mathbf{G})(I,\sigma)$ the subspace of elements of $A_0(G(k)\backslash\mathbf{G})$ killed by I and on which $\mathbf{K}$ acts via σ. Then $A_0(G(k)\backslash\mathbf{G})(I,\sigma)$ is finite-dimensional.

Proof The ideal $I\mathfrak{z}^0$ of $\mathfrak{z}^0$ is of finite codimension. Set $\mathbf{K}^0 = \mathbf{K} \cap \mathbf{G}^0$. Denote by Σ the set of irreducible representations σ' of $\mathbf{K}^0$ such that, for $d \in D$, if σ'_d and σ_d are the representations of $\mathbf{K} \cap d^{-1}\mathbf{K}^0 d$ defined by

$$\sigma'_d(k) = \sigma'(dkd^{-1}), \ \sigma_d(k) = \sigma(k),$$

then

$$\mathrm{Hom}_{\mathbf{K}\cap d^{-1}\mathbf{K}^0 d}(\sigma'_d, \sigma_d) \neq \{0\}.$$

The set Σ is finite. It is clear that if $\phi \in A_0(G(k)\backslash\mathbf{G})(I,\sigma)$ and $k \in D\mathbf{K}$, the function ϕ^0_k belongs to

$$V = \bigoplus_{\sigma' \in \Sigma} A_0(G^0(k)\backslash\mathbf{G}^0)(I\mathfrak{z}^0, \sigma').$$

Fix a finite set $S \supset S_0$ of places of k such that K_v can be lifted to a subgroup $\tilde{K}_v$ for every $v \notin S$, and large enough so that σ is trivial on $\tilde{K}_v$ for all $v \notin S$. The quotient

$$\mathbf{K}^0\backslash\mathbf{K}^0 D\mathbf{K}/\prod_{v\notin S}\tilde{K}_v$$

is finite. Fix a set $\{k_i;\ i = 1,\dots,n\} \subset D\mathbf{K}$ of representatives of the quotient. The map $\phi \mapsto (\phi^0_{k_1},\dots,\phi^0_{k_n})$ is injective on $A_0(G(k)\backslash\mathbf{G})(I,\sigma)$. Thus this space is mapped injectively into V^n which is finite-dimensional. □

7. Cuspidal components

The definitions and statements of I.3.4 and I.3.5 remain valid. The proofs are identical if G satisfies condition $(*)$ of Section 5. Otherwise we introduce the commutator G' of T_{G^0}. This commutator satisfies $(*)$ and the statements for G' are immediate consequences of the properties of G.

8. Eisenstein series

Let $P = MU$ be a standard parabolic subgroup of G and $\pi \in \Pi_0(\mathbf{M})$ an irreducible cuspidal automorphic representation of $\mathbf{M}$. We define $\mathrm{Re}\,\pi \in \mathrm{Re}\,X_M \simeq \mathrm{Re}\,\mathfrak{a}_M^*$ and the space $A_0(U(\mathbb{A})M(k)\backslash\mathbf{G})_\pi$ as in the connected case.

First suppose $\langle\mathrm{Re}\,\pi, \check{\alpha}\rangle > \langle\rho_P, \check{\alpha}\rangle$ for all $\alpha \in \Delta(T_{M^0}, G)$. Let $\phi \in A_0(U(\mathbb{A})M(k)\backslash\mathbf{G})_\pi$. Define a function on $\mathbf{G}$ by

$$E(\phi,\pi)(g) = \sum_{\gamma\in P(k)\backslash G(k)} \phi(\gamma g).$$

This series converges and defines an automorphic form.

Proof Let us write $g = g^0k$, with $g^0 \in \mathbf{G}^0$, $k \in D\mathbf{K}$. We identify the set B of Section 4 with a system of representatives in $\mathrm{Norm}_{G(k)}(T_{\min}, P^0_{\min})$. We can write

$$E(\phi,\pi)(g) = [P(k) : P^0(k)]^{-1} \sum_{\gamma\in P^0(k)\backslash G(k)} \phi(\gamma g^0 k)$$

$$(1) \qquad E(\phi,\pi)(g) = [P(k) : P^0(k)]^{-1}\sum_{b\in B} \sum_{\gamma\in P^0(k)\ G^0(k)} \phi(\gamma b^{-1}g^0k).$$

Fix $b \in B$. The group bPb^{-1} is a standard parabolic subgroup of G. Denote it by ${}^bP = {}^b M {}^b U$ and write $b\pi$ for the representation of ${}^b\mathbf{M}$ deduced from π by conjugation. The function ${}^b\phi$ defined by ${}^b\phi(g) = \phi(bg)$ belongs to $A_0({}^bU(\mathbb{A}){}^bM(k)\backslash\mathbf{G})_{b\pi}$. As $b \in \mathrm{Norm}_{G(k)}(T_{\min}, P^0_{\min})$, b sends $\Delta(T_{M^0}, G)$ on $\Delta(T_{{}^bM^0}, G)$: thus we still have $\langle\mathrm{Re}\,b\pi, \check{\alpha}\rangle > \langle\rho_{{}^bP}, \check{\alpha}\rangle$ for all $\alpha \in \Delta(T_{{}^bM^0}, G)$. As we saw in Section 6, there exists a finite set of irreducible cuspidal representations of ${}^b\mathbf{M}^0$, denoted by $\pi'_1, \ldots, \pi'_t$, such that the function $({}^b\phi)^0_k$ belongs to

$$\bigoplus_{i=1}^{t} A_0({}^bU(\mathbb{A}){}^bM^0(k)\backslash\mathbf{G}^0)_{\pi_i}.$$

Moreover $\mathrm{Re}\,\pi_i = \mathrm{Re}\,b\pi$ for every i since $\mathrm{Re}\,\pi_i$ and $\mathrm{Re}\,b\pi$ are determined by their restriction to $A_{{}^b\mathbf{M}^0}$ and $A_{{}^b\mathbf{M}}$, respectively: these restrictions are equal on $A_{{}^b\mathbf{M}}$ and the latter is of finite index in $A_{{}^b\mathbf{M}^0}$. The series defining $E(({}^b\phi)^0_k, g^0)$ is thus convergent. Now, by (1), we have the equality

$$(2) \qquad \begin{aligned} E(\phi,\pi)(g) &= [P(k) : P^0(k)]^{-1}\sum_{b\in B} \sum_{\gamma\in {}^bP^0(k)\backslash G^0(k)} \phi(b^{-1}\gamma g^0k) \\ &= [P(k) : P^0(k)]^{-1}\sum_{b\in B} E(({}^b\phi)^0_k, g^0). \end{aligned}$$

This gives the result. □

It is a consequence of equality (2) above that the properties of analytic continuation proved in Chapter IV in the connected case remain valid in the present situation.

Let $P' = M'U'$ be a standard parabolic subgroup of G. For every $w \in W$ such that $wMw^{-1} = M'$, we define as in the connected case the representation $w\pi$ of $\mathbf{M}'$ and the intertwining operator $M(w,\pi)$: $A_0(U(\mathbb{A})M(k)\backslash\mathbf{G})_\pi \to A_0(U'(\mathbb{A})M'(k)\backslash\mathbf{G})_{w\pi}$. Let $W_1(M,M') = \{w \in W \mid wMw^{-1} = M'\}$, $B_1(M,M') = W_1(M,M') \cap B$. In the connected case, we fixed a system of representatives $W(M,M')$ of $W_1(M,M')/W_M$: it was the set of elements w of minimal length in their class wW_M. Here there does not seem to be a natural system of representatives for $B_1(M,M')/B_M$. Rather than fixing an arbitrary system of representatives, we will keep all of $B_1(M,M')$, which obliges us to divide all the formulae by $|B_M| = [M(k) : M^0(k)] = [P(k) : P^0(k)]$. We thus set:

$$W(M,M') = \bigcup_{M''} B_1(M'',M')W^0(M^0,M''^0)$$
$$= \bigcup_{M''} W^0(M''^0,M'^0)B_1(M,M'')$$

where the union is over all the standard Levis M'' of G. This set represents the quotient

$$W_1(M,M')/W^0_{M^0}.$$

We then have the equality

$$E(\phi,\pi)_{P'} = |B_M|^{-1} \sum_{w\in W(M,M')} M(w,\pi)\phi.$$

This is a consequence of the equality (2) and of Proposition II.1.7.

9. The functions m_P and trivialisation of fibre bundles

Let M be a standard Levi subgroup of G; we denote by U the unipotent radical of the standard parabolic Levi M. Also let $\mathfrak{P}$ be an orbit under $X_M^{\mathbf{G}}$ of irreducible cuspidal representations of M (automorphic would actually suffice here). Frequently, in the connected case, we used the function m_P defined in I.1.4. The introduction of this function has two uses: on the one hand it is used to obtain upper bounds (see II.1.5, II.2.5) and on the other to trivialise the fibre bundle over $\mathfrak{P}$ whose fibre over $\pi \in \mathfrak{P}$ is the space $A(U(\mathbb{A})M(k)\backslash\mathbf{G})_\pi$. The sections of this fibre bundle are thus functions ϕ on $\mathfrak{P}$ with values in the functions on $\mathbf{G}$ satisfying:

(1) $$\phi(\pi) \in A(U(\mathbb{A})M(k)\backslash\mathbf{G})_\pi,\ \forall\pi \in \mathfrak{P}.$$

We thus generalise the definition of m_P without supposing that $\mathbf{G}= U(\mathbb{A})\mathbf{MK}$, asking only that m_P be a map of $\mathbf{G}$ into $\mathbf{M}/\mathbf{M}^1$ satisfying:

(2) $$\forall g \in \mathbf{G},\ \forall u \in U(\mathbb{A}),\ \forall m \in \mathbf{M},\ \forall k \in \mathbf{K},\ \ m_P(umgk) = \overline{m}m_P(g),$$

where $\overline{m}$ is the image of m in $\mathbf{M}/\mathbf{M}^1$. It is easy to construct all the maps having property (2). Indeed, fix a set D_P of representatives in $\mathbf{G}$ of double cosets $U(\mathbb{A})\mathbf{M}\backslash\mathbf{G}/\mathbf{K}$. Then the set of maps of $\mathbf{G}$ into $\mathbf{M}/\mathbf{M}^1$ having the property (2) is in bijection with the set of maps of D_P in $\mathbf{M}/\mathbf{M}^1$. We sketch this rapidly. Let f be a map of D_P into $\mathbf{M}/\mathbf{M}^1$. Then f is the restriction to D_P of a map m_P of $\mathbf{G}$ into $\mathbf{M}/\mathbf{M}^1$ satisfying (2) if and only if for every $d \in D_P$ and for every $g \in U(\mathbb{A})\mathbf{M}d\mathbf{K}$:

(a) $(\mathbf{M}^1\backslash(U(\mathbb{A})\mathbf{M}^1 g\mathbf{K}d^{-1}) \cap \mathbf{M})$ is reduced to a point, $\overline{m}_g$,

(b) $m_P(g) = \overline{m}_g m_P(d)$.

It is clear that condition (b) defines m_P. Thus we need only prove (a) which is independent of f. Let us prove it. Write $g = umdk$ with $u \in U(\mathbb{A})$, $m \in \mathbf{M}$, $k \in \mathbf{K}$. Then:

$$\begin{aligned}(U(\mathbb{A})\mathbf{M}^1 g\mathbf{K}d^{-1}) \cap \mathbf{M} &= \mathbf{M}^1(U(\mathbb{A})g\mathbf{K}d^{-1} \cap \mathbf{M}) \\ &= \mathbf{M}^1 m(U(\mathbb{A})d\mathbf{K}d^{-1} \cap \mathbf{M}).\end{aligned}$$

We must thus prove that every element λ of X_P with values in $\mathbb{R}_{>0}$, extended to a character of $U(\mathbb{A})\mathbf{M}$ trivial on $U(\mathbb{A})$, is trivial on $d\mathbf{K}d^{-1} \cap U(\mathbb{A})\mathbf{M}$. Now, $d\mathbf{K}d^{-1} \cap U(\mathbb{A})\mathbf{M}$ is a compact subgroup of $U(\mathbb{A})\mathbf{M}$. Its image under λ is thus a compact subgroup of $\mathbb{R}_{>0}$, so it is reduced to 1. This concludes the proof.

We deduce from this description the following property: let m_P and m'_P be maps of $\mathbf{G}$ into $\mathbf{M}/\mathbf{M}^1$ satisfying (2). Then there exists $r \in \mathbb{N}$ and a finite set of elements, $(\overline{m}_i)_{i\in[1,r]}$, of $\mathbf{M}/\mathbf{M}^1$ such that:

$$\forall g \in \mathbf{G},\ \exists\, i \in [1,r],\ \text{such that } m_P(g) = m'_P(g)\overline{m}_i.$$

We also deduce from this the following

Corollary *Let $M \subset M'$ be standard Levi subgroups. Let m_P (resp. $m_{P'}$) be a map of $\mathbf{G}$ into $\mathbf{M}/\mathbf{M}^1$ or $\mathbf{M}'/\mathbf{M}'^1$ satisfying (2), or the analogue of (2) for M', respectively. Then there exists $r \in \mathbb{N}$ and a finite set $(\overline{m}'_i)_{i\in[1,r]}$, of elements of $\mathbf{M}'/\mathbf{M}'^1$ such that, writing* proj *for the natural map of $\mathbf{M}/\mathbf{M}^1$ to $\mathbf{M}'/\mathbf{M}'^1$, we have:*

$$\forall g \in \mathbf{G}, \exists i \in [1,r],\ \textit{such that}\ \operatorname{proj} m_P(g) = m_{P'}(g)\overline{m}'_i.$$

Let us return to the trivialisation of the fibre bundle over $\mathfrak{P}$ whose fibre over π is the space $A(U(\mathbb{A})M(k)\backslash\mathbf{G})_\pi$. Fix $\pi_0 \in \mathfrak{P}$. In the function field case, we note that $\operatorname{Fix}_{X_M^G}\mathfrak{P}$ operates on $A(U(\mathbb{A})M(k)\backslash\mathbf{G})_{\pi_0}$, since we fixed a map m_P of $\mathbf{G}$ into $\mathbf{M}/\mathbf{M}^1$ satisfying (2), by:

$$\forall \mu \in \operatorname{Fix}_{X_M^G}\mathfrak{P},\ \forall \phi \in A(U(\mathbb{A})M(k)\backslash\mathbf{G})_{\pi_0},\quad \mu.\phi = (m_P)^\mu \phi.$$

Thus the choice of m_P allows us to construct a fibre bundle over

$X_M^{\mathbf{G}}/\mathrm{Fix}_{X_M^{\mathbf{G}}}\mathfrak{P}$ of fibre $A(U(\mathbb{A})M(k)\backslash\mathbf{G})_{\pi_0}$ which we denote by:

$$(3) \qquad X_M^{\mathbf{G}} \times_{m_P}^{\mathrm{Fix}_{X_M^{\mathbf{G}}}\mathfrak{P}} A(U(\mathbb{A})M(k)\backslash\mathbf{G})_{\pi_0}.$$

In the number field case, (3) is the trivial fibre bundle on $X_M^{\mathbf{G}}$ of fibre $A(U(\mathbb{A})M(k)\backslash\mathbf{G})_{\pi_0}$. Thanks to the choice of m_P, we construct a map of the set of sections ϕ of the fibre bundle over $\mathfrak{P}$ described above into the set of sections $\tilde{\phi}$ of the fibre bundle (3), by:

$$\forall\lambda \in X_M^{\mathbf{G}},\ \forall g \in \mathbf{G},\ \ \tilde{\phi}(\lambda)(g) = m_P(g)^{-\lambda}\phi(\pi \otimes \lambda)(g).$$

Lemma (a) *We use the same notation as above, in particular π_0 and m_P. The map $\phi \mapsto \tilde{\phi}$ defines an isomorphism between sections of the two fibre bundles we are considering. If ϕ is $\mathbf{K}$-finite, then ϕ is of finite rank in the following sense: there exists a finite set $r \in \mathbb{N}$ $(\phi_i)_{i\in[1,r]}$ of functions on $X_M^{\mathbf{G}}$ and a finite set $(f_i)_{i\in[1,r]}$ of elements of $A(U(\mathbb{A})M(k)\backslash\mathbf{G})_{\pi_0}$ such that:*

$$(4) \qquad \tilde{\phi} = \sum_{i=1}^{r} \phi_i \otimes f_i.$$

The proof is the same as in the connected case; see II.1.2.

Let us calculate the change of trivialisation when m_P is replaced by another map m_P' also satisfying (2). We construct a map $\phi \mapsto \tilde{\phi}'$ analogous to $\phi \mapsto \tilde{\phi}$.

Lemma (b) *With the same notation as above, there exists a finite set J of elements of $\mathbf{M}/\mathbf{M}^1$ such that for every $\mathbf{K}$-finite ϕ, there exists an expression $\tilde{\phi} = \sum_{i=1}^{r} \phi_i \otimes f_i$ satisfying (4) for which $\tilde{\phi}' = \sum_{i=1}^{r} \phi_i' \otimes f_i$ with ϕ_i', for $i \in [1,r]$, of the form:*

$$\phi_i'(\lambda) = \phi_i(\lambda)\overline{m}^{\lambda},\ \forall\lambda \in X_M^{\mathbf{G}},$$

where $\overline{m}$ is an element of J independent of λ.

We recall the notation D_P introduced earlier. It suffices to prove the lemma for the set of sections ϕ for which there exists $d \in D_P$ such that for every $\pi \in \mathfrak{P}$ the support of the function $\phi(\pi)$ lies in $U(\mathbb{A})\mathbf{M}d\mathbf{K}$. Let ϕ be of this type and $d \in D_P$ as above. We easily check that we can write $\tilde{\phi}$ in the form $\sum_{i=1}^{r} \phi_i \otimes f_i$ satisfying (4) with the f_i supported in $U(\mathbb{A})\mathbf{M}d\mathbf{K}$. The definitions then show that:

$$\tilde{\phi}'(\lambda) = \sum_{i=1}^{r} \phi_i(\lambda) m_P(d)^{\lambda} m_P'(d)^{-\lambda} f_i,$$

i.e.

$$\tilde{\phi}' = \sum_{i=1}^{r} \phi_i' \otimes f_i,$$

with $\phi_i'(\lambda) = \phi_i(\lambda)m_P(d)^{-\lambda}m_P'(d)^{-\lambda}$, for every $\lambda \in X_M^{\mathbf{G}}$. This concludes the proof.

10. Paley-Wiener functions, $P_{(M,\mathfrak{P})}$ and $P^R_{(M,\mathfrak{P})}$

Let $(M, \mathfrak{P})$ be a cuspidal datum, i.e. M is a standard Levi of G and $\mathfrak{P}$ is an orbit under $X_M^{\mathbf{G}}$ of cuspidal representations of $\mathbf{M}$. As in II.1.2, we denote by $P(X_M^{\mathbf{G}})$ the set of Paley-Wiener functions; for R a large positive real number, we define $P^R(X_M^{\mathbf{G}})$ as in II.1.4.

Definition Let $P_{(M,\mathfrak{P})}$ or $P^R_{(M,\mathfrak{P})}$ be the set of $\mathbf{K}$-finite sections f of the fibre bundle on $\mathfrak{P}$ whose fibre over π is the space $A(U(\mathbb{A})M(k)\backslash\mathbf{G})_\pi$, such that, in the trivialisation of Lemma 9(b), the section $\tilde{f}$ belongs to $P(X_M^{\mathbf{G}})\otimes A(U(\mathbb{A})M(k)\backslash\mathbf{G})_{\pi_0}$ or $P^R(X_M^{\mathbf{G}})\otimes A(U(\mathbb{A})M(k)\backslash\mathbf{G})_{\pi_0}$, *respectively.*

Note that it is a consequence of 9(*c*) that these definitions are independent of the choice of trivialisation, i.e. of the choice of m_P. It is also easy to check that these definitions are independent of the choice of $\pi_0 \in \mathfrak{P}$. As in II.1.13, we define a Fourier transform denoted by F, and the evaluation ϵ at the unit element of $\mathbf{M}$. For $\phi \in P_{(M,\mathfrak{P})}$ or $P^R_{(M,\mathfrak{P})}$, we want to consider $\epsilon F(\phi)$. Let $\lambda_0 \in X_M$ be an element of $\operatorname{Re}\mathfrak{P}$: then

$$\epsilon F(\phi) = \int_{\pi\in\mathfrak{P},\ \operatorname{Re}\pi=\lambda_0} \phi(\pi)\, d\pi. \tag{1}$$

Fix m_P and π_0 as in Section 9 and write

$$\tilde{\phi} = \sum_{i=1}^{r} \phi_i \otimes f_i \in P(X_M^{\mathbf{G}}) \otimes A(U(\mathbb{A})M(k)\backslash\mathbf{G})_{\pi_0}.$$

As in II.1.3, we check that we have:

$$\forall g \in \mathbf{G},\ (\epsilon F(\phi))(g) = \sum_{i=1}^{r} (F\phi_i)(m_P(g))f_i(g) \tag{2}$$

where F in the right-hand side is the Fourier transform on $\operatorname{Im} X_M^{\mathbf{G}}$. We deduce from this that if $\phi \in P_{(M,\mathfrak{P})}$, there exists a compact set ω of $U(\mathbb{A})\mathbf{M}^1 Z_{\mathbf{G}}\backslash\mathbf{G}$ such that $\epsilon F(\phi)(g) = 0$ if the image of g in $U(\mathbb{A})\mathbf{M}^1 Z_{\mathbf{G}}\backslash\mathbf{G}$ is not contained in ω.

11. Pseudo-Eisenstein series

Let $(M, \mathfrak{P})$ be a cuspidal datum as in Section 10. We denote by ξ the restriction to $Z_{\mathbf{G}}$ of the central character of elements of $\mathfrak{P}$. Let R be a large positive real number and let $\phi \in P^R_{(M,\mathfrak{P})}$. We define as in the connected case the pseudo-Eisenstein series (the notation ϵ and F are as in Section 10):

$$(1) \qquad \theta_\phi(g) = \sum_{\gamma \in U(k)M(k)\backslash G(k)} (\epsilon F(\phi))(\gamma g), \forall g \in \mathbf{G}.$$

The proof of the connected case can be applied word for word to prove:

Proposition (a) *(i) The series (1) converges absolutely and uniformly on every compact set of* $\mathbf{G}$. *It defines a* $\mathbf{K}$*-finite function on* $G(k)\backslash\mathbf{G}$ *on which* $Z_{\mathbf{G}}$ *acts via the character* ξ. *(ii) Suppose that* ξ *is unitary. Then* θ_ϕ *is square integrable modulo* $Z_{\mathbf{G}}$.

We also have:

Proposition (b) *Let* $\phi \in P_{(M,\mathfrak{P})}$ *and let* f *be a slowly increasing* $\mathbf{K}$*-finite function on* $G(k)\backslash\mathbf{G}$ *on which* $Z_{\mathbf{G}}$ *acts via* $\overline{\xi}^{-1}$. *Then* $|\theta_\phi \overline{f}|$ *is integrable on* $G(k)Z_{\mathbf{G}}\backslash\mathbf{G}$.

To avoid absolute values, we can suppose from the start that $f = |f|$ and $\phi = |\phi|$, i.e. $\phi(\pi)(g) = |\phi(\pi)(g)|$, for every $\pi \in \mathfrak{P}$, and for every $g \in \mathbf{G}$. We denote by f_U the constant term of f along U; it is a slowly increasing, $\mathbf{K}$-finite function on $\mathbf{G}$. We then have:

$$\int_{G(k)Z_{\mathbf{G}}\backslash\mathbf{G}} \theta_\phi(g)\overline{f}(g)\, dg < +\infty \iff \int_{U(\mathbb{A})M(k)Z_{\mathbf{G}}\backslash\mathbf{G}} (\epsilon F(\phi))(g)\overline{f}_U(g)\, dg < +\infty.$$

The right-hand integral converges by (2) in Section 10 and the arguments just after that equation.

12. Constant terms of pseudo-Eisenstein series; scalar product

Lemma II.2.2 is valid under the hypotheses of this appendix with the unique difference that in (iv) we must divide the right-hand side by the cardinal of the set:

$$B_M := \{w \in W_M; w(\alpha) > 0, \forall \text{ positive roots } \alpha \text{ of } M\}.$$

Let us rewrite (iv) in the simplest case which is also the most important one.

Let $(M, \mathfrak{P})$ be a cuspidal datum and let P' be a standard parabolic Levi subgroup M'. Let $\phi \in P^R_{(M,\mathfrak{P})}$.

Proposition

(i) $(\theta_\phi)_{P'} = 0$ *if M is not conjugate to a Levi subgroup M'.*
(ii) *Suppose that M' is conjugate to M. Then:*

$$(\theta_\phi)_{P'} = |B_M|^{-1} \sum_{w \in W(M,M')} \int_{\pi \in \mathfrak{P},\ \mathrm{Re}\,\pi = \lambda_0} M(w,\pi)\phi(\pi)\, d\pi,$$

where λ_0 is a 'very positive' element of $\mathrm{Re}\,\mathfrak{P}$*, a subset of* $\mathrm{Re}\,X_M$*).*

Let ξ be the restriction to Z_G of the central character of the elements of $\mathfrak{P}$ and suppose that ξ is unitary; we then have $\mathfrak{P} = -\overline{\mathfrak{P}}$. We denote by $\mathfrak{X}$ the equivalence class of $(M, \mathfrak{P})$, i.e. $\{(wMw^{-1}, w\mathfrak{P})\}$ where w runs through $W(M)$. For $(M, \mathfrak{P}), (M', \mathfrak{P}') \in \mathfrak{X}$, we denote by $W((M, \mathfrak{P}), (M', \mathfrak{P}'))$ the subset of $W(M)$ which conjugates $(M, \mathfrak{P})$ and $(M', \mathfrak{P})$. We then have:

Proposition *Let $(M, \mathfrak{P})$ and $(M', \mathfrak{P}')$ be cuspidal data the restriction of whose unitary character to Z_G is ξ. Let R be a large positive real number and let $\phi' \in P^R_{(M',\mathfrak{P}')}, \phi \in P^R_{(M,\mathfrak{P})}$. We have:*

$$\langle \theta_{\phi'}, \theta_\phi \rangle = |B_M|^{-1} \int_{\pi \in \mathfrak{P},\ \mathrm{Re}\,\pi = \lambda_0} A(\phi', \phi)(\pi)\, d\pi,$$

where

$$A(\phi', \phi)(\pi) = \sum_{w \in W((M,\mathfrak{P}),(M',\mathfrak{P}'))} \langle M(w^{-1}, -w\overline{\pi})\phi'(-w\overline{\pi}), \phi(\pi) \rangle,$$

where B_M is defined above and where λ_0 is a 'very positive" element of $\mathrm{Re}\,\mathfrak{P}$.

For every equivalence class of cuspidal data $\mathfrak{X}$ the restriction of whose unitary character to Z_G is ξ, we denote by $L^2(G(k)\backslash \mathbf{G})_{\mathfrak{X}}$ the closed subspace of $L^2(G(k)\backslash \mathbf{G})_\xi$ generated by the pseudo-Eisenstein series θ_ϕ with $\phi \in P_{(M,\mathfrak{P})}$, where $(M, \mathfrak{P})$ runs through $\mathfrak{X}$. As in the connected case we have:

Theorem $L^2(G(k)\backslash \mathbf{G})_\xi = \hat{\oplus} L^2(G(k)\backslash \mathbf{G})_{\mathfrak{X}}$, *where $\mathfrak{X}$ runs through the set of equivalence classes of cuspidal data relative to the unitary central character ξ.*

13. The operators $\Delta(f)$

We define algebras $H^R_{\mathfrak{X}}, H^R_\xi$ and so on as in the connected case. The definitions and results of III.1 remain valid, as does Proposition III.2.

Indeed, if G is one of its own Levi subgroups, the proof is the same as in the connected case. Otherwise, let G' be the Levi subgroup G associated with the Levi G° of G°; it is the largest Levi subgroup. The proposition is thus true for $\mathbf{G}'$. Let $\phi \in A(G(k)\backslash\mathbf{G})_\xi$. Fix $h \in \mathbf{G}$ and denote by $\phi_h : G(k)\backslash\mathbf{G}' \to \mathbb{C}$ the function defined by $\phi_h(g) = \phi(gh)$. Let us choose a finite set

$$D_h \subset \mathbb{C}[\mathrm{Re}\,\mathfrak{a}^{\mathbf{G}}_{G'}] \times \mathrm{Hom}(Z_{\mathbf{G}'}, \mathbb{C}^*) \times A(G(k)\backslash\mathbf{G})$$

such that:

(i) for every $(Q, \eta, \psi) \in D_h, \psi \in A(G'(k)\backslash\mathbf{G}')_\eta$;

(ii) for every $g \in \mathbf{G}', \phi_h(g) = \sum_{(Q,\eta,\psi)\in D_h} Q(\log^{\mathbf{G}}_{G'} g)\psi(g)$.

Let $f \in H^R_\xi$, $\eta \in \mathrm{Hom}(Z_{\mathbf{G}'}, \mathbb{C}^*)$ such that $\|\mathrm{Re}\,\eta\|^2 + R'^2 < R^2$ and $Q \in \mathbb{C}[\mathrm{Re}\,\mathfrak{a}^{\mathbf{G}}_{G'}]$. We define $f_{Q,\eta} \in H^{G',R'}_\eta$ (the analogue of H^R_ξ for G') as follows: let $(M, \mathfrak{P}')$ be a pair consisting of a standard Levi and an orbit under $X^{\mathbf{G}'}_M$ of cuspidal representations of $\mathbf{M}$ the restriction of whose central character to $Z_{\mathbf{G}'}$ is η; for $\pi \in \mathfrak{P}'$ and $\phi \in A_0(M(k)\backslash\mathbf{M})_\pi$, we set:

$$f_{Q,\eta}(\pi)\phi = \partial_Q(\lambda \mapsto (-\lambda) \circ f(\pi \otimes \lambda \circ \psi))_{|\lambda=0},$$

where the variable λ belongs to $X^{\mathbf{G}}_{G'}$.

For every $Q \in \mathbb{C}[\mathrm{Re}\,\mathfrak{a}^{\mathbf{G}}_{G'}]$, let us fix a finite set $\Delta(Q) \subset \mathbb{C}[\mathrm{Re}\,\mathfrak{a}^{\mathbf{G}}_{G'}]^2$ such that for all functions A, B on $X^{\mathbf{G}}_{G'}$, we have the formula

$$\partial_Q(AB) = \sum_{(Q',Q'')\in\Delta(Q)} \partial_{Q'}(A)\partial_{Q''}(B).$$

We then set for $g \in \mathbf{G}'$;

$$\phi'_h(g) = \sum_{(Q,\eta,\psi)\in D_h} \sum_{(Q',Q'')\in\Delta(Q)} Q'(\log^{\mathbf{G}}_{G'} g)(\Delta(f_{Q'',\eta})\psi)(g).$$

It is fairly easy to check that

(i) this term does not depend on the choice of D_h and $\Delta(Q)$;

(ii) if $gh = g'h'$, $\phi'_h(g) = \phi'_{h'}(g')$.

This allows us to define a function $\Delta(f)\phi$ by $\Delta(f)\phi(gh) = \phi'_h(g)$. Then

(iii) $\Delta(f)\phi$ is left-invariant by $G(k)$;

note the construction gives invariance *a priori* only under $G'(k)$; we must use the properties of invariance of f under all of $G(k)$. We then show as in Chapter III that $\Delta(f)\phi$ is an automorphic form on $G(k)\backslash\mathbf{G}$ which satisfies the condition of Proposition III.2.1.

If G is one of its own Levi subgroups, Proposition III.3.1 remains true; we set by definition $G_{\mathrm{der}} = G^0_{\mathrm{der}}$ and $N(G) = N(G^0)$. If G is not one of its Levi subgroups, there is no discrete spectrum, i.e. $A^2(G(k)\backslash\mathbf{G})_\xi = \{0\}$. Indeed, we see that if $\phi \in A(G(k)\backslash\mathbf{G})_\xi$ and $g \in \mathbf{G}$ are such that $\phi(g) \neq 0$, the integral

$$\int_{Z_{\mathbf{G}}(Z_{G^0}(k)\cap Z_{\mathbf{G}^0})\backslash Z_{\mathbf{G}^0}} |\phi(zg)|^2 \, dz$$

cannot be convergent. This result will appear as a special case of spectral decomposition; see below.

14. Spectral decomposition

Chapter V can be generalised with no changes to the non-connected case. Indeed, the object of Chapter V is the spectral decomposition of the scalar product of pseudo-Eisenstein series. The method we follow uses the following:

the explicit formula for the scalar product, which is the same in the connected case and the non-connected case up to the scalar factor $|B_M|^{-1}$;

analysis (essentially the residue theorem) on the orbits under $X_M^{\mathbf{G}}$ of cuspidal representations of $\mathbf{M}$ (for a standard Levi M of G) for (scalar-valued) meromorphic functions with polynomial singularities along singular hyperplanes for the intertwining operators. These orbits and these hyperplanes are of the same nature in the connected and the non-connected cases;

in the number field case, the spectral decomposition of the operator $\Delta(f_0)$, where f_0 is defined as in the connected case, which is a standard result of analysis.

We do not rewrite the theorems of V but they do remain entirely valid here.

Chapter VI also generalises; we recall that its object is the description of the spaces $A_{\mathfrak{C},\pi}$, using notation as in the connected case. However, in Chapter VI we did use induction on the semi-simple rank of G and the generalisation is thus immediate only if G is one of its own standard Levis. In the general case, we denote by G' the largest standard Levi subgroup of G; it is also a standard parabolic subgroup. To simplify, we suppose here that k is a number field. We denote by ξ' the character of $Z_{\mathbf{G}}A_{G'}^{\mathbf{G}}$ extending ξ to $Z_{\mathbf{G}}$ and trivial on $A_{G'}^{\mathbf{G}}$. The generalisation of Chapter VI to G' gives the spectral decomposition of $L^2(G'(k)\backslash\mathbf{G})_{\xi'}$ as a function of the discrete spectrum of the standard Levi subgroups of G', i.e. of G. The result for $\mathbf{G}$ is then:

Proposition *$L^2(G(k)\backslash\mathbf{G})_\xi$ is isomorphic to the set of functions f defined almost everywhere on $\operatorname{Im} X^{\mathbf{G}}_{G'}$ with values in $L^2(G'(k)\backslash\mathbf{G})_{\xi'}$ satisfying, for every element w of the Weyl group of G of minimal length modulo the Weyl group of $(G')^0$:*

$$M(w^{-1}, w\lambda)f(w\lambda) = f(\lambda)$$

almost everywhere, and also satisfying:

$$\int_{\operatorname{Im} X^{\mathbf{G}}_{G'}} \|f(\lambda)\| \, d\lambda < +\infty,$$

where $\|f(\lambda)\|$ is the norm in $L^2(G'(k)\backslash\mathbf{G})_{\xi'}$. The isomorphism is given using the Eisenstein series from G' to G; note that this series is just a finite sum over $G'(k)\backslash G(k)$.

The proof is the same as the proof of the last theorem of VI, once we have proved the equivalent of VI.1.4 and VI.2.4, i.e.

Lemma *Let $(M, \mathfrak{P})$ be a cuspidal datum relative to the character ξ of $Z_{\mathbf{G}}$ and let $\mathfrak{S}$ be an affine subspace of $\mathfrak{P}$. We denote by $\mathfrak{S}'$ the set of elements of $\mathfrak{S}$ the restriction of whose central character to $A^{\mathbf{G}}_{G'}$ is trivial. Then:*

(i) *$\mathfrak{S} \in \operatorname{Sing}^G$ if and only if $\mathfrak{S}' \in \operatorname{Sing}^{G'}$ and $\mathfrak{S} = \mathfrak{S}' \otimes X^{\mathbf{G}}_{G'}$. If these conditions are fulfilled, then:*

$$\operatorname{Res}^G_{\mathfrak{S}} A(\phi', \phi) = \operatorname{Res}^{G'}_{\mathfrak{S}'} A(\phi', \phi),$$

for every $\phi', \phi \in P_{\mathfrak{X}}$; the notation is as in the connected case.

(ii) *Let $\mathfrak{C}$ be the conjugacy class of $\mathfrak{S}$ and $\mathfrak{C}'$ that of $\mathfrak{S}'$. Let $\pi' \in \mathfrak{S}'$ be a general point and let $\mu \in X^{\mathbf{G}}_{G'}$. The Eisenstein series of G' to G induces an isomorphism of $A_{\mathfrak{C}',\pi'}$ onto $A_{\mathfrak{C},\pi\otimes\mu}$ which is an isometry (up to a scalar) if μ is unitary.*

(i) We recall the calculation of VI.1.4(b), with $L = G'$. We fix $\lambda_0 = \lambda_0^{G'} + \mu_0$ as there, but there are no positivity conditions to impose on μ_0. Thus we can (and do) suppose that $\mu_0 = 0$. We do the aforementioned calculation, to obtain VI.1.4(b)(6)$_n$. Note that in (4), we replace 'τ^L minimal in $\tau^L W_L$' by '$\tau^{G'}$ minimal in its right class modulo the Weyl group of $(G')^0$' and we divide the right-hand side by $|B|^{-1}$ where B is the intersection of the set of these elements with the Weyl group of G'. The factor $|B|$ appears at the end of the calculation which follows (4) because an element τ of $W(M)$ has $|B|$ decompositions $\tau = \tau^{G'}\tau_{G'}$. Formula (6)$_n$ being proved, we obtain (7) with no vanishing hypotheses on ϕ, simply because $\mu_0 = 0$. We also have VI.1.4(b)(8) and we conclude as in VI.1.4(b).

(ii) Recall the notation B introduced in the proof of (i). We check that for every element of $P_{\mathfrak{X}}^{R}$, for every $w \in W_{G'}(M)$ and for every $\mu \in X_{G'}^{\mathbf{G}}$:

$$\mathrm{Res}_{w\mathfrak{S}}^{G} E(\phi, w\pi' \otimes \mu) = E_{G'}^{G}(\mu)\mathrm{Res}_{w\mathfrak{S}'}^{G'} E(\phi, w\pi').$$

The injectivity of $E_{G'}^{G}(\mu)$ can be proved by a calculation of the scalar product as in VI.2.4. Surjectivity is easy: we know that $A_{\mathfrak{C},\pi'\bigotimes\mu}$ is generated by the set of elements

$$\mathrm{Res}_{w\mathfrak{S}}^{G} E(\phi, w(\pi' \otimes \mu)),$$

where w runs through $W(M)$. According to (i), we know that the element written above is also:

$$\mathrm{Res}_{w\mathfrak{S}'}^{G'} E(\phi, w(\pi' \otimes \mu)).$$

With what precedes, this means that $E_{G'}^{G}(\mu)$ induces a surjection of $\oplus_{w\in B} A_{w\mathfrak{C}',w\pi'}$ onto $A_{\mathfrak{C},\pi'\otimes\mu}$. It thus suffices to check that for every $w \in B$:

$$E_{G'}^{G}(\mu)A_{w\mathfrak{C}',w\pi'} = E_{G'}^{G}(\mu)A_{\mathfrak{C}',\pi'}.$$

This is done as at the end of VI.2.4.

Corollary *The statements of VI.2 are true up to the following change: the functions on* $\mathrm{Im}\, X_L^{\mathbf{G}}$ *with values in* $\mathrm{ind}_{\mathbf{K}\cap L}^{\mathbf{K}} L^2(L(k)\backslash\mathbf{L})$ *must be replaced by the sections of the fibre bundle on* $\mathrm{Im}\, X_L^{\mathbf{G}}$ *whose fibre over* μ *is the space* $L^2(U_L(\mathbb{A})L(k)\backslash\mathbf{G})_{\xi_L\otimes\mu}$*, where* U_L *is the radical unipotent of the parabolic standard of Levi L and where* $\xi_L \otimes \mu$ *is the character of* $Z_{\mathbf{G}}A_L^{\mathbf{G}}$ *which extends* ξ *to* $Z_{\mathbf{G}}$ *and is equal to* μ *on* $A_L^{\mathbf{G}}$.

Bibliography

[A1] **Arthur, J.**, A trace formula for reductive groups I. terms associated to classes in $G(Q)$, *Duke Math. J.* **45** (1978), 911–52.

[A2] **Arthur, J.**, A trace formula for reductive groups II: applications of a truncation operator, *Comp. Math.* **40** (1980), 87–121.

[A3] **Arthur, J.**, Eisenstein series and the trace formula, in *Automorphic forms, representations and L-functions, Part 1, Proc. of Symp. in Pure Math.* **33**, AMS 1979, 253–74.

[A4] **Arthur, J.**, On some problems suggested by the trace formula, in Lie group representations II, *LN* **1041**, Springer Verlag 1983, 1–49.

[A5] **Arthur, J.**, *Lectures at the University of Paris* **7**, 1987.

[A6] **Arthur, J.**, Unipotent automorphic representations: global motivations, in *Automorphic Forms, Shimura Varieties and L-functions*, II, Academic Press 1988.

[BDKV] **Bernstein, J., Deligne, P.**, Le 'centre' de Bernstein, in Représentations des groupes réductifs sur un corps local, Hermann, Paris 1985.

[Bor] **Borel, A.**, Groupes linéaires algébriques, *Ann. Math.* **64** (1956), 20–82.

[BJ] **Borel, A., Jacquet, H.**, Automorphic forms and automorphic representations, in *Automorphic forms, representations and L-functions, Part 1, Proc. of Symp. in Pure Math.* **33**, AMS 1979, 189–202.

[BT] **Borel, A., Tits, J.,** Groupes réductifs, *Publ. Math. de l'IHES* **27** (1965), 55–150.

[B] **Bourbaki, N.,** *Groupes et algèbres de Lie,* chapters 4, 5 and 6, Hermann 1968.

[B2] **Bourbaki, N.,** *Algèbre commutative,* chapters 1 to 4, Masson, Paris 1985.

[C] **Colin de Verdières, Y.,** Une nouvelle démonstration du prolongement méromorphe des séries d'Eisenstein *CRAS* **293** (1981), 361–63.

[D] **Diximier, J.,** *Les algèbres d'opérateurs dans l'espace hilbertien,* Paris 1957.

[E] **Efrat, I.,** The Selberg trace formula for $PSL_2(\mathbb{R})^n$, *Memoirs AMS* **359** (1987).

[G] **Godement, R.,** Introduction à la théorie de Langlands, *Séminaire Bourbaki* **321** (1967/68).

[H] **Harder, G.,** Chevalley groups over function fields and automorphic forms, *Annals of Math.* **100** (1974), 249–306.

[HC] **Harish-Chandra,** *Automorphic forms on semi-simple Lie groups,* LN **62,** Springer Verlag, 1968.

[HC2] **Harish-Chandra,** Two theorems on semi-simple Lie groups, *Ann. of Math.* **83** (1966), 74–128.

[HC3] **Harish-Chandra,** Discrete series for semi-simple Lie groups II, *Acta Math.* **116** (1966), 1–111.

[J] **Jacquet, H.,** On the residual spectrum of $GL(N)$, in *Lie group representations II,* LN **1041,** Springer Verlag 1983, 185–208.

[L] **Langlands, R. P.,** *On the functional equations satisfied by Eisenstein series,* LN **544,** Springer Verlag 1976.

[M1] **Morris, L. E.,** Eisenstein series for reductive groups over global function fields I: the cusp form case, *Can. J. Math.* **34** (1982), 91–168.

[M2] **Morris, L. E.,** Eisenstein series for reductive groups over global function fields II: the general case, *Can. J. Math.* **34** (1982), 1112–82.

[M3] **Morris, L. E.,** Some remarks on Eisenstein series for metaplectic coverings, *Can. J. Math.* **35** (1983), 974–85.

[OW] **Osborne, M S., Warner, G.,** *The theory of Eisenstein systems,* Academic Press 1981.

[Ro] **Rogawski, J.,** On modules over the Hecke algebra of a p-adic group, *Invent. Math.* **79** (1985), 443–65.

[Ru] **Rudin, W.,** *Functional analysis,* McGraw-Hill, 1974.

[Sp] **Springer, T.,** Reductive groups, in Automorphic forms, representations and L-functions, Part 1, *Proc. of Symp. in pure Math.* **32,** AMS 1979, 3–28.

[S] **Steinberg, R.,** Générateurs, relations et relèvements, *Colloque sur la théorie des groupes algébriques,* Bruxelles, 1962, (CBRM), 113–28.

[St] **Stone, M. H.,** *Linear transformations in Hilbert space and their applications to analysis,* AMS, New York 1932.

[W] **Waldschmidt, M.,** *Un 1er cours sur les nombres transcendants,* Orsay 1973.

Index